Handbook of High Entropy Alloys

Handbook of High Entropy Alloys (HEAs) offers a comprehensive and multidisciplinary overview of these advanced materials from foundational principles to synthesis techniques, advanced characterizations, and preparation methods. It also explores conventional as well as novel and emerging applications.

- Provides a brief introduction to HEAs, offering essential background for readers new to the field
- Covers various metallurgical techniques, starting from the traditional to advanced techniques used to prepare HEAs
- Discusses the chances of forming various possible phases and their predictions using different methods
- Details advanced instruments used to characterize microstructural, mechanical, corrosion, wear, and oxidative resistance
- Investigates corrosion and catalytic, high-temperature, and high-strength properties
- Describes an array of applications such as hydrogen splitting, energy conversion, electrochemical sensors, fuel cells, and batteries

This wide-ranging reference offers researchers, engineers, and engineering students a complete understanding of the preparation of high entropy alloys for modern and advancing applications.

Emerging Materials and Technologies

Series Editor: Boris I. Kharissov

The Emerging Materials and Technologies series is devoted to highlighting publications centered on emerging advanced materials and novel technologies. Attention is paid to those newly discovered or applied materials with potential to solve pressing societal problems and improve quality of life, corresponding to environmental protection, medicine, communications, energy, transportation, advanced manufacturing, and related areas.

The series takes into account that, under present strong demands for energy, material, and cost savings, as well as heavy contamination problems and worldwide pandemic conditions, the area of emerging materials and related scalable technologies is a highly interdisciplinary field, with the need for researchers, professionals, and academics across the spectrum of engineering and technological disciplines. The main objective of this book series is to attract more attention to these materials and technologies and invite conversation among the international R&D community.

Organic-Inorganic Hybrid Nanomaterials: Energy Harvesting, Storage, and Advanced Applications
Edited by Srikanta Moharana, Bibhuti B. Sahu and Suresh Sagadevan

Nanocarbon to Polymer-Nanocarbon Composites: From Synthesis to Breakthrough Performance
Edited by Srikanta Moharana, Bibhuti B. Sahu and Suresh Sagadevan

Materials for Dye Degradation
Edited by Tarun Parangi and Soyebkhan Pathan

Circularity of Cementitious Materials: A Practical Approach
Christian Paglia

Machine Learning for Semiconductor Materials
Edited by Neeraj Gupta, Rashmi Gupta, Rekha Yadav, Sandeep Dhariwal and Rajkumar Sarma

Multilayer Nanostructured Wear-Resistant Coatings for Metal-Cutting Tools
by A.A. Vereschaka, S.N. Grigoriev and V.P. Tabakov

Multifunctional Inorganic Nanomaterials: Synthesis, Properties and Applications
Edited by H.C. Ananda Murthy, Velumani Subramaniam, Mika Sillanpää, R. Balachandran and Gurumurthy Hedge

Handbook of High Entropy Alloys: Fundamentals to Applications
Edited by Shashanka Rajendrachari

For more information about this series, please visit: www.routledge.com/Emerging-Materials-and-Technologies/book-series/CRCEMT

Handbook of High Entropy Alloys

Fundamentals to Applications

Edited by
Shashanka Rajendrachari

CRC Press
Taylor & Francis Group
Boca Raton London New York

CRC Press is an imprint of the
Taylor & Francis Group, an **informa** business

Designed cover image: © 2026 Shutterstock

First edition published 2026
by CRC Press
2385 NW Executive Center Drive, Suite 320, Boca Raton FL 33431

and by CRC Press
4 Park Square, Milton Park, Abingdon, Oxon, OX14 4RN

CRC Press is an imprint of Taylor & Francis Group, LLC

ISBN: 978-1-032-85557-8 (hbk)
ISBN: 978-1-041-12307-1 (pbk)
ISBN: 978-1-003-66409-3 (ebk)

DOI: 10.1201/9781003664093

Typeset in Times
by KnowledgeWorks Global Ltd.

Contents

Preface

In recent years, high entropy alloys (HEAs) have emerged as one of the most promising classes of advanced materials, drawing the attention of researchers, engineers, and industries worldwide. Their unique composition, involving multiple principal elements in near-equiatomic ratios, offers exceptional combinations of mechanical strength, thermal stability, corrosion resistance, and functional versatility, opening up new frontiers in materials science and engineering. This edited book brings together the contributions of leading researchers and experts from various institutions to provide a comprehensive, multidisciplinary exploration of the subject. The chapters are thoughtfully organized to guide readers through the foundational principles, synthesis techniques, advanced characterizations, and wide-ranging applications of HEAs. The book begins with a brief historical perspective and introduction to HEAs, offering essential background for readers new to the field. It then delves into various metallurgical and mechanical alloying methods for the preparation of HEAs, followed by discussions on thermodynamic design strategies and advanced characterization techniques that reveal their complex structures and properties. Further, the book addresses the mechanical, structural, thermal, corrosion, and radiation resistance properties of HEAs, providing valuable insights into their performance under diverse conditions. In addition to conventional applications, several chapters explore novel and emerging applications, such as their use in aerospace, electrochemical sensing, catalysis, and as alternatives to traditional Ni-based superalloys. The final chapters of the book reflect on other advanced and future applications of HEAs, offering a forward-looking view on the potential impact of these materials in next-generation technologies. This collection aims to serve as a valuable resource for graduate students, researchers, and professionals working in the fields of materials science, metallurgy, mechanical engineering, and applied physics. By providing both fundamental knowledge and practical insights, this book hopes to inspire new ideas, collaborations, and innovations in the ever-evolving landscape of HEA research.

I extend my heartfelt gratitude to all the contributing authors for their expertise, dedication, and valuable contributions, which have made this book a meaningful addition to the scientific literature.

Preface

Editor

Shashanka Rajendrachari is an Associate Professor in the Department of Basic Sciences, Global Academy of Technology, Bengaluru, India. He was previously an Assistant Professor in the Department of Metallurgical and Materials Engineering, Bartin University, Turkey. He earned a PhD at the National Institute of Technology, Rourkela, India. He is the first recipient of the prestigious Powder Metallurgy Student of the Year Award by the Powder Metallurgy Association of India for 2015 at IIT Bombay. He was awarded the prestigious Young Scientist 2020 Award at the Fourth International Scientist Awards ceremony on engineering, science, and medicine in Chennai, India. He is listed in the top 2% of scientists in the world for 2020, 2021, 2022, and 2023, as reported by Stanford University, USA, and published by Elsevier. He was awarded third place for the project competition in the Fifth R&D Project Market held in Turkey in 2022. He is experienced in preparing nano-structured stainless-steel powders, high entropy alloys, and shape memory alloys by the mechanical alloying method. He has more than eight years of research and teaching experience and has published more than 80 research articles, 1 German patent, 18 books and 15 book chapters in various peer-reviewed international publications. He has more than 3000 citations. His areas of research include powder metallurgy, mechanical alloying, nanomaterials, corrosion, electrochemical sensors, etc.

Contributors

Santhy A
Department of Chemistry
School of Physical Sciences
Amrita Vishwa Vidyapeetham
Amritapuri, Kerala, India

Syed Nasimul Alam
National Institute of Technology, Rourkela
Rourkela, Odisha, India

Serkan Bati
Faculty of Engineering and Architecture
Batman University
Batman, Turkey

Ismail Bayar
Department of Mechanical Engineering
Batman University
Batman, Turkey

Tapas Kumar Bhanj
Advanced Materials Technology Department
CSIR – Institute of Minerals and Materials Technology
Bhubaneswar, Odisha, India

Vatsala Chaturvedi
Department of Metallurgical Engineering
O.P. Jindal University
Raigarh, Chhattisgarh, India

Jason Daza
Department of Physics
University of Girona
Girona, Spain

Krishna Dutta
National Institute of Technology, Rourkela
Rourkela, Odisha, India

Parvathi J
Department of Chemistry
School of Physical Sciences
Amrita Vishwa Vidyapeetham
Amritapuri, Kerala, India

Avala Lava Kumar
Department of Metallurgical and Materials Engineering
Indian Institute of Technology – Ropar
Bara Phool, Punjab, India

Hareesha M
Department of Mechanical Engineering
Malnad College of Engineering – Hassan
Belagavi, Karnataka, India

Simi M
Department of Chemistry
School of Physical Sciences
Amrita Vishwa Vidyapeetham
Amritapuri, Kerala, India

Rajesh M
Department of Mechanical Engineering
Malnad College of Engineering – Hassan
Belagavi, Karnataka, India

Naveen Kumar Mindi
National Institute of Technology, Rourkela
Rourkela, Odisha, India

D. Mishra
Department of Metallurgical and Materials Engineering
Indian Institute of Technology – Jodhpur
Jheepasani, Rajasthan, India

Rahul Mitra
Department of Materials and Metallurgical Engineering
Indian Institute of Technology – Kharagpur
Kharagpur, West Bengal, India

Sharath B N
Department of Mechanical Engineering
Malnad College of Engineering – Hassan
Belagavi, Karnataka, India

Dharavathu Naresh
International Advanced Research Centre for Powder Metallurgy and New Materials
Hyderabad, Telangana, India

D. Narsimhachary
School of Engineering and Technology
G H Raisoni University
Saikheda, Madhya Pradesh, India

S. S. Nene
Department of Metallurgical and Materials Engineering
Indian Institute of Technology – Jodhpur
Jheepasani, Rajasthan, India

Ajit Panigrahi
Advanced Materials Technology Department
CSIR – Institute of Minerals and Materials Technology
Bhubaneswar, Odisha, India

Ghananshu Patil
National Institute of Technology, Rourkela
Rourkela, Odisha, India

Anshuman Patra
Metallurgical and Materials Engineering
National Institute of Technology, Rourkela
Rourkela, Odisha, India

K. Pramod
Osmania University
Hyderabad, Telangana, India

Shashanka R
Department of Basic Sciences
Global Academy of Technology,
Bengaluru, India

Raghavendra Rao
Department of Mechanical Engineering
Malnad College of Engineering – Hassan
Belagavi, Karnataka, India

Navya Roby
Department of Chemistry
School of Physical Sciences
Amrita Vishwa Vidyapeetham
Amritapuri, Kerala, India

Debdas Roy
Department of Materials and Metallurgical Engineering
National Institute of Advanced Manufacturing Technology
Ranchi, Jharkhand, India

Pradeep S
Department of Mechanical Engineering
Malnad College of Engineering – Hassan
Belagavi, Karnataka, India

Nityananda Sahoo
National Institute of Technology, Rourkela
Rourkela, Odisha, India

Pradyut Sengupta
Advanced Materials Technology Department
CSIR – Institute of Minerals and Materials Technology
Bhubaneswar, Odisha, India

Koushik Sikdar
Department of Materials and Metallurgical Engineering
National Institute of Advanced Manufacturing Technology
Ranchi, Jharkhand, India

J. Singh
Department of Metallurgical and Materials Engineering
Indian Institute of Technology – Jodhpur
Jheepasani, Rajasthan, India

Ayush Sourav
Defence Institute of Advanced Technology
Pune, Maharashtra, India

Joan-Josep Suñol
University of Girona
Girona, Spain

Lakshmi Krishnaa Suresh
Department of Chemistry
School of Physical Sciences
Amrita Vishwa Vidyapeetham
Amritapuri, Kerala, India

Shanmugasundaram T.
Defence Institute of Advanced Technology
Pune, Maharashtra, India

Sudeep Kumar T.
Defence Institute of Advanced Technology
Pune, Maharashtra, India

Trinath Talapaneni
Department of Metallurgical Engineering
O.P. Jindal University
Raigarh, Chhattisgarh, India

Asma Wederni
University of Girona
Girona, Spain

Sandeep Kumar Yadav
Defence Institute of Advanced Technology
Raigarh, Chhattisgarh, India

Section I

Introduction to High Entropy Alloys

1 A Brief History and the Advances of High Entropy Alloys

Serkan Bati

1.1 THE EFFECTS OF TECHNOLOGY AND MATERIALS SCIENCE ON HUMANITY

The scientific and historical development of high entropy alloys (HEAs) is discussed in this section. A single principal element has traditionally served as the basis for alloy design; however, because of the limitations of this strategy, scientists have suggested an alternative approach for materials design. This alternate technique aims to increase configurational entropy in order to contribute to the generation of thermodynamically stable solid solutions over complicated multiphase structures by combining five or more primary components in about equiatomic proportions. Early experimental studies showed that single-phase HEAs with face-centered cubic (FCC) or body-centered cubic (BCC) structures could be successfully manufactured, with high hardness and excellent corrosion resistance. Over time, this approach has evolved beyond purely metallic systems, encompassing a wide variety of compositions and microstructures with potential for various applications. This expansion has been accelerated by advances in processing technologies and the growing demand for multifunctional, high-performance materials. This review outlines the origins of HEAs, the fundamental scientific principles underlying their development, and the key milestones that have shaped their role within the field of advanced materials science.

Throughout history, humanity has continuously sought to advance technology to make life easier, meet needs, and sustain societal progress. Indeed, this technological advancement has always been closely tied to developments in materials technology. Materials science and technology, which have played one of the most significant roles in the development of modern humanity, have exhibited an uninterrupted process of advancement from the hunter-gatherer era to the present day.

The ancestors of modern humans, the hunter-gatherers, initially produced simple tools from natural materials, such as stone, bone, or wood. Later, with the transition to agricultural societies, an increasing accumulation of knowledge expanded the range of tools and equipment used, leading to the production of more functional implements [1]. The transition to an agricultural society relieved people from the trouble of finding food and allowed them to spend more time thinking. With growing knowledge, humanity began to utilize a wider range of materials, and the simple but functional

DOI: 10.1201/9781003664093-2

tools made from these materials represented significant technological progress for their time.

Copper, one of the few metals found in its metallic form in nature, was the first metal to be put to practical use by humans. Subsequently, bronze, regarded as the first alloy in history, was produced by combining copper and tin. This alloy, with its harder and more durable structure, was widely utilized. The discovery of iron enabled humans to create stronger tools and weapons, and iron played a crucial role in the development of civilizations [2]. By processing materials obtained from their surroundings, humans increased their chances of survival, advanced their societies, and, over time, focused on areas like agriculture, art, and trade, further advancing their civilizations.

This situation led humanity to a new concept: technology. The origin of the word technology is based on the Greek words "techne" and "logos". "Techne" refers to the knowledge behind the creation of a product; that is, the process of producing art or craft. While "logos" is attributed to craft work by some academic circles, another view is that it is rumored to be derived from the Greek word "mind" [3]. This etymological root demonstrates that technology is based both on practical applications and on theoretical knowledge and the mind.

However, throughout history, one of the greatest obstacles to societal development has been the inability of materials science and technology to meet the demands of the time. New inventions were either rendered nonfunctional due to the limitations of existing materials or had to await the development of new materials. As a result, materials science and technology have historically been driven by a constant need for innovation and progress.

1.2 FROM THE EVOLUTION OF ALLOYS TO HIGH ENTROPY ALLOYS

Alloying is the process of adding a relatively small amount of a secondary element to a primary element to impart desired properties to materials. The concentration of the primary phase, known as the solvent, typically ranges between 60% and 90% [4]. With a few exceptions (e.g., bronze, steel, brass, Wood's metal, Inconel), the alloying process has proceeded in this way, leading to alloys being commonly named after their primary constituent materials, such as iron alloys, aluminum alloys, titanium alloys, and nickel alloys [5].

Alloys exhibit superior properties compared to pure metals, making them more prevalent in industrial applications. As is well known, alloys have been a part of human life since the Bronze Age, marked by the use of the copper-tin alloy known as bronze. Later, the Iron Age began with the discovery of iron, which is very rare in nature, and its ability to be processed. During this era, tools made from iron alloys facilitated advancements in agriculture and construction, contributing to population growth. Moreover, developments in military technology enabled the production of more durable weapons.

The development of alloys progressed slowly until the Industrial Revolution. The main reason for this was that the existing alloys sufficiently met the general needs of society prior to industrialization. Additionally, the limited understanding of chemistry and physics, which are the basis of modern metallurgy, and therefore the lack of production techniques and advanced material knowledge that could create homogeneous structures, constrained the pace of alloy development. Nonetheless, alloys such as Indian Wootz and Damascus steel, as well as various alloys such as gold-silver, lead-tin, and brass, were produced during this period. From today's perspective, we can argue that alloys developed slowly. However, when we consider the needs and technological limitations of the era, this progression is understandable.

The prosperity of today's society is dependent on industrial advancements, which means that new materials must be continuously discovered, existing materials must be improved, and these resources must be used in appropriate applications. With the Industrial Revolution, materials science and technology gained new momentum. The creation of new materials tailored to specific needs in industries such as automotive, agriculture, food, textile, medical, computer, space, and aviation has become one of the cornerstones of technological progress. In particular, the increase in production brought about by industrialization has increased the demand for new materials, and in this process, materials science has taken shape as an important research area with the integration of disciplines such as metallurgy, polymer science, ceramics, solid-state physics, and semiconductors in the 1950s and 1960s [6].

Before the Industrial Revolution, steel production was limited to specialized applications. Later, during the Industrial Revolution, the Bessemer process was developed by the British inventor Henry Bessemer (1813–1898). This process involved oxidizing excess carbon and impurities in liquid iron and removed in gaseous form by an air blowing process, and a more homogeneous steel with lower carbon was obtained. However, this method was only effective with iron ores low in phosphorus and sulfur content [7].

Subsequently, the British inventor Sidney Gilchrist Thomas (1850–1885) introduced improvements that enabled the use of lower-quality, phosphorus-rich ores for steel production [7]. This innovation made the abundant, high-phosphorus, low-grade ores in Europe usable, reducing costs and facilitating large-scale steel production. These advancements significantly contributed to the realization of projects, such as railways, bridges, and factories, thereby accelerating and elevating the Industrial Revolution.

The advances in alloy production continued rapidly after the Industrial Revolution. With new manufacturing techniques and technological advancements, materials science was established, resulting in revolutionary innovations in fields such as automotive, aviation, energy, and defense. These advancements led to the production of more efficient, durable, and environmentally friendly materials.

Duralumin, Monel, high-speed steels, stainless steels, superalloys (nickel or cobalt based), titanium alloys, nitinol, magnetic alloys, metallic glasses, aluminum foams, and titanium foams were among the numerous groundbreaking alloys developed during the 1900s.

Thanks to the production of high-speed steels, the performance of cutting tools was enhanced, thereby increasing the speed and efficiency of manufacturing processes. Stainless steels offered resistance to corrosion, revolutionizing the food and chemical industries.

Duralumin has contributed greatly to the development of the aviation sector due to its lightness and durability [8]. The high-temperature resistance of nickel superalloys played a crucial role in the success of advanced engineering applications such as jet engines and gas turbines [9].

Titanium alloys, known for their light weight and high strength, found widespread use in the aviation industry and provided innovative solutions in biomedical applications, particularly in implant production [10]. Nitinol has revolutionized stents and other biomedical devices and opened new horizons in the medical world with its shape memory, superelasticity, biocompatibility, and corrosion-resistant properties [11, 12].

Metallic glasses (amorphous metals), with their superior magnetic properties, high strength and hardness, low elasticity, and resistance to corrosion and wear, have found applications in diverse sectors, including electronics, implants, defense, and aviation sectors, enabling significant advancements in these fields [13].

Alloys have always had a very critical place among these materials. Alloy systems, as mentioned before, consist of a base metal or primary metal (solvent) and an additional or secondary (solute) metal or non-metal element. However, alloying processes have generally been limited to this basic approach. In 2004, innovative alloy systems called high entropy alloys (HEAs) were independently proposed by research teams led by Jien-Wei Yeh (Taiwan) and Brian Cantor (UK). This approach created a new paradigm in materials science.

1.3 BRIEF HISTORY OF HIGH ENTROPY ALLOYS

In the late 1990s and the early 2000s, materials science experienced a significant leap forward with the discovery of a series of innovative materials, including carbon nanotubes, graphene, metallic glasses, and perovskite materials. During this period, developments in nanocomposites, spintronic materials, and nanostructured metals also played a crucial role, leading to groundbreaking advancements in fields such as electronics, energy, biomedical applications, and aerospace. One such material that emerged during this era was HEAs.

Among advanced materials, HEAs have attracted great attention in recent years [14, 15]. Unlike conventional alloys, HEAs are composed of five or more elements combined in equal or nearly equal proportions. This innovative design concept has given the materials superior properties, such as high strength, corrosion resistance, thermal stability, and wear resistance [16].

HEAs are materials formed by combining more than one element (not less than 5) in equal or approximate amounts, compared to traditional alloys. In fact, the configurational entropy increases in multicomponent systems for both the liquid and solid phases, which raises the glass transition temperature and facilitates the formation of metallic glasses at lower cooling rates. The diversity of elements increases configurational entropy, resulting in the formation of HEAs—a unique class of materials that are a transition material between amorphous metallic glasses and crystalline conventional alloys [4].

In 2004, an alloy system, not yet officially named but soon to be defined, had been developed by a group of scientists led by Brian Cantor [17] and Jien-Wei Yeh [18].

While Cantor and his team referred to these materials as "equiatomic multicomponent alloys," Yeh and his team named them "high entropy alloys," thereby becoming the originators of the term. So, what led the researchers to invent this alloy system? To answer this question, we must look back into history. While the invention of alloys can be traced to the Bronze Age, it is not necessary to delve that far for HEAs. For possibly the earliest work on HEAs, we only need to look to the late 18th century which is much more recent.

In 1788, Karl Franz Achard (1753–1821), a German scientist, examined alloys containing five to seven elements in a French-language book titled *Recherches sur les Propriétés des Alliages Métalliques.* Achard conducted an extensive study on more than 900 alloys composed of 11 metallic elements. His work involved various mechanical and physical tests, such as wear resistance, hardness, strength, density, and impact resistance, as well as a series of chemical tests exposing the alloys to dry air, humid air, humid air containing hydrochloric acid (HCl) vapor, and humid air containing hydrogen sulfide (H_2S) vapor. However, this research was largely overlooked until it was featured in Cyril Stanley Smith's book [19], *Four Outstanding Researches in Metallurgical History* [20].

The first known study after Achard's was conducted in 1981 by Brian Cantor and an undergraduate student, Alain Vincent, at the University of Sussex. In this study, equiatomic alloys composed of various elements were produced. The researchers even succeeded in creating an alloy composed of 20 different elements. In a specific alloy consisting of five elements, $Fe_{20}Cr_{20}Ni_{20}Mn_{20}Co_{20}$, they observed the formation of a face-centered cubic (FCC) structure. However, this work was not published internationally, aside from being documented as an undergraduate thesis.

After a hiatus, Cantor resumed his research in 1998 with Peter Knight, another undergraduate student at the University of Oxford. Once again, this work did not materialize into an international publication, apart from the undergraduate thesis. Eventually, Isaac Chang, a graduate student at Oxford, continued this research and presented it at an international conference in 2002, marking the first time it entered the scientific literature. Ultimately, the research was published in 2004 in the journal *Materials Science and Engineering A* under the title "Microstructural Development in Equiatomic Multicomponent Alloys".

This study [17] focused on identifying and developing potential new alloys located in the central regions of phase diagrams, as opposed to the traditional alloys at the endpoints. The researchers aimed to transcend the boundaries of conventional alloy development strategies and explore broader alloy systems. Using casting and melt spinning techniques, they produced alloys with 16 and 20 components. The alloys were found to be multiphase, crystalline, and brittle. In an alloy containing five components ($Fe_{20}Cr_{20}Ni_{20}Mn_{20}Co_{20}$), they observed dendritic solidification and the formation of an FCC structure. Additionally, alloys with six to nine components, predominantly transition metals, were found to possess a primary FCC dendritic phase. In the study, they stated that there were no metal glasslike structures and that different factors were effective for their formation.

Jien-Wei Yeh, a Taiwanese scientist, started to work on HEAs, which did not have a name yet, in 1995, independently of Cantor. By 2004, when the first international study [18] on the topic was published, Yeh and his team at National Tsing Hua

University had already conducted master's thesis projects with six students (K. H. Huang in 1996, K. T. Lai in 1998, Y. H. Hsu in 2000, Y. D. Huang in 2001, T. J. Tung in 2002, and C. Y. Chen in 2002) [21].

Yeh and his team initially defined HEAs as alloys composed of five or more elements in equal or nearly equal proportions [22]. Later studies suggested that the proportions of these elements could vary between 5% and 35%.

In the first study conducted for this purpose in 1996, a student named K. H. Huang prepared 20 different alloys with 40 distinct equiatomic compositions, consisting of 5 to 9 elements, using the arc melting method. These alloys were subjected to characterization processes such as microstructure analysis, hardness testing, and corrosion-resistance evaluation [20]. Their initial focus was on moving beyond conventional alloy systems—such as iron-, aluminum-, and nickel-based systems—and exploring multicomponent systems. The researchers emphasized that, according to traditional phase diagram knowledge, various intermetallic compounds were expected to form in this system. However, their investigations revealed that the new system did not behave as anticipated. They explained this phenomenon by noting that the high mixing enthalpy of the included elements reduced the tendency for phase segregation and ordering, thereby facilitating the formation of solid solutions [18].

Although Cantor and Yeh's work is considered the starting point for HEAs, the contributions of Professor Srinivasa Ranganathan cannot be overlooked. In his research [23], Ranganathan emphasized that metallurgists traditionally preferred systems with few alloying elements due to the increasing complexity of phase diagrams and processing variables with a higher number of components. However, he noted a shift in the early 2000s, highlighting three major developments: bulk metallic glasses, superelastic and superplastic alloys, and HEAs.

In 2003, before Yeh's publication, Ranganathan visited Yeh and wrote an article titled "Alloyed Pleasures: Multimetallic Cocktails" which introduced HEAs to the literature. He described Yeh's innovative idea of creating a new alloy system by combining multiple metals in equal proportions. Highlighting the significance of Yeh's work, Ranganathan noted that these alloys retained high hardness levels, even after annealing at 1000°C for 12 hours. He gave information that several patents were obtained from these studies and that the publication processes were ongoing.

The first-generation HEAs primarily focused on mechanical properties and were simpler to produce, featuring single-phase FCC/BCC (body-centered cubic) structures. In contrast, second-generation HEAs, designed to optimize not only mechanical properties but also electrical, magnetic, and other functional properties, involved more complex manufacturing techniques. These were characterized by multiphase and nanostructured compositions, with more than four principal elements [24].

HEAs have significant potential in advanced technological sectors such as automotive, aviation, electronics, energy, biomedical applications, and defense [25]. These materials address a wide range of applications, from structural components to energy storage systems, and are recognized as a novel approach in materials science.

1.4 HIGH ENTROPY EFFECT

To effectively explain the high entropy effect, it is first necessary to define entropy. Entropy can be described in various disciplines, including physics, chemistry, statistical mechanics, statistics, economics, ecology, information theory, and even theology and philosophy. However, a common theme across all these fields is that entropy represents disorder, uncertainty, and randomness. In thermodynamics, an increase in entropy signifies a transition of a system into a less ordered state. This does not imply a loss of energy but rather indicates that the energy becomes less available for useful work. While energy is conserved (according to the first law of thermodynamics), the potential of energy to perform work decreases as entropy increases.

Boltzmann expressed the entropy of a system in statistical thermodynamics as follows:

$$S = k \ln \Omega \tag{1.1}$$

Where:

- S: Entropy of the system
- k: Boltzmann constant (1.38×10^{-23} J·K^{-1})
- Ω: The number of microstates or possible configurations of the system.

This equation indicates that entropy is directly proportional to the natural logarithm of the number of microstates.

For solid solutions, the increase in system entropy resulting from the mixing of atoms of different alloying elements in various configurations is referred to as mixing entropy or configurational entropy. In multiprincipal element HEAs, the arrangement of atoms has the most significant influence on entropy. If the impact of other configurations on entropy is ignored, the atomic arrangement's mixing entropy dominates the system's entropy [26]. The entropy of HEAs composed of multiple elements is calculated using the configurational or mixing entropy formula:

$$\Delta S_{mix} = -R\sum_{i=1}^{n} x_i \ln\left(x_i\right) \tag{1.2}$$

If the fractions of the elements are equal, this can be simplified as:

$$\Delta S_{mix} = R \ln(n) \tag{1.3}$$

Where:

- ΔS_{mix}: Mixing entropy J·mol^{-1}·K^{-1}
- R: Universal gas constant 8.314 J·mol^{-1}·K^{-1}
- x_i: Mole fraction of each element in the alloy.
- n: Number of elements in the alloy.

High mixing entropy is evaluated together with the Gibbs free energy (G):

$$\Delta G_{mix} = \Delta H_{mix} - \Delta TS_{mix} \tag{1.4}$$

Where:

- ΔG_{mix}: Gibbs free energy of the mixture
- ΔH_{mix}: Mixing entalpy
- T: Temperature
- ΔS_{mix}: Mixing entropy.

High entropy causes the $-T\Delta S_{mix}$ term to dominate, reducing the free energy and promoting the formation of homogeneous solid solutions in HEAs.

Alloys are classified based on their mixing entropies into three categories:

1. Low-entropy alloys ($\Delta S \leq 1.0R$)
2. Medium-entropy alloys ($1.0R \leq \Delta S \leq 1.5R$)
3. High entropy alloys ($\Delta S \geq 1.5R$)

1.5 DEFINITION, EVOLUTION, AND CLASSIFICATION OF HEAs

HEAs are typically named by listing the constituent elements in alphabetical order. For example, AlCoCrFeNi refers to an alloy consisting of aluminum (Al), cobalt (Co), chromium (Cr), iron (Fe), and nickel (Ni). If the elements are present in different ratios, the atomic proportions are specified after the element symbols. For instance, notations like $Al_xCoCrFeNi$ are used, where x represents the molar ratio of aluminum.

As is well known, various definitions exist for HEAs. While establishing a universal definition is challenging, Miracle and Senkov [27], in their review, have proposed three distinct definitions for HEAs.

The first one is the composition-based definition. According to this definition, HEAs consist of five or more principal elements. Moreover, the concentration of each principal element must be between 5% and 35% atomic percentage [28]. In practical applications, the number of components often stays between five and nine since the growth in configurational entropy slows down when the number of elements exceeds nine [29].

The second one is the entropy-based definition. This definition asserts that the configurational entropy of HEAs must exceed 1.5R. There appears to be a consensus on this criterion [30]. For instance, when composed of five elements in equal proportions, the configurational entropy reaches 1.61R, thereby satisfying both the composition- and entropy-based definitions.

The third definition highlights the complexity and potential contradictions between the previous two criteria. For example, an alloy composed of five elements in proportions of 5-5-20-35-35% has a configurational entropy of 1.36R, which qualifies it as an HEA according to the composition-based definition but not the entropy-based one. Conversely, an alloy comprising four elements in equal molar ratios, with a configurational entropy of 1.39R, is recognized as an HEA by some researchers [31, 32] due to its higher entropy value compared to the five-element alloy. However, it does not satisfy the composition-based definition.

In addition to HEAs, other terminologies such as multi-principal element alloy (MPEA) and complex concentrated alloy (CCA) have gained acceptance in the literature [27]. These alloys are also referred to by terms like multi-component alloy (MCA) and concentrated solid solution alloy (CSS). However, as shown in Figure 1.1, which illustrates the number of publications over the years, despite the existence of these alternative names, research in this field predominantly continues to be published under the HEA designation.

Although there have been many innovations and groundbreaking developments in HEAs, researchers have different opinions on whether these developments are more important than others. However, each of these developments can be considered an important milestone in the evolution of HEAs. In this context, Musicó et al. [33] listed the evolution of HEAs in 2020 as follows:

2004: High Entropy Alloys (HEAs)
2005: High Entropy Nitrides (HENs)
2015: High Entropy Oxides (HEOs) [34]
2016: High Entropy Metal Diborides (HEBs) [35]
2018: High Entropy Carbides (HECs) [36]
2018: High Entropy Sulfides (HESs) [37]
2019: High Entropy Fluorides (HEFs) [38]
2019: High Entropy Alumina Silicides or High Entropy Ceramics (HECs) [39].

In addition to these, several HEA types have been developed to date, some of which are listed below:

Eutectic High Entropy Alloys (EHEAs)
High Entropy Alloy Films (HEAFs)
High Entropy OxideMicro/Nano Materials (HEO MNMs)
High Entropy Shape Memory Alloys (HESMAs)
High Entropy SuperAlloys (HESAs)
Metastable High Entropy Alloys (MHEAs)
Refractory High Entropy Alloys (RHEAs)

Miracle and Senkov [27] analyzed 37 elements used in 408 studies on HEAs and classified them into different alloy families. It was observed that more than 70% of the studies utilized Co (301), Cr (301), Fe (348), and Ni (341). The most frequently used element groups were transition metals, with 22 occurrences. Their classification is shown in Figure 1.2.

Li et al. [40] made a more comprehensive and different classification. They listed the definitions made by other researchers in different categories and saw that five different classifications were taken as the basis. The first is composition-based (two different), the second is microstructure-based (six different), the third is microstructure and property-based, the fourth is mechanical property-based, and the fifth is phase

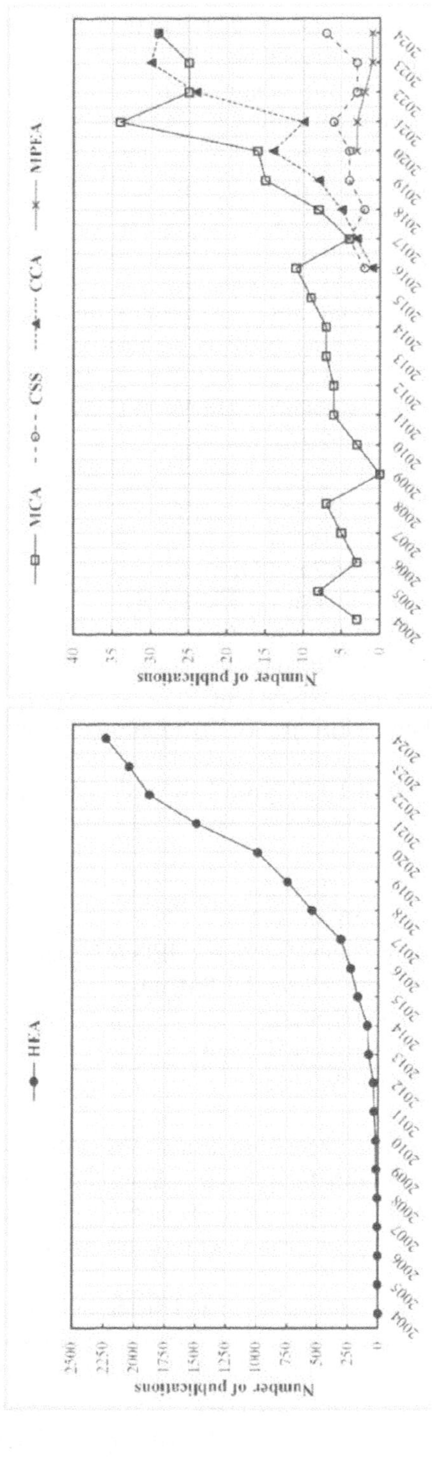

FIGURE 1.1 Comparison of the number of research articles with HEA and alternative names (surveyed in WoS up to Dec. 2024).

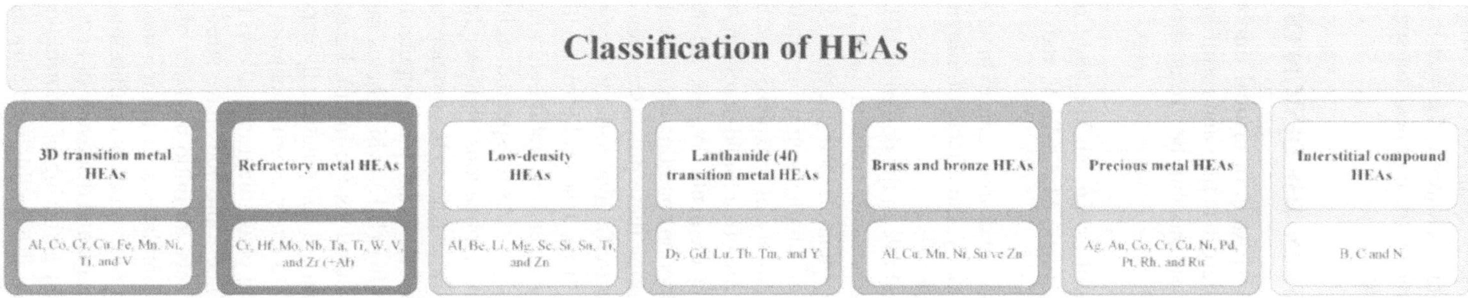

FIGURE 1.2 Classification of HEAs sorted by elemental groupings by Miracle and Senkov [27].

stability–based. Of course, all these definitions can perhaps be expanded more easily and comprehensively in the future. However, the fact is that there are many alloys that have not yet been discovered in the composition space, which may make these classifications more complicated.

1.6 EFFECT OF ELEMENTS IN THE DESIGN OF HEAs

The selection of elements for HEAs is of critical importance. Adding a specific element can alter the alloy's microstructure and result in significant changes to its properties. In general, HEA elements can be grouped based on their tendency to form certain crystal structures:

Elements forming FCC structures: Cu, Ni, Mn, and Co.
Elements forming BCC structures: Al, Cr, W, V, and Ti.

Molybdenum, depending on conditions, can exhibit a tendency to form either BCC or FCC structures. Essentially, the effects of alloying elements are similar to those observed in stainless steel [41].

These elements play a critical role in the design of HEAs, and additional properties provided by some other elements are listed in Table 1.1.

Although HEAs have great potential, their unexplored, vast compositional space makes discovering new materials challenging using traditional methods and material diversity. Despite the availability of inexpensive transition metals (Fe, Ni, Cr) that facilitate discovery, HEAs are often found in regions that are difficult to explore due to the presence of refractory elements (W, Nb, Ta) that increase material costs. Additionally, variations in production conditions can affect the alloys' microstructure and mechanical properties, making these processes even more complex. Therefore, the use of more efficient and cost-effective methods is of great importance.

Alloy design has evolved from trial-and-error methods to more systematic approaches, with HEAs largely extending traditional alloy strategies. Application-driven design focuses on tailoring HEAs for specific real-world uses, such as refractory alloys for high-temperature applications, lightweight alloys for transportation, and high entropy superalloys as alternatives to commercial superalloys [40].

Mooraj and Chen [43] discuss HEAs and the computational methods used for their design. It covers machine learning (ML), which helps rapidly explore large design spaces, and first-principles calculations (ab-initio), based on quantum mechanics, which provide high accuracy but are computationally expensive. Molecular dynamics (MD) simulations are used to study atomic behavior under various conditions, while CALPHAD (calculation of phase diagrams) allows for the efficient exploration of phase formation in alloys. These methods have strengths and weaknesses: ML is fast but relies on large data sets, ab-initio methods offer physical insights but at high computational costs, MD simulates deformation mechanisms, and CALPHAD is computationally efficient but only predicts equilibrium phases. Combining these approaches can enhance the design and discovery of new alloys.

Although the most important gathering of information for scientists is through literature, some researchers can also do this with various ML applications. One of

TABLE 1.1
Effect of the Alloying Elements on HEAs [24, 26, 42]

Element	Symbol	Effect of the Properties
Titanium	Ti	• Increases lattice distortion due to large atomic radius. • Improves hardness and strength.
Zirconium	Zr	• It may cause the formation of intermetallic phase and therefore reduce strength and plasticity.
Vanadium	V	• Improves the microstructure by creating homogeneous structures in dendritic regions. • Improves hardness and strength. • Reduces oxidation resistance.
Chromium	Cr	• Reduces hardness and strength. • Improves plasticity.
Molybdenum	Mo	• Increases lattice distortion due to large atomic radius. • The alloy's microstructure is improved by the inclusion of BCC. • Improves plasticity, hardness, and strength.
Manganese	Mn	• Reduces oxidation resistance. • Reduces cost.
Cobalt	Co	• Exhibits ferromagnetic properties. • Increases plasticity and wear resistance when added in small amounts.
Nickel	Ni	• Increases paramagnetic properties.
Copper	Cu	• Provides superior mechanical properties by forming Cu-rich nanophases.
Gold	Au	• Provides a link between the primary elements (Cr, Mo, Fe, Ni, and Cu).
Silver	Ag	• Causes the formation of Ag-Cu hypotactic phases.
Tin	Sn	• FCC phase may be shown at low rates. • When the Sn content is low (x = 0.01), there is a significant increase in plasticity and strength with rising Sn content. • When the Sn content is high (x = 0.05), the elongation reaches its maximum. • If the content increases further, the alloy precipitates intermetallic compounds of hard and brittle phases, reducing elongation.
Niobium	Nb	• May lead to intermetallic formation. • Reduces strength and hardness. • Can enhance oxidation resistance when moderately added to the alloy.
Aluminum	Al	• Causes a transition from FCC to BCC. • The crystal structure transforms from porous to dendritic. • Increases microhardness. • Enhances wear resistance.
Iron	Fe	• Provides magnetic properties. • Has no effect on phase and microstructure in the solid solution.
Boron	B	• May improve wear resistance. • Can increase compressive strength at high temperatures.
Silicon	Si	• Increases compressive strength. • Improves plasticity. • Reduces density.
Germanium	Ge	• Causes instability in the dendritic structure within the FCC phase. • Tends to cluster within the dendritic structure.

the best examples in this regard is the way Pei et al. [44] used to suggest new materials from 6 million texts in the literature, which has somehow facilitated this difficult process. Researchers who aim to improve existing materials or discover new materials have suggested 494 promising alloys out of approximately 2.6 million candidate alloys. This study has shown a path that can be used in future material discovery studies.

1.7 PROPERTIES AND APPLICATIONS OF HEAs

As in traditional alloys, the common crystal structures in HEAs include face-centered cubic (FCC and body-centered cubic (BCC). HEAs with predominantly FCC crystal structures exhibit high ductility, impact resistance, and low hardness, whereas HEAs with predominantly BCC crystal structures demonstrate high hardness but are brittle, which reduces ductility and toughness [45]. HEAs that contain both phases enable versatile mechanical properties, and the mechanical characteristics can be optimized through the ratio and compatibility of these two structures. Additionally, HEAs can exhibit amorphous and hexagonal close-packed (HCP) structures [46]. These crystal structures are fundamental factors that determine the physical, chemical, and mechanical properties of HEAs [26].

HEAs exhibit superior mechanical properties compared to traditional alloys. Examples include strength, elongation percentage, hardness, corrosion resistance, wear resistance, high-temperature stability, fatigue resistance, radiation resistance, and magnetic properties. For instance, HEAs can display a wide range of hardness values, from 140 HV to 900 HV, as processing parameters significantly influence hardness [47]. Although coating other materials with HEAs can be challenging, successful coatings with high hardness, corrosion resistance, and wear resistance have been achieved using hot forming methods such as laser cladding and thermal spraying, as well as cold forming methods like electrical discharge machining (EDM), deposition and magnetron sputtering [48]. There are also studies in the literature on corrosion resistance [49–52].

Some studies on HEAs report compressive strengths exceeding 2 GPa and yield strengths above 2.7 GPa [53]. Other studies have achieved hardness values around 6 GPa and compressive yield strength near 2 GPa [54, 55]. These values demonstrate that HEAs can achieve strength levels far surpassing those of traditional alloys.

Certain refractory HEAs exhibit higher yield strength than superalloys and retain this performance up to approximately 1600°C. For example, Senkov et al. [32] found that the yield strength of the $Nb_{25}Mo_{25}Ta_{25}W_{25}$ alloy at 1000°C is approximately twice that of Inconel 718 or Haynes 230, and for $V_{20}Nb_{20}Mo_{20}Ta_{20}W_{20}$, it is nearly four times higher. They attributed this superior performance to slow elemental diffusion, high melting temperatures, and the preservation of the single-phase BCC structure.

Although there are studies on the fatigue testing of HEAs, the available data is limited. Under comparable conditions, some HEAs exhibit superior fatigue resistance compared to conventional metallic materials [40].

Electrocatalysts play a critical role in energy storage and conversion, acting as catalysts in electrochemical reactions. HEAs are considered to have great potential for electrocatalyst design. To develop eco-friendly and sustainable energy solutions, scientists are investigating the electrochemical conversion of abundant substances such as water (H_2O), carbon dioxide (CO_2), and nitrogen (N_2) into high-value products like hydrogen (H_2), alcohols, and ammonia. These products can be used in fuel cells to convert chemical energy into energy sources [56].

It is also thought that HEAs can be used as biomaterials. What is expected from biomaterials is that they have high yield strength, low elasticity, high fatigue resistance, and sufficient ductility, as well as corrosion resistance and biocompatibility. Studies conducted with HEAs show that although they have higher strength than other metals and are promising for the future, they need more in-depth analysis [57]. Furthermore, high entropy nitrides and carbides have been reported to have potential as coating materials for biomedical applications [58, 59].

High entropy carbides, with melting points above 3000°C, belong to the category of materials suitable for high-temperature applications such as rocket nozzles and nuclear reactors [58]. Studies on HEAs are also available for nuclear applications [60, 61] and hydrogen storage [62–64], which is expected to become even more significant in the future. Many studies are being conducted on the hydrogen storage kinetics of HEAs, and it is anticipated that research in this area will continue. However, challenges such as high mass density, low storage capacities at room temperature, and safety issues (e.g., evaporation and embrittlement) necessitate further investigation [65].

It is not possible to limit HEA applications to a specific field, as there are still many unexplored and unresearched areas related to HEAs. According to the current knowledge, Kumar et al. [58] summarized the application areas of some HEAs as follows:

1. Electronic devices: ACuSnTi, FeCoNiSiAl, CuAlCoNi
2. High-temperature conditions: (aerospace): AlCrMgTiV, AlCoCrFeTiNi
3. High-temperature conditions: (refractory): TiZrNbHfTa, NbWMoTaV
4. Marine applications: CoCrFeNiCu, AlCoCrFeNiCu
5. Cryogenic conditions: CrMnFeCoNi, CrFeCoNiPd, CoCrFeMnNi
6. Coatings: CrMnFeCoNi-WC, FeCoNiCrMn, $MgMoNbFeTi_2Y_x$
7. Welding: CoCrFeNi
8. Nuclear equipment: TiVCrTa, VCrMnTi, TiHfMnCr, TiHfVMoZr

   Additionally, some specific alloys related to other application areas have been identified in other studies [66]:
9. Hydrogen storage: HfNbTiVZr, MoNbTiVZr, NbTiVZr
10. Molten salt storage: $CoCrFeNiTi_{0.5}$, $Al_5Co_{18}Cr_7Fe_9Ni_{51}Ti_5$

HEAs have garnered significant interest from researchers due to their unique properties, which surpass those of conventional alloys. HEAs are still regarded as an emerging science, even though Web of Science data indicates that there are about 11,000 SCI-indexed studies on the subject at present. Challenges such as difficulties in mass production, high manufacturing costs, complexities in phase prediction and determining optimal compositions, a lack of detailed studies on structural properties, and limited industrial applications have hindered the realization of their full potential [67]. Nevertheless, HEAs are anticipated to play a critical role in the development of next-generation materials and in meeting the evolving demands of modern industries. Given its relatively short history of about 20 years, there remains substantial progress to be made in this field.

REFERENCES

1. Golding E. A history of technology and environment: From stone tools to ecological crisis: Routledge; 2016.
2. Kundig KJ, Cowie JG. Copper and copper alloys. In: Kutz M, editor. Mechanical Engineers' Handbook, Volume 1: Materials and Mechanical Design. Hoboken, NJ. John Wiley & Sons; 2006. p. 117–220
3. Tulley RJ. Is there techne in my logos? On the origins and evolution of the ideographic term–technology. International Journal of Technology, knowledge and society 2008;4:93–104.
4. Nene S, Sinha S, Yadav D, Dutta A. Metallurgical aspects of high entropy alloys. Journal of Alloys and Compounds 2024;1005:175849.
5. George EP, Raabe D, Ritchie RO. High-entropy alloys. Nature Reviews Materials 2019;4:515–34.
6. Ikeda S, Kotani M. A new direction in mathematics for materials science: Springer; 2015.
7. Williams AM. The geography of iron and steel: Routledge; 2015.
8. Ouissi T. Evolution of Light Alloys in Aeronautics: The Case of Duralumin from its Discovery to the End of WWII. Nacelles [Online], 8 | 2020, Online since 20 mai 2020, connection on 15 juin 2025. http://interfas.univ-tlse2.fr/nacelles/923.
9. Schafrik R, Sprague R. Superalloy technology-a perspective on critical innovations for turbine engines. Key Engineering Materials 2008;380:113–34.
10. Marin E, Lanzutti A. Biomedical applications of titanium alloys: A comprehensive review. Materials 2023;17:114.
11. Kilic M, Yenigun B, Bati S, Balalan Z, Kirik I. Effect of Cu addition on porous NiTi SMAs produced by self-propagating high-temperature synthesis. Materials Testing 2019;61:1140–4.
12. McKelvey A, Ritchie R. Fatigue-crack propagation in Nitinol, a shape-memory and superelastic endovascular stent material. Journal of Biomedical Materials Research: An Official Journal of The Society for Biomaterials, The Japanese Society for Biomaterials, and The Australian Society for Biomaterials and the Korean Society for Biomaterials 1999;47:301–8.
13. Jabed A, Bhuiyan MN, Haider W, Shabib I. Distinctive features and fabrication routes of metallic-glass systems designed for different engineering applications: A review. Coatings 2023;13:1689.
14. Zhang J, Wang C, Huang S, Xiang X, Xiong Y, Xu B, et al. Design high-entropy electrocatalyst via interpretable deep graph attention learning. Joule 2023;7:1832–51.
15. Liu X, Zhang J, Pei Z. Machine learning for high-entropy alloys: Progress, challenges and opportunities. Progress in Materials Science 2023;131:101018.
16. Menghani J, Vyas A, Patel P, Natu H, More S. Wear, erosion and corrosion behavior of laser cladded high entropy alloy coatings – A review. Materials Today: Proceedings 2021;38:2824–9.
17. Cantor B, Chang I, Knight P, Vincent A. Microstructural development in equiatomic multicomponent alloys. Materials Science and Engineering: A 2004;375:213–8.
18. Yeh JW, Chen SK, Lin SJ, Gan JY, Chin TS, Shun TT, et al. Nanostructured high-entropy alloys with multiple principal elements: Novel alloy design concepts and outcomes. Advanced Engineering Materials 2004;6:299–303.

19. Smith CS. Four outstanding researches in metallurgical history. Lecture on Outstanding Research Series. American Society for Testing and Materials, Philadelphia, 1963, 35 pp.
20. Murty BS, Yeh J-W, Ranganathan S, Bhattacharjee PP. High-entropy alloys: Elsevier; 2019.
21. Hsu C-Y, Yeh J-W, Chen S-K, Shun T-T. Wear resistance and high-temperature compression strength of Fcc CuCoNiCrAl 0.5 Fe alloy with boron addition. Metallurgical and Materials Transactions A 2004;35:1465–9.
22. Yeh J-W. Alloy design strategies and future trends in high-entropy alloys. Jom 2013;65:1759–71.
23. Ranganathan S. Alloyed pleasures: Multimetallic cocktails. Current Science 2003; 85:1404–6.
24. Zhang Y. High entropy materials: Springer; 2023.
25. Arun S, Radhika N, Saleh B. Exploring the potential of high entropy alloys: A comprehensive review on microstructure, properties and applications: Part II. Johnson Matthey Technology Review 2024;68:567–82.
26. He B, Zu Y, Mei Y. Design of advanced electrocatalysts for the high-entropy alloys: Principle, progress, and perspective. Journal of Alloys and Compounds 2023;958:170479.
27. Miracle DB, Senkov ON. A critical review of high entropy alloys and related concepts. Acta materialia 2017;122:448–511.
28. Kumar D. Recent advances in tribology of high entropy alloys: A critical review. Progress in Materials Science 2023;136:101106.
29. Han X, Zhang L, Wang X, Song S, Zhang H. Recent progress on the synthesis and applications of high-entropy alloy catalysts. Nano Research Energy 2023;2:e9120084.
30. Miracle DB, Miller JD, Senkov ON, Woodward C, Uchic MD, Tiley J. Exploration and development of high entropy alloys for structural applications. Entropy 2014;16: 494–525.
31. Senkov ON, Wilks GB, Miracle DB, Chuang CP, Liaw PK. Refractory high-entropy alloys. Intermetallics 2010;18:1758–65.
32. Senkov ON, Wilks GB, Scott JM, Miracle DB. Mechanical properties of Nb25Mo25Ta25W25 and V20Nb20Mo20Ta20W20 refractory high entropy alloys. Intermetallics 2011;19:698–706.
33. Musicó BL, Gilbert D, Ward TZ, Page K, George E, Yan J, et al. The emergent field of high entropy oxides: Design, prospects, challenges, and opportunities for tailoring material properties. APL Materials 2020;8.
34. Rost CM, Sachet E, Borman T, Moballegh A, Dickey EC, Hou D, et al. Entropy-stabilized oxides. Nature Communications 2015;6:8485.
35. Gild J, Zhang Y, Harrington T, Jiang S, Hu T, Quinn MC, et al. High-entropy metal diborides: A new class of high-entropy materials and a new type of ultrahigh temperature ceramics. Scientific Reports 2016;6:37946.
36. Zhou J, Zhang J, Zhang F, Niu B, Lei L, Wang W. High-entropy carbide: A novel class of multicomponent ceramics. Ceramics International 2018;44:22014–8.
37. Zhang R-Z, Gucci F, Zhu H, Chen K, Reece MJ. Data-driven design of ecofriendly thermoelectric high-entropy sulfides. Inorganic Chemistry 2018;57:13027–33.
38. Chen X, Wu Y. High-entropy transparent fluoride laser ceramics. Journal of the American Ceramic Society 2020;103:750–6.
39. Wen T, Liu H, Ye B, Liu D, Chu Y. High-entropy alumino-silicides: A novel class of high-entropy ceramics. Science China Materials 2020;63:300–6.
40. Li W, Xie D, Li D, Zhang Y, Gao Y, Liaw PK. Mechanical behavior of high-entropy alloys. Progress in Materials Science 2021;118:100777.
41. Soni VK, Sinha AK. Effect of alloying elements, phases and heat treatments on properties of high-entropy alloys: A review. Transactions of the Indian Institute of Metals 2023;76:897–914.

42. Yu B, Ren Y, Zeng Y, Ma W, Morita K, Zhan S, et al. Recent progress in high-entropy alloys: A focused review of preparation processes and properties. Journal of Materials Research and Technology 2024; 29: 2689–2719.
43. Mooraj S, Chen W. A review on high-throughput development of high-entropy alloys by combinatorial methods. Journal of Materials Informatics 2023; 3: 4.
44. Pei Z, Yin J, Liaw PK, Raabe D. Toward the design of ultrahigh-entropy alloys via mining six million texts. Nature Communications 2023;14:54.
45. Sinha AK, Rathore RK, Soni VK, Rao KR, Upadhyay N, Pamnani G. Chapter 8 Effect of alloying elements on the phases of high-entropy alloys. In: Anil K, Rituraj C, Vikas D, Marta M-D, editors. High-entropy alloys. Berlin, Boston: De Gruyter; 2023. p. 89–98.
46. Hu J, Li X, Zhao Q, Chen Y, Yang K, Wang Q. An overview on fatigue of high-entropy alloys. Materials 2023;16:7552.
47. V B, M AX. Development of high entropy alloys (HEAs): Current trends. Heliyon 2024;10:e26464.
48. Lu K, Zhu J, Ge W, Hui X. Progress on new preparation methods, microstructures, and protective properties of high-entropy alloy coatings. Coatings 2022.
49. Cheng H, Liu Z, Luo H, Pan Z, Wang X, Zhao Q, et al. Tuning the microstructure to improve corrosion resistance of additive manufacturing high-entropy alloy in proton exchange membrane fuel cells environment. Corrosion Science 2023;213:110969.
50. Fu Y, Li J, Luo H, Du C, Li X. Recent advances on environmental corrosion behavior and mechanism of high-entropy alloys. Journal of Materials Science & Technology 2021;80:217–33.
51. Qiu Y, Thomas S, Gibson MA, Fraser HL, Birbilis N. Corrosion of high entropy alloys. npj Materials Degradation 2017;1:15.
52. Shuang S, Ding Z, Chung D, Shi S, Yang Y. Corrosion resistant nanostructured eutectic high entropy alloy. Corrosion Science 2020;164:108315.
53. Zhang Y, Wen C, Dang P, Lookman T, Xue D, Su Y. Toward ultra-high strength high entropy alloys via feature engineering. Journal of Materials Science & Technology 2024;200:243–52.
54. Fu Z, Chen W, Fang S, Zhang D, Xiao H, Zhu D. Alloying behavior and deformation twinning in a CoNiFeCrAl0. 6Ti0. 4 high entropy alloy processed by spark plasma sintering. Journal of Alloys and Compounds 2013;553:316–23.
55. Ji W, Wang W, Wang H, Zhang J, Wang Y, Zhang F, et al. Alloying behavior and novel properties of CoCrFeNiMn high-entropy alloy fabricated by mechanical alloying and spark plasma sintering. Intermetallics 2015;56:24–7.
56. Zhang Y, Wang D, Wang S. High-entropy alloys for electrocatalysis: Design, characterization, and applications. Small 2022;18:2104339.
57. Castro D, Jaeger P, Baptista AC, Oliveira JP. An overview of high-entropy alloys as biomaterials. Metals 2021;11:648.
58. Kumar A, Singh A, Suhane A. Mechanically alloyed high entropy alloys: Existing challenges and opportunities. Journal of Materials Research and Technology 2022;17:2431–56.
59. Bhaskaran Nair R, Supekar R, Morteza Javid S, Wang W, Zou Y, McDonald A, et al. High-entropy alloy coatings deposited by thermal spraying: A review of strengthening mechanisms, performance assessments and perspectives on future applications. Metals 2023;13:579.
60. Kozlovskiy AL, Zdorovets MV. Study of hydrogenation processes in radiation-resistant nitride ceramics. Journal of Materials Science: Materials in Electronics 2020;31:11227–37.
61. Pickering EJ, Carruthers AW, Barron PJ, Middleburgh SC, Armstrong DEJ, Gandy AS. High-entropy alloys for advanced nuclear applications. Entropy 2021;23:98.

62. Fu Y, Luo H, Pan Z, Wei Y, Gan B, Bi Z, et al. Hydrogen induced microstructure, mechanical properties and cracking evolution in a novel CoCrNiMo medium-entropy alloy. Journal of Alloys and Compounds 2023;939:168790.
63. Luo H, Pan Z, Wang X, Cheng H, Nazarov AA, Li X. Hydrogen-assisted failure in partially recrystallized carbon alloyed equiatomic CoCrFeMnNi high-entropy alloy. Corrosion Science 2022;203:110357.
64. Pan Z, Luo H, Cheng H, Zhao Q, Wang X, Fu Y, et al. A nano-sized precipitate strengthened medium entropy alloy with superior strength-ductility synergy and exceptional corrosion resistance. Materials Today Physics 2023;33:101062.
65. Han L, Zhu S, Rao Z, Scheu C, Ponge D, Ludwig A, et al. Multifunctional high-entropy materials. Nature Reviews Materials 2024;9:846–865.
66. Pineda F, Martínez C, Martin P, Aguilar C. High-entropy alloys: A review of their performance as promising materials for hydrogen and molten salt storage. Reviews on Advanced Materials Science 2023;62:20230150.
67. Menon A, Rajak DK, Singh DK. High entropy alloys: Next-generation material for space exploration. Materials Chemistry and Physics 2025;329:130091.

2 An Introduction to High Entropy Alloys

Anshuman Patra

2.1 INTRODUCTION

The chapter contains comprehensive overview of high entropy alloys. The evolution of high entropy alloys, formation mechanisms, properties, and applications will be discussed. With the progress of technological innovation, multicomponent high entropy alloys demonstrate an interesting microstructure-property relationship. The sustainable applications of high entropy alloys demand understanding the basis of alloy designing and microstructure development with the deliberation of thermodynamics. The chapter's emphasis is on the fundamental thermodynamic principles involves in the development of high entropy alloys, several prospective high entropy alloys, and several essential effects which show significant impacts on phase and thermal stability of high entropy alloys.

2.2 CONCEPTS OF HIGH ENTROPY ALLOYS DEVELOPMENT

Conventional alloys, including binary, ternary, or even more elements based the system, comprise a base constituent which possesses a higher concentration compared to others. However, in high entropy alloys (HEAs), the number of elements can vary to five or more. The composition of the elements can vary from 5 to 35 atomic % [1]. The concept of HEAs was developed from the publications in 2004 [1, 2]. Apart from composition-based interpretation of HEAs, another most important aspect in comprehending HEAs is mixing entropy. Mixing entropy (ΔS_{mix}) involves several entropy contributions, such as vibrational entropy (ΔS_{mix}^{vib}), configurational entropy (ΔS_{mix}^{conf}), electronic randomness entropy (ΔS_{mix}^{elec}), and magnetic dipole entropy (ΔS_{mix}^{mag}) [3]. Among all entropies, the configurational entropy is significant in the HEAs' formation. ΔS_{mix}^{conf} can be explained from the following equation:

$$\Delta S_{mix}^{conf} = -R\sum_{i=1}^{n} c_i \ln c_i \tag{2.1}$$

R represents the universal gas constant, n denotes the number of elements contained in the HEA, and c_i signifies the *i*th constituent's mol fraction [1]. The plot between composition and configurational entropy is displayed in Figure 2.1, which illustrates that the highest entropy (9.15 J/mol K) is attained at the core region of the plot, which suggests the zone of high entropy effect [4].

DOI: 10.1201/9781003664093-3

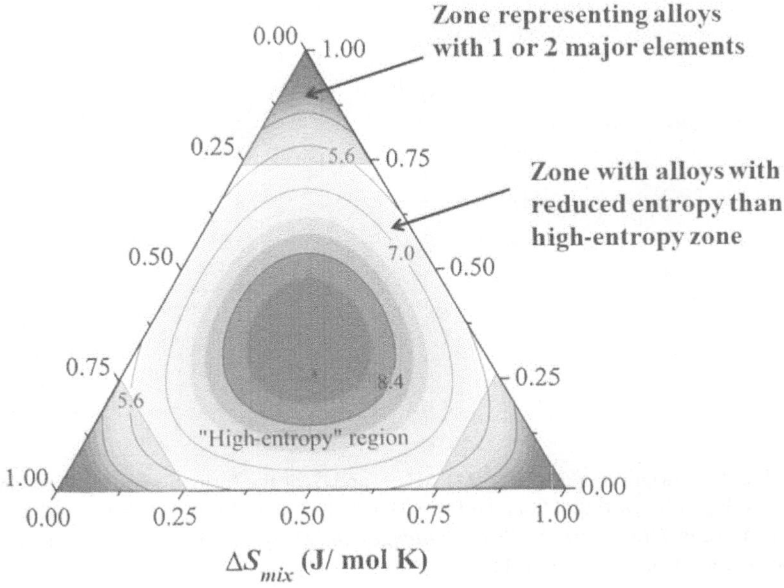

FIGURE 2.1 Plot representing entropy zone for different alloys. Corner area: Alloys with one or two major elements. Central area: High entropy zone. (Adapted from Materials Today, 19(6), Y.F. Ye, Q. Wang, J. Lu, C.T. Liu, Y. Yang, High-entropy alloy: Challenges and prospects, 349-362, Copyright (2015), with permission from Elsevier [4].)

If the HEA is equiatomic, the mixing entropy is as follows:

$$\Delta S_{mix} = R \ln n \tag{2.2}$$

The solid solution formation according to equation 2.2 can be enhanced by a combination of large participating elements in HEAs and therefore enormous mixing entropy [1]. The configurational entropies can vary from zero to 2.57R, respectively, for 1 and 13 elements [5]. For more than one element in an alloy, all constituents are regarded as equimolar. The HEAs are defined as alloys possessing configurational entropy ≥ 1.5R (R: gas constant) [5]. Other than the mixing entropy, two other parameters decide the nature of alloy formation, such as mixing enthalpy (ΔH_{mix}) and size difference of atoms (δ). Based on the two parameters, the final output can be a single-phase solid solution, multiphase, or amorphous. Figure 2.2 shows the effect of ΔS_{mix}, ΔH_{mix} and δ on the different phase formation ability in cast HEAs [4]. The condition for the formation of solid solution (single phase) is -15 kJ/mol $< \Delta H_{mix} <$ 5 kJ/mol and $0 < \delta < 5$ [6]. The metastable phase of HEA is also possible when $\Delta H_{mix} < -15$ kJ/mol and $\delta > 8$ [7].

Other research cites that both root mean square (RMS) residual strain and dimensionless elastic energy decide the phase formation and stability. As per the Lindemann's condition, a crystalline structure is unstable once RMS residual strain exceeds about 10% [8]. The relationship between RMS residual strain and the dimensionless elastic energy with diverse phases is presented in Figure 2.3 [9]. According

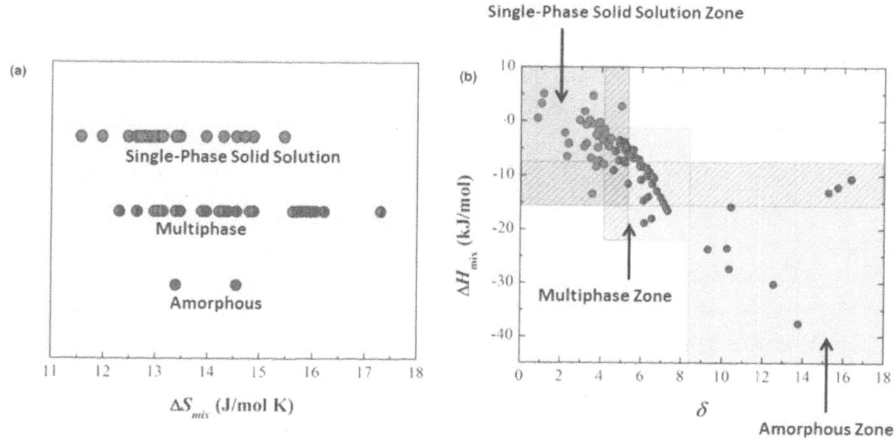

FIGURE 2.2 Variation of (a) ΔS_{mix}, (b) ΔH_{mix}, and δ for the formation of various phases in HEAs developed by casting. (Adapted from Materials Today, 19(6), Y.F. Ye, Q. Wang, J. Lu, C.T. Liu, Y. Yang, *High-entropy alloy: Challenges and prospects*, 349–362, Copyright (2015), with permission from Elsevier" [4].)

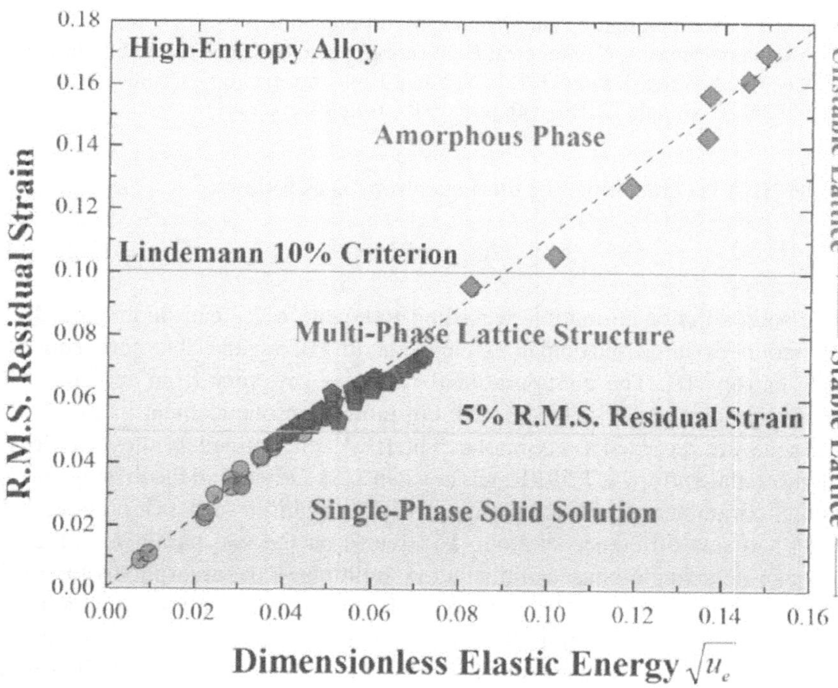

FIGURE 2.3 Variation of RMS residual strain with elastic energy (dimensionless) for HEAs. (Reprinted from Acta Mater., 94, Y.F. Ye, C.T. Liu, Y. Yang, A geometric model for intrinsic residual strain and phase stability in high entropy alloys, 152-161, Copyright (2015), with permission from Elsevier" [9].)

to the representation, the solid solution (single phase) is dominant when the RMS residual strain < 5%, amorphous structure forms if the strain is > 10% and multi-phase forms when the strain is within 5–10% [9].

Valence electron concentration (VEC) indicates the structure of the phase in HEAs. VEC can be represented as:

$$VEC = \sum_{i=1}^{n} C_i (VEC)_i \tag{2.3}$$

C_i: element's atomic fraction and $(VEC)_i$: concentration of valence electron in individual element (denoted by i). As per the VEC criteria, the face-centered cubic (FCC) phase forms when VEC ≥ 8 and the body-centered cubic (BCC) phase forms if VEC < 6.87 and FCC+BCC forms when 6.87 ≤ VEC < 8 [10]. In another investigation, the solid solution formation capacity of an alloy has been defined by Ω and can be related with melting point (T_m), ΔH_{mix}, and ΔS_{mix} as shown [11]:

$$\Omega = \frac{T_m \Delta S_{mix}}{|\Delta H_{mix}|} \tag{2.4}$$

The melting point of the alloy can be evaluated as:

$$T_m = \sum_{i=1}^{n} c_i (T_m)_i \tag{2.5}$$

$(T_m)_i$ is the melting temperature of the *i*th element. The condition for a solid solution based HEA formation is Ω > 1, where $T_m \Delta S_{mix}$ will be dominant over ΔH_{mix}. If Ω < 1, ΔH_{mix} will be the dominant factor to form intermetallic phases in HEAs [11]. Other literature has discussed that the ratio of mismatch entropy (S_σ) and Boltzmann constant (k_B) regulate the crystal structure variation. The FCC structure formation that is favourable is S_σ/k_B is less and a BCC structure forms if S_σ/k_B is high due to the adjustment of large lattice strain in BCC which is less packed than FCC [12]. Additionally, the ratio of configurational entropy and fusion entropy also decides the solid solution formation rule in equiatomic and non-equiatomic alloys. As the ratio > 1 facilitates a solid solution formation in equiatomic alloys, the non-equiatomic formation is assisted if the ratio > 1.2 [12]. However, as per the discussion, cumulatively all the factors contribute in the formation of specific phases in HEAs.

2.3 CLASSIFICATION OF HIGH ENTROPY ALLOYS

HEAs can be classified into various groups based on phase, such as ordered solid-solution (BCC), disordered solid solution (BCC, FCC, HCP), combination of ordered and disordered solid solution (FCC + BCC, BCC + B2, FCC + B2), completely intermetallic (α, β, σ, μ, laves phase), and intermetallic and solid solution (BCC+laves, FCC+ intermetallic, etc.) [13]. The report shows that the number of exclusive arrangement of major elements can be estimated from the following equation:

$$C_n^r = \frac{n!}{r!(n-r)!} \tag{2.6}$$

n: quantity of elements, r: major elements in alloy.

TABLE 2.1
High Entropy Alloy Groups and Corresponding Elements

Group	Major Elements in Group	Reference
3d transition HEA	Al, Co, Cr, Cu, Fe, Mn, Ni, Ti, and V	[15, 16]
Refractory HEA	Cr, Hf, Mo, Nb, Re, Ta, V, W, Zr, Co, Ni, Al, C, N, Si, Ti	[17]
4f transition HEA	Yb, Eu, Dy, Gd, Tb, Ho, Tm, Lu	[18, 19]

For n = 75 and $3 \leq r \leq 6$, the total novel systems will be in the millions [14]. The area of HEA is too extensive and, overall, the classification can be guided by the major groups, as mentioned in Table 2.1.

A combination of five or more elements can be combined from the above group for HEA development. Some of the investigated series of combination of HEAs is illustrated in Figure 2.4 [1, 2, 20–42]. Diverse crystal structures, as indicated by different colours of rectangular boxes at the bottom of Figure 2.4, are apprehended from the combinations [42].

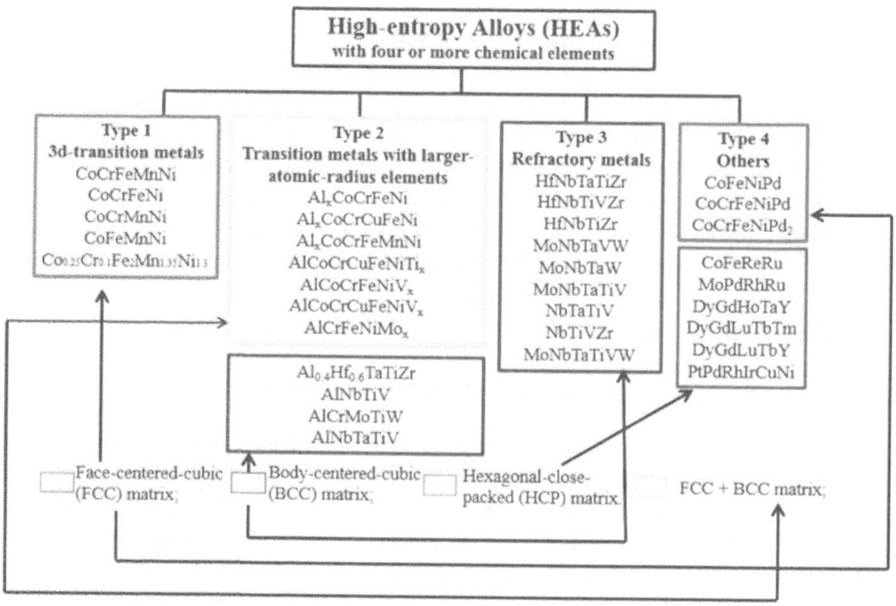

FIGURE 2.4 Structure-wise categories of HEAs. (Adapted from Curr. Opin. Solid State Mater. Sci., 21(5), H.Y. Diao, R. Feng, K.A. Dahmen, P.K. Liaw, Fundamental deformation behavior in high-entropy alloys: An overview, 252-266, Copyright (2017), with permission from Elsevier" [42].)

2.4 INFLUENCING FACTORS IN HIGH ENTROPY ALLOYS

Several effects which are prevalent in HEAs are cocktail effect, high entropy effect, lattice distortion effect, and sluggish diffusion effect. These effects influence the thermal and phase stabilities and contribute excellent mechanical properties. Manifestation of these effects results in replacing many established alloys for cutting-edge applications.

2.4.1 Cocktail Effect

The cocktail effect indicates that the elements in HEAs complement each other to improve the final properties. However, no individual element possesses the target property. The effect can be illustrated as combined contributions of composition, structure, and microstructure [43]. Unanticipated interactions among elements in HEAs result in remarkable properties of materials, which is the outcome of the cocktail effect. Attainment of groups of properties (significant strength and fracture toughness, thermoelectric outcome/photovoltaic change/piezoelectricity) can be the deliverables from such an effect [14, 44]. Inductors and transformers require enhanced magnetism and reduced coercivity which is another illustration of the cocktail effect [44]. As per other literature [45], the HEA is the consequence of mixing of different components in a 'multi-metallic cocktail'. Various impacts in cocktail alloying have been reported in literature such as (a) crystalline metastable structure formation in HEA [46, 47], (b) deliver driving force for the development of intermetallic phases [48], and (c) addition of interstitial element results in significant lattice distortion and short range ordering (SRO) is developed among the base and interstitial elements [49, 50]. Figure 2.5 shows the representation of various cocktail impacts, which facilitate martensitic transformation, microtwinning driven by mechanical perturbations, and strengthening through precipitation of intermetallic and interstitial addition [50]. It is reported that the stacking fault energy (SFE) of five-element HEAs (NiCoCrFeMn) is higher compared to alloys containing three elements (NiCoCr) [46]. Consequently, deformation-induced twining is observed at a later stage (strain around 25%) in NiCoCrFeMn HEAs compared to NiCoCr [51]. Twinning is detected at a much lower strain (around 7.5%) in cryogenic condition [51].

The cocktail effect in CrMnFeCoNi HEAs shows superlative catalytic impact related with MgH_2 hydrogen storage [52]. In $Nb_{25}Mo_{25}Ta_{25}W_{25}$ and $V_{20}Nb_{20}Mo_{20}Ta_{20}W_{20}$ HEAs, the melting points evaluated are greater than 2650°C, which is higher than the melting point of V, Nb, and Mo and the phenomenon is due to the cocktail effect [38]. Chang et al. have reported that the influence of the cocktail effect is significant on the mechanical and tribological properties and oxidation resistance of Ta-high entropy nitride coatings [53]. The cocktail effect is realized from the contribution of Al and Cr on the enhanced oxidation resistance of $(AlCrTa_2TiZr)N$ and $(Al_2Cr_2HfMoNbTa)N$. In addition, the impact of the cocktail effect is also comprehended from the improved wear resistance in (MoNbTaTiZr)N and $(Al_2Cr_2HfMoNbTa)N$ coatings compared to $(AlCrTa_2TiZr)N$ and (HfNbTaTiZr)N, supported by Mo addition [53]. Ishizu and Kitagawa have anticipated that the

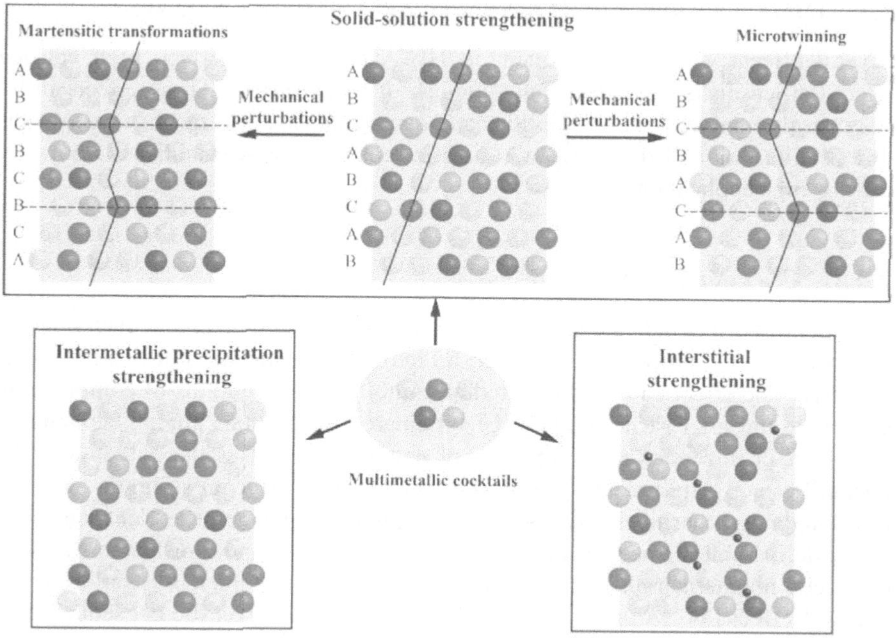

FIGURE 2.5 Several cocktail effects in multimetallic system. (Reprinted from Scr. Mater., 187, B. X. Cao, C. Wang, T. Yang, C. T. Liu, Cocktail effects in understanding the stability and properties of face-centered-cubic high-entropy alloys at ambient and cryogenic temperatures, 250-255, Copyright (2020), with permission from Elsevier" [50].)

superconducting critical temperature resulting from the cocktail effect in HEAs containing the composition of $Hf_{21}Nb_{25}Ti_{15}V_{15}Zr_{24}$ (BCC phase) is regulated by mixing entropy and VEC [54].

2.4.2 High Entropy Effect

The high entropy effect has significant importance to stabilize the solid solution in a multielement system at an elevated temperature [43]. In a multielement system, the formation of the brittle phase in terms of intermetallics can be decreased by this effect [1, 6]. For the formation of the solid solution, the mutual solubility among the elements should be higher, which is assisted by the high entropy effect. In traditional alloys, the enthalpy of mixing (ΔH_{mix}) is much higher than $-T\Delta S_{mix}$ and therefore, the free energy of mixing (ΔG_{mix}) is nearly equivalent with ΔH_{mix}. However, in HEAs, the high ΔS_{mix} results in decreasing the ΔG_{mix} and consequently a solid solution forms [6, 43]. Literature clarifies that in an equimolar FeCoCrNi system no long-range chemical ordering is present and a developed FCC structure is disordered [23]. In another study, equiatomic as cast AlCoCrCuFeNi alloy produced by low cooling rate exhibits the evidence of various phases with diverse composition. However, when the alloys were produced by splat quenching, a single BCC phase forms [55]. It is concluded in the research that a high mixing entropy forms a cubic structure phase irrespective of the cooling rate [55]. Luan et al. have also explained that the competition between a single-phase solid solution and intermetallic phases is governed by a configurational and vibrational entropy effect [56]. The free energy minimization from a high vibrational entropy from a combined effect of the number of elements and temperature can increase the propensity of a single-phase solid solution formation [56]. Even though the HEA exhibits disordered atomic organization, several researches indicate the evidence of SRO in HEAs [57–59].

Otto et al. have explained that in CoCrFeMnCu, TiCrFeMnNi, CoMoFeMnNi, CoVFeMnNi, and CoCrVMnNi alloys multiple phases are evident, which direct the incidence of other driving forces (non-configurational entropy) apart from the configurational entropy which too regulate the stability of phases [60]. It is therefore summarized that HEAs are related to a true solid solution and multiphase alloys with reduced entropy can be denoted as 'multicomponent alloys' [60].

2.4.3 Lattice Distortion Effect

In HEAs, there is variation in size and electronegativity among atoms which results in wide deviation in interatomic angle and distance [61]. Therefore, the atoms are located at a position different from an ideal location, which causes serious distortion in the lattice. The difference in crystal structure and bond strength also influences the lattice distortion. The lattice distortion effect generates stress or strain fields and dislocation motion is hindered when the dislocation passes through the field. The phenomenon generates solid-solution strengthening in HEAs [43]. However, the lattice distortion effect in different alloys can be due to the contribution of different sets of parameters, such as the atomic size difference between Co and Ni, which is very small; still the lattice distortion effect is enormous due to different crystal structures (Co: HCP, Ni: FCC) [43]. Zaddach et al. have reported that increasing the number of elements with Ni reduces the stacking fault energy due to intense lattice distortion. Consequently, the formation of nanotwins and the strain hardening effect improve the strength and elongation [43, 62]. The effect of increased addition of elements (Ti→Ti-Nb→Ti-Nb-Ta-Hf-Zr) in the unit cell of a BCC crystal structure is schematically described in Figure 2.6 [63]. It is clearly evident from the illustration that the extent of lattice distortion enhances with elevated element addition.

The strengthening effect σ_f^i imposed by an element (i) in HEAs is due to a mismatch, presented as [64]:

$$\sigma_f^i = AGc_i^{2/3}\delta_i^{4/3} \tag{2.7}$$

A: constant (non-dimensional) (0.04) [65], G: material's shear modulus, c_i: atomic fraction of each major element of HEA, δ_i: mismatch parameter. δ_i is represented as [66]:

$$\delta_i = \xi\left(\delta G_i^2 + \beta\delta r_i^2\right)^{1/2} \tag{2.8}$$

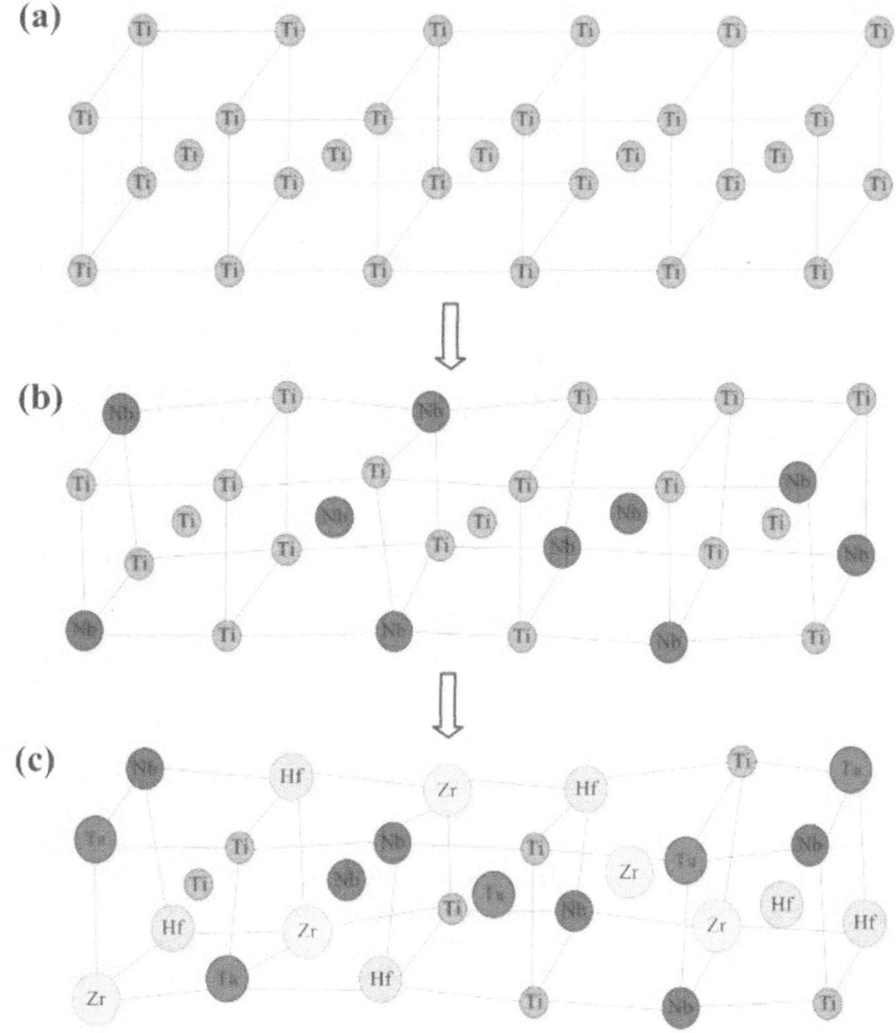

FIGURE 2.6 Structure of (a) β-Ti, lattice distortion in (b) Ti-Nb system and (c) HEA of TiNbTaZrHf. (Reprinted from J. Mater. Sci. Technol., 34(2), Z. Wang, Q. Fang, J. Li, B. Liu, Y. Liu, Effect of lattice distortion on solid solution strengthening of BCC high-entropy alloys, 349-354, Copyright (2017), with permission from Elsevier" [63].)

δG_i: disparity of elastic property, δr_i: mismatch in atomic size, ξ: 1 (FCC), 2.5 (BCC), $\beta > 16$ (edge dislocation), $3 < \beta < 16$ (edge dislocation) [66]. The strengthening of HEAs based on solid solution and contributed by element *i* can be defined as:

$$\sigma_f = \sum c_i \sigma_f^i \tag{2.9}$$

The δr_i and δG_i can be calculated as following where ijklm of HEAs comprises a jklm matrix and element i [67, 68]:

$$\delta r_i = \frac{\delta r_{ijklm}^{ave} - \delta r_{jklm}^{ave}}{\delta c_i} \tag{2.10}$$

$$\delta G_i = \frac{\delta G_{ijklm}^{ave} - \delta G_{jklm}^{ave}}{\delta c_i} \tag{2.11}$$

δc_i: variation of atomic fraction of i among HEAs of ijklm and jklm. δr_{ijklm}^{ave}, δr_{jklm}^{ave} and δG_{ijklm}^{ave}, δG_{jklm}^{ave} are calculated from subsequent equations (2.12, 2.13) [67, 68].

$$\delta r^{ave} = \sum_{i}^{n} \sum_{j}^{n} c_i c_j \delta r_{ij} \tag{2.12}$$

$$\delta G^{ave} = \sum_{i}^{n} \sum_{j}^{n} c_i c_j \delta G_{ij} \tag{2.13}$$

δr_{ij}, δG_{ij} signify atomic size and elastic difference among i and j atoms.

Wang et al. have prepared an intermetallic of $(CoNi)_{50}(TiZrHf)_{50}$ (high entropy system) (in atomic %) by the arc melting route [69]. The authors have achieved enriched strength and flow properties in the intermetallic which is often challenging to attain in conventional intermetallics. Lattice distortion assisted strengthening and additionally an amorphous phase at low temperature, elevated temperature dynamic hardening facilitates the flow property at increased homologous temperature [69]. Huang et al. have described the relationship between mean square atomic displacement (MSAD) with phase stability in a CoNiCrVMo system and elucidate that with the enhancement in the Cr, V, and Mo addition, the MSAD increases and such elevated lattice distortion reduces the phase stability [70]. HEAs experience both local short-range ordering and lattice distortion, which may impact the mechanical property in a different way; however, to achieve sustainable mechanical properties (high strength with plasticity), synergy between distortion and ordering has to be established [70, 71].

2.4.4 Sluggish Diffusion Effect

The poor diffusion rate in HEAs compared to traditional alloys give rise to several outstanding properties, such as strength and structural stability at elevated temperature and generation of nanostructure [39, 72–74]. Several authors have reported diverse factors for the sluggish diffusion effect in HEAs. Tsai et al. have reported that the variation in lattice potential energy at the atomic position facilitates sluggish diffusion [75]. The potential energy variation is associated with chemical and elastic factors. The chemical factor is regulated by mixing enthalpy and the elastic aspect is governed by the atomic misfit [76]. The enhancement in potential energy variations offers difficulties in vacancy mobility and impedes atomic diffusion [75]. Dąbrowa et al.

have explained that the crystal structure largely influences the coefficient of diffusion compared to composition at an elevated temperature [77]. In another study, the inter-diffusion coefficients among the elements of HEAs have been accredited for the comprehension of the diffusion effect [78]. The sluggish effects originated from the lattice distortion and therefore atomic migration is impeded [1, 79, 80]. Even the hypothesis of sluggish diffusion has been introduced as an essential effect in HEAs; there are many diverse explanations regarding its existence. Dąbrowa et al. [77] have reported reduced diffusion coefficients Al-Co-Cr-Fe-Ni HEAs compared to traditional alloys. It is also summarized in recent literature that HEAs with a higher number of elements may not qualify for lower diffusion rates. In this context, the most influential impact on the diffusion coefficient is contributed by the elements. To substantiate the statement, it is also mentioned that diffusion of Ni is at a minimum among CoCrFeNi and CoCrFeMnNi; however, Co diffusion rate depends on a particular temperature [81]. The diffusion of Co in CoCrFeMnNi is large compared to CoCrFeNi above the temperature [81]. Jones et al. explained that the formation of the B2 phase of primarily NiAl in $Al_{0.5}$CrFeCoNiCu HEA occurs inside 1 h at 700°C, < 0.1 h at 800 and 900°C [82]. Based on the phase formation time of < 0.1 h, the authors conclude that sluggish diffusion kinetics is not manifested in the HEAs [82]. The effect of sluggish diffusion in HEAs can also be explained by percolation, as studied by several researchers [83, 84]. The percolation is this regard refers to the association developed by constituents, which possesses rapid diffusion kinetics [84]. The verge of site percolation concept provides an understanding about the compositional effect on diffusion kinetics. At the boundary of site percolation, the diffusion coefficient is lowest and for HEAs the reported boundary value is 20 at.% for FCC (five-element system) and 25 at.% for BCC (four-element system) [85, 86]. Sen et al. have reported that MSADs are higher in a Hexagonal closed packed (HCP) system (HfTiZr, $Al_5Hf_{25}Sc_{20}Ti_{25}Zr_{25}$ and $Al_{15}Hf_{25}Sc_{10}Ti_{25}Zr_{25}$) [87] compared to a FCC system (CrMnFeCoNi) [88] which indicates contrary effect of "sluggish diffusion" in HEA. Sen et al. explained that the atomic misfit and MSAD are responsible for such a phenomenon in HCP HEAs [87]. The interdiffusion study of Co-Cr-Fe-Mn-Ni, Co-Fe-Mn-Ni, Co-Cr-Fe-Ni, Co-Cr-Mn-Ni, and Al-CO-Cr-Fe-Ni HEAs by Dąbrowa et al. directs that diffusion may not be necessary sluggish in all HEAs; rather they observed rapid diffusion kinetics in Co-Fe-Mn-Ni and Co-Cr-Mn-Ni when the temperature scale was absolute [89]. Such high diffusion is attributed to elevated concentration of vacancies in equilibrium [90]. In the melting point normalized scale, Al-Co-Cr-Fe-Ni and Co-Cr-Fe-Ni alloys considering Fe diffusion and Cr diffusion show similar diffusion, as in traditional alloys [89], which concludes that the manifestation of sluggish diffusion effect is not valid in every HEA. According to Daw and Chandross, the sluggish diffusion arises when the lattice mismatch of equimolar FCC alloys is 1–3% and beyond that the diffusion kinetics is faster [91]. Literature illustrates that the Zr diffusion coefficients in TiZrHfNbV and TiZrHfNbTa HEAs are higher compared to Hf-Nb binary alloys on an opposite homologous temperature scale [92]. Figure 2.7 displays the tendency of sluggish and opposite-sluggish diffusion of Ti in several metal and alloys. It is suggested that at higher configuration entropy, which a characteristic of HEA, the diffusion of Ti in HfTiZr, $Al_5Hf_{25}Sc_{20}Ti_{25}Zr_{25}$, and $Al_{15}Hf_{25}Sc_{10}Ti_{25}Zr_{25}$ does not follow poor diffusion kinetics [87].

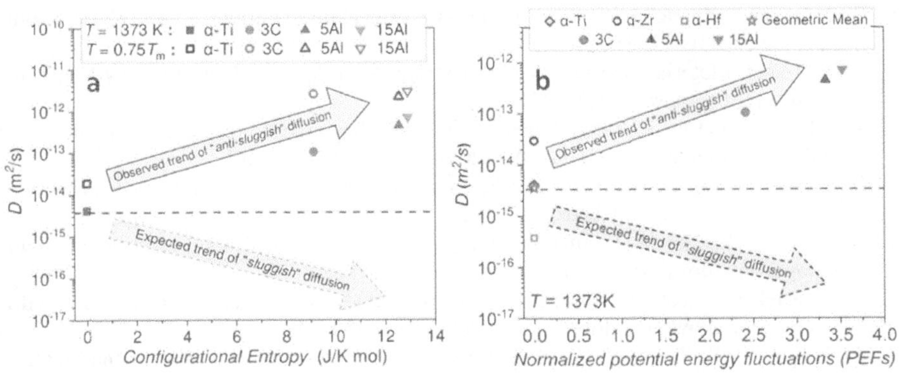

FIGURE 2.7 Variation of diffusion coefficient of Ti with (a) configurational entropy and (b) normalized potential energy fluctuations in HCP system. (Reprinted from Scr. Mater., 224, S. Sen, X. Zhang, L. Rogal, G. Wilde, B. Grabowski, S. V. Divinski, 'Anti-sluggish' Ti diffusion in HCP high-entropy alloys: Chemical complexity vs. lattice distortions, 115117, Copyright (2022), with permission from Elsevier [87].)

2.5 INFERENCES

The genesis of HEAs with respect to thermodynamic consideration is highlighted in the chapter. The central effects regulating the properties of HEAs are further discussed. However, all the effects may be valid under a specific parameter set. The synergistic effect provided by the elements in HEAs possesses vital significance due to outstanding contribution to final properties. Higher increase in configurational entropy, significant lattice distortion, and poor diffusion kinetics depict the phase formation in HEAs. Simply mixing larger numbers of elements does not permit the evidence of the above effects in HEAs. The structure, atomic size, and thermodynamic parameters of elements need to be considered to comprehend the sustainability of HEAs.

REFERENCES

1. J.-W. Yeh, S.-K. Chen, S.-J. Lin, J.-Y. Gan, T.-S. Chin, T.-T. Shun, C.-H. Tsau, S.-Y. Chang, Nanostructured high-entropy alloys with multiple principal elements: Novel alloy design concepts and outcomes, Adv. Eng. Mat., 6 (2004) 299–303.
2. B. Cantor, I.T.H. Chang, P. Knight, A.J.B. Vincent, Microstructural development in equiatomic multicomponent alloys, Mater. Sci. Eng. A, 375-377 (2004) 213–218.
3. X. Wang, W. Guo, Y. Fu, High-entropy alloys: Emerging materials for advanced functional applications, J. Mater. Chem. A, 9 (2021) 663–701.
4. Y.F. Ye, Q. Wang, J. Lu, C.T. Liu, Y. Yang, High-entropy alloy: Challenges and prospects, Materials Today. 19 (6) (2016) 349–362.
5. J.W. Yeh, Alloy design strategies and future trends in high-entropy alloys, JOM. 65 (12) (2013) 1759–1771.
6. Y. Zhang, T.T. Zuo, Z. Tang, M.C. Gao, K.A. Dahmen, P.K. Liaw, Z.P. Lu, Microstructures and properties of high-entropy alloys, Prog. Mater. Sci., 61 (2014) 1–93.

7. A. Inoue, T. Zhang, T. Masumoto, Glass-forming ability of alloys, J. Non-Cryst. Solids, 156-158 (1993) 473–480.
8. Y. Zhang, Materials Design of High-Entropy Materials. In: High-Entropy Materials. Springer, Singapore (2019), p. 46.
9. Y.F. Ye, C.T. Liu, Y. Yang, A geometric model for intrinsic residual strain and phase stability in high entropy alloys, Acta Mater., 94 (2015) 152–161.
10. S. Guo, C. Ng, J. Lu, C.T. Liu, Effect of valence electron concentration on stability of fcc or bcc phase in high entropy alloys, J. Appl. Phys., 109 (10) (2011) 103505.
11. X. Yang, Y. Zhang, Prediction of high-entropy stabilized solid-solution in multi-component alloys, Mater. Chem. Phys., 132 (2012) 233–238.
12. R. Raghavan, K.C.H. Kumar, B.S. Murty, Analysis of phase formation in multi-component alloys, J. Alloys Compd., 544 (2012) 152–158.
13. S. Singh, N.K. Katiyar, S. Goel, S.N. Joshi, Phase prediction and experimental realisation of a new high entropy alloy using machine learning, Sci. Rep., 13 (2023) 4811.
14. D.B. Miracle, High entropy alloys as a bold step forward in alloy development, Nat. Commun., 10 (2019) 1805.
15. D.B. Miracle, O.N. Senkov, A critical review of high entropy alloys and related concepts, Acta Mater., 122 (2017) 448–511.
16. S. Gorsse, D.B. Miracle, O.N. Senkov, Mapping the world of complex concentrated alloys, Acta Mater., 135 (2017) 177–187.
17. O.N. Senkov, D.B. Miracle, K.J. Chaput, J.P. Couzinie, Development and exploration of refractory high-entropy alloys – A review, J. Mater. Res., 33 (2018) 3092–3128.
18. R. Soler, A. Evirgen, M. Yao, C. Kirchlechner, F. Stein, M. Feuerbacher, D. Raabe, G. Dehm, Microstructural and mechanical characterization of an equiatomic YGdTbDyHo high entropy alloy with hexagonal close-packed structure, Acta Mater., 156 (2018) 86–96.
19. A. Takeuchi, K. Amiya, T. Wada, K. Yubuta, W. Zhang, High-entropy alloys with a hexagonal close-packed structure designed by equi-atomic alloy, strategy and binary phase diagrams, JOM. 66 (2014) 1984–1992.
20. Z. Wu, H. Bei, F. Otto, G.M. Pharr, E.P. George, Recovery, recrystallization, grain growth and phase stability of a family of FCC-structured multi-component equiatomic solid solution alloys, Intermetallics, 46 (2014) 131–140.
21. Y. Wu, W.H. Liu, X.L. Wang, D. Ma, A.D. Stoica, T.G. Nieh, Z.B. He, Z.P. Lu, In-situ neutron diffraction study of deformation behavior of a multi-component high-entropy alloy, Appl. Phys. Lett., 104 (5) (2014) 051910.
22. B. Gludovatz, A. Hohenwarter, D. Catoor, E.H. Chang, E.P. George, R.O. Ritchie, A fracture-resistant high-entropy alloy for cryogenic applications, Science. 345 (6201) (2014) 1153–1158.
23. M.S. Lucas, G.B. Wilks, L. Mauger, J.A. Muñoz, O.N. Senkov, E. Michel, J. Horwath, S.L. Semiatin, M.B. Stone, D.L. Abernathy, E. Karapetrova, Absence of long-range chemical ordering in equimolar FeCoCrNi, Appl. Phys. Lett., 100 (25) (2012) 251907.
24. M.J. Yao, K.G. Pradeep, C.C. Tasan, D. Raabe, A novel, single phase, non-equiatomic FeMnNiCoCr high-entropy alloy with exceptional phase stability and tensile ductility, Scr. Mater., 72–73 (2014) 5–8.
25. L.J. Santodonato, Y. Zhang, M. Feygenson, C.M. Parish, M.C. Gao, R.J. Weber, J.C. Neuefeind, Z. Tang, P.K. Liaw, Deviation from high-entropy configurations in the atomic distributions of a multi-principal-element alloy, Nat. Commun., 6 (2015) 5964.
26. M.A. Hemphill, T. Yuan, G.Y. Wang, J.W. Yeh, C.W. Tsai, A. Chuang, P.K. Liaw, Fatigue behavior of $Al_{0.5}CoCrCuFeNi$ high entropy alloys, Acta Mater., 60 (16) (2012) 5723–5734.

27. Z. Tang, T. Yuan, C.-W. Tsai, J.-W. Yeh, C.D. Lundin, P.K. Liaw, Fatigue behavior of a wrought $Al_{0.5}CoCrCuFeNi$ two-phase high-entropy alloy, Acta Mater., 99 (2015) 247–258.
28. S.G. Ma, S.F. Zhang, M.C. Gao, P.K. Liaw, Y. Zhang, A successful synthesis of the $CoCrFeNiAl_{0.3}$ single-crystal, high-entropy alloy by Bridgman solidification, JOM, 65 (12) (2013) 1751–1758.
29. T.-T. Shun, Y.-C. Du, Microstructure and tensile behaviors of FCC Al0.3CoCrFeNi high entropy alloy, J. Alloys Compd., 479 (1–2) (2009) 157–160.
30. J.Y. He, W.H. Liu, H. Wang, Y. Wu, X.J. Liu, T.G. Nieh, Z.P. Lu, Effects of Al addition on structural evolution and tensile properties of the FeCoNiCrMn high-entropy Alloy system, Acta Mater., 62 (2014) 105–113.
31. S. Guo, C. Ng, C.T. Liu, Anomalous solidification microstructures in Co-free $AlxCrCuFeNi_2$ high-entropy alloys, J. Alloys Compd., 557 (2013) 77–81.
32. Z. Tang, M.C. Gao, H. Diao, T. Yang, J. Liu, T. Zuo, Y. Zhang, Z. Lu, Y. Cheng, Y. Zhang, K.A. Dahmen, P.K. Liaw, T. Egami, Aluminum alloying effects on lattice types, microstructures, and mechanical behavior of high-entropy alloys systems, JOM, 65 (12) (2013) 1848–1858.
33. O.N. Senkov, S.V. Senkov, C. Woodward, Effect of aluminum on the microstructure and properties of two refractory high-entropy alloys, Acta Mater., 68 (2014) 214–228.
34. B.S. Li, Y.P. Wang, M.X. Ren, C. Yang, H.Z. Fu, Effects of Mn Ti and V on the microstructure and properties of AlCrFeCoNiCu high entropy alloy, Mater. Sci. Eng. A, 498 (1–2) (2008) 482–486.
35. Y. Dong, K. Zhou, Y. Lu, X. Gao, T. Wang, T. Li, Effect of vanadium addition on the microstructure and properties of AlCoCrFeNi high entropy alloy, Mater. Des., 57 (2014) 67–72.
36. M.-R. Chen, S.-J. Lin, J.-W. Yeh, M.-H. Chuang, S.-K. Chen, Y.-S. Huang, Effect of vanadium addition on the microstructure, hardness, and wear resistance of $Al_{0.5}CoCrCuFeNi$ high-entropy alloy, Metall. Mater. Trans. A, 37 (5) (2006) 1363–1369.
37. Y. Dong, Y. Lu, J. Kong, J. Zhang, T. Li, Microstructure and mechanical properties of multi-component $AlCrFeNiMo_x$ high-entropy alloys, J. Alloys Compd., 573 (2013) 96–101.
38. O.N. Senkov, G.B. Wilks, D.B. Miracle, C.P. Chuang, P.K. Liaw, Refractory high-entropy alloys, Intermetallics, 18(9) (2010) 1758–1765.
39. O.N. Senkov, G.B. Wilks, J.M. Scott, D.B. Miracle, Mechanical properties of $Nb_{25}Mo_{25}Ta_{25}W_{25}$ and $V_{20}Nb_{20}Mo_{20}Ta_{20}W_{20}$ refractory high entropy alloys, Intermetallics, 19 (5) (2011) 698–706.
40. M.S. Lucas, L. Mauger, J.A. Muñoz, Y. Xiao, A.O. Sheets, S.L. Semiatin, J. Horwath, Z. Turgut, Magnetic and vibrational properties of high-entropy alloys, J. Appl. Phys., 109 (7) (2011) 07E307.
41. S. Sohn, Y. Liu, J. Liu, P. Gong, S. Prades-Rodel, A. Blatter, B.E. Scanley, C.C. Broadbridge, J. Schroers, Noble metal high entropy alloys, Scr. Mater., 126 (2017) 29–32.
42. H.Y. Diao, R. Feng, K.A. Dahmen, P.K. Liaw, Fundamental deformation behavior in high-entropy alloys: An overview, Curr. Opin. Solid State Mater. Sci., 21 (5) (2017) 252–266.
43. J. W. Yeh, Physical metallurgy of high-entropy alloys, JOM, 67 (2015) 2254–2261.
44. V. Balaji, M.X. Anthony, Development of high entropy alloys (HEAs): Current trends, Heliyon, 10 (7) (2024) e26464.
45. S. Ranganathan, Alloyed pleasures: Multimetallic cocktails, Curr. Sci., 85 (2003) 1404–1406.
46. G. Laplanche, A. Kostka, C. Reinhart, J. Hunfeld, G. Eggeler, E.P. George, Reasons for the superior mechanical properties of medium-entropy CrCoNi compared to high-entropy CrMnFeCoNi, Acta Mater., 128 (2017) 292–303.

47. S. Huang, W. Li, S. Lu, F. Tian, J. Shen, E. Holmström, L. Vitos, Temperature dependent stacking fault energy of FeCrCoNiMn high entropy alloy, Scr. Mater., 108 (2015) 44–47.
48. W.H. Liu, Z.P. Lu, J.Y. He, J.H. Luan, Z.J. Wang, B. Liu, Y. Liu, M.W. Chen, C.T. Liu, Ductile CoCrFeNiMox high entropy alloys strengthened by hard intermetallic phases, Acta Mater., 116 (2016) 332–342.
49. Z. Lei, X. Liu, Y. Wu, H. Wang, S. Jiang, S. Wang, X. Hui, Y. Wu, B. Gault, P. Kontis, D. Raabe, L. Gu, Q. Zhang, H. Chen, H. Wang, J. Liu, K. An, Q. Zeng, T.G. Nieh, Z. Lu, Enhanced strength and ductility in a high-entropy alloy via ordered oxygen complexes, Nature, 563 (7732) (2018) 546–550.
50. B.X. Cao, C. Wang, T. Yang, C.T. Liu, Cocktail effects in understanding the stability and properties of face-centered-cubic high-entropy alloys at ambient and cryogenic temperatures, Scr. Mater., 187 (2020) 250–255.
51. G. Laplanche, A. Kostka, O. Horst, G. Eggeler, E.P. George, Microstructure evolution and critical stress for twinning in the CrMnFeCoNi high-entropy alloy, Acta Mater., 118 (2016) 152–163.
52. L. Wang, L. Zhang, X. Lu, F. Wu, X. Sun, H. Zhao, Q. Li, Surprising cocktail effect in high entropy alloys on catalyzing magnesium hydride for solid-state hydrogen storage, Chem. Eng. J., 465 (2023) 142766.
53. C.C. Chang, Y.T. Hsiao, Y.L. Chen, C.Y. Tsai, Y.J. Lee, P.H. Ko, S.Y. Chang, Lattice distortion or cocktail effect dominates the performance of tantalum-based high-entropy nitride coatings, Appl. Surf. Sci., 577 (2022) 151894.
54. N. Ishizu, J. Kitagawa, New high-entropy alloy superconductor $Hf_{21}Nb_{25}Ti_{15}V_{15}Zr_{24}$, Results Phys., 13 (2019) 102275.
55. S. Singh, N. Wanderka, B.S. Murty, U. Glatzel, J. Banhart, Decomposition in multi-component AlCoCrCuFeNi high-entropy alloy, Acta Mater., 59 (1) (2011) 182–190.
56. H.-W. Luan, Y. Shao, J.-F. Li, W.-L. Mao, Z.-D. Han, C. Shao, K.-F. Yao, Phase stabilities of high entropy alloys, Scr. Mater., 179 (2020) 40–44.
57. R. Zhang, S. Zhao, J. Ding, Y. Chong, T. Jia, C. Ophus, M. Asta, R.O. Ritchie, A.M. Minor, Short-range order and its impact on the CrCoNi medium-entropy alloy, Nature, 581 (2020) 283–287.
58. S. Ma, J. Zhang, B. Xu, Y. Xiong, W. Shao, S. Zhao, Chemical short-range ordering regulated dislocation cross slip in high-entropy alloys, J. Alloys Compd., 911 (2022) 165144.
59. D. Li, P.K. Liaw, L. Xie, Y. Zhang, W. Wang, Advanced high-entropy alloys breaking the property limits of current materials, J. Mater. Sci. Technol., 186 (2024) 219–230.
60. F. Otto, Y. Yang, H. Bei, E.P. George, Relative effects of enthalpy and entropy on the phase stability of equiatomic high-entropy alloys, Acta Mater., 61 (7) (2013) 2628–2638.
61. C. Tandoc, Y.-J. Hu, L. Qi, P.K. Liaw, Mining of lattice distortion, strength, and intrinsic ductility of refractory high entropy alloys, npj Comput. Mater., 9 (2023) 53.
62. A.J. Zaddach, C. Niu, C.C. Koch, D.L. Irving, Mechanical properties and stacking fault energies of NiFeCrCoMn high-entropy alloy, JOM, 65 (12) (2013) 1780–1789.
63. Z. Wang, Q. Fang, J. Li, B. Liu, Y. Liu, Effect of lattice distortion on solid solution strengthening of BCC high-entropy alloys, J. Mater. Sci. Technol., 34 (2) (2018) 349–354.
64. O.N. Senkov, J.M. Scott, S.V. Senkova, D.B. Miracle, C.F. Woodward, Microstructure and room temperature properties of a high-entropy TaNbHfZrTi alloy, J. Alloys Compd., 509 (2011) 6043–6048.
65. H.W. Yao, J.W. Qiao, J.A. Hawk, H.F. Zhou, M.W. Chen, M.C. Gao, Mechanical properties of refractory high-entropy alloys: Experiments and modelling, J. Alloys Compd., 696 (2017) 1139–1150.
66. R. Labusch, A statistical theory of solid solution hardening, Phys. Stat. Sol., 41 (2) (1970) 659–669.

67. H.A. Moreen, R. Taggart, D.H. Polonis, A model for the prediction of lattice parameters of solid solutions, Metall. Trans., 2 (1971) 265–268.
68. L. Li, Q. Fang, J. Li, B. Liu, Y. Liu, P.K. Liaw, Lattice-distortion dependent yield strength in high entropy alloys, Mater. Sci. Eng. A, 784 (2020) 139323.
69. H. Wang, P.Y. Yang, W.J. Zhao, S.H. Ma, J.H. Hou, Q.F. He, C.L. Wu, H.A. Chen, Q. Wang, Q. Cheng, B.S. Guo, J.C. Qiao, W.J. Lu, S.J. Zhao, X.D. Xu, C.T. Liu, Y. Liu, C.W. Pao, Y. Yang, Lattice distortion enabling enhanced strength and plasticity in high entropy intermetallic alloy, Nat. Commun., 15 (2024) 6782.
70. J. Huang, W. Fang, C. Xue, T. Peng, H. Yu, J. Li, L. Sun, X. He, B. Liu, Y. Yang, F. Yin, Exploring the relationship between lattice distortion and phase stability in a multi-principal element alloy system based on machine learning method, Comput. Mater. Sci., 221 (2023) 112089.
71. H. Wang, Q. He, X. Gao, Y. Shang, W. Zhu, W. Zhao, Z. Chen, H. Gong, Y. Yang, Multifunctional high entropy alloys enabled by severe lattice distortion, Adv. Mater., 36 (2024) 2305453.
72. D.L. Beke, G. Erdelyi, On the diffusion in high-entropy alloys, Mater. Lett., 164 (2016) 111–113.
73. M.H. Tsai, C.W. Wang, C.W. Tsai, W.J. Shen, J.W. Yeh, J.Y. Gan, W.W. Wu, Thermal stability and performance of NbSiTaTiZr high-entropy alloy barrier for copper metallization, J. Electrochem. Soc., 158 (2011) H1161.
74. T.T. Shun, C.H. Hung, C.F. Lee, Formation of ordered/disordered nanoparticles in FCC high entropy alloys, J. Alloys Compd., 493 (2010) 105–109.
75. K.-Y. Tsai, M.-H. Tsai, J.-W. Yeh, Sluggish diffusion in Co-Cr-Fe-Mn-Ni high-entropy alloys, Acta Mater., 61 (2013) 4887–4897.
76. Q.F. He, Y.F. Ye, Y. Yang, The configurational entropy of mixing of metastable random solid solution in complex multicomponent alloys, Appl. Phys., 120 (2016) 154902.
77. J. Dąbrowa, W. Kucza, G. Cieślak, T. Kulik, M. Danielewski, J.W. Yeh, Interdiffusion in the FCC-structured Al-Co-Cr-Fe-Ni high entropy alloys: Experimental studies and numerical simulations, J. Alloys Compd., 674 (2016) 455–462.
78. V. Verma, A. Tripathi, K.N. Kulkarni, On interdiffusion in FeNiCoCrMn high entropy alloy, J. Phase Equilib. Diffus., 38 (2017) 445–456.
79. J.W. Yeh, Recent progress in high-entropy alloys, Ann. Chimie Sci. Matériaux, 31 (2006) 633–648.
80. C.J. Tong, Y.L. Chen, J.W. Yeh, S.J. Lin, S.K. Chen, T.T. Shun, C.H. Tsau, S.Y. Chang, Microstructure characterization of $Al_xCoCrCuFeNi$ high-entropy alloy system with multiprincipal elements, Metall. Mater. Trans. A, 36A (2005) 881–893.
81. M. Vaidya, K.G. Pradeep, B.S. Murty, G. Wilde, S.V. Divinski, Bulk tracer diffusion in CoCrFeNi and CoCrFeMnNi high entropy alloys, Acta Mater., 146 (2018) 211–224.
82. N.G. Jones, R. Izzo, P.M. Mignanelli, K.A. Christofidou, H.J. Stone, Phase evolution in an $Al_{0.5}CrFeCoNiCu$ high entropy alloy, Intermetallics, 71 (2016) 43–50.
83. H. Ouyang, B. Fultz, Percolation in alloys with thermally activated diffusion, J. Appl. Phys., 66 (1989) 4752–4755.
84. B. Xu, S. Ma, Y. Xiong, J. Zhang, S. Huang, J.J. Kai, S. Zhao, Exploring the influence of percolation on vacancy-mediated diffusion in CoCrNi multi-principal element alloys, Mater. Des., 223 (2022) 111238.
85. X. Xu, J. Wang, J.P. Lv, Y. Deng, Simultaneous analysis of three-dimensional percolation models, Front. Phys., 9 (2014) 113–119.
86. Y.N. Osetsky, L.K. Béland, A.V. Barashev, Y. Zhang, On the existence and origin of sluggish diffusion in chemically disordered concentrated alloys, Curr. Opin. Solid State Mater. Sci., 22 (3) (2018) 65–74.
87. S. Sen, X. Zhang, L. Rogal, G. Wilde, B. Grabowski, S.V. Divinski, 'Anti-sluggish' Ti diffusion in HCP high-entropy alloys: Chemical complexity vs. lattice distortions, Scr. Mater., 224 (2023) 115117.

88. M. Mizuno, K. Sugita, H. Araki, Defect energetics for diffusion in CrMnFeCoNi high-entropy alloy from first-principles calculations, Comput. Mater. Sci., 170 (2019) 109163.
89. J. Dąbrowa, M. Zajusz, W. Kucza, G. Cieślak, K. Berent, T. Czeppe, T. Kulik, M. Danielewski, Demystifying the sluggish diffusion effect in high entropy alloys, J. Alloys Compd., 783 (2019) 193–207.
90. M. Vaidya, S. Trubel, B.S. Murty, G. Wilde, S.V. Divinski, Ni tracer diffusion in CoCrFeNi and CoCrFeMnNi high entropy alloys, J. Alloys Compd., 688 (Part B) (2016) 994–1001.
91. M.S. Daw, M. Chandross, Sluggish diffusion in random equimolar FCC alloys, Phys. Rev. Mater., 5 (2021) 043603.
92. J. Zhang, C. Gadelmeier, S. Sen, R. Wang, X. Zhang, Y. Zhong, U. Glatzel, B. Grabowski, G. Wilde, S.V. Divinski, Zr diffusion in BCC refractory high entropy alloys: A case of 'non-sluggish' diffusion behaviour, Acta Mater., 233 (2022) 117970.

Section II

Preparation and Characterization of High Entropy Alloys

3 Preparation of High Entropy Alloys by Various Metallurgical Methods

Ismail Bayar

3.1 INTRODUCTION

High entropy alloys (HEAs) are typically prepared as massive bulk ingots, coatings, or powders. These alloys are obtained by various preparation methods depending on the required properties and applications. Preparation methods can be classified according to the dimension or state of the HEA. According to the dimensions, the preparation methods are mechanical alloying, gas atomization for HEA particles, laser cladding, magnetron sputtering for coating alloys, additive manufacturing, casting, arc melting, vacuum arc melting, and Bridgman solidification for bulk alloys. HEA particles are prepared by mechanical alloying and then sintered by powder metallurgy processes to form a dense structure. HEA element powders are produced spherically by gas atomization and used in other processes. Mechanical alloying is used for solid-state, vacuum arc melting, arc melting, Bridgman solidification, laser cladding for liquid-state, and magnetron sputtering for gas-state according to the state of materials. Figure 3.1 illustrates common HEA preparation methods depending on the state of materials.

HEAs are prepared by processes similar to conventional alloy preparation methods, but the high entropy characteristic of these alloys requires higher concentrations of each element. The preparation of HEAs requires a detailed investigation of the basic principles of melting and casting theory, solidification and crystallization theory, and heat treatment [1]. Alloy preparation processes play an important role in creating microstructures and stabilizing phase structures. Once the alloys are prepared, a heat treatment process can be performed to further improve the alloys' properties [2]. This chapter describes commonly used HEA preparation methods.

3.2 SOLID PREPARATION METHODS

This section describes the mechanical alloying and sintering processes used to prepare solid-state HEAs. Mechanical alloying is followed by sintering to achieve a dense alloy structure. Spark plasma sintering is a commonly used method for the preparation of HEAs.

DOI: 10.1201/9781003664093-5

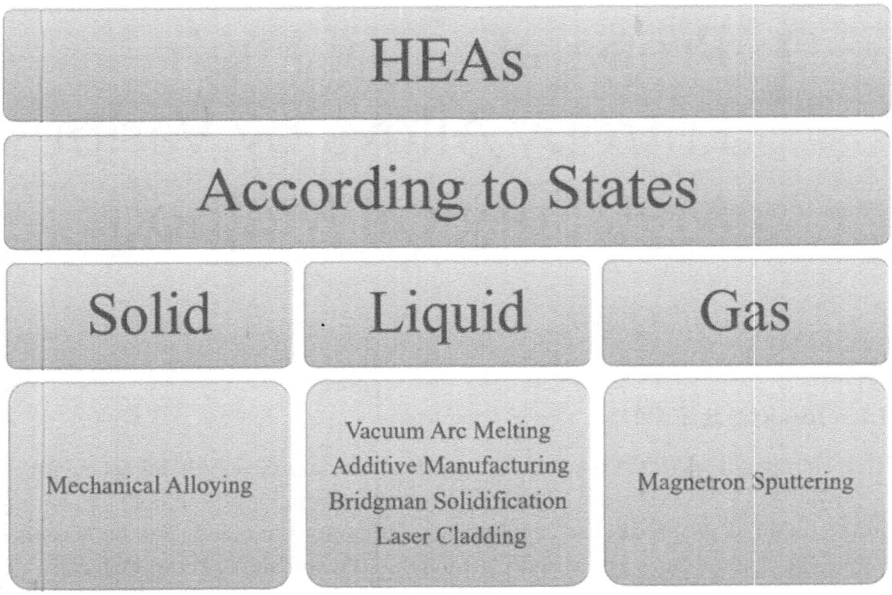

FIGURE 3.1 HEA preparation methods depending on the state of materials.

3.3 MECHANICAL ALLOYING

Mechanical alloying (MA) is a method based on the mixing of alloy powders in a mill by physical interactions. In this process, the powder particles interact with mechanisms such as plastic deformation, fracture, and cold welding as a result of collisions that occur during grinding, allowing the materials to blend at the atomic level and form a homogeneous structure. The fact that HEAs contain five or more elements, and their homogeneous mixing is critical in the production process, this makes the MA method an essential technique in HEA production.

In order for HEAs to be a dense component, they must be sintered following the MA process. This sintering process is usually carried out by the Spark Plasma Sintering (SPS) method. In the SPS method, the powders are densified in a mold under the influence of high temperature and mechanical pressure in a vacuum chamber to form a solid structure. Figure 3.2 shows an illustration of the spark plasma sintering process. The vacuum chamber is established by pumping argon or a similar inert gas into the system in a controlled manner. As a result, HEAs homogenously blended by the MA method are formed into a dense structure by the SPS method.

The parameters MA milling time and SPS sintering temperature are important in the preparation of HEAs. By changing these parameters, mechanical properties can be improved. Silas Anamu et al. [3] investigated the effects of MA milling time and the SPS sintering temperature process parameters on relative density and microhardness in $Ti_{20}Al_{20}Cr_5Nb_5Ni_{19}Cu_{12}Co_{19}$ HEA prepared by MA and SPS methods. In the study, it was stated that milling time and sintering temperature are critical in achieving high density, which contributes to densification as well as superior

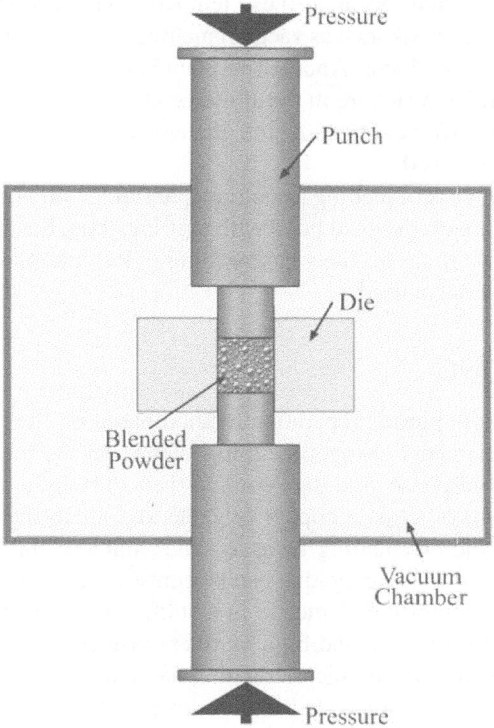

FIGURE 3.2 Schematic illustration of spark plasma sintering method.

mechanical properties. According to the optimization model, 99.9% relative density, 0.099999994% porosity, and 835.21 HV microhardness values can be obtained at a milling time of 7.2 hours and a sintering temperature of 899.7°C. In another study [4], the effects of sintering temperatures in the range of 550-1100°C on phase formation and microstructure were investigated. In $Al_{0.3}CoCrFeNi$ HEA alloy, the microstructure was observed to consist of face-centered cubic (FCC) solid solutions containing B2 precipitates. It was found that the size and content of these precipitates decreased with increasing temperature. It is also stated that the microstructure and especially the size, distribution and growth of large grains can be controlled by milling and sintering parameters. The sintering temperature is a critical parameter in the preparation of HEAs by the SPS process. In addition, sintering time, applied pressure, and heating rate also play an important role.

In addition to the SPS method, hot-pressing [5, 6] and hot isostatic pressing [7, 8] techniques are used to sinter HEAs to form a dense structure.

3.4 LIQUID PREPARATION METHODS

HEAs are prepared in coating (film) and bulk form by liquid phase methods. To prepare bulk HEAs in the liquid phase, all elements in the composition must be completely melted, and this process usually requires high temperatures. However, evaporation problems may occur at these temperatures since some elements have low melting points. Methods such as vacuum melting and additive manufacturing are used to overcome this problem. Another problem is the encountered to the inability of a homogeneous microstructure in the alloys produced. To overcome this problem, the alloy is re-melted after cooling and the process is repeated until the desired level of homogeneity is achieved.

Coating is the process of melting a coating material, in wire or powder form, onto a substrate to form a metallurgical bond with that substrate. Laser cladding is widely used in this process. Tungsten inert gas welding (TIG) and plasma transferred arc (PTA) techniques are also used.

3.5 ARC MELTING

Arc melting is a liquid phase preparation method based on the principle of melting different alloying elements in ingot form together in an arc melting furnace, mixing them in the liquid phase, and then cooling them. The system consists of an arc torch for the melting process, a copper crucible and a cooling system. Schematic diagrams showing the arc melting furnace under ambient and vacuum conditions are provided in Figure 3.3. Arc melting can be considered as a suitable technique for forming HEAs of some of the elements. In a study [9] on AlCoCuFeTi HEA alloy with excellent wear resistance and high hardness properties prepared by arc melting, the effects of annealing on microstructure, nanomechanical behavior, tribological properties, and corrosion resistance were investigated. The results showed that AlCoCuFeTi alloy consists of a Co-enriched L21 phase, Cu-enriched FCC phase, and (Fe, Ti) enriched Laves phase. The annealing process promoted the formation of the FCC and Laves phases while reducing the volume fraction of the L21 phase. The high hardness value of the alloy was attributed to the presence of L21 and Laves phases. The highest hardness value (14.1 ± 1.3 GPa) was obtained in the sample annealed at 1100°C, while the lowest Young's modulus (256 ± 11 GPa) was

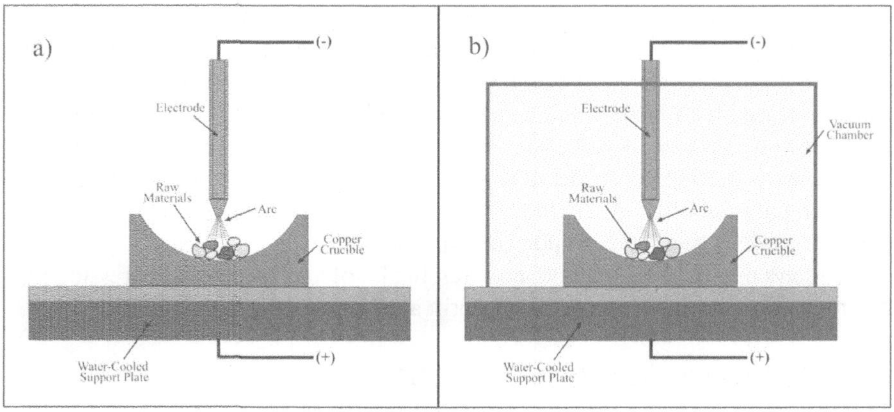

FIGURE 3.3 Schematic diagrams of the arc melting furnace under (a) ambient and (b) vacuum conditions.

obtained in the sample annealed at 900°C. All specimens exhibited superior wear resistance compared to cast-HEAs due to the mild oxidational wear mechanism. In particular, the sample annealed at 1100°C showed the best wear resistance and high corrosion resistance. Chen et al. [10] prepared $CrFeCoNiAl_{0.6}$ HEA by arc melting and sub-rapid cooling with controlled low cooling rates. The study showed that ultra-thin FCC+BCC dual-phase structures can be obtained at appropriate cooling rates, and the cooling rate can adjust the mechanical properties such as yield strength (~580–1460 MPa) and hardness (~260–550 HV) with the fraction of BCC phase. The sub-rapid solidification process in arc melting offers a method for the development of high-performance HEAs.

The arc torch temperature over 3000°C can make arc melting unsuitable for elements with low melting temperatures. Temperature control is critical to prevent or minimize the evaporation of such elements. Therefore, vacuum or vacuum induction heating methods may be more appropriate for materials with low melting temperatures.

3.6 VACUUM ARC MELTING

Vacuum arc melting (VAM) is one of the preferred techniques among liquid phase preparation methods. Although this method is systematically similar to arc melting, it involves creating a vacuum environment using argon gas in the furnace. An arc torch, copper crucible, and cooling system are also included in the system. The vacuum environment provides a protective atmosphere, eliminating the effects of reactive gases such as oxygen, nitrogen and water vapour in the air. In this way, the homogeneity of the prepared alloys is increased, and their microstructure is improved. Parameters such as arc intensity, cooling rate, crucible size and solidification rate are critical in the VAM method. Controlling these parameters allows different phases and microstructures to be obtained. The VAM method is widely preferred for the preparation of HEAs due to its precise control over composition and purity. This method is especially appropriate for metals with high melting points and metals with high susceptibility to oxidation. In a study [11] examining the microstructure, mechanical and wear properties of FeCoNiCrMnTix HEAs prepared by VAM method, it was determined that the crystal structure of the alloys consists of FCC and BCC phases, and the precipitation of Ni3Ti reinforcement phase suppresses the FCC phase and promotes the formation of the BCC phase. The dendrites of the FCC phase are enriched in Fe, Co, and Cr, while Ni, Mn, and Ti elements are concentrated in the inter-dendrite regions of the BCC phase. The yield strength of FeCoNiCrMnTix HEAs varies between 296–601 MPa depending on the Ti content. Tensile test results showed that the fracture mechanism was characterized by ductile and brittle fracture properties. It has been reported that the oxide layer formed protects the alloy matrix during friction and the wear mechanism gradually changes from adhesive wear to abrasive wear.

The VAM method is also used to prepare refractory high entropy alloys (RHEA). RHEAs differ from normal HEAs in terms of the elements used in their composition. While normal HEAs are generally synthesized from metals such as Fe, Co, Ni, Cu, and Al, RHEAs are prepared using refractory materials such as Mo, Nb, Cr, Ta, Ti, V, W, and Hf [12]. RHEAs generally have many features compared to normal

HEAs, such as excellent resistance to thermal softening, good ductility, oxidation resistance, excellent corrosion and wear resistance [13]. In a study prepared by the NbZrTiCrAl RHEA alloy using the VAM method, the microstructure and oxidation behavior of the alloy were investigated [14]. It was found that the alloy showed different levels of oxidation resistance at 800°C and 1000°C but exhibited inadequate resistance to long-term oxidation at 1200°C. It was also reported that the alloy has a single-phase BCC solid solution structure and exhibits an inhomogeneous microstructure consisting of dendrite and inter-dendrite regions.

3.7 VACUUM INDUCTION MELTING

The vacuum induction melting (VIM) method is used to melt metallic materials with different melting temperatures in a vacuum environment with an electromagnetic induction heating system. In the process, mixing is carried out to obtain a homogeneous alloy from different elements. The melted material is then cooled and poured into a mold. These features make it a preferred method in the production of HEAs [15]. The macro, microstructure, and mechanical properties of the cast single-phase $Al_{0.2}CoCrFeNi$ HEA alloy prepared by the VIM method were investigated [16]. Classical solidification properties were observed in the macrostructure of the alloy, and no macrosegregation was detected. However, Al and Ni microsegregation was observed between the dendritic and interdendritic regions in the upper part. Differences in tensile strength were recorded between the upper center (330 MPa) and lower center (464 MPa) of the alloy, and this was attributed to the changes in the dendrite arm spacing. Nanoindentation tests confirmed the relatively high hardness in the upper center and showed that this was due to the composition of the dendritic regions. The study showed that this structure produced by VIM can play a leading role in the development of structural materials.

The microstructure and mechanical properties are improved by applying heat treatment to the HEAs prepared with VIM. In this context, the effects of heat treatment temperatures on the microstructure, hardness, and tensile properties of the CoCuFeMnNi dual-phase HEAs prepared with the VIM method were investigated in the research conducted [17]. The presence of Cu-rich and Cu-poor phases in the cast sample was determined; it was observed that the Cu-rich content decreased with increasing heat treatment temperatures. In mechanical tests, while the tensile strength slightly decreased with increasing heat treatment temperatures, a significant improvement was observed in the failure stress and ductility properties. For example, after heat treatment at 891°C, the failure strain increased by 84.2%, while the tensile strength decreased by only 5.28%. In addition, the fracture behavior of the alloy changed from brittle to ductile as the heat treatment temperatures increased.

3.8 BRIDGMAN SOLIDIFICATION METHOD

The Bridgman solidification method is a technique generally used in the production of materials with a single crystal or orientated solidification structure. In this method, the temperature gradient and the direction of heat flow generated during the solidification of a molten material are precisely controlled, allowing the growth

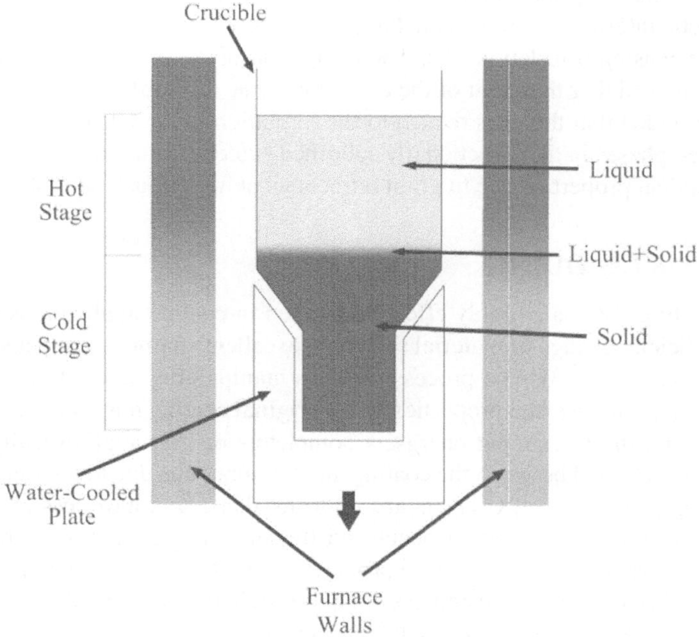

FIGURE 3.4 Schematic of Bridgman solidification technique.

rate and direction of the crystal to be regulated. Schematically, the Bridgman solidification method is given in Figure 3.4. This controlled solidification process offers the possibility of improving the mechanical properties of the alloy by optimizing its microstructure. In a crucible, alloy elements are melted, and the molten alloy is gradually solidified. A water cooling system is used to cool the outer surface of the molten alloy [18]. The mechanical properties of single-crystal materials differ from those of polycrystals, and this difference is generally beneficial. Therefore, the Bridgman solidification method is preferred in cases where HEAs are required to have a single crystal structure.

With this method it is possible to obtain a variety of microstructures and mechanical properties at different rates of translation. $Al_{0.8}CrFeNi_{2.2}$ alloy, prepared by arc melting method and subjected to directed solidification by Bridgman solidification method in the range of 1.4–14 μm/s translation rates, exhibited a planar solid-liquid interface growth at low rates and a regular lamellar array was formed. With the increase in the translation rates, a transition from planar to cellular structure was observed at the interface. Mechanical behavior analyses showed that the translation rate increased the yield strength without compromising ductility. A maximum yield strength of 558 MPa was obtained with a total strain of approximately 6.5% [19]. $Al_{0.9}CoCrNi_{2.1}$ eutectic high entropy alloy (EHEA) was prepared by the Bridgman solidification method at translation rates ranging from 6 to 120 μm/s. During orientated solidification, a lamellar eutectic structure containing FCC ($L1_2$) and BCC (B2) phases was formed due to different translation rates. The character of the solid-liquid interface of the alloy showed a transition from an almost planar structure to a cellular eutectic interface, with the lamella spacing decreasing from 7.17 μm to 1.65 μm with an increasing translation rate. The elongation of $Al_{0.9}CoCrNi_{2.1}$ eutectic HEA was two times higher than that of the as-cast alloy at 120 μm/s translation rate, and it was concluded that this was related to the formation of the $L1_2$ phase and Cr-rich precipitated phases in the directionally solidified eutectic structure, which increased the mechanical properties due to grain refinement at high translation rates [20].

3.9 LASER CLADDING

Laser cladding (LC) is a highly effective method in coating applications due to its coating efficiency, range of material selection, excellent coating quality and controllable properties [21–23]. The process involves manipulating a substrate and particles. In the LC process, the properties of the original coating material are preserved by controlling the laser input energy; a completely dense metallurgically bonded layer can be obtained between the coating and the substrate; due to the rapid heating and cooling process, the LC layers are uniform, dense and have less microscopic defects and little thermal effect occurs on the substrate [24]. Cracks, holes, and deformation are minimized by these properties. The LC system consists of a laser beam as a heat source, a focusing system and nozzle to focus and direct the beam, and a manipulator for moving the substrate. The LC system is schematically shown in Figure 3.5. In this system, there are two types of configurations: the particles can either be pre-placed onto the substrate (Figure 3.5a) or fed into the system during the process using a particle feeding apparatus (Figure 3.5b). In the process, the laser beam creates a melt pool. When the laser beam moves away, the melt pool solidifies rapidly. The melt pool consists of substrate and particles and has a solid metallurgical bond after solidification. Due to its local heat-affected zone and rapid heating-cooling process, LC technology is widely used for the preparation of HEAs.

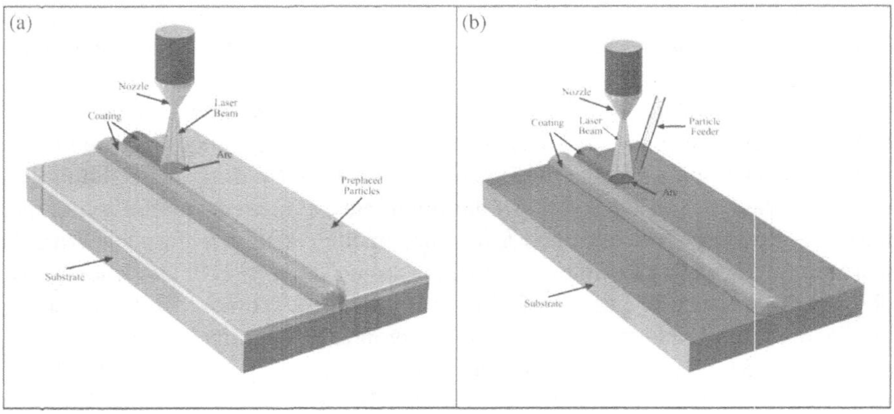

FIGURE 3.5 Schematic diagrams of laser cladding processes: (a) pre-placed particles and (b) with a particle feeder.

The effects of process parameters on the geometric and metallurgical properties of HEA alloys prepared by the LC method are investigated. In a study in which AlCoCrFeNiMo HEA coating layer was fabricated at different pulse frequencies (0.5 Hz, 5 Hz, 50 Hz, 500 Hz, and 5000 Hz) using continuous wave (CW) and pulsed wave (PW) laser parameters, the microstructure revealed the existence of BCC, σ, and B2 phases. The study findings reveal that PW coatings exhibit finer grain sizes, reduced σ phase ratio, superior wear resistance, corrosion resistance and corrosion-wear resistance compared to CW coatings. In particular, an impact frequency of 50 Hz was identified as the most suitable parameter to achieve the smallest grain size and optimum performance. The reduction of grain size and the presence of an appropriate proportion of the σ phase significantly improve wear and corrosion resistance by limiting plastic deformation and diffusion of oxygen atoms, while increasing the cutting resistance during friction [25]. Feng et al. [26] response surface method (RSM) was used to optimize the LC parameters for $Fe_{50}Mn_{30}Co_{10}Cr_{10}$ HEA cladding. Mathematical models were developed to analyse the relationships between processing parameters such as laser power, scanning speed and particle feed rate and results such as dilution ratio, microhardness, W/H and porosity. The results showed that dilution ratio and microhardness are positively correlated with laser power and scanning speed and negatively correlated with particle feed rate. Scanning speed was found to be the most influencing factor on porosity. The optimum parameters were determined as 1750 W laser power, 8.0 mm/s scanning speed, and 11.5 g/min particle feed rate. Under these parameters, the coating exhibited excellent ductility and strength with 0.07% porosity, 208.32 $HV_{0.2}$ microhardness, 55.93% deformation, and 2153.7 MPa compressive strength. The coating was composed of 15.6% HCP and 84.4% FCC phases. The fracture mechanism was reported to be transcrystalline fracture and toughness fractures. It was reported that the model showed a strong agreement with the experimental results.

Various processes are used to improve the properties of HEA coatings, such as ultrasonic impact treatment (UIT) and heat treatment following the LC process. Gu et al. [27] investigated CrMnFeCoNi HEA coatings by ultrasonic impact treatment (UIT) following LC process. The findings revealed that UIT optimized the grain size of the laser-coated CrMnFeCoNi HEA coatings, improved the surface tension, decreased the FeOOH content while increasing the proportion of Cr compounds, thus resulting in a significant improvement in the corrosion resistance and mechanical performance of the coatings. Using the LC method, Liu et al. [28] prepared $AlCoCrFeNiTi_{0.8}$ coatings and then investigated the evolution of microstructure and wear resistance after different temperatures of heat treatment. The coating consists of a BCC-FeCr solid solution matrix, a B2-AlNi precipitate phase, and an in-situ TiC particle phase. The microstructure was found to be stable below 700°C. At 900°C, the coarsening behavior of AlNi precipitates after heat treatment was observed, and at 1200°C, the Ostwald ripening phenomenon of the precipitates appeared. A close relationship was found between microstructure evolution and wear resistance, and it was determined that the coarsening and maturation of the precipitates caused a decrease in wear resistance. The coating exhibited a significant decrease in wear resistance after heat treatment at 1200°C.

3.10 ADDITIVE MANUFACTURING

Additive manufacturing (AM) is an innovative technology that allows material to be deposited in layers to create a three-dimensional part. Each layer is created by adding on top of the previous layer. Unlike conventional machining methods, AM technology is based on the principle of material addition rather than material removal. This method enables the direct production of three-dimensional models, provides high efficiency in manufacturing parts with complex geometries and offers design freedom. In addition, localized melting and subsequent rapid solidification processes are observed in AM methods. These features make AM technology a highly suitable method for preparing HEAs. In AM processes, materials in powder or wire form are used as feedstock. In HEA preparation, these feedstocks, prepared according to the alloy's composition, are brought together by local melting and rapid cooling processes to form the required alloy microstructure. Powder fluidity plays an important role in obtaining homogeneous dispersion and high-density alloys in AM processes, contributing to the improvement of mechanical properties and the formation of a clear shape. Powders with spherical or nearly spherical morphology and good fluidity properties reduce defects such as porosity and cracks during production; therefore, fine spherical powders are preferred in AM techniques. Commonly used methods for spherical powder production include ball milling (mechanical alloying), gas atomization, and inductively integrated thermal plasma spheroidization [29]. Studies on HEA have been carried out using AM methods such as powder bed fusion (PBF) [30, 31], selective laser melting (SLM) [32, 33], selective electron beam melting (SEBM) [34, 35], and wire arc additive manufacturing (WAAM) [36, 37].

To improve mechanical properties and microstructures, AM parameter studies are carried out. In the study [38] where HEA alloy was prepared using the GTA-WAAM method using $Al_{0.1}CoCrFeNi$ wire as stock material, the mechanical properties of the alloy were investigated. HEA thin-walled, multilayered HEA alloy structures were deposited with two heat input conditions, low and high, with the same energy density. In both, large columnar grains with cellular dendritic substructures and a homogeneous composition are observed; 200 A current 2000 mm/min (wire feed speed) and 200 mm/min (travel speed) parameters were defined as near-optimal processing conditions due to good surface quality, higher ductility, and higher deposition rate. In the studies of Bardo et al. [39], Ni-Co-Fe-Mn-Ti HEAs were successfully prepared using the PBF process using elemental powders. The AM process was carried out by changing the laser power and laser exposure time. Changes in process parameters and the content of alloying elements directly affected the machinability of the alloys and the sample properties. In the study, it was shown that Ni-Co-Fe-Mn-Ti HEAs with low porosity and high chemical homogeneity could be produced from elemental powders, especially for five-component HEAs; the best results were obtained with 250–300 W laser power and 145 μs laser exposure time.

The mechanical properties of HEAs prepared by AM processes were effectively enhanced by heat treatment. In the study conducted on CrMnFeCoNi HEA with SiC addition, after 2 hours of heat treatment at 900°C, the yield strength, tensile strength, and elongation of the 5% SiC alloy increased from 343 MPa to 431 MPa, from 631 MPa to 717 MPa, and from 33.2% to 36.4%, respectively. In addition, significant improvements were achieved in tensile strength and ductility (706 MPa and 30.3%) in the 10% SiC alloy after 2 hours of heat treatment at 700°C. These improvements were associated with the transformation of carbide particles into a fine and uniform structure, precipitation strengthening, and optimization of stress distribution [40]. The study presents an innovative heat treatment approach that optimizes the strengthening phases. Although HEAs exhibit high strength under ambient conditions, their tensile properties deteriorate at high temperatures. The aging process applied to $(CoCrNi)_{94}(TiAl)_6$ HEA prepared by the laser PBF method revealed the temperature-dependent changes in microstructure and mechanical properties. The initial microstructure contained cell boundaries reinforced with Al_2O_3 nanoparticles and ultra-fine cellular structures, and this structure provided high yield strength (725~750 MPa) and elongation (> 28%) at room temperature. Cellular structures were preserved up to 700°C but disappeared at 800°C. Aging at 600°C and above caused the formation of γ' precipitates. Samples aged at 700°C showed the highest yield (1227 MPa) and tensile strength (1539 MPa) at room temperature. Shen et al. [41] investigated the effect of W content on the microstructure and mechanical properties of CoCrFeNiWx HEA alloys successfully prepared by powder plasma arc additive manufacturing (PPA-AM) and found that increasing W content increased hardness and strength along with phase transformations but decreased ductility. Shen [37] examined the effects of heat input on the forming quality, microstructure, and mechanical properties of $CoCrFeNiMo_{0.2}$ HEAs prepared by WAAM using cold metal transfer (CMT) technology; it was noted that increased heat input deteriorated the forming quality, caused phase growth, and negatively affected the mechanical properties, but improved the tensile properties perpendicular to the deposition direction. Due to their stable structure and high melting temperatures, AM techniques are also used to prepare RHEA [42] and austenitic HEAs [36] in addition to other HEAs.

The AM process's microstructural variations can be explained by contrasting it with the casting method. In the study [43] where AlCoCrFeNi HEA was prepared by induction melting and AM methods, microstructure, and mechanical properties were investigated. B2 and BCC phases were detected in both methods, but the microstructure showed significant differences. A coarse-grained dendritic structure was obtained with induction melting, while a finer-grained and dendritic-like structure was obtained with the AM method. This difference was attributed to the differences in cooling rates. While FeCr and NiAl-rich regions were observed with induction melting, only Cr fluctuations were detected in SLM. This study shows that microstructure and elemental separation can be controlled depending on the preparation method, thus enabling the design of new alloys with different mechanical properties. It has not been possible to prepare crack- and porosity-free samples of some elements in HEA alloys using the AM method due to thermal stresses and phase transformations. To overcome these problems, preheating or change in alloy composition is recommended [43]. The AM methods are characterized by their ability to prepare complex alloys and their design freedom compared to other HEA preparation methods.

3.11 GAS PREPARATION METHODS

This section includes HEA coatings in the form of thin coatings prepared by steam. Coatings are deposited on a workpiece surface, usually called the substrate, by various deposition methods.

3.12 MAGNETRON SPUTTERING

Magnetron sputtering is the most widely used method for the preparation of HEAs from the gas phase. Magnetron sputtering is a physical vapour deposition (PVD) technique in which magnetic fields concentrate plasma on the target material, atoms are plucked from the target surface by ion collisions, and these atoms are transported onto the substrate to form a thin coating. This method allows HEAs to be prepared in a thin coating form. The superior performance properties of HEAs in bulk form have led researchers to study the protective properties of HEAs in coating form on materials [44]. Magnetron sputtering is the most widely preferred method due to its advantages, such as rapid coating formation, easy adjustability of the process parameters, the use of multiple targets, and strong adhesion of the coatings with the substrate [45]. Also, the properties of the coatings can be optimized by varying the combinations of different elements in the HEA, allowing various properties to be obtained [46].

In the preparation of HEAs by magnetron sputtering, the bias voltage is an important parameter of the efficiency of the process and coating properties. In a study where different bias voltages were investigated [44], CoCrFeNi thin-coating HEAs were prepared on 304 stainless-steel and single-crystal Si substrates by magnetron sputtering. The results obtained revealed that all coatings exhibited equiaxed nanoscale grains and demonstrated a columnar microstructure. Fe, Co, Cr, and Ni elements exhibited a homogeneous distribution in all coatings, and all coatings had a face-centered cubic (FCC) crystal structure. At 0 V bias voltage, the CoCrFeNi HEA film formed a single-phase FCC structure, while the crystallinity first decreased and then increased with increasing bias voltage. In particular, the lowest crystallinity was observed at –200 V bias voltage due to the FCC nanocrystalline and amorphous dual-phase structure of the films. However, at –200 V bias voltage, the thin films reached the maximum density and minimum surface roughness, while the grain size decreased to the lowest value of about 10 nm. Furthermore, higher corrosion resistance was achieved by optimizing the bias voltage. [47].

Radiofrequency (RF) power in magnetron sputtering is another critical parameter used to control the physical and chemical properties of HEA thin coatings. RF magnetron sputtering offers significant advantages, such as ease of use, high deposition rate, high coating purity, and strong coating-substrate metallurgical bonding [48]. In a study where (TiVCrNbSiTaBY)N HEA coatings were deposited on silicon wafers and stainless-steel plates by RF magnetron sputtering, the effects of RF power on the elemental composition, morphology, and microstructure of the coatings were investigated. The coatings exhibit a surface roughness of less than 1 nm, indicating that a highly uniform surface is obtained. Moreover, the coatings contain nanocomposite structure of amorphous Si_3N_4 with the FCC phase. An increase in the amorphous phase content was observed with increasing RF power, which resulted in the refinement of the crystal grains. The structural changes led to an initial increase in the nano hardness and elastic modulus of the coatings, followed by a decreasing trend. In particular, the coating grown with 800 W RF power reached the highest mechanical properties with a nano hardness of ~38.1 GPa and an elastic modulus of ~451.2 GPa, respectively [49].

REFERENCES

1. Y. Zhang, C.C. Koch, S.G. Ma, H. Zhang, and Y. Pan, Fabrication Routes. In: Gao, M., Yeh, JW., Liaw, P., Zhang, Y. (eds) High-Entropy Alloys. (2016) pp. 151–179. Springer, Cham. https://doi.org/10.1007/978-3-319-27013-5_5
2. S. Ragunath, N. Radhika, and B. Saleh, Advancements and future prospects of additive manufacturing in high-entropy alloy applications, Journal of Alloys and Compounds. 997 (2024) 174859.
3. U.S. Anamu, O.O. Ayodele, E. Olorundaisi, B.J. Babalola, K. Ukoba, N.M. Kibambe, M.C. Umba, T.-C. Jen, and P.A. Olubambi, Response surface methodology and process optimization of spark plasma sintered parameters of Ti20Al20Cr5Nb5Ni19Cu12Co19 high entropy alloy, Manufacturing Letters. 41 (2024) 1708.
4. A. Saviot, P. Sallamand, J.-P. Monchoux, C. Marcelot, R. Cours, N. Geoffroy, J.-M. Jouvard, and S. Le Gallet, Phase and microstructure formation during reactive spark plasma sintering of AlxCoCrFeNi (x=0.3 and 1) high entropy alloys, Journal of Materials Research and Technology. 32 (2024) 3047.
5. M. Çelebi, O. Güler, A.H. Karabacak, A. Çanakçı, S.B. Akçay, M. Kocaman, H. Çuvalcı, and T. Varol, Investigation of microstructural, hardness, and wear properties of AlCrCuFeNi high entropy alloys produced by hot-pressing with the enhancement of manufacturability through electroless Ag incorporation, Tribology International. 201 (2025) 110277.
6. B. Liu, H. Chen, J. Zhou, W. Wang, S. Xi, and X. Chen, Interfacial bonding behavior of WC/AlCoCrFeNi2.1 eutectic high-entropy alloy matrix composites fabricated by fast hot pressing sintering, Vacuum. 217 (2023) 112574.
7. X. Pan, X. Wang, and C. Qiu, Microstructural development and mechanical behavior of a near-eutectic high entropy alloy Al1.8CrCuFeNi2 fabricated by hot isostatic pressing, Intermetallics. 171 (2024) 108361.
8. J. Fiocchi, C.A. Biffi, M. Elnemr, J. Shipley, A. Tuissi, and R. Casati, Hot isostatic pressing and heat treatments of LPBFed CoCuFeMnNiTi0.13 high-entropy alloy: Microstructure and mechanical properties, Journal of Materials Research and Technology. 26 (2023) 9127.
9. X. Lai, Y. Ren, Q. Wang, S. Xing, C. Xu, J. Hou, I. Baker, and H. Wu, Effect of annealing on microstructure and properties of AlCoCuFeTi high-entropy alloy fabricated by arc melting, Intermetallics. 170 (2024) 108327.
10. C. Chen, Y. Fan, W. Wang, H. Zhang, J. Hou, R. Wei, T. Zhang, T. Wang, M. Li, S. Guan, and F. Li, Synthesis of ultrafine dual-phase structure in CrFeCoNiAl0.6 high entropy alloy via solid-state phase transformation during sub-rapid solidification, Journal of Materials Science & Technology. 113 (2022) 253.
11. Y. Li, Z. Wang, L. Sun, and X. Wang, Mechanical properties and wear behaviors of FeCoNiCrMnTix high-entropy alloys manufactured by vacuum arc melting, Journal of Materials Research and Technology. 30 (2024) 1962.
12. S.H. Guler, A. Yakin, O. Guler, A.K. Chattopadhyay, and T. Simsek, A critical review of the refractory high-entropy materials: RHEA alloys, composites, ceramics, additively manufactured RHEA alloys, Current Applied Physics. 70 (2025) 87.

13. J. Yi, S. Tang, C. Zhang, M. Xu, L. Yang, L. Wang, and L. Zeng, Microstructure and mechanical properties of a new refractory equiatomic CrHfNbTaTi high-entropy alloy, JOM. 73 (3) (2021) 934.
14. P. Zhang, Y. Li, Z. Chen, J. Zhang, and B. Shen, Oxidation response of a vacuum arc melted NbZrTiCrAl refractory high entropy alloy at 800–1200°C, Vacuum. 162 (2019) 20.
15. Y.-F. Yang, F. Hu, T. Xia, R.-H. Li, J.-Y. Bai, J.-Q. Zhu, J.-Y. Xu, and G.-F. Zhang, High entropy alloys: A review of preparation techniques, properties and industry applications, Journal of Alloys and Compounds. 1010 (2024) 177691.
16. S. Dudala, S.C. Krishna, and R. Korla, Macrostructure, microstructure, and mechanical properties of Al0.2CoCrFeNi high-entropy alloy produced by vacuum induction melting, Transactions of the Indian Institute of Metals. 77 (6) (2024) 1489.
17. S. Pourmorad Kaleybar, and H. Khorsand, Effects of annealing temperature on the microstructure and mechanical properties of the As-cast dual-phased CoCuFeMnNi high entropy alloys fabricated by vacuum induction melting method, Journal of Materials Engineering and Performance. (2024).
18. X. Li, Y. Fautrelle, and Z. Ren, Influence of an axial high magnetic field on the liquid–solid transformation in Al–Cu hypoeutectic alloys and on the microstructure of the solid, Acta Materialia. 55 (4) (2007) 1377.
19. K.N. Campo, M. Wischi, J.F.Q. Rodrigues, L.F. Starck, M.C. Sangali, and R. Caram, Directional solidification of the Al0.8CrFeNi2.2 eutectic high-entropy alloy, Journal of Materials Research and Technology. 30 (2024) 8874.
20. Y. Xu, S. Liu, J. Chang, Y. He, S. Liu, R. Hua, J. Zou, Q. Zhao, and Y. Chen, Formation of lamellar eutectic structure and improved mechanical properties of directional solidified Al0.9CoCrNi2.1 high-entropy alloy, Intermetallics. 173 (2024) 108430.
21. X.-M. Xie, X.-B. Liu, B.-M. He, F.-Z. Zhang, J.-X. Liang, X.-Y. Liu, and J. Zheng, Laser cladding Ni60-SiC/Ti3SiC2 self-lubricating composite coatings on IN718 alloy: Wear mechanisms and oxidation behaviors, Wear. 560-561 (2025) 205611.
22. A.M. Sedighi, S.F. Nabavi, and A. Farshidianfar, A review on effect of cooling rate on metallurgical, mechanical, geometrical characteristics and defects of laser cladding process, Lasers in Manufacturing and Materials Processing. 11 (3) (2024) 677.
23. D. Xiang, D. Wang, S. Wang, L. Yan, and Y. Chen, Analysis and optimization of laser cladding Ti-B4C composite coatings based on the interaction and GABP-NSGAII algorithm, Surface and Coatings Technology. 493 (2024) 131186.
24. X.-W. Qiu, and C.-G. Liu, Microstructure and properties of Al2CrFeCoCuTiNix high-entropy alloys prepared by laser cladding, Journal of Alloys and Compounds. 553 (2013) 216.
25. W. Zhao, X. Li, J. Bi, S. Li, C. Fang, H. Zhang, S. Gao, Y. Lv, and G. Xiao, Effect of pulse frequency on the microstructure and properties of laser cladding layer of AlCoCrFeNiMo high entropy alloy, Journal of Materials Research and Technology. 33 (2024) 9044.
26. L. Feng, G. Jin, X. Cui, B. Lu, Z. Dong, Y. Guan, X. Li, X. Yan, and M. Liu, Optimizing laser cladding process parameters for transformation-induced plasticity FeMnCoCr high entropy alloy: A multi-objective approach, Surface and Coatings Technology. 495 (2025) 131536.
27. B. Gu, J. Chu, Y. Wang, G. Xu, L. Gao, Y. Yang, Y. Hu, and H. Zhang, Effects of ultrasonic impact treatment on the corrosion resistance of laser-cladded CrMnFeCoNi high-entropy alloy coatings, Surface and Coatings Technology. 489 (2024) 131102.
28. H. Liu, J. Liu, X. Li, P. Chen, H. Yang, and J. Hao, Effect of heat treatment on phase stability and wear behavior of laser clad AlCoCrFeNiTi0.8 high-entropy alloy coatings, Surface and Coatings Technology. 392 (2020) 125758.

29. A. Ostovari Moghaddam, N.A. Shaburova, M.N. Samodurova, A. Abdollahzadeh, and E.A. Trofimov, Additive manufacturing of high entropy alloys: A practical review, Journal of Materials Science & Technology. 77 (2021) 131.
30. L. Jinlong, C. Yuqing, L. Di, and Z. Zhiping, Effect of passivation potential on corrosion behaviors of laser powder bed fusion FeCrCoNi high-entropy alloy, Colloids and Surfaces A: Physicochemical and Engineering Aspects. 702 (2024) 135181.
31. S. Wu, H. Wang, S. Wang, and W. Liu, Strong yet ductile (FeCoNi)86Al7Ti7 high-entropy alloy via laser powder bed fusion, Materials Today Communications. 41 (2024) 110285.
32. J.Q. Shi, C.P. Huang, S.H. Deng, Z.J. Tan, H.L. Lu, J.Z. Hao, F.R. Shen, Y.F. Jia, J. Chen, Q. Wang, L.H. He, and G. Wang, In-situ neutron diffraction study of the strengthening mechanism and deformation behavior of cellular structure in high-entropy alloys by additive manufacturing, International Journal of Plasticity. 181 (2024) 104081.
33. X. Hu, Z. Xu, X. Jia, S. Li, Y. Zhu, and A. Xia, Microstructure and magnetic properties of AlCoCrCuFeNi high-entropy alloy prepared by selective laser melting, Journal of Alloys and Compounds. 1010 (2025) 177740.
34. B. Xiao, W. Jia, J. Wang, and L. Zhou, Selective electron beam melting of WMoTaNbVFeCoCrNi refractory high-entropy alloy, Materials Characterization. 193 (2022) 112278.
35. M. Zhang, X. Zhou, D. Wang, L. He, X. Ye, and W. Zhang, Additive manufacturing of in-situ strengthened dual-phase AlCoCuFeNi high-entropy alloy by selective electron beam melting, Journal of Alloys and Compounds. 893 (2022) 162259.
36. Y. Huang, X. Chen, W. Wu, Y. Wang, and Y. Wang, Effect of laser remelting on the microstructure and properties of an AlCoFeNi eutectic high-entropy alloy fabricated through double-wire arc additive manufacturing, Journal of Alloys and Compounds. 1010 (2025) 177509.
37. Q. Shen, J. Xue, Z. Zheng, X. Yu, and N. Ou, Effects of deposition strategies on microstructure and mechanical properties of wire arc additive manufactured CoCrFeNiMo0.2 high-entropy alloy, Materials Science and Engineering: A. 918 (2024) 147486.
38. M.R.U. Ahsan, G.-J. Seo, X. Fan, P.K. Liaw, S. Motaman, C. Haase, and D.B. Kim, Effects of process parameters on bead shape, microstructure, and mechanical properties in wire + arc additive manufacturing of Al0.1CoCrFeNi high-entropy alloy, Journal of Manufacturing Processes. 68 (2021) 1314.
39. R. Bardo, R. Dziurka, K. Fryzowicz, G. Cios, M. Marciszko-Wiąckowska, M. Gajewska, C. Hain, and P. Bała, Influence of process parameters on the quality of powder bed fusion-fabricated Ni-Co-Fe-Mn-Ti high entropy alloy prints using elemental powders, Journal of Alloys and Compounds. 972 (2024) 172862.
40. C. Ren, X. Zhang, Y. Zhou, D. Wang, X. Yang, and J. Li, Effect of heat treatment on microstructure and mechanical properties of CrMnFeCoNi high-entropy alloy with SiC addition fabricated via additive manufacturing, Materials Science and Engineering: A. 895 (2024) 146229.
41. Q. Shen, J. Xue, X. Yu, Z. Zheng, and N. Ou, Powder plasma arc additive manufacturing of CoCrFeNiWx high-entropy alloys: Microstructure evolution and mechanical properties, Journal of Alloys and Compounds. 922 (2022) 166245.
42. Y. Zhang, H. Wang, B. Su, Y. Zhu, C. Shen, Z. Li, B. Liu, and J. Yang, High-temperature mechanical behavior and microstructural destabilization evolution of the lightweight refractory high-entropy alloy prepared by laser additive manufacturing, Materials Science and Engineering: A. 915 (2024) 147280.
43. M.-L. Yi, H.-L. Xia, L. Zhang, W.-J. Li, J. Tu, S.-K. Zheng, L.-Z. Wang, and Y.-B. Du, Comparison of microstructures, mechanical and tribological properties of Al0.3CoCrFeNi

high-entropy alloy prepared by induction melting and selective laser melting, Materials Today Communications. 37 (2023) 106942.
44. K. Zhao, X. Hao, B. Huang, X. Zhao, J. Ma, and C. Wang, Effect of bias voltage on the structure and properties of CoCrFeNi high entropy alloy thin films prepared by magnetron sputtering, Materials Today Communications. 40 (2024) 110217.
45. A. Obeydavi, A. Shafyei, and J.-W. Lee, Effect of sputtering power and substrate bias on microstructure, mechanical properties and corrosion behavior of CoCrFeMnNi high entropy alloy thin films deposited by magnetron sputtering method, Intermetallics. 172 (2024) 108369.
46. Y.-F. Yang, F. Hu, T. Xia, R.-H. Li, J.-Y. Bai, J.-Q. Zhu, J.-Y. Xu, and G.-F. Zhang, High entropy alloys: A review of preparation techniques, properties and industry applications, Journal of Alloys and Compounds. 1010 (2025) 177691.
47. X.K. Zeng, Y.T. Li, X.D. Zhang, M. Liu, J.Z. Ye, X.L. Qiu, X. Jiang, and Y.X. Leng, Effect of bias voltage on the structure and properties of CuNiTiNbCr dual-phase high entropy alloy films, Journal of Alloys and Compounds. 931 (2023) 167371.
48. L. Zhang, H. Liu, H. Chen, Y. Chen, N. Li, C. Tan, and H. Ma, Preparation of amorphous TiO2 films by RF magnetron sputtering: Process optimization and effect of sputtering pressure on electrochromic properties, Physica B: Condensed Matter. 697 (2025) 416726.
49. X. Zhang, X. Zeng, Y. Liu, J. Liu, A. Pogrebnjak, V. Pelenovich, Q. Wan, X. Liu, H. Wang, Y. Lei, and B. Yang, Effects of RF magnetron sputtering power on the structure and nanohardness of high-entropy alloys (TiVCrNbSiTaBY)N Hard coatings, Ceramics International. 49 (20) (2023) 33418.

4 Thermodynamic Designing of High Entropy Alloys

Naveen Kumar Mindi, Krishna Dutta, Nityananda Sahoo, Ghananshu Patil and Syed Nasimul Alam

4.1 INTRODUCTION

Since the Bronze Age, the designing of alloys has followed the development of a system with one or two main elements and adjusting their properties by adding smaller amounts of other elements. A few examples of alloy systems include those based on aluminum, nickel, iron, and copper [1–3]. A huge demand is there for materials developing in structural applications that show better mechanical properties having balanced strength and ductility [4]. The concept of high entropy alloys (HEAs) was first introduced by Yeh et al. [5] and Cantor et al. [6], which involves equiatomic multicomponent alloys comprising a disordered solid solution phase. These alloys are named HEAs as their liquid states show significantly higher mixing entropy compared to conventional alloy systems. As a result, the entropy effect is more prominent in HEAs. Earlier, the focus of researchers in alloy design was on the corners of the phase diagram for traditional alloys that use a minor design space. However, for the design of HEAs, the focus must be mainly on the central portion of the phase diagram shown in Figure 4.1 (a). In conventional alloys, we generally design by taking the base element and n types of alloys can be acquired with n different base elements. Moreover, if a combination of p elements was chosen from total n elements to form an equiatomic alloy ($p = 2, 3, 4, \ldots, n$) the total potential alloys increased from n to $N = 2^n - n - 1$. If $n < 3$, the considerable difference cannot be observed but with an increase in n, this difference can raise due to the exponential dependence of N on n. The distribution of individual elements in conventional and HEAs is shown in Figure 4.1 (b). Yeh et al. [5] defined HEAs as alloys having five or more primary elements, where the concentration of each element ranges between 5 and 35 atomic %. In addition to these major elements, HEAs may consist of minor elements with concentrations less than 5 atomic % [7, 8]. The advantage of HEAs when compared with traditional alloys because they achieve excellent properties such as hardness, impressive strength, wear resistance, corrosion resistance, and unique mechanical performance at elevated and cryogenic temperatures [9–12]. The most important observation is that HEAs usually show excellent physical, mechanical, and chemical properties. Due to their versatility, ease of fabrication, and environmental safety, HEAs hold many potential applications as structural

DOI: 10.1201/9781003664093-6

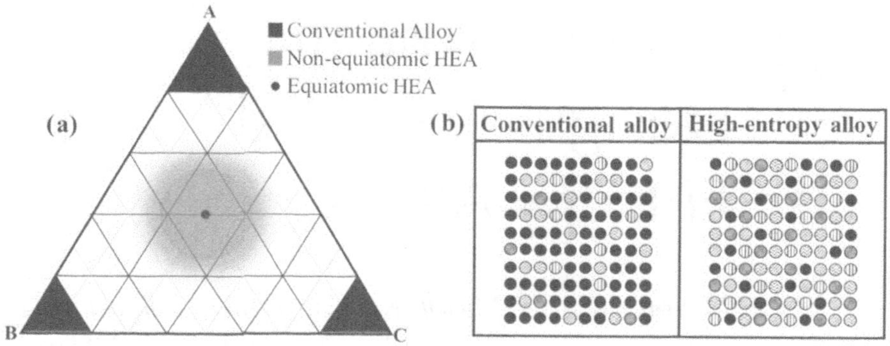

FIGURE 4.1 (a) Ternary phase diagram of both conventional and high entropy alloys and (b) individual elements distribution of conventional and high entropy alloys.

and functional materials, including thermoelectric devices, magnetic materials, and superconductors.

Cantor et al. [6] studied the FeCoNiMnCr multicomponent alloy, which formed a face-centered cubic (FCC) single-phase solid solution. Meanwhile, Yeh et al. [5] examined various combinations of elements like Ti, V, Al, Ta, Mo, Zr, and Nb, which exhibited superior mechanical properties compared to conventional alloy systems. HEAs generally exhibit body-centered cubic (BCC) or FCC solid solution structures, dual phase, and sometimes hexagonal close-packed (HCP) structures as well. Table 4.1 summarizes the experimentally reported some of the HEAs forming BCC, FCC, HCP, and dual-phase structures. The slip planes are closely packed in an FCC structure, whereas the BCC structure lacks such close-packed slip planes. The critical slip stress in FCC-structured materials is less than the BCC-structured materials. Because of that reason, HEAs with a BCC structure tend to show higher yield strength but low ductility and FCC structure exhibit lower yield strength and higher ductility. Therefore, an alloy consisting of both BCC and FCC phases is probable to attain a balance of high strength and sufficient ductility.

This chapter overviews the different methods used for forecasting the various phases formed in HEAs. Some of the important phase prediction techniques discussed in this chapter are the parametric approach, CALPHAD, machine learning approach, and density functional theory (DFT) shown in Figure 4.2. All these methods application will be discussed in detail with some examples.

The parametric approach mainly explains the thermodynamic parameters and their limits to obtain a solid solution. Also, it predicts the different phase formations concerning the values of some of the main thermodynamic parameters such as enthalpy of mixing (ΔH_{mix}), entropy of mixing (ΔS_{mix}), atomic size difference (δ), electronegativity difference (ΔN), solid solution parameter (Ω), valence electron concentration (VEC). Gibbs free energy plays a crucial role in finding the thermodynamic phase stability with enthalpy of mixing and entropy of mixing. It helps to forecast the phase stability and equilibrium of HEAs. The lower value of Gibbs free energy indicates a more stable phase.

TABLE 4.1
A Few Results of Single-Phase and Dual-Phase Formation After Alloying

Alloys	Structure	References
CoCrFeNi	FCC	[13]
CoCrMnNi	FCC	[14]
FeCrMnNiCo	FCC	[6]
CoCrFeNiPd	FCC	[15]
NbTaTiV	BCC	[16]
MoNbTaW	BCC	[17, 18]
NbTiVZr	BCC	[19]
MoNbTaVW	BCC	[17, 18]
MoNbReTaVW	BCC	[20]
MoNbReTaTiVW	BCC	[20]
DyGdHoTbY	HCP	[21]
DyGdLuTbTm	HCP	[21]
NiMnCrMoW	BCC+FCC	[22]
CrMnFeVTi	BCC+FCC	[23]
AlCoCrCuFeNi	BCC+FCC	[2]
AlCoCrFeMnNi	BCC+FCC	[24]

Phase diagrams are generally established through equilibration experiments, thermal analysis, and identification of phase using X-ray diffraction and scanning electron microscopy methods. To conduct these experiments is expensive and time-consuming, especially for more complex systems. The CALPHAD (acronym of CALculation of PHAse Diagram) method [25, 26] allows us to get phase diagrams

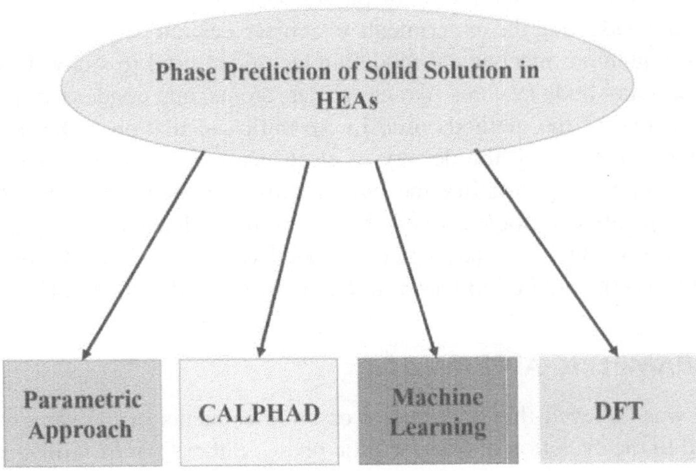

FIGURE 4.2 Various types of phase prediction methods in high entropy alloys.

of multicomponent alloy systems, which are valuable for basic investigation in areas like solidification and solid-state transformations as well as for alloy design and process development. It is dependent on the established thermodynamic principle that the equilibrium is attained in a system with the least Gibbs energy at a particular composition, pressure, and temperature. The main benefit of CALPHAD is to help in predicting phase diagrams of higher order by extrapolation from the systems of lower-order, such as binary and ternary systems. This makes it highly valuable for the design of commercial alloys that contain multiple components. Utilizing the CALPHAD approach necessitates the utilization of specialized computational software and thermodynamic data sets. In the last 30 years, several commercial software packages, such as Thermo-Calc, PandatTM, and FactSage, were developed to assist in implementing this approach. There is an increasing demand for accurate and dependable databases that contain information on the thermodynamic properties of HEAs. Two primary databases used are PanHEA, which comprises comprehensive binary and ternary phase data for components such as Al, Cu, Co, Cr, Ni, Fe, Mn, and TCNI7, which specifically concentrates on binary and select ternary systems about nickel-based alloys. These data sets serve as essential for precisely forecasting phase diagrams and alloy behaviors.

HEAs with exceptional microstructures and properties are in high demand at present. Discovering such HEAs with traditional designing methods is time-consuming and laborious. To resolve these hurdles, ML has emerged with efficient solution. Machine learning is a computational technique to identify statistical patterns and obtain insights from data without specific programming. It has become a widely used tool in the investigation of the materials. Huang et al. [27] used ML to find solid solution-hardened HEAs, emphasizing the complexity of local atomic environments and interactions. They established a systematic method for developing new HEAs, which was confirmed through experimental results. Jaiswal et al. [28] discovered various ML models to forecast solid solution phases in HEAs, considering the prominence of input parameters and accuracy determination. In the present study, one of the ML models was used to predict the phases which is economical and time-saving compared to conducting the experiments to achieve desired outputs.

DFT is a quantum mechanical modeling technique used to study the electronic structure of many-body systems like molecules, atoms, and condensed matter. DFT is known as one of the methods used in Ab-initio calculations. The main benefit of DFT is that employing the density of electrons as the core variable instead of the wave function may simplify the intricate many-electron problem. Because of its great computational efficiency and ability to fairly balance accuracy and computational cost, it is ideal for studying complicated systems such as HEAs. The main focus will be on the prediction phases in HEAs by using the method DFT.

4.2 PARAMETRIC APPROACH

Earlier, it was believed that adding five or more elements in the same proportion would lead to the evolution of intermetallic phases rather than forming a solid solution, because of the differences in the formation of heat and atomic sizes among two elements. However, Cantor and Yeh found that certain HEAs form phases of random solid solution, which can be explained by thermodynamic parameters that help in solid solution formation. In traditional binary alloys, the development of solid solutions follows the Hume-Rothery rules, which describe the conditions for solubility between elements. These rules indicate that solid solubility majorly depends upon crystal structure, atomic size, electronegativity, and valence electron concentration.

The Hume-Rothery rules are stated as follows:

a. The atomic size difference between solute and solvent should not be more than 15%.
b. Both elements' crystal structures should be the same.
c. The electronegativity of the two elements must be similar.
d. The valency should match for both solute and solvent.

In case of HEAs, the important thermodynamic parameters such as entropy of mixing (ΔS_{mix}), enthalpy of mixing (ΔH_{mix}), atomic size difference (δ), electronegativity difference (ΔN), VEC, solid solution parameter (Ω), geometrical parameter ($\wedge$), atomic packing factor (γ), d-orbital energy level parameter ($\overline{M_d}$), and Φ were used for predicting the formation of solid solution and distinct phases. To know the formation of solid solution phase in HEAs, different methods were developed and proposed by many researchers. Zhang et al. proposed that formation of solid solution in HEAs was determined with the thermodynamic parameters like entropy of mixing (ΔSmix), enthalpy of mixing (ΔHmix), and the atomic size difference (δ). These are determined by using the equations (4.1), (4.2), and (4.3), respectively.

$$\Delta S_{mix} = -R\sum_{i=1}^{n} X_i \ln X_i. \tag{4.1}$$

where, R = universal gas constant (8.314 J-K^{-1}-mol^{-1}) and X_i = mole fraction of the i_{th} element,

$$\Delta H_{mix} = 4 \sum_{i=1,\, i \neq j}^{n} \left(\Delta H_{ij}^{mix} \right) X_i X_j \tag{4.2}$$

where ΔH_{ij}^{mix} = enthalpy of mixing for the binary alloy i-j, and X_i, X_j represent the mole fraction of the elements.

$$\delta = \sqrt{\sum_{i=1}^{n} X_i \left(1 - \frac{r_i}{\overline{r}} \right)} \tag{4.3}$$

where, $\overline{r} = \sum_{i=1}^{n} X_i r_i$; r_i = radius of the i_{th} element and $\overline{r}$ = mean radius

They established that the solid solution formation happens when ΔS_{mix}, ΔH_{mix}, and δ fall within the following ranges: $12 \leq \Delta S_{mix} \leq 17.5$ *J/(Kmol)*, $-20 \leq \; \leq 5$ kJ/mol, and $\delta \leq 6.4\%$. The ΔH_{mix} should not be too large, because higher positive ΔH_{mix} typically leads to the formation of intermetallics. The difference in atomic size should be small. If it is large the strain energy will be increased beyond which it destabilizes the phases of solid solution. Furthermore, the ΔS_{mix} value must be high enough as it is a crucial factor in stabilizing the solid solution phases. The 'electronegativity difference (ΔN)' parameter was useful in predicting the topological closed packed (TCP) phases [29]. It was noted that the TCP phase is stable when $\Delta N > 0.133$ except in the case of high concentrations of HEAs containing aluminum. For the formation of a solid solution, a low value of electronegativity difference must be there and it can be calculated by equation (4.4).

$$\Delta N = \sqrt{\sum_{i=1}^{n} X_i \left(N_i - \bar{N}\right)^2} \tag{4.4}$$

Where $\bar{N} = \sum_{i=1}^{n} X_i N_i$, ($N_i$ is the electronegativity of i_{th} element and $\bar{N}$ = mean electronegativity).

Guo et al. [30] stated that VEC is a crucial factor for determining the phase formation in HEAs. The BCC solid solution phase forms when VEC < 6.87, while the FCC solid solution occurs when VEC ≥ 8. This criterion is valuable for designing HEAs. The value of VEC is determined by using equation (4.5).

$$\text{VEC} = \sum_{i=1}^{n} X_i \left(\text{VEC}\right)_i \tag{4.5}$$

A dimensionless parameter 'Ω' that connects the effect of entropy and enthalpy is used to forecast solid solution formation in HEAs [31]. The value of 'Ω' is calculated by using the equation (4.6).

$$\Omega = \frac{T_m \Delta S_{mix}}{\left|\Delta H_{mix}\right|} \tag{4.6}$$

Here, $T_m = \sum_{i=1}^{n} X_i \left(T_m\right)_i$; where $(T_m)_i$ and T_m are the melting temperatures of the i_{th} element, and the alloy.

It was found that the required condition for the stable solid solution formation in HEAs is $\Omega \geq 1.1$ and $\delta \leq 6.6$. The smaller atomic size difference (δ) and higher Ω value help in forming a stable solid solution. Additionally, a novel geometric parameter ($\wedge$) was introduced to assess disordered solid solution formation and its value must be greater than 0.96. Therefore, a high value of $\wedge$ is important for solid solution formation. The geometric parameter is determined using equation (4.7).

$$\wedge = \Delta S_{mix} / \delta^2 \tag{4.7}$$

The relation between Ω and δ was shown by Yang et al. [31] in Figure 4.3. The triangular zone marked by S represents the distribution of simple solid solution phase formation of different alloys. Here, the HEAs attain a small δ value and larger values of Ω. This suggests that the strain energy produced due to lattice distortion is less in the alloy and the effect of ΔS_{mix} is enough to exceed that of ΔH_{mix} for the solid solution formation. The zone marked with S+I indicates the solid solution formation and intermetallic phases. The intermetallic compounds are formed in zone I which has the larger atomic size and the effect of ΔH_{mix} for HEAs. The bulk-metallic glasses

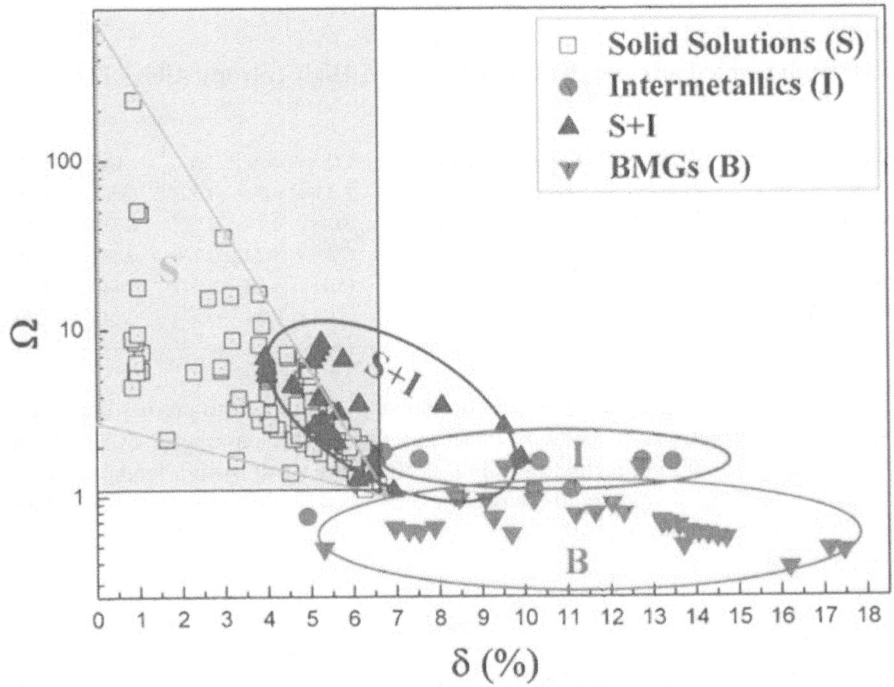

FIGURE 4.3 The relation between thermodynamic parameters Ω and δ representing the solid solution [31].

(BMGs) are marked in zone B as having the larger value of δ and the lowest values of Ω. Thus, the limits of Ω and δ were fixed as discussed earlier based on this analysis.

Wang et al. [32] suggested a novel atomic packing parameter (γ) useful for random atomic packing in HEAs and superalloys. The author gives the condition for the formation of single-phase solid solution as $\gamma \leq 1.175$. Thus, γ tends for solid solution formation in HEAs when its value is low. of. The γ value is calculated by using the equation (4.8).

$$\gamma = \frac{W_s}{W_l} \tag{4.8}$$

Lu et al. [33] introduced a d-orbital energy level parameter to forecast the stability of the TCP phase in HEAs. They proposed that the TCP phase remains stable when $\overline{M_d} > 1.09$. The elements such as Cr, Co, Mo, Cu, Ni, Fe, Si, Ti, Nb, and other transition metals the $\overline{M_d}$ value is valid, whereas this parameter does not apply to HEAs with a higher concentration of V and Al. This parameter value is determined by equation (4.9).

$$\overline{M_d} = \sum_{i=1}^{n} X_i \left(M_d\right)_i \tag{4.9}$$

TABLE 4.2
Thermodynamic Parameters of $NiMnCrMoW_x$ High Entropy Alloys [22]

Alloy	T_m(K)	ΔS_{mix} ($J\text{-}K^{-1}mol^{-1}$)	ΔH_{mix} ($kJ\text{-}mol^{-1}$)	Ω	δ%	Λ	ΔN	VEC
$NiMnCrMoW_{0.2}$	2156.99	12.57	−3.22	8.43	4.23	0.703	0.202	7.19
$NiMnCrMoW_{0.4}$	2227.09	13.01	−2.77	10.47	4.24	0.725	0.219	7.13
$NiMnCrMoW_{0.6}$	2290.63	13.24	−2.38	12.74	4.24	0.738	0.231	7.08
$NiMnCrMoW_{0.8}$	2349.48	13.35	−2.05	15.31	4.23	0.747	0.240	7.04
$NiMnCrMoW_{1.0}$	2403.80	13.38	−1.76	18.28	4.21	0.754	0.248	7.00

In addition, King et al. [34] proposed a novel parameter Φ to predict the single-phase HEA formation. They found that the value of this parameter Φ >1 for the disordered solid solution formation. Thus, if the value of Φ is high tends to form a solid solution. The Φ value with the data generated by an online tool can calculated and given by the equation (4.10).

$$\Phi = \frac{\Delta G_{ss}}{-\left|\Delta G_{max}\right|} \tag{4.10}$$

Mindi et al. [22] investigated $NiMnCrMoW_x$ (x = 0.2, 0.4, 0.6, 0.8, 1.0) HEAs processed through the powder metallurgy route. The thermodynamic parameters calculated theoretically using different equations satisfy each parameter's limits. The list of calculated thermodynamic parameters is presented in Table 4.2. The phase prediction of these alloys can be done with the help of VEC and its values lie between 6.87 and 8, indicating the dual-phase (BCC + FCC) formation. The development of phases in AlCoCrCuFe and CoCrCuFeNi HEAs synthesized by mechanical alloying and then sintered by spark plasma sintering was investigated by Praveen et al. [35]. They confirmed that configurational entropy is not enough to reduce the Cu-rich FCC and σ phase formation, and highlighted the importance of ΔH_{mix} in analyzing the stability of phases in HEAs.

4.3 CALPHAD MODELING

HEAs aim to maximize configurational entropy by incorporating multiple major elements in approximately equal proportions. This approach differs from standard alloy design, which usually emphasizes one or two important elements. Although the theory indicates that high configurational entropy should promote the development of single-phase solid solutions, the majority of known HEAs consist of several phases. This demonstrates that the mere combination of various components does not ensure the formation of a single-phase alloy. It is important to meticulously choose the compositions [36]. For example, Cantor et al. [6] conducted a study on alloys consisting of 20 and 16 elements in identical proportions. However, their findings revealed that these alloys exhibited many phases and were prone to brittleness. While the authors were able to develop a CoCrFeMnNi alloy with a single-phase FCC structure, the introduction of additional elements resulted in the formation

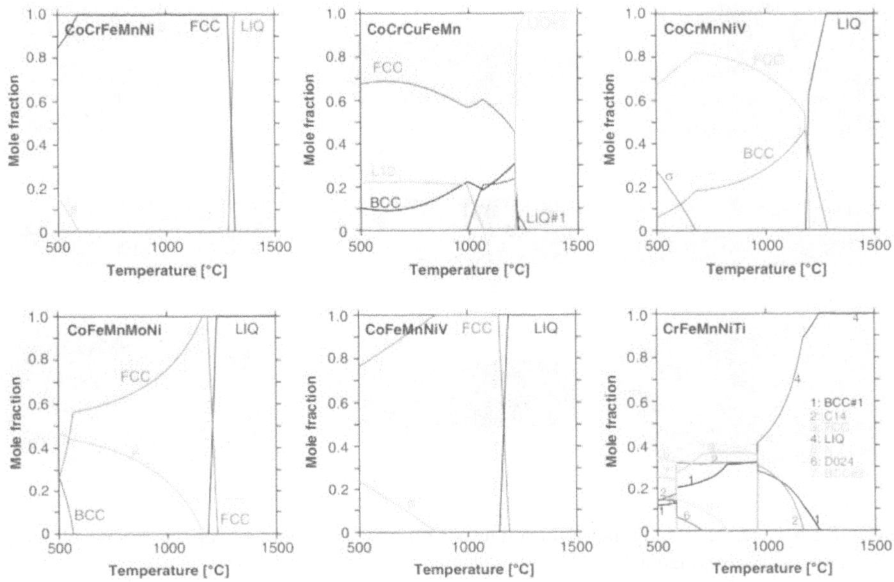

FIGURE 4.4 The phases formed under equilibrium cooling conditions in mole fraction versus temperature plots for CoCrFeMnNi, CoCrCuFeMn, CoCrMnNiV, CoFeMnMoNi, CoFeMnNiV, and CrFeMnNiTi HEAs studied experimentally by Otto et al. [37].

of several phases. Similarly, the investigation carried out by Otto et al. [37] to find the phase stability in six alloys CoCrFeMnNi, CoCrCuFeMn, CoCrMnNiV, CoFeMnMoNi, CoFeMnNiV, and CrFeMnNiTi. From this study, they have found a single-phase FCC solid solution formed only in CoCrFeMnNi, while the other alloys comprised of more than one phase after annealing at 1000°C or 800°C. Using the TCNI7 database, which includes all edge binaries and some ternaries of these alloys (except the TTNI8 database for CoCrCuFeMn and CoFeMnNiV), the findings by Otto et al. are qualitatively reproduced by the CALPHAD calculations illustrated in Figure 4.4. Using the database single FCC phase was predicted in CoCrFeMnNi HEA, whereas in CoFeMnMoNi an FCC matric phase with a minor amount of μ phase was predicted. A complex microstructure with four phases was anticipated in CrFeMnNiTi HEA. For CoFeMnNiV, the FCC phase was predicted to be stable over a limited temperature range along with a minor σ phase forming at lower temperatures. It looks like both databases underestimate the thermal stability of the σ phase in the systems as it was seen in CoCrCuFeMn and CoCrMnNiV by Otto et al. [37]. This indicates that both databases require further development, especially in descriptions of all ternary systems.

4.3.1 Methodologies of CALPHAD

The CALPHAD method uses a phenomenological approach to construct a coherent thermodynamic model of a system with numerous components shown in Figure 4.5. This requires the establishment of a "thermodynamic database" that contains

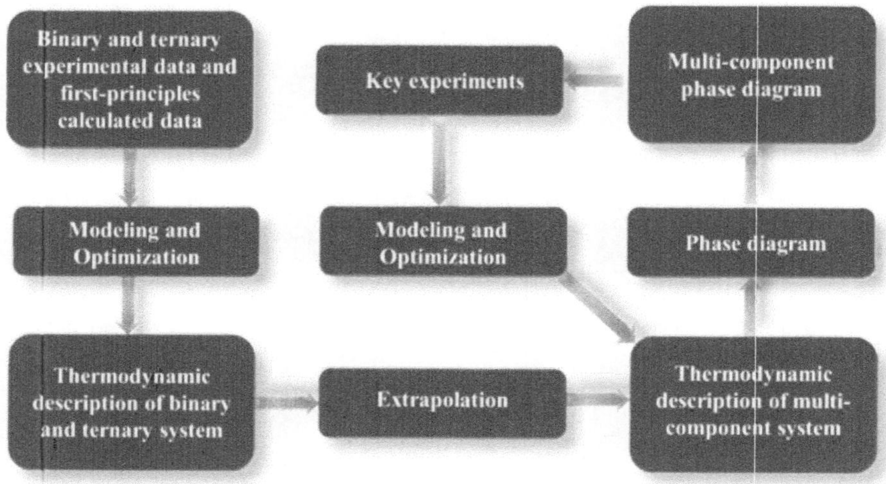

FIGURE 4.5 CALPHAD method used to obtain a thermodynamic representation of a system with multi-components.

parameters characterizing the Gibbs energies of diverse binary (two-component) and ternary (three-component) systems. These factors are essential for the precise composition ranges under investigation. The method commences by collecting existing thermochemistry and phase equilibrium data from the literature for the more straightforward binary and ternary systems. If the data is unavailable, it is necessary to develop and carry out experiments to acquire it. In addition, theoretical calculations based on fundamental physical principles, known as first-principles calculations, can be employed to complement the experimental results. After collecting enough data, the subsequent stage involves determining the model parameters for the phases involved using an optimization technique. This entails fine-tuning the settings to optimize their alignment with the collected data. These parameters must adhere to physical authenticity rather than solely being numerically optimized, particularly in situations where there is a lack of experimental evidence. This implies that the thermodynamic descriptions are reliable and precise. Once credible thermodynamic models have been established for smaller systems, often binaries and ternaries, the subsequent task involves constructing a comprehensive thermodynamic database for more complicated systems, such as quaternary, by extrapolating from these simpler models. Empirical evidence has demonstrated that this approach typically yields favorable results.

Nevertheless, if a novel phase emerges inside the quaternary system, it is imperative to accurately define and align its Gibbs free energy with the existing database via optimization. Occasionally, modifications to the binary and ternary models are required when the extrapolated data does not align with experimental findings. Extra caution should be exercised when dealing with "hypothetical" phases, which refer to those that are unstable or metastable in some simpler systems. Historically, databases dedicated to certain materials such as Al alloys, steels, or

Ni-based superalloys have mostly concentrated on a limited set of essential constituents. Conversely, databases for HEAs must include the complete spectrum of compositions, as HEAs possess compositions that are situated at the center of a multidimensional composition space. Developing a HEA database presents significant difficulties and necessitates the careful use of thermodynamic models that can accurately depict the Gibbs energies of all phases within the system.

i. **Determination of Gibbs Free Energy for Elements**

The equations provided describe the Gibbs energy of pure elements with distinct structures and how they are adjusted to incorporate magnetic ordering.

The Gibbs energy of a pure element *i* with a particular structure ϕ, referred to as the enthalpy of its standard state at room temperature, is given by:

$$^{0}G_i^{\varphi}(T) = a + bT + cT\,lnT + dT^2 + eT^3 + fT^{-1} + gT^7 + hT^{-9} \qquad (4.11)$$

The coefficients *a* through *h* are empirical parameters derived from experimental data and tabulated values, such as those presented by Dinsdale [38].

a: A constant term representing the base value of the Gibbs energy.
bT: A linear term in temperature, capturing the linear dependence of Gibbs energy on temperature.
cT lnT: A term involving the product of temperature and the natural logarithm of temperature, accounting for more complex temperature dependence.
dT^2: A quadratic term, capturing non-linear temperature effects.
eT^3: A cubic term, further refining the temperature dependence at higher temperatures.
fT^{-1}: An inverse temperature term, important for modeling behavior at low temperatures.
gT^7 and hT^{-9}: Higher-order terms in temperature, allowing for very fine adjustments to the model to capture subtle physical effects.

For elements that demonstrate magnetic ordering, such as nickel (Ni), the molar Gibbs energy is modified by the introduction of an additional factor, $^{mag}G_m^{\varphi}$. This phrase quantifies the influence of magnetic ordering on the total Gibbs energy. The formula representing this magnetic ordering is:

$$^{mag}G_m^{\varphi} = RT\,ln\left(\beta^* + 1\right) f\left(T/T^*\right) \qquad (4.12)$$

R: The universal gas constant.
T: The absolute temperature.
β^*: The average magnetic moment per atom of the element, expressed in Bohr magnetons.
T^*: The critical temperature for magnetic ordering, also known as the Curie temperature for ferromagnetic materials.

f(T/T)*: A function that describes how the magnetic contribution evolves with temperature relative to the critical temperature T^*. The specific form of this function f is also adapted from Dinsdale.

Physical interpretation:

- *ln(β*+1)*: This logarithmic term captures the dependence of the Gibbs energy on the magnetic moment of the element.
- *f(T/T*)*: This function modifies the magnetic ordering by considering the ratio between the present temperature and the critical temperature. It accounts for the changes in magnetic properties as the material undergoes a transition at its critical temperature.

Equation (4.11) fundamentally explains the relationship between temperature and the Gibbs energy for non-magnetic factors. On the other hand, Equation (4.12) incorporates the required adjustment for magnetic influences in components that experience magnetic ordering. These equations, when combined, offer a comprehensive description of the thermodynamic properties of pure elements at various temperatures.

ii. **Substitutional Solution Model**

The Substitutional Solution Model is a core idea in the CALPHAD technique, which is employed to describe the thermodynamic properties of solid (BCC, FCC, and HCP) and liquid solutions. This model is especially valuable for illustrating phases in which various atomic species can replace each other on a shared sublattice, which is prevalent in metallic alloys and certain ceramic systems. The Substitutional Solution Model involves the occupation of crystal lattice sites by atoms or molecules, where different elements can replace one another on these sites. The model assumes ideal mixing on the lattice sites, with atoms distributed randomly based on their mole fractions. To account for deviations from ideality, excess terms are incorporated into real systems to adapt to randomness. The molar Gibbs energy of a substitutional solid solution phase (φ) can be expressed as:

$$G_m^{\varphi} = {}^{ref}Gm_m^{\varphi} + {}^{id}G_m^{\varphi} + {}^{ex}Gm_m^{\varphi} + {}^{mag}Gm_m^{\varphi} \tag{4.13}$$

The expression " ${}^{ref}Gm_m^{\varphi}$ " refers to the baseline state of the Gibbs energy. In this state, each component in the solution acts perfectly according to its attributes as a pure element.

$${}^{ref}Gm_m^{\varphi} = \sum_i x_i . {}^{0}Gm_m^{\varphi} \tag{4.14}$$

The molar Gibbs energy of a substitutional solid solution phase, denoted as ϕ, incorporates several factors and serves as a complete metric. The mole fraction of element i, represented as x_i, indicates the relative amount of each element present in the solution. The excess Gibbs energy term, ${}^{ex}G_m^{\varphi}$, quantifies the deviations from ideal behavior caused by interactions between various elements in their nonmagnetic state, which can result from factors like atomic size mismatch or chemical interactions. The term ${}^{mag}Gm^{\varphi}_{m}$, known as the magnetic contribution term, takes into consideration the impact of magnetic interactions inside the alloy. This is especially significant for alloys that contain magnetic elements. The equation for Gibbs free energy is given by the formula

$$^{id}G^{\varphi}_{m} = RT\sum_{i} x_i . \, ln \; x_i \tag{4.15}$$

Real solutions frequently exhibit deviations from ideal behavior as a result of interactions among several atoms. The deviations are quantified by the excess Gibbs energy ${}^{ex}Gm^{\varphi}_{m}$ for a quinary system given below

$$\begin{aligned} ^{ex}G^{\varphi}_{m} &= \sum_{i\neq j} x_i \; x_j L^{\varphi}_{i,j} + \sum_{i\neq j\neq k} x_i \; x_j \; x_k \; L^{\varphi}_{i,j,k} + \sum_{i\neq j\neq k\neq l} x_i \; x_j \; x_k \; x_l \; L^{\varphi}_{i,j,k,l} \\ &+ \sum_{i\neq j\neq k\neq l\neq m} x_i \; x_j \; x_k \; x_l x_m \; L^{\varphi}_{i,j,k,l,m} \end{aligned} \tag{4.16}$$

The nonmagnetic excess Gibbs energy (${}^{ex}G^{\varphi}_{m}$) for a quinary system incorporates interactions that go beyond ideal mixing, effectively portraying the complex structure of multi-component interactions. The expression is a combination of contributions from binary, ternary, quaternary, and quinary interactions. Each interaction is defined by its respective interaction parameters $L^{\varphi}_{i,j}$, $L^{\varphi}_{i,j,k}$, $L^{\varphi}_{i,j,k,l}$, $L^{\varphi}_{i,j,k,l,m}$. The parameters can be expressed using temperature-dependent polynomials, such as the Redlich-Kister polynomial, which is used to describe binary interactions ($L^{\varphi}_{i,j}$) given below [39].

$$L^{\varphi}_{i,j} = \sum_{v=0}^{n} L^{\varphi}_{i,j}\left(x_i - x_j\right)^{v} \tag{4.17}$$

$$^{v}L^{\varphi}_{i,j} = {}^{v}a^{\varphi}_{i,j} + {}^{v}b^{\varphi}_{i,j} + {}^{v}c^{\varphi}_{i,j} \; T \; ln\left(T\right) + \cdots \tag{4.18}$$

The variables mentioned are the coefficients ${}^{v}a^{\varphi}_{i,j}$, ${}^{v}b^{\varphi}_{i,j}$, ${}^{v}c^{\varphi}_{i,j}$ ... are variables that require optimization using experimental data.

The interaction parameter $L^{\varphi}_{i,j,k}$ applies to ternary systems, and the expression is given by

$$L^{\varphi}_{i,j,k} = v_i L^{\varphi}_{i} + v_j L^{\varphi}_{j} + v_k L^{\varphi}_{k} \tag{4.19}$$

where $v_i = x_i - (1\text{-}x_i\text{-}x_j\text{-}x_k)/3$, $v_j = x_j - (1\text{-}x_i\text{-}x_j\text{-}x_k)/3$, and $v_k = x_k - (1\text{-}x_i\text{-}x_j\text{-}x_k)/3$.

These L^{φ}_{i}, L^{φ}_{j}, L^{φ}_{k} parameters represent the interaction contributions from the corresponding corners of the Gibbs energy surface of the ternary system.

If the solution phase has magnetic ordering, the magnetic component of the Gibbs energy, denoted as ${}^{mag}G^{\varphi}_{m}$, is considered. The temperature at which ordering becomes critical (T^{*}_{m}) and the magnetic moment (β^{*}_{m}) are both dependent on the composition and can be stated using equations that involve the mole fractions (x_i) and expressed by

$$T_m^* = \sum_i x_i T_i^* + {}^{ex}T_m^* \tag{4.20}$$

The parameter β_m^* represents a distinct attribute, most likely related to magnetism, of a system composed of several components. It is mathematically stated as:

$$\beta_m^* = \sum_i x_i \, \beta_i^* + {}^{ex}\beta_m^* \tag{4.21}$$

The excess terms ${}^{ex}T_m^*$ and ${}^{ex}\beta_m^*$ are described by equations similar to the previously discussed Equations 4.17 and 4.18. The use of quaternary and quinary excess terms, as explained in Equation 4.16, is limited due to their significantly lesser impact compared to binary and ternary interaction parameters. The reason behind this can be attributed to the inherent characteristics of their mathematical expression. These higher orders include the multiplication of several molar fractions. The variables x, j, k, and l represent values in a quaternary system, while the variables $x_i x_j x_k x_l$ and $x_i x_j x_k x_l x_m$ represent values in a quinary system. When handling combinations including multiple components, the molar compositions of these products get significantly reduced. This is particularly applicable to systems in which the components are present in equimolar ratios, as seen in Figure 4.6.

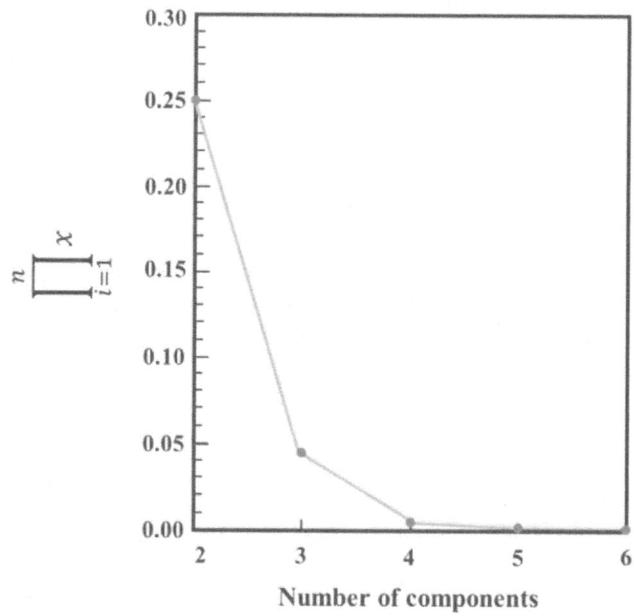

FIGURE 4.6 The relationship between the product of equimolar compositions and the number of components.

iii. **Stoichiometric Compound Model**

The stoichiometric compound model is used to describe intermetallic phases that have very narrow or no compositional range, often appearing in binary systems. In this model, the Gibbs free energy of the compound A*p*B*q*, G_m^{φ}, depends solely on temperature:

$$G_m^{\varphi} = \sum_i x_i^0 G_i^{\varphi} + \Delta_f G(\text{A}p\text{B}q) \tag{4.22}$$

The formula for G_m^{φ} includes the Gibbs energies of the individual components G_i^{φ} weighted by their mole fractions x_i, plus the Gibbs energy of formation $\Delta_f G$(A*p*B*q*) for the stoichiometric compound. The term $\Delta_f G$(A*p*B*q*), which typically varies with temperature, can be expressed as a linear function of temperature:

$$\Delta_f G(\text{A}p\text{B}q) = \Delta_f H(\text{A}p\text{B}q) - T.\,\Delta_f S(\text{A}p\text{B}q) \tag{4.23}$$

Here, $\Delta_f H$(A*p*B*q*) represents the enthalpy of formation, and $\Delta_f S$(A*p*B*q*) represents the entropy of formation. Equations (4.22) and (4.23) can be extended to describe stoichiometric compounds in systems with more than two components.

iv. **Optimization**

The goal of optimization in thermodynamics is to refine the parameters of Gibbs energy functions for various phases such that the models can anticipate behaviors outside of the tested temperature and composition ranges and accurately replicate experimental data. An overview of the CALPHAD technique optimization procedure for HEA systems is provided below:

Database Collection and Evaluation: To start, compile all thermodynamic databases that are available for binary and ternary systems that are associated with your materials. Examine these databases' compatibility and dependability to make sure the information they offer is correct and consistent.

Model Selection: Select suitable thermodynamic models by considering the crystallographic details of the system's phases. For accurate phase representations and predictions, the right model is essential.

Data Collection for New Systems: If you are dealing with a system that does not have a preexisting database or requires optimization, gather and arrange pertinent thermodynamic and phase equilibrium data. The data may encompass any experimental findings pertaining to Gibbs energy. DFT calculations can be employed to provide the required thermodynamic data for hypothetical phases.

Critical Data Evaluation: Thoroughly examine and scrutinize the gathered data, eliminating those that are flawed or present contradictions. Proficiency in phase diagrams and experimental procedures is essential for this procedure. It is crucial to carefully consider the specificities of the experimental settings and the dependability of the data, as any mistakes in these areas can lead to complications during the optimization process.

Initial Parameter Setting: Commence the optimization procedure by utilizing initial parameter values derived from a limited yet essential data set. For binary systems, this may entail utilizing crucial facts such as phase equilibria or enthalpy of formation. Excessive data usage at the outset can complicate the procedure.

Refinement and Validation: After optimizing the binary and ternary systems, utilize the models on HEAs and evaluate the outcomes by comparing them to existing experimental data. Enhance the accuracy by making adjustments to the models, particularly for hypothetical stages. Continue enhancing the model by incorporating additional experimental data as it becomes accessible.

The database will be updated to ensure its accuracy as new experimental data becomes available. The PanHEA database has a component that enables the estimation of phase diagrams and characteristics for quinary and quaternary systems within a specified composition range. Subsequent sections will analyze the comparisons between CALPHAD forecasts and experimental results.

4.4 MACHINE LEARNING

Solid solution phase prediction in HEAs using ML approaches involves huge data sets and advanced algorithms for identifying stable phases based on thermodynamic principles and alloy composition. The important steps are discussed below:

i. **Data Collection and Preparation**

   The data need to be gathered from experimental studies, computational results (like CALPHAD and DFT), and existing databases that comprise compositions, phase stability, and related properties. Understand and extract features from the data, such as elemental properties (atomic radius, electronegativity), mixing enthalpies, mixing entropies, and specific alloying behaviors.

ii. **Machine Learning Model Selection**

   Implementation of regression techniques is to forecast the possibility of phase formation based on the features. Algorithms such as linear regression, random forests, and gradient boosting can be used to get better outputs. The models that can be used are support vector machines, decision trees, and neural networks for predicting the phases.

iii. **Training the Model**

   Split the data set into training and test sets to calculate model performance. Use k-fold cross-validation to confirm the model generalizes well to hidden data.

iv. **Model Evaluation**

   Consider model performance using metrics such as precision, accuracy, recall, and F1 score for classification tasks or regression tasks (mean absolute error and R^2). The analysis needs to be done to find the most influential feature in predicting solid solution formation, helping to refine the phase stability understanding.

v. **Phase Prediction**

The trained model must be used to predict solid solution phases for novel or untested alloy compositions. To quantify uncertainty in phase prediction, use the methods that help to assess the results' reliability.

vi. **Validation and Iteration**

To confirm the presence of predicted phases validation can be done by experimental synthesis and characterization (e.g., XRD, SEM). Refine the model continuously with new data from computational studies or experiments to enhance accuracy and predictive power.

Huang et al. [40] implemented three distinct ML for the prediction of phases in HEAs such as K-nearest neighbors' (KNNs), support vector machine (SVM), and artificial neural network (ANN). The ML model has been discovering the phase selection rule by using a broad experimental data set algorithm of 401 distinct HEAs, such as 174 solid solutions, 54 intermetallic compounds, and 173 solid solutions with intermetallic phases. The respective MLFFNN (multi-layer feed forward neural network) architecture for these HEAs is shown in Figure 4.7. KNN, SVM, and ANN model accuracy is 68.6%, 64.3%, and 74.3%, respectively. Figure 4.8 shows the mean testing accuracy values represented in scatter plots for different combinations

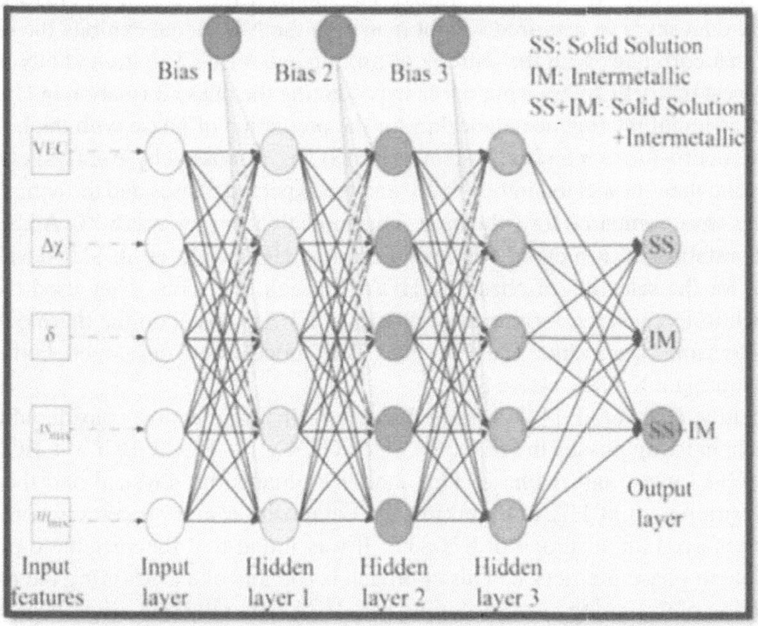

FIGURE 4.7 MLFFNN representation consisting input layer, three hidden layers, and one output layer [40].

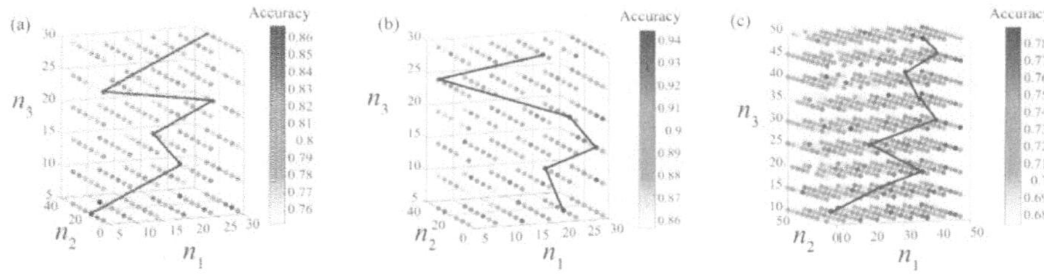

FIGURE 4.8 A 3D scatter plot presenting how the mean testing accuracy of categorizing two phases varies with the number of neurons in each hidden layer, denoted as n_1, n_2, and n_3, respectively: (a) SS and IM; (b) SS + IM and IM; (c) SS and SS + IM [40].

of neurons in the three hidden layers. The testing accuracy values using ANN can be observed in classification composed of a pair of the three phases: the solid solution phase the solid solution and Intermetallic phases, the solid solution + intermetallic and Intermetallic phases, as well as the solid solution and solid solution + Intermetallic phases, are 86.7%, 94.3%, and 78.9% respectively. Among the three ML models, the trained ANN model exhibits well and is useful in predicting phases of new HEAs.

Islam et al. [41] applied an ML approach for phase prediction in HEAs. The correlation was established between five parameters that led to the selection of phase in various HEAs. A neural networking (NN) model was trained to categorize the subsequent phase based totally on the input variable. The model achieved 99% in the learning of the total data set. The acquired weight matrix in the NN model exhibits the essential aspect that correlates with the entropy of mixing and VEC. The trained NN parameters suggest that VEC plays a main role in predicting the phase formation in HEAs. An applied artificial intelligence algorithm for the prediction of phase with the help of an adaptive neuro-fuzzy interface system (ANFIS) was discussed by Agarwal et al. [42]. The testing data set was fed into ANFIS, and the expected phases and the actual phases of HEAs were compared for validation. This model's accuracy was 80%. Additionally, ANFIS established a method for designing new HEAs. Qu et al. [43] developed a method for the selection of phases in HEAs through ML tools. They used the SVM approach to forecast the formation of phases in HEAs. Based on the thermodynamic parameters and composition of HEAs, two SVM models were developed, both models demonstrating reliability above 85%.

Recently, Dixit et al. [44] presented a new ANN architecture considered to forecast eight existing phases in HEAs: BCC, FCC, FCC1 + FCC2, BCC1 + BCC2, B2, Laves, C14 Laves, and σ Phase. This model evaluates the physical and thermodynamic parameters of HEAs by taking the composition and processing route. This model achieved an accuracy of 87.083%. It was found that the variable δ plays the main role in phase prediction. This approach is capable of ascertaining the potential application of designing and developing new HEAs, thereby selecting the processing routes.

Zhang et al. [45] suggested a correlation between different phases and nine thermodynamic parameters, including the formation enthalpy of intermetallic compounds

(H_{IM}), the enthalpy of mixing for amorphous phase (H_{AM}), the formation enthalpy of solid solutions (H_{ss}), the mixing enthalpy of the liquid phase (H_L), the entropy of mixing (S_{mix}), the elastic energy of the alloy (H_E), atomic size difference (δ), the Ω parameter, and the melting temperature of the alloy (T_m). They utilized an ML approach along with an extended Miedema model to predict various phases in HEAs. Kernel principal component analysis (KPCA) was suggested for feature selection and variable transformation. The thermodynamic parameters for HEAs were assessed using a geometric model and Miedema theory, providing feature data for the ML model. The results suggested that atomic size difference and elastic energy play important roles in phase formation. The accuracy of the testing set using a SVM with four features and KPCA reached 0.9743. The extended Miedema model provides clear thermodynamic parameters necessary for designing HEAs. Also, the combination of SVM and KPCA models proves effective in the prediction of phases in HEAs.

4.5 DENSITY FUNCTIONAL THEORY

The fundamental ideas of density functional theory (DFT) are derived from two theorems published by Hohenberg and Kohn in 1964 [46]. DFT establishes the foundation for using the density of electrons to determine the properties of many electron systems.

i. **Hohenberg-Kohn Theorem I**

The first Hohenberg-Kohn (HK) theorem states that the ground-state properties of a many-electron system are uniquely determined by its electron density, ρ(r). This theorem establishes that all observable features of a system, including total energy, can be expressed as a function of the electron density. In other words, knowing the electron density alone is sufficient to determine all system characteristics, such as energy and external potential.

Mathematically, the ground-state energy E_0 is represented as a functional of the ground-state electron density, $\rho_0(r)$:

$$E_0 = E\left[\rho_0\right]$$

To prove this, we assume that two different external potentials, $V_{ext}(r)$ and V_{ext}' (r), give rise to the same electron density $\rho_0(r)$. These potentials would correspond to different Hamiltonians H^ and H^′, leading to different ground-state wave functions, Ψ and Ψ′. According to the variational principle, the ground-state energy of any Hamiltonian is minimized by its corresponding wave function, leading to:

$$E_0 < \left\langle \Psi' \middle| H^\wedge \middle| \Psi' \right\rangle$$

$$E_0' < \left\langle \Psi \middle| H^{\wedge\prime} \middle| \Psi \right\rangle$$

ii. **Hohenberg-Kohn Theorem II**

The second Hohenberg-Kohn theorem introduces the existence of a universal energy functional, F[ρ], which, when minimized, provides the exact ground-state energy of the system. This functional incorporates the kinetic energy of the electrons and their interaction energy.

The total energy of the system is given by:

$$E[\rho] = F[\rho] + \int V_{ext}(r)\rho(r)dr$$

Here, F[ρ] represents the sum of the kinetic energy functional T[ρ] and the electron-electron interaction energy functional $E_{int}[\rho]$:

$$F[\rho] = T[\rho] + \text{Eint}[\rho] F[\rho] = T[\rho] + \text{Eint}[\rho]$$

The ground-state density is found by minimizing this energy functional with respect to the electron density:

$$\delta_E[\rho]\delta_\rho(r) = 0$$

iii. **The Kohn-Sham Equations**

Although the HK theorems provide a theoretical foundation for DFT, practical applications require a simplification of the electron-electron interactions. To make DFT computationally feasible, Kohn and Sham introduced a method that replaces the many-electron problem with a set of non-interacting electrons that produce the same ground-state density as the original system [47].

The Kohn-Sham equations, derived by minimizing the total energy functional, are given by:

$$\left[-1/2\nabla^2 + V_{eff}(r)\right]\psi_i(r) = \epsilon_i\psi_i(r)$$

where $\psi_i(r)\psi_i(r)$ are the Kohn-Sham orbitals, ϵ_i are the corresponding eigenvalues, and $V_{eff}(r)$ is the effective potential defined as:

$$V_{eff}(r) = V_{ext}(r) + \int \rho(r')|r - r'|dr' + V_{xc}(r)$$

Here, $V_{xc}(r) = \delta E_{xc}[\rho] / \delta\rho(r)$ is the exchange-correlation potential.

The accuracy of DFT calculations depends on the choice of the exchange-correlation function. Two common approximations are the local density approximation (LDA) and the generalized gradient approximation (GGA).

4.5.1 Case Studies and Experimental Results

The utilization of DFT in the thermodynamic modeling of HEAs has attracted considerable interest because of its capacity to forecast the stability, phase formation, and mechanical characteristics of these intricate materials. This section explores multiple case studies and experimental findings that demonstrate the effectiveness of DFT in both academic and industrial settings.

a. **Computational Parameters and Setup**

Within the domain of highly efficient algorithms, laboratory investigations frequently commence with meticulous computational simulations employing DFT. Accurate findings are highly dependent on the careful selection of exchange-correlation functionals, pseudopotentials, and k-point mesh densities. The Perdew-Burke-Ernzerhof (PBE) generalized gradient approximation (GGA) functional is commonly employed in the study of HEAs because it offers a good trade-off between accuracy and computing expense [48]. The plane-wave cut-off energy is a vital characteristic that is usually determined within the range of 400–600 eV, depending on the intricacy of the alloy and the constituent elements. The Monkhorst-Pack grids are commonly utilized to sample the Brillouin zone. For more precise total energy calculations, finer grids such as 12 × 12 × 12 are required [49]. The selection of pseudopotentials, such as projector augmented-wave (PAW) or ultrasoft pseudopotentials, is also a crucial factor in determining the correctness of the simulation. PAW potentials are favored because of their superior efficiency in handling core electrons, resulting in improved convergence properties in complicated alloys [50].

b. **Thermodynamic Properties and Phase Stability**

DFT has played a crucial role in forecasting the thermodynamic characteristics of HEAs. For example, the enthalpy of mixing is a crucial factor that affects the stability of different phases. Research has indicated that HEAs, such as CoCrFeMnNi, have negative enthalpies of mixing, which implies a propensity for phase separation at lower temperatures [7]. This has been empirically verified using X-ray diffraction (XRD) and transmission electron microscopy (TEM) investigations. The configurational entropy is another important thermodynamic feature that is maximized in HEAs because of the nearly equal concentration of numerous main components. The phenomenon of entropy-driven stabilization of single-phase solid solutions in HEAs has been extensively studied and supported by both DFT calculations and experimental evidence [5]. CoCrFeMnNi has been demonstrated to exhibit a stable FCC structure, as supported by both DFT predictions and experimental phase diagrams [51].

c. **Mechanical Properties**

DFT simulations have played a crucial role in comprehending the mechanical characteristics of HEAs. For instance, the elastic constants (C_{ij}) of HEAs can be calculated using the stress-strain method inside DFT, which offers valuable information about the material's rigidity and directional dependence. Research conducted on NbTaTiV has demonstrated high elastic moduli, indicating exceptional mechanical strength. This finding has been supported by nanoindentation research conducted by Gludovatz et al. in 2014 [52]. In addition, DFT has been employed to forecast the optimal shear strength of HEAs. The CoCrFeMnNi alloy has been discovered to have shear strength similar to typical alloys but with greater ductility. This has been confirmed through micro-compression testing, according to Gao et al. in 2013 [53].

d. **Electronic and Magnetic Properties**

The utilization of DFT is also employed to investigate the electrical and magnetic characteristics of HEAs. The electronic density of states (DOS) offers valuable information regarding the electrical conductivity and stability of the alloy. FeCoCrNiCu has a metallic character, as evidenced by DFT calculations that show a high DOS at the Fermi level. This observation aligns with its observed electrical conductivity, as reported by Sohn et al. in 2020 [54]. DFT has also been utilized to predict magnetic characteristics, including magnetic moment and Curie temperature. The magnetic properties of CoCrFeMnNi were significantly affected by the nearby atomic arrangement, resulting in an intricate magnetic structure that was subsequently verified by neutron diffraction investigations [55].

4.5.2 Industrial Applications

HEAs have garnered significant attention from the aerospace industry due to their exceptional mechanical qualities, especially under elevated temperatures. DFT has been widely employed to simulate the phase stability of HEAs at elevated temperatures, a critical factor for their utilization in turbine blades and other components operating under high-temperature conditions. The AlCoCrFeNi alloy has been recognized as a potential material for high-temperature uses because of its stable BCC structure, which has been predicted by DFT and validated by high-temperature X-ray diffraction (XRD) experiments. Furthermore, DFT investigations have played a crucial role in fine-tuning the composition of HEAs to suit certain uses. For example, the incorporation of refractory elements like Nb and Mo into HEAs has been demonstrated to improve their strength at high temperatures. This improvement has been confirmed through creep tests conducted in industrial environments [56].

Within the nuclear sector, materials must endure harsh conditions, such as intense radiation exposure and corrosive surroundings. HEAs are considered promising materials for nuclear applications because of their exceptional resistance to radiation and corrosion. DFT simulations have been employed to simulate the energies required for defect formation and the barriers for defect migration in HEAs, offering valuable insights into their ability to withstand radiation. For instance, research using DFT on the FeCrNiMn HEA has demonstrated that the alloy has lower vacancy formation energies in comparison to conventional stainless steels. This indicates that the alloy has a stronger resistance to radiation, as suggested by Granberg et al. in 2017 [57]. Ion irradiation studies have confirmed that HEA has exceptional resistance to swelling and phase separation caused by radiation.

The automobile industry is progressively turning to HEAs as a solution for developing lightweight materials with enhanced strength, which can contribute to improved fuel efficiency and safety. The utilization of DFT has played a crucial role in accurately forecasting the mechanical characteristics of HEAs that are of utmost importance in the field of automobile engineering. The AlCoCrFeNi alloy has been recognized as a promising option for automotive components because of its excellent strength-to-weight ratio, as projected by DFT and verified by conducting tensile tests [58]. In addition, DFT investigations have been employed to optimize the selection of alloying elements in HEAs to get the desired qualities. For instance, the incorporation of titanium (Ti) into the AlCoCrFeNi HEA has been demonstrated to improve its hardness and resistance to wear, rendering it appropriate for applications in automobile gears and bearings [59]. The validity of these findings has been confirmed through large-scale testing conducted in an industrial setting, where the alloy exhibited higher performance in comparison to conventional materials.

4.5.3 Challenges in Industrial Implementation

Although DFT offers valuable insights into the characteristics of HEAs, there are difficulties in applying these discoveries to practical industrial uses. A significant obstacle is the expansion of HEA production due to the extensive processing techniques needed to handle the high melting temperatures and complex phase diagrams of these alloys. In addition, the expense of raw materials and the absence of established supply chains for HEAs pose further challenges. Notwithstanding these difficulties, some sectors have started to embrace HEAs in specialized uses, leveraging their distinct characteristics to gain substantial benefits. Further industrial application of HEAs is anticipated in the future due to ongoing progress in DFT models and experimental methodologies.

4.5.4 Comparative Analysis with Traditional Materials

HEAs, as compared to standard alloys, demonstrate enhanced thermodynamic stability, especially at high-temperature conditions. Research using DFT has demonstrated that the presence of a large amount of configurational entropy in HEAs helps to stabilize the formation of single-phase solid solutions. This prevents the occurrence of phase separation and enhances the performance of the material at high temperatures. Conversely, conventional alloys frequently experience phase separation and deterioration when exposed to high temperatures, which restricts their use in challenging situations. For example, DFT calculations on the CoCrFeMnNi alloy have demonstrated its stability in the FCC phase at elevated temperatures. This finding has been verified by conducting differential scanning calorimetry (DSC) and high-temperature XRD studies [7]. By contrast, conventional stainless steels like 304 and 316 demonstrate phase instability and the creation of carbides when exposed to high temperatures, resulting in a decline in mechanical characteristics.

HEAs exhibit superior mechanical properties in comparison to conventional alloys. The results of DFT simulations indicate that HEAs, such as NbTaTiV, demonstrate superior elastic moduli and yield strengths in comparison to standard alloys such as Inconel 718 and titanium alloys [52]. The unusual atomic configurations and the presence of numerous primary elements are responsible for impeding dislocation motion and enhancing mechanical strength. The validity of these findings has been confirmed by experimental research, where HEAs have exhibited exceptional strength and ductility in a range of mechanical tests, such as tensile, compressive, and impact tests [53]. Conversely, conventional alloys frequently experience a compromise between strength and ductility, which restricts their effectiveness in challenging applications.

HEAs exhibit superior corrosion resistance compared to conventional materials. DFT research has been employed to simulate the generation and durability of passive films in HEAs, offering a valuable understanding of their corrosion characteristics. For instance, DFT simulations conducted on the FeCoCrNiCu alloy have demonstrated the formation of a durable passive coating in corrosive surroundings, akin to conventional stainless steels. Experimental studies have demonstrated that HEAs possess enhanced corrosion resistance in harsh environments, such as marine and acidic conditions when compared to conventional alloys [60]. HEAs are very desirable for use in the maritime and chemical processing sectors due to their exceptional corrosion resistance, which is of utmost importance in these industries.

Although HEAs possess improved qualities in comparison to conventional materials, their higher costs are attributed to the utilization of numerous primary elements, some of which are both expensive and rare. DFT investigations have been employed to optimize the composition of HEAs to attain specific qualities while simultaneously reducing expenses. For instance, the replacement of costly materials such as cobalt with more readily available elements like iron has been demonstrated to decrease expenses without significantly impacting performance [61]. The importance of environmental factors in material selection is growing. HEAs, because of their exceptional characteristics, have the potential to extend the life span of products and decrease the amount of materials needed, therefore compensating for the higher upfront expenses and environmental consequences. Nevertheless, the recycling and management of HEAs at the end of their life span remain subjects that necessitate additional investigation.

REFERENCES

1. Q. C. Fan, B. S. Li, Y. Zhang, Influence of Al and Cu elements on the microstructure and properties of (FeCrNiCo)Al_xCu_y high-entropy alloys. J. Alloys Compd. 614 (2014) 203–210. https://doi.org/10.1016/j.jallcom.2014.06.090.
2. C. J. Tong, Y. L. Chen, S. K. Chen, J. W. Yeh, T. T. Shun, C. H. Tsau, S. J. Lin, S. Y. Chang, Microstructure characterization of AlxCoCrCuFeNi high-entropy alloy system with multiprincipal elements. Metall Mater Trans A. 36 (2005) 881–893. https://doi.org/10.1007/s11661-005-0283-0
3. W. Chen, Z. Fu, S. Fang, H. Xiao, D. Zhu, Alloying behavior, microstructure and mechanical properties in a FeNiCrCo0.3Al0.7 high entropy alloy. Mater. Des. 51 (2013) 854–860. https://doi.org/10.1016/j.matdes.2013.04.061.
4. A. Y. Ivannikov, I. K. Grebennikov, Y. A. Klychevskikh, A. V. Mikhailova, K. V. Sergienko, M.A. Kaplan, Fabrication, microstructure, and physico-mechanical properties of Fe–Cr–Ni–Mo–W high-entropy alloys from elemental powders. Metals (Basel). 12 (2022) 1764. https://doi.org/10.3390/met12101764.

5. J. W. Yeh, S. K. Chen, S. J. Lin, J. Y. Gan, T. S. Chin, T. T. Shun, C. H. Tsau, S. Y. Chang, Nanostructured high-entropy alloys with multiple principal elements: Novel alloy design concepts and outcomes. Adv. Eng. Mater. 6 (2004) 299–303. https://doi.org/10.1002/adem.200300567.
6. B. Cantor, I. T. H. Chang, P. Knight, A. J. B. Vincent, Microstructural development in equiatomic multicomponent alloys. Mater. Sci. Eng. A 375–377 (2004) 213–218. https://doi.org/10.1016/j.msea.2003.10.257.
7. Y. Zhang, T. T. Zuo, Z. Tang, M. C. Gao, K. A. Dahmen, P. K. Liaw, Z. P. Lu, Microstructures and properties of high-entropy alloys. Prog. Mater. Sci. 61 (2014) 1–93. https://doi.org/10.1016/j.pmatsci.2013.10.001.
8. B. S. Murty, J. W. Yeh, S. Ranganathan, High Entropy Alloys, Butterworth-Heinemann, London (UK) (2014).
9. C. Suryanarayana, Mechanical alloying and milling. Prog. Mater. Sci. 46 (2001) 1–184. https://doi.org/10.1016/S0079-6425(99)00010-9.
10. S. K. Dewangan, V. Kumar, Heat treatment and processing route consequences on the microstructure and hardness behavior of tungsten-containing high-entropy alloys. J. Alloys Compd. 923 (2022) 166167. https://doi.org/10.1016/j.jallcom.2022.166167.
11. K. R. Rao, S. K. Sinha, Effect of sintering temperature on microstructural and mechanical properties of SPS processed CoCrCuFeNi based ODS high entropy alloy. Mater. Chem. Phys. 256 (2020) 123709. https://doi.org/10.1016/j.matchemphys.2020.123709.
12. D. Kumar, O. Maulik, S. Kumar, V. K. Sharma, Y. V. S. S. Prasad, V. Kumar, Impact of tungsten on phase evolution in nanocrystalline AlCuCrFeMnWx (x = 0, 0.05, 0.1 and 0.5 mol) high entropy alloys. Mater. Res. Express 4 (2017) 114004. https://doi.org/10.1088/2053-1591/aa96df.
13. M. S. Lucas, G. B. Wilks, L. Mauger, J. A. Muñoz, O. N. Senkov, E. Michel, J. Horwath, S. L. Semiatin, M. B. Stone, D. L. Abernathy, E. Karapetrova, Absence of long-range chemical ordering in equimolar FeCoCrNi. Appl. Phys. Lett. 100 (2012) 1–5. https://doi.org/10.1063/1.4730327.
14. Z. Wu, H. Bei, F. Otto, G. M. Pharr, E. P. George, Recovery, recrystallization, grain growth and phase stability of a family of FCC-structured multi-component equiatomic solid solution alloys. Intermetallics 46 (2014) 131–140. https://doi.org/10.1016/j.intermet.2013.10.024.
15. M. S. Lucas, L. Mauger, J. A. Muoz, Y. Xiao, A. O. Sheets, S. L. Semiatin, J. Horwath, Z. Turgut, Magnetic and vibrational properties of high-entropy alloys. J. Appl. Phys. 109 (2011). https://doi.org/10.1063/1.3538936.
16. X. Yang, Y. Zhang, P. K. Liaw, Microstructure and compressive properties of NbTiVTaAlx high entropy alloys. Procedia Eng. 36 (2012) 292–298. https://doi.org/10.1016/j.proeng.2012.03.043.
17. O. N. Senkov, G. B. Wilks, D. B. Miracle, C. P. Chuang, P. K. Liaw, Refractory high-entropy alloys. Intermetallics 18 (2010) 1758–1765. https://doi.org/10.1016/j.intermet.2010.05.014.
18. J. Pan, T. Dai, T. Lu, X. Ni, J. Dai, M. Li, Microstructure and mechanical properties of Nb25Mo25Ta25W25 and Ti8Nb23Mo23Ta23W23 high entropy alloys prepared by mechanical alloying and spark plasma sintering. Mater. Sci. Eng. A 738 (2018) 362–366. https://doi.org/10.1016/j.msea.2018.09.089.
19. O.N. Senkov, S. V. Senkova, D.B. Miracle, C. Woodward, Mechanical properties of low-density, refractory multi-principal element alloys of the Cr-Nb-Ti-V-Zr system. Mater. Sci. Eng. A 565 (2013) 51–62. https://doi.org/10.1016/j.msea.2012.12.018.
20. O. Ridge, (12) Patent Application Publication (10) Pub. No.: US 2013/0108502 A1, 1 (2013).

21. A. Takeuchi, K. Amiya, T. Wada, K. Yubuta, W. Zhang, High-entropy alloys with a hexagonal close-packed structure designed by equi-atomic alloy strategy and binary phase diagrams. JOM. 66 (2014) 1984–1992. https://doi.org/10.1007/s11837-014-1085-x.
22. N. K. Mindi, S. N. Alam, K. Dutta, Effect of tungsten on microstructure and mechanical properties of novel NiMnCrMoW x high-entropy alloys developed by mechanical alloying. J. Mater. Eng. Perform. (2024). https://doi.org/10.1007/s11665-024-10107-3.
23. R. K. Song, L. J. Wei, C. X. Yang, S. J. Wu, Phase formation and strengthening mechanisms in a dual-phase nanocrystalline CrMnFeVTi high-entropy alloy with ultrahigh hardness. J. Alloys Compd. 744 (2018) 552–560. https://doi.org/10.1016/j.jallcom.2018.02.029.
24. C. Wang, W. Ji, Z. Fu, Mechanical alloying and spark plasma sintering of CoCrFeNiMnAl high-entropy alloy. Adv. Powder Technol. 25 (2014) 1334–1338. https://doi.org/10.1016/j.apt.2014.03.014.
25. L. Kaufman, H. Bernstein, Computer Calculation of Phase Diagrams. Academic, New York (1970).
26. M. Hillert, Phase Transformations. ASM, Cleveland (1968).
27. X. Huang, C. Jin, C. Zhang, H. Zhang, H. Fu, Machine learning assisted modelling and design of solid solution hardened high entropy alloys. Mater. Des. 211 (2021), 110177. https://doi.org/10.1016/j.matdes.2021.110177.
28. U. K. Jaiswala, V. K. Yegi, M. R. Rahula, P. K. Gandham, Machine learning-enabled identification of new medium to high entropy alloys with solid solution phases. Comput. Mater. Sci 197 (2021), 110623. https://doi.org/10.1016/j.commatsci.2021.110623.
29. Y. Dong, Y. Lu, L. Jiang, T. Wang, T. Li, Effects of electro-negativity on the stability of topologically close-packed phase in high entropy alloys. Intermetallics. 52 (2014) 105–109. https://doi.org/10.1016/j.intermet.2014.04.001.
30. S. Guo, C. Ng, J. Lu, and C.T. Liu, Effect of valence electron concentration on stability of fcc or bcc phase in high entropy alloys. J. Appl. Phys. 109 (2011) 103505. https://doi.org/10.1063/1.3587228.
31. X. Yang, Y. Zhang, Prediction of high-entropy stabilized solid solution in multi-component alloys. Mater. Chem. Phys 132 (2012) 233–238. https://doi.org/10.1016/j.matchemphys.2011.11.021.
32. Z. Wang, Y. Huang, Y. Yang, J. Wang, C.T. Liu, Atomic-size effect and solid solubility of multicomponent alloys. Scripta Mater. 94 (2015) 28–31, https://doi.org/10.1016/j.scriptamat.2014.09.010.
33. Y. Lu, Y. Dong, L. Jiang, T. Wang, T. Li, Y. Zhang, A criterion for topological close-packed phase formation in high entropy alloys. Entropy (2015) 2355–2366, https://doi.org/10.3390/e17042355.
34. D. J. M. King, S. C. Middleburgh, A. G. McGregor, M. B. Cortie, Predicting the formation and stability of single-phase high-entropy alloys. Acta Mater. 104 (2016) 172–179, https://doi.org/10.1016/j.actamat.2015.11.040.
35. S. Praveen, B. S. Murty, R. S. Kottada, Alloying behavior in multi-component AlCoCrCuFe and NiCoCrCuFe high entropy alloys. Mater. Sci. Eng. A. 534 (2012), 83. https://doi.org/10.1016/j.msea.2011.11.044.
36. M. Gao, D. Alman, Searching for next single-phase high-entropy alloy compositions. Entropy 15 (2013) 4504–4519. https://doi.org/10.3390/e15104504.
37. F. Otto, Y. Yang, H. Bei, E. P. George Relative effects of enthalpy and entropy on the phase stability of equiatomic high-entropy alloys. Acta Mater. 61 (2013) 2628–2638. doi:10.1016/j.actamat.2013.01.042
38. A. T. Dinsdale, SGTE data for pure elements. Calphad. 15 (4) (1991) 317–425. https://doi.org/10.1016/0364-5916(91)90030-N.

39. O. Redlich, A. T. Kister, Algebraic representation of thermodynamic properties and the classification of solutions. Ind. Eng. Chem. 40 (1948) 345–348. https://doi.org/10.1021/ie50458a036.
40. W. Huang, P. Martin, H. L. Zhuang. Machine learning phase prediction of high-entropy alloys. Acta Mater. 169 (2019) 225–236. https://doi.org/10.1016/j.actamat.2019.03.012
41. N. Islam, W. Huang, H. L. Zhuang. Machine learning for phase selection in multi-principal element alloys. Comput Mater Sci. 150 (2018) 230–235. https://doi.org/10.1016/j.commatsci.2018.04.003
42. A. Agarwal, A. K. P. Rao. Artificial intelligence predicts body-centered-cubic and face-centered-cubic phases in high-entropy alloys. JOM. 71 (2019) 3424–3432. https://doi.org/10.1007/s11837-019-03712-4.
43. N. Qu, Y. Chen, Z. Lai, Y. Liu, J. Zhu, The phase selection via machine learning in high entropy alloys. Procedia Manuf. 37 (2019) 299–305. https://doi.org/10.1016/j.promfg.2019.12.051.
44. S. Dixit, V. Singhal, A. Agarwal, A. K. Prasada Rao, Multi-label phase-prediction in high-entropy-alloys using. Mater Lett. 268 (2020) 127606. https://doi.org/10.1016/j.matlet.2020.127606.
45. L. Zhang, H. Chen, X. Tao, et al., Machine learning reveals the importance of the formation enthalpy and atom-size difference in forming phases of high entropy alloys. Mater Des. 193 (2020) 108835. https://doi.org/10.1016/j.matdes.2020.108835.
46. P. Hohenberg, W. Kohn: Inhomogeneous Electron Gas. Physics (College Park Md). (1964)
47. W. Kohn, J. S. Lu, Self-consistent equations including exchange and correlation effects. Phys Rev. 140 (1965) A1133.
48. P. P. John, B. Kieron, E. Matthias, Generalized gradient approximation made simple. Phys Rev Lett. 77 (1996) 3865.
49. H. J. Monkhorst, J. D. Pack, Special points for brillouin-zone integrations. Phys Rev B. 13 (1976) 5188–5192. https://doi.org/10.1103/PhysRevB.13.5188
50. P. E. Blöchl, Projector augmented-wave method, Phys Rev B. 50 (1994) 17953–17979. https://doi.org/10.1103/PhysRevB.50.17953
51. F. He, Z. Wang, Y. Li et al., Kinetic ways of tailoring phases in high entropy alloys. Sci Rep. 6 (2016) 1–8. https://doi.org/10.1038/srep34628
52. B. Gludovatz, A. Hohenwarter, D. Catoor, et al., A fracture-resistant high-entropy alloy for cryogenic applications. Science. 345 (2014) 1153–1158. https://doi.org/10.1126/SCIENCE.1254581.
53. M. C. Gao, P. K. Liaw, J. W. Yeh, Y. Zhang, High-entropy alloys: Fundamentals and applications. 1–516 (2016). https://doi.org/10.1007/978-3-319-27013-5
54. D. G. Kim, Y. H. Jo, J. M. Park, et al., Effects of annealing temperature on microstructures and tensile properties of a single FCC phase CoCuMnNi high-entropy alloy. J Alloys Compd. 812 (2020) 152111. https://doi.org/10.1016/J.JALLCOM.2019.152111.
55. Y. Ikeda, B. Grabowski, F. Körmann, Ab initio phase stabilities and mechanical properties of multicomponent alloys: A comprehensive review for high entropy alloys and compositionally complex alloys. Mater Charact.147 (2019) 464–511. https://doi.org/10.1016/J.MATCHAR.2018.06.019
56. O. N. Senkov, S. Gorsse, D. B. Miracle, High temperature strength of refractory complex concentrated alloys. Acta Mater. 175 (2019) 394–405. https://doi.org/10.1016/J.ACTAMAT.2019.06.032
57. F. Granberg, K. Nordlund, M. W. Ullah, Mechanism of radiation damage reduction in equiatomic multicomponent single phase alloys. Phys Rev Lett. 116 (2016) 135504. https://doi.org/10.1103/PhysRevLett.116.135504.

58. N. Tabassum, Y. S. Sistla, R. G. Burela, A. Gupta, Structural, electronic, mechanical and thermal properties of AlxCoCrFeNi ($0 \leq x \leq 2$) high entropy alloy using density functional theory. Met Mater Int. 30 (2024) 3349–3369. https://doi.org/10.1007/S12540-024-01709-6
59. C. O. Ujah, D. V. V. Kallon, V. S. Aigbodion, High entropy alloys prepared by spark plasma sintering: Mechanical and thermal properties. Mater Today Sustain. 25 (2024) 100639. https://doi.org/10.1016/J.MTSUST.2023.100639
60. D. B. Miracle, O. N. Senkov, A critical review of high entropy alloys and related concepts. Acta Mater. 122 (2017) 448–511. https://doi.org/10.1016/J.ACTAMAT.2016.08.081
61. J. Kumar, S. Jha, A. Raturi, A. Bajpai, et al. Novel alloy design concepts enabling enhanced mechanical properties of high entropy alloys. Front Mater. 9 (2022) 868721. https://doi.org/10.3389/fmats.2022.868721

5 Advanced Characterization of High Entropy Alloys

D. Narsimhachary, K. Pramod and Dharavathu Naresh

High entropy alloys (HEAs) have gained significant attention due to their exceptional mechanical properties, such as high strength, corrosion resistance, and thermal stability. Their complex microstructures, often involving multiphase or nano-structured phases, necessitate advanced characterization methods to unravel their fundamental behaviour. This chapter reviews the application of key characterization techniques—scanning electron microscopy (SEM), electron backscatter diffraction (EBSD), transmission electron microscopy (TEM), atom probe tomography (APT), and nanoindentation—in the analysis of HEAs. SEM provides insights into surface morphology, while EBSD maps crystallographic orientations and phase distributions. TEM reveals atomic-scale structures and dislocations, and APT offers precise compositional mapping at the atomic level. Nanoindentation serves as a pivotal technique for measuring hardness, and elastic modulus. Despite their unique advantages, these techniques face challenges related to sample preparation, resolution limitations, and data interpretation, especially for multi-phase HEAs. By integrating these advanced methods, we can achieve a comprehensive understanding of HEA behaviour, which is critical for tailoring these materials for specific high-performance applications.

5.1 INTRODUCTION

The design and development of advanced metallic alloys are crucial for meeting the growing demand for high-performance materials in industries such as aerospace, energy, and automotive. Traditionally, alloys have been designed using single-principal-element systems, where one dominant element forms the base (e.g., Fe-C alloys, Al-Mg, Al-Mn, Al-Cu, Al-Zn, Cu alloys), with smaller amounts of other elements added to adjust specific properties. While this approach has successfully produced a wide range of alloys for industrial use, it limits improvements in key mechanical properties such as toughness and strength [1–3]. This limitation has driven researchers to develop new materials, such as high entropy alloys (HEAs), which exhibit enhanced physical and mechanical properties, thereby expanding their industrial applications [4].

DOI: 10.1201/9781003664093-7

HEAs represent an innovative class of materials that offer significant improvements in mechanical properties compared to conventional alloy systems. These alloys are composed of multiple principal elements mixed in equimolar or near-equimolar ratios. The high configurational entropy in HEAs stabilizes solid solution phases, reducing phase segregation, and improving homogeneity. This unique composition imparts remarkable mechanical properties, excellent thermal stability, and strong resistance to corrosion, making HEAs suitable for extreme environmental conditions [5].

However, characterizing HEAs presents several challenges. Their complex compositions make characterization difficult, as the wide range of alloying possibilities leads to overlapping phases. HEAs typically exhibit a multiphase microstructure, further complicating their analysis. Achieving a single-phase material is crucial for studying the fundamental behaviour of HEAs. By focusing on microstructure, researchers can identify those that offer unique properties, advancing the understanding of HEAs for targeted applications.

To analyze HEAs, traditional characterization techniques such as powder X-ray diffraction (XRD) are commonly used. However, XRD, which typically operates at a wavelength of 1.5418 Å, cannot effectively resolve nano-sized secondary phases and lacks spatial resolution, making it difficult to distinguish localized composition variations [6]. Additionally, amorphous or very fine-grained HEAs may yield broad peaks, complicating phase identification [7–9].

Other techniques, including scanning electron microscopy (SEM), transmission electron microscopy (TEM), and X-ray photoelectron spectroscopy (XPS), are frequently employed to analyse phase structures, particle morphology, and chemical states. However, these methods often fall short in detecting detailed multi-elemental mixing or subtle structural differences in HEAs. Advanced synchrotron X-ray techniques, typically operating at much shorter wavelengths, such as 0.2113 Å, overcome these limitations by providing higher resolution. Synchrotron XRD is particularly effective in detecting overall phase structures, immiscible phases, and impurities that are difficult to identify with other techniques, ensuring precise confirmation of high entropy mixing.

Additional techniques such as X-ray absorption spectroscopy (XAS) provide detailed insights into atomic coordination, helping researchers understand local ordering and short-range interactions. Hard X-ray photoelectron spectroscopy (HAXPES) offers valuable information about electronic structures, including valence band configurations, which are directly linked to catalytic properties [10].

This chapter provides fundamental insights into selecting appropriate techniques for handling immiscible phases, a critical challenge in the development of HEAs. By integrating advanced tools with traditional methods, researchers can effectively address this challenge. These combined approaches not only aid in understanding the underlying mechanisms governing the behaviour of immiscible phases but also enhance the ability to fabricate HEAs with tailored and improved properties. Furthermore, the chapter emphasizes the importance of choosing the right methodology to optimize the microstructural characteristics of HEAs, which directly influence their performance in various applications.

5.2 XRD

XRD is the primary and most widely used technique for determining the phases formed in any alloy system. This section primarily provides insights into how XRD graphs change concerning (1) composition percentage, (2) powder size (i.e., milled HEAs), (3) strain, (4) heat treatment, and (5) alloy processing (i.e., pre- and post-fabrication) [3–5]. The microstructural complexity of HEAs complicates the identification of overlapping phases, which play a crucial role in their strengthening mechanisms. To explore this, Yesh et al. and his team used the XRD technique to analyze the peaks and understand how designed alloys (i.e., Cu–Ni–Al–Co–Cr–Fe–Si alloy) contribute to mechanical properties in relation to their microstructures in HEAs. However, several issues arise during the analysis of XRD peaks. For instance, in Figure 5.1, the XRD patterns of the Cu–Ni–Al–Co–Cr–Fe–Si alloy series show a gradual reduction in peak intensity and variations in peak positions due to alloying elements. The reason for this variation is lattice distortion, which induces strain leading to modification of peaks.

Few HEAs such as 1. Co-Cr-Fe-Mn-Ni system (Cantor alloys). This alloy is known for its single-phase face-centered cubic (FCC) structure, which contributes to its exceptional ductility and fracture resistance [6, 7]. XRD studies reveal that Co-Cr-Fe-Mn-Ni exhibits a distinct FCC diffraction pattern, characterized by the presence of (111), (200), (220), and (311) peaks, with no additional peaks corresponding to secondary phases [9]. But when Cantor alloys are exposed to high temperatures, phase separation occurs, leading to the formation of secondary phases, such as Cr-rich intermetallic compounds or σ-phase, can occur. Studies have shown that non-equiatomic variations, such as Co-Cr-Fe-Ni (without Mn), tend to form the σ-phase, which is detrimental to ductility. Similarly, the addition of Mo to Co-Cr-Fe-Mn-Ni has been found to promote intermetallic formation, shifting the microstructure away from a stable FCC solid solution. These observations highlight the need for precise composition control to maintain a truly single-phase FCC structure in this system. One must be cautious when working with these alloys because the minimal atomic size differences among their constituent elements are very close [8]. This can pose limitations for the XRD technique [10, 11].

Nb-Mo-Ta-W (classified as a refractory alloy) favours a body-centred cubic (BCC) structure due to the significant atomic size mismatches and high cohesive energies of its constituent elements [12]. These factors promote the formation of a BCC solid solution rather than an FCC phase, leading to increased hardness and high-temperature stability at the expense of ductility [13]. XRD analysis of Nb-Mo-Ta-W displays characteristic BCC reflections at (110), (200), and (211), with no FCC-related peaks, confirming the formation of a stable BCC phase [14–16]. These alloys also show the absence of secondary phases, indicating that the alloy maintains a single-phase BCC structure under equilibrium conditions. Some studies have demonstrated that adding Ti to Nb-Mo-Ta-W leads to the precipitation of Laves phases, modifying the alloy's mechanical properties and potentially limiting its applicability in high-temperature environments. In order to resolve these, lave phase one needs a high-resolution XRD [16].

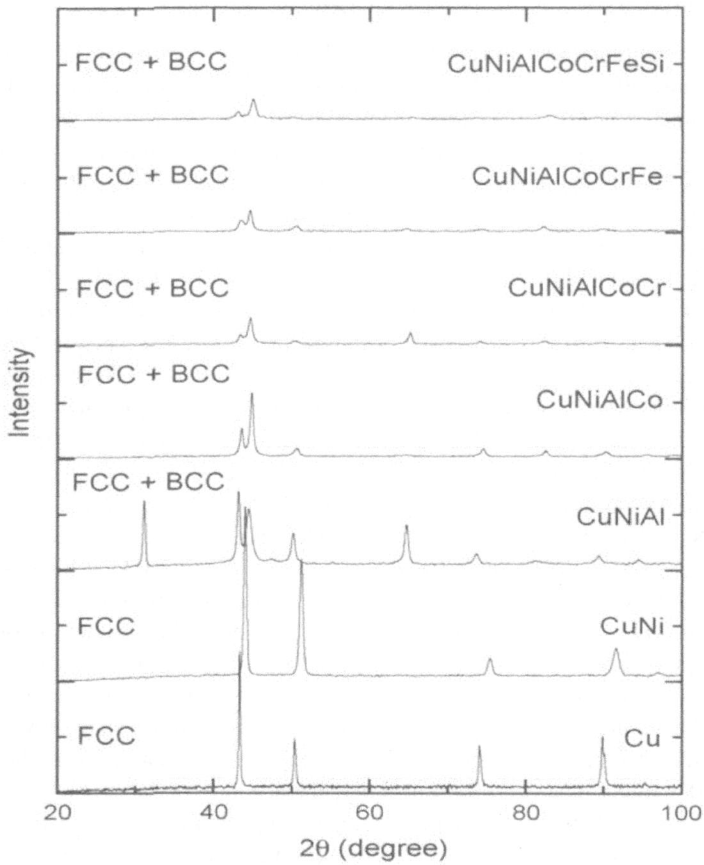

FIGURE 5.1 XRD patterns of Cu–Ni–Al–Co–Cr–Fe–Si alloy series with an increasing number of incorporated principal elements (intensity scale between two major ticks is 1000 cps) [11].

Most equiatomic HEAs described in the literature do not only exist as single-phase solid solutions under all conditions; rather, they often exhibit multiphase microstructures consisting of secondary solid solution phases and intermetallic compounds.

But resolving these phases and compounds through XRD is difficult due to its resolving [17, 18].

Case 1: CoCrFeNi + Al → FCC + BCC Dual-Phase Structure

Introducing aluminium into the CoCrFeNi system leads to phase separation, with FCC and BCC domains coexisting in the microstructure. XRD analysis of this alloy reveals both FCC peaks at (111) and (200) and BCC peaks at (110) and (200), indicating the formation of a dual-phase structure [19]. The coexistence of these phases results in a balance between strength and ductility, where the FCC phase contributes to plasticity, while the BCC phase enhances hardness [20].

Case 2: CoCrFeMnNi + Mo → FCC + σ-Phase Formation

When molybdenum is added to CoCrFeMnNi, a brittle σ-phase emerges, disrupting the single-phase FCC structure. This is evident from XRD patterns, where additional peaks at (102), (110), and (201) confirm the presence of σ-phase intermetallics [21]. The formation of this phase significantly reduces ductility but enhances wear resistance, making it suitable for coatings but less ideal for structural applications [22].

Case 3: NbMoTaW + Ti → BCC + Laves Phase

Alloying NbMoTaW with titanium promotes the formation of Laves phases, which appear in XRD patterns as additional peaks alongside the primary BCC reflections. These intermetallic compounds further increase hardness but reduce overall toughness, limiting their use in applications requiring impact resistance [23].

1. Single-Phase HEAs:
   - FCC HEAs (e.g., CoCrFeMnNi) → High ductility, excellent toughness.
   - BCC HEAs (e.g., NbMoTaW) → High strength, superior thermal stability.
2. Dual-Phase and Multi-Phase HEAs:
   - FCC + BCC (e.g., CoCrFeNi + Al) → Balanced strength and ductility.
   - FCC + σ (e.g., CoCrFeMnNi + Mo) → Enhanced wear resistance, lower plasticity.
   - BCC + Laves (e.g., NbMoTaW + Ti) → Extreme hardness, reduced impact resistance.

5.2.1 Reason for Advanced Techniques for Structural Insights

The unusual behaviour observed in phase detection of HEAs can be attributed to the limitations of conventional XRD techniques. if we consider the CuKα radiation is used while XRD has limited penetration, particularly in alloys containing high atomic number transition elements such as Ta or W. This can result in partial absorption of the primary X-ray beam by the majority phase before it reaches any minority phases. Consequently, secondary phases could go undetected, especially when the counting statistics are not ideal. In high-quality samples, detection limits for minor phases are approximately 1% in a lab setup, and as low as 0.0001% using synchrotron radiation. However, HEAs generally possess short correlation lengths for their periodic structures (typically 10–20 nm), which leads to broad Bragg peaks, increasing the detection limits. Moreover, if phases within the HEA share similar lattice parameters, XRD peaks might overlap, making it difficult to identify all phases. This could lead to mischaracterizing the material as single phase [24].

5.2.2 High-Resolution X-Ray Technique

To resolve the secondary peaks, scientists used a high-resolution X-ray technique to study the differences in the crystal structures (called lattice misfit) of 5Al-0.01B and 10Fe-0.01B alloys which tend to form two phases, γ and γ', in aged alloys at room

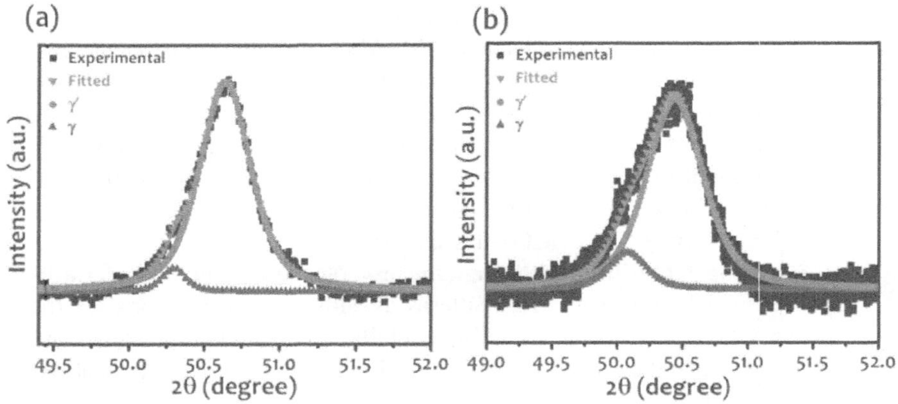

FIGURE 5.2 XRD peaks of (200) plane from the (a) 5Fe-0.01B and (b) 10Fe-0.01B alloys, subjected to aging at 900°C for 20 hours. The XRD peaks are asymmetric and consisting of two peaks. Peak fitting using the Pseudo-Voigt function gives two peaks pertaining to the γ and γ' phases [26].

temperature. They observed that the X-ray peaks for the (200) plane in the 5Al-0.01B and 10Fe-0.01B alloys were not symmetrical (see Figure 5.2 (a, b)). This asymmetry occurs because the γ and γ' phases have different crystal sizes, and the γ phase experiences slight distortion due to stress caused by these differences.

To analyze this, they used a mathematical method (Pseudo-Voigt function) to split the asymmetric (200) peak into two overlapping peaks, representing the γ and γ' phases. The intensity of the X-ray peaks depends on factors like the material's structure, the amount of each phase present, the texture from solidification, and changes in ordering. Because of this, the peak intensity does not directly indicate the exact amount of each phase. However, if we assume that the peak intensity relates to the phase volume and that the stronger peak comes from the phase with more volume (γ'), we can calculate the lattice misfit (δ) between the γ and γ' phases using a specific formula [25].

The crystal sizes (lattice parameters) of the γ and γ' phases, as well as the lattice misfit, were measured for three alloys: 5Fe-0.01B, 10Fe-0.01B, and Fe-free 0Fe-0.01B. The lattice misfit is negative because the γ phase has a smaller crystal size than the γ' phase. Adding iron (Fe) increases the crystal size of both phases, but it does not significantly change the lattice misfit. For example, adding 5% and 10% Fe changed the misfit to –0.63% and –0.67%, respectively, compared to –0.64% for the Fe-free alloy.

5.3 POWDER NEUTRON SCATTERING

Neutron scattering is another powerful technique for probing the atomic structure, phase composition, and microstructural features of materials, including HEAs. Unlike XRD, neutron scattering is highly sensitive to light elements, isotopic variations, and magnetic structures, making it particularly useful for studying complex alloys with multiple phases. This technique is crucial for analysing residual stress, dislocations, and lattice strain, providing insights into the mechanical behaviour and thermal stability of HEAs.

However, neutron scattering requires specialized expertise due to its complex data interpretation, reliance on nuclear scattering cross-sections, and the need for sophisticated computational modelling. Additionally, it demands access to large-scale facilities such as nuclear reactors or spallation neutron sources, making it less accessible compared to conventional XRD. Despite these challenges, neutron scattering remains an indispensable tool for characterizing HEAs at an atomic level, offering unique insights into their microstructural evolution [27, 28].

For an accurate characterization of polycrystalline refractory high entropy alloys (R-HEAs), high-resolution powder X-ray diffraction (PXRD) using synchrotron radiation or powder neutron scattering (PNS) is recommended. A quantitative Rietveld refinement is essential to validate the structural model. Additionally, analysing peak broadening, particularly the full-width at half-maximum (FWHM), provides valuable information on lattice strain and the correlation length of the material's structure. High-resolution PXRD or PNS also enables the determination of the atomic pair distribution function (PDF), which is useful for studying lattice distortions and local atomic ordering. Temperature-dependent PDF analysis is particularly beneficial for tracking variations in lattice distortions and local ordering at different temperatures. Moreover, 3D-PDF calculations utilizing single-crystal data offer a more comprehensive understanding of local ordering phenomena [29].

Tong et al. [30] investigated how the number of elements in low-entropy alloys (LEAs), medium-entropy alloys (MEAs), and HEAs influences the lattice distortions, peak shifts, and peak intensities in PXRD. Their findings indicated that increasing the number of components leads to noticeable distortions, affecting peak behaviour. Owen et al. [31] further observed that while the local lattice strain in HEAs, such as the Cantor alloy (FCC Co-Cr-Fe-Mn-Ni), was comparable to that in simpler alloys, the extent of local distortion fluctuations was significantly higher, especially for Cr and Mn.

Differentiating HEAs with partially disordered B2 (CsCl-type) structures from those with fully disordered body-centered cubic (BCC) structures presents a challenge. In BCC structures, specific reflections are absent, making the detection of superstructure peaks difficult when there is minimal scattering contrast between atoms in different Wyckoff positions. A prime example of this is tungsten (W), where the low contrast in BCC structures complicates superstructure peak identification in XRD.

In alloys such as (Hf, Zr) and (Nb, Ta), the most prominent superstructure peak observed is the (100) reflection, with an intensity ratio of about 1.3×10^{-4} relative to the strongest (011) peak when using CuKα radiation. Due to its low intensity, this peak is only detectable in high-quality HEAs with narrow diffraction peaks through synchrotron PXRD. When employed neutron scattering with the same wavelength, the intensity ratio of the (100) peak to the strongest (123) reflection increases slightly to around 4.4×10^{-4}. Neutron diffraction also provides better scattering contrast for elements with similar X-ray scattering factors, such as vanadium and titanium, chromium, manganese and iron, or copper and nickel [1, 4].

For in-depth studies of ordering phenomena, single-crystal XRD remains the preferred technique. When used with a pixel detector, this method can capture intensity ratios spanning up to nine orders of magnitude, making it highly effective for analysing disorder and diffuse scattering, thereby identifying local ordering features. If HEA single crystals can be successfully grown, this approach provides the most precise structural and property characterization, as demonstrated by studies from researchers such as Mishra, Ma, and Yasuda. Additionally, small-angle scattering techniques, including small-angle X-ray scattering (SAXS) and small-angle neutron scattering (SANS), can be particularly useful in investigating local clustering effects or nanoparticle precipitation in HEAs.

5.3.1 Scanning Electron Microscopy (SEM)

SEM is a crucial technique for the microstructural characterization of HEAs. It provides high-resolution imaging of surface morphology, grain structures, and phase distributions, making it indispensable for analysing microstructural features and fracture surfaces. As a fundamental tool in research laboratories, SEM is particularly valuable for studying HEAs with complex phase compositions, offering critical insights into their structural characteristics and phase formed. It enables the visualization of key microstructural features such as phase segregation, particle size, and the distribution of alloying elements. Additionally, when combined with energy-dispersive X-ray spectroscopy (EDX), SEM facilitates elemental mapping and semi-quantitative chemical analysis, providing deeper insights into compositional heterogeneity within HEAs. In this section, authors have focused on only two HEA systems for discussion: Co-Cr-Fe-Ni-Nbx (FCC) and Al-Co-Cr-Fe-Ni_2 (BCC).

Co-Cr-Fe-Ni-Nbx alloy was scanned under an electron microscope which had a varying Nb concentrations (x = 0.1, 0.25, 0.5, and 0.8), leading to a change from hypo-eutectic to hyper-eutectic structure. Compositions with x values of 0.1, 0.25, and 0.5 exhibited hypo-eutectic characteristics, while x = 0.8 represented a hyper-eutectic structure. The eutectic structure was achieved at x = 0.63, corresponding to a Nb mole fraction of 14%. The incorporation of Nb, which has a larger atomic radius than the other constituent elements, promotes eutectic structure formation and introduces lattice distortion.

In hypo-eutectic compositions, the proeutectic FCC phase predominantly forms, whereas in hyper-eutectic compositions, the proeutectic Laves phase becomes dominant. Figure 5.3 depicts the eutectic lamellar structure, where an increase in Nb content leads to a higher eutectic volume fraction and a reduction in lamellar spacing. As a result, hyper-eutectic compositions are expected to exhibit a denser microstructure with a greater eutectic volume fraction. In the Co-Cr-Fe-Ni-$Nb_{0.1}$, Co-Cr-Fe-Ni-$Nb_{0.25}$, and Co-Cr-Fe-Ni-$Nb_{0.5}$ alloys, the proeutectic region is relatively depleted in Nb, while the eutectic structure is Nb-enriched. Conversely, the $CoCrFeNiNb_{0.8}$ alloy is characterized by a Nb-rich proeutectic Laves phase.

During the solidification process, hypo-eutectic compositions first undergo the nucleation and growth of the proeutectic FCC phase, followed by the simultaneous development of both phases during the eutectic reaction, resulting in a lamellar

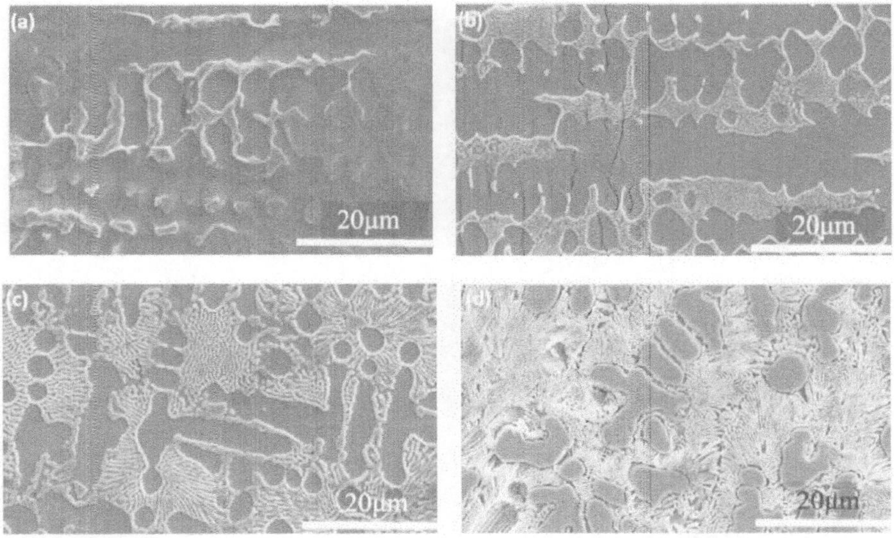

FIGURE 5.3 SEM images of Co-Cr-Fe-Ni-Nbx alloys for hypo-eutectic (a) x = 0.1, (b) x = 0.25, (c) x = 0.5 and hyper-eutectic, and (d) x = 0.8 structures [32].

microstructure. A similar mechanism occurs in hyper-eutectic compositions; however, the growth process is delayed due to interactions between eutectic FCC phase particles. Among these alloys, Co-Cr-Fe-Ni-$Nb_{0\cdot 8}$ exhibits the highest hardness, primarily due to the increased volume fraction of the Laves phase at higher Nb concentrations. Since the intermetallic Laves phase is inherently harder than the FCC phase, the reduced lamellar spacing further enhances the alloy's strength. On the other hand, as Nb content decreases, the alloy's ductility tends to improve. A comprehensive understanding of the eutectic transformation allows for precise control over Nb content to tailor the alloy's mechanical properties.

Shi et al. [33] has used eutectic high entropy alloys (EHEAs), such as Al-Co-Cr-Fe-Ni_2, that have emerged as a distinct category of multi-principal-element eutectic materials, offering extensive in situ composite possibilities. By employing tailored thermo-mechanical processing, an ultrafine-grained duplex microstructure can be engineered to retain its intrinsic lamellar nature, enabling unique property combinations that were previously unattainable through conventional reinforcement techniques. Similar to the as-cast EHEA depicted in Figure 5.4a, the developed dual-phase hierarchical lamellar (DPHL) HEA exhibits a characteristic lamellar morphology. The inter-lamellar spacing within different lamellar domains varies between 1.5 and 5 μm (Figure 5.4b), demonstrating that the structure is inherited from the as-cast EHEA. Furthermore, precipitated BCC phases of varying sizes are dispersed within the FCC lamellae (Figure 5.4c).

To achieve a comprehensive understanding of the microstructure, a detailed analysis using TEM combined with energy-dispersive spectroscopy (EDS) was performed. The investigation revealed that the lamellae were composed of recrystallized grains rather than simple phase bands, with occasional annealing twins present

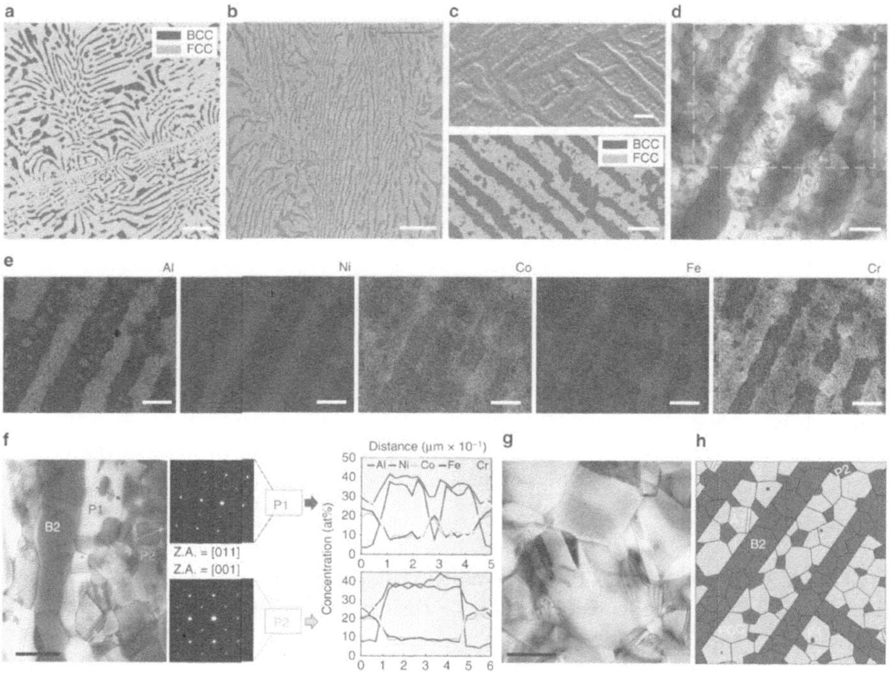

FIGURE 5.4 Microstructures of the as-cast EHEA and the hierarchical DPHL700. (a) Electron-backscatter-diffraction (EBSD) phase image of the as-cast EHEA. (b, c) SEM image, high-magnification SEM image, and EBSD-phase image of the DPHL HEA. RD, rolling direction; TD, transverse direction. (d) Scanning TEM (STEM) image exhibiting a more detailed DPHL structure. (e) EDS maps of the identical region marked in d showing the distribution of Al, Ni, Co, Fe, and Cr. (f) Enlarged STEM image showing the distribution of P_2 (the intergranular B_2 Phase, marked by blue arrows) and P_1 (the intragranular B2 phase, marked by red arrow), and the corresponding SADPs and EDS composition profiles. (g) TEM image showing annealing twins. (h) Microstructural schematic sketch of the DPHL structure. AT, annealing twin. Scale bars, 20 μm in a, b, 2 μm in c, 1 μm in d–f and 500 nm in g [34].

within the FCC grains (Figure 5.4d, g). Additionally, EDS mapping and selected-area diffraction patterns (SADPs) indicated that the NiAl-rich lamellae, approximately 1 μm thick, consisted of B_2 grains, while the Fe- and Cr-enriched lamellae corresponded to FCC grains (Figure 5.4e). The average grain size of both the FCC and B_2 phases was approximately 0.71 μm. Interestingly, Cr-enriched BCC nano-precipitates, which are typically observed within the B_2 lamellae of as-cast EHEAs, were absent in the dual-phase hierarchical lamellar (DPHL) HEA [34–36].

Furthermore, as depicted in Figure 5.4e and f, a significant presence of the BCC phase precipitates were identified within the FCC lamellae. These precipitates were classified into two types of NiAl-rich phases: small, sparsely distributed intragranular B_2 grains (P_1), ranging from 50–180 nm in size, and larger intergranular B_2 grains (P_2) with an average size of around 350 nm (Figure 5.4f). These observations suggest a complex phase decomposition process originating from the initial FCC lamellae, a phenomenon now recognized in the EHEA category. Similar decomposition behaviour has been previously reported in certain single-phase HEAs, such as the equiatomic Cr-Mn-Fe-Co-Ni HEA, where it is attributed to limited entropic stabilization. While these alloys were traditionally considered thermally stable single-phase solid solutions due to their high mixing entropy, recent findings challenge this assumption [37, 38].

5.3.2 Electron Backscatter Diffraction (EBSD)

Electron backscatter diffraction (EBSD) is a powerful technique for mapping crystallographic orientations, grain boundary characteristics, and phase distributions in HEAs. It is particularly useful for understanding deformation mechanisms and microstructural evolution in these alloys. Unlike TEM, which transmits electrons through a thin sample to produce high-resolution images, diffraction patterns, and spectroscopic data, EBSD operates within the SEM to analyze surface microstructures. This makes it an essential tool for studying both grain morphology and phase transformations in HEAs. For further insights, readers can refer to the sources provided, as this section focuses solely on explaining the role of EBSD in HEA characterization.

The single-phase, Co-Cr-Fe-Mn-Ni HEA which exhibited FCC structure, EBSD maps show a random grain orientation, consistent with isotropic mechanical properties. In contrast, the Al-Co-Cr-Fe-Ni HEA, which exhibits a dual-phase FCC + BCC microstructure, demonstrates texture variations, where FCC grains align preferentially along specific crystallographic directions. For refractory HEAs like Nb-Mo-Ta-W, EBSD has identified the presence of high-angle grain boundaries (HAGBs), which are reported to improve creep resistance at high temperatures [5, 39–41].

However, EBSD struggles with poor diffraction pattern quality in some nanostructured or severely deformed HEAs, such as those processed via high-pressure torsion (HPT). In such cases, advanced EBSD techniques like transmission Kikuchi diffraction (TKD) help in resolving the sub-micron grain orientations. Additionally, EBSD was performed on metastable HEAs like Co-Cr-Fe-Ni-Ti, based on the crystallographic information obtained from EBSD, including grain orientation, misorientation, and strain accumulation, which played a crucial role in validating the experimental results and formulating the strain-hardening mechanism study deformation-induced phase transformations in metastable HEAs like Co-Cr-Fe-Ni-Ti, providing insights into their strain-hardening mechanisms [39].

A critical observation in EBSD studies of cold-rolled Co-Cr-Fe-Mn-Ni is the occurrence of strain-induced phase transformation. At higher strains (above 50% cold reduction), deformation twins and hexagonal close-packed (HCP) phase fractions start appearing. This phase transformation occurs due to a reduction in stacking fault energy (SFE), which promotes the transition from the FCC to HCP phase. This phenomenon has been extensively reported in the literature, demonstrating that increased cold work drives a higher volume fraction of the HCP phase, influencing mechanical behaviour by enhancing work hardening and ductility simultaneously [39–41].

5.4 TRANSMISSION ELECTRON MICROSCOPY (TEM)

TEM is a critical tool for resolving nanoscale structures, dislocations, and phase transformations in materials. By transmitting electrons through a thin sample, TEM generates high-resolution images, diffraction patterns, and spectroscopic data, enabling atomic-scale imaging and diffraction analysis. This capability makes TEM indispensable for investigating the fundamental properties of HEAs, which often exhibit complex microstructures with multiple phases and fine-scale features. Additionally, TEM facilitates the study of both surface morphology and internal microstructures, providing comprehensive insights into the material's characteristics."

Single-phase FCC HEAs/MEAs, such as Cr-Mn-Fe-Co-Ni alloys, experience significant ductility loss at intermediate temperatures (400°–800°C). This phenomenon is linked to the rapid segregation of principal elements (Cr, Mn, Ni) at grain boundaries (GBs), leading to decohesion and premature failure. Unlike intermetallic compounds, which cause embrittlement at room temperature but mitigate stress concentration at high temperatures, GB segregation remains a dominant factor in ductility loss. Studies on Cr20-Mn20-Fe20-Co20-Ni20 HEA confirm that preventing GB segregation through preexisting precipitates can mitigate this issue, providing a strategy to preserve ductility in HEAs/MEAs under intermediate-temperature deformation [5, 35, 42, 43].

A TEM image of a specimen was tested at 700°C, depicted in Figure 5.5a, illustrates a high-angle GB, indicated by an arrow. Comprehensive structural analyses, including XRD, high-resolution TEM, electron diffraction, and scanning electron microscopy with energy-dispersive spectroscopy (SEM-EDS), confirmed the absence of secondary phases along the GB. The atom probe tomography (APT) atom map in Figure 5.5b highlights the segregation of trace elements carbon (C) and boron (B) at the GB, present in impurity concentrations of 0.2 atomic % (at%) and 0.008 at%, respectively, as detailed in Figure 5.5. Notably, no segregation of elements known to contribute to embrittlement, such as phosphorus (P), sulphur (S), or oxygen (O), was observed at the GBs.

Figures 5.5c and 5.5d display is composition surfaces, while Figures 5.5e and 5.5f provide one-dimensional (1D) compositional profiles, both indicating localized enrichment of chromium (Cr), nickel (Ni), and manganese (Mn) in nanometre-scale regions along the GB, extending approximately 5 nm in thickness. The observed segregation patterns suggest that Ni and Mn tend to cluster together at GBs, whereas Cr and Ni exhibit independent segregation behaviour, implying a repulsive interaction between them during GB segregation, as illustrated in Figure 5.5d [35, 43, 44].

APT analysis also confirms that the embrittled GBs exhibit enrichment of Cr, Ni, and Mn, in addition to minimal traces of C and B (Figures 5.5b–f). Prior studies on coarse-grained Cr20-Mn20-Fe20-Co20-Ni20 HEAs annealed at temperatures exceeding 1000°C have reported uniform distribution of principal elements at GBs, with the only detectable segregation involving trace elements C and B. This indicates that, there is a segregation of Cr, Ni, and Mn at GBs occurred during tensile testing at 700°C. It has been established that residual strain from cold rolling can facilitate Cr diffusion into GBs during annealing, leading to the formation of

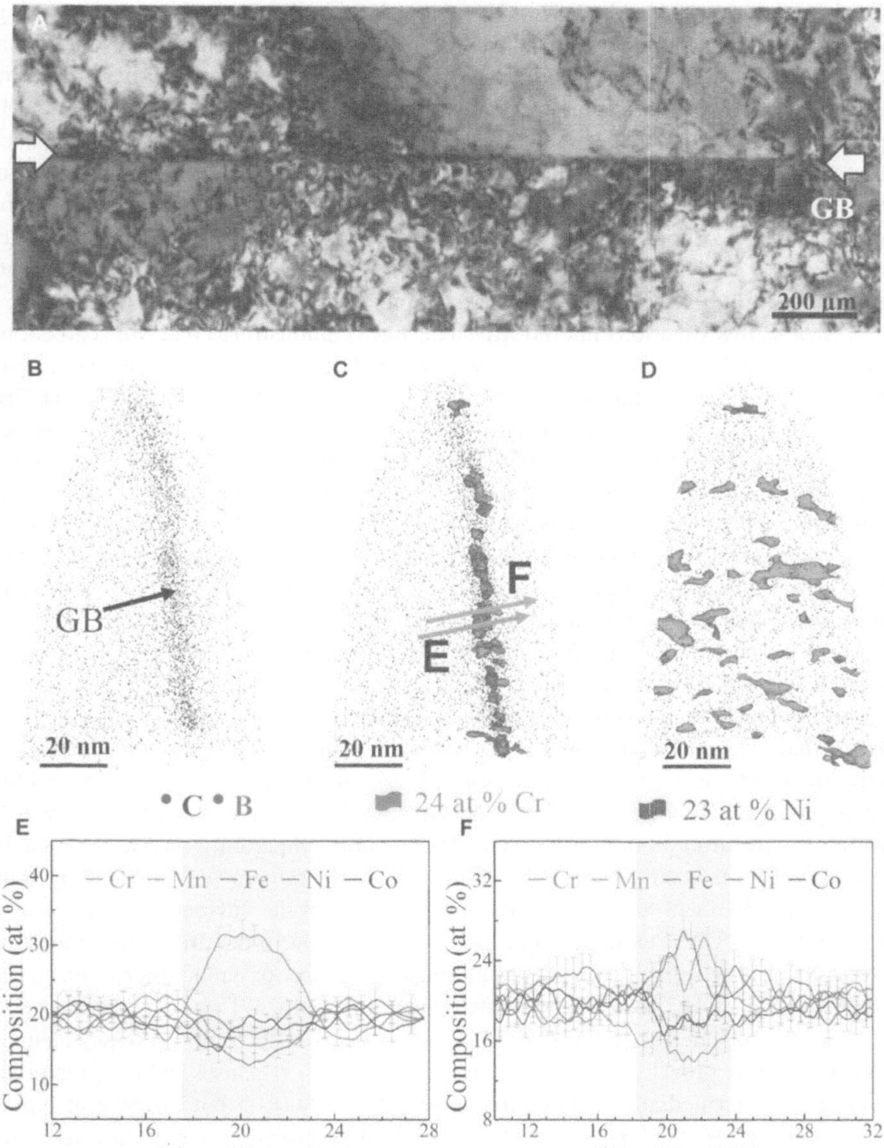

FIGURE 5.5 (a) TEM image of the single-phase coarse-grained $Cr_{20}Mn_{20}Fe_{20}Co_{20}Ni_{20}$ sample after a tensile test at 700°C. (b) Atom map of the tip showing GB segregation of impurities C and B. Cr (24 at %) and Ni (23 at %) is composition surfaces (c) viewed with the GB edge-on and (d) viewed from the GB normal. 1D compositional profiles (e) along the cyan arrow and (f) along the orange arrow indicated in (c); the shaded areas represent the approximate GB regions in thickness direction [35].

Cr-rich σ phases in Cr-Mn-Fe-Co-Ni system HEAs. Based on this understanding, it is reasonable to conclude that plastic deformation HEA promotes the accumulation of Cr, Ni, and Mn at GBs. However, the low concentrations of these elements are insufficient to trigger precipitation of Cr-rich σ phases, BCC Cr, or L10-NiMn phases. Additionally, experiments were conducted by the researcher and noticed that short exposure time of 10 minutes at 700°C is insufficient for the formation of σ phase precipitates, as verified by TEM analysis in Figure 5.5a. The composition of potential GB precipitates in $Cr_{20}Mn_{20}Fe_{20}Co_{20}Ni_{20}$ HEAs further corroborates the tendency of Ni and Mn to co-segregate at GBs, whereas Cr and Ni segregate separately, aligning with the APT observations in Figure 5.5 [34, 43, 44].

Previous investigations into Fe-Mn steels have demonstrated that Mn segregation at GBs can induce embrittlement by weakening GB cohesion due to its local magnetic contributions. In the presented study, the co-segregation of Ni and Mn at GBs (Figure 5.5f) raises the possibility of forming an ordered L10-NiMn phase, which could contribute to reduced GB cohesion. Additionally, Cr competes for GB sites and repels Ni (Figures 5.5c–e), potentially facilitating the development of brittle Cr-rich, Ni-depleted σ phases along GBs. This formation could significantly impair ductility due to weakened interfacial bonding. However, the presence of trace elements C and B at GBs may enhance GB cohesion, provided that M6C carbide precipitation does not occur. The 1D compositional profiles in Figure 5.5 indicate that the concentrations of C and B at Cr-enriched and Ni (Mn)-enriched GBs are too low to support carbide or boride formation.

Some other studies has also discussed the GB segregation such as $CoFeNi_2V_{0.5}Mo_{0.2}$ HEAs have revealed that the absence of Cr and Mn can help maintain ductility at intermediate temperatures. Their finding supports the hypothesis that Cr, Ni, and Mn segregation at GBs in Cr-Mn-Fe-Co-Ni system HEAs may contribute to embrittlement [45], resulting in ductility loss as deformation temperature increases. Notably, this segregation occurs even at relatively low temperatures (< 700°C), contributing to embrittlement at intermediate temperatures. In the current study, the specimen tested at 700°C exhibited a pronounced reduction in ductility while retaining strain-hardening capacity, serving as an illustrative case for nanocluster-induced GB embrittlement in HEAs at intermediate temperatures [46, 47].

It was concluded that the GB decohesion was due to multi-principal element nanoclusters is prevalent in HEAs. Similar to Cr-rich σ phases, which precipitate within the intermediate temperature range (400°C–800°C) but dissolve at higher temperatures (> 800°C), nanoclusters at GBs may also dissolve at elevated temperatures, thereby preventing ductility loss. This aligns with reports indicating that Cr-Mn-Fe-Co-Ni HEAs display improved ductility and even super plasticity when tested at temperatures above 900°C.

Wu et al. has used the TEM to confirm this perspective. They performed detailed TEM analyses. As illustrated in Figure 5.6a, dense planar slip bands are observed in the FCC phase of the fractured as-cast (AC) alloy, while significantly few dislocations are detected in the B2 phase under a nominal strain of 15%. Additionally, dislocations in the B2 phase tend to concentrate near the phase boundaries (PBs). In contrast, the phase-separated and refined (PSR) alloy, subjected to a similar nominal strain of 18% (Figure 5.6b), exhibits a high density of dislocations in both the FCC

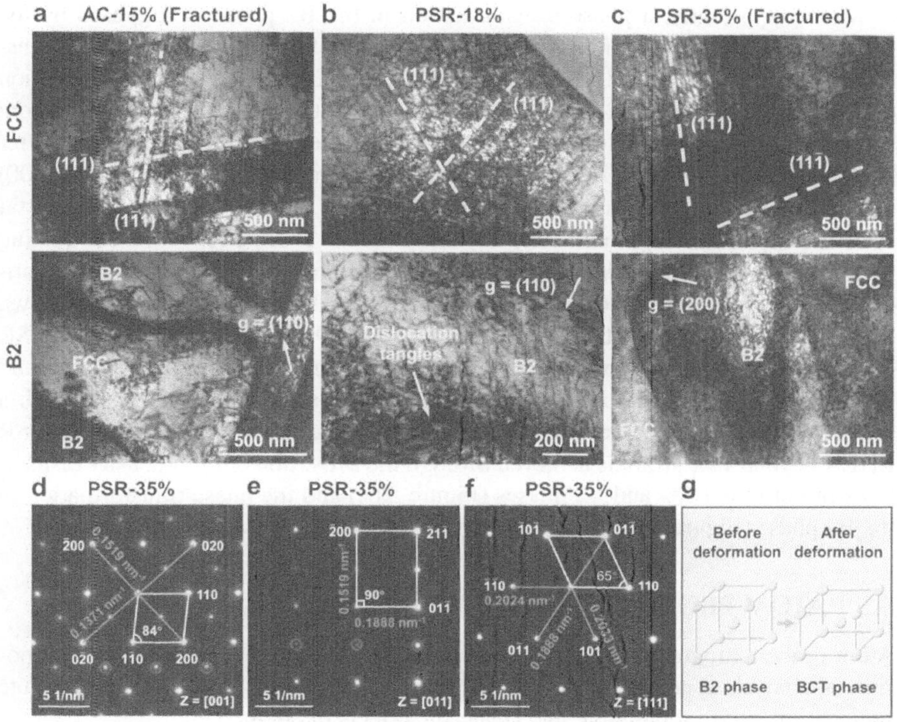

FIGURE 5.6 (a) Dislocation substructure in the fractured AC EHEA. (b, c) Dislocation substructures in the 18% and 35% stretched PSR EHEAs, respectively. (d–f) SAED patterns of the original B2 phase in the fractured PSR specimen taken from the zone axis of [001], [011], [$\bar{1}$11], respectively, revealing the BCT crystal structure. The superlattices are marked by the red circles. (g) Schematic diagram illustrating the crystal structure transformation before and after deformation [48].

and B2 phases. We also investigated the dislocation structures of the PSR eutectic high entropy alloy (EHEA) at nominal strains of 0.5% and 8%. Well-developed planar slip bands in both phases indicate that the FCC and B2 phases in the PSR EHEA contribute to strain hardening simultaneously from the onset of plastic deformation. As the strain increases further, the dislocation density rises significantly in both phases (Figure 5.6c), demonstrating their exceptional capacity for dislocation storage.

Particularly in the B2 phase, although no evidence of slip transfer across the PBs is observed, its inherent superior deformability enables the activation of a high density of dislocations. Consequently, geometrically necessary dislocations multiply significantly due to the heterogeneous microstructure, enhancing the sustainable increase in back stress strengthening. Through loading-unloading-reloading experiments, the back stress near failure is quantitatively measured to be approximately 900 MPa, explaining the high strain hardening capacity of the PSR EHEA [48–50].

More importantly, a phase transformation in the B2 phase is triggered by the increased strain. We identified a B2-to-body-centered tetragonal (BCT) phase transformation in the PSR EHEA during deformation. Selected area electron diffraction (SAED) patterns of the B2 phase, taken along the [001] zone axis, transition from square-arranged spots before deformation to parallelogram-arranged spots after deformation (Figure 5.6d), indicating distinct interplanar spacings between the (200) and (020) planes. Quantitative calculations (Figure 5.6 d–f) reveal that the crystal structure expands by approximately 6% along the (200) direction while contracting by about 5% along the (002) and (020) directions after deformation, as schematically depicted in Figure 5.6g. The ductile-transformable B2 phase has been shown to significantly enhance fatigue life in duplex HEAs. In the presented study on PSR EHEA, the phase transformation improves the deformability of the B2 phase and promotes coordinated deformation between the soft and hard phases, even with a low transformation content. Therefore, the sustainable strain hardening of the PSR EHEA beyond 18% strain is achieved through the activation of high-density dislocations in both the FCC and B2 phases (Figure 5.6c) and the phase transformation in the B2 phase (Figure 5.6g).

5.5 ATOM PROBE TOMOGRAPHY (APT)

Atom probe tomography (APT) is an advanced technique that provides 3D compositional analysis at atomic resolution. It is particularly useful for investigating solute segregation, phase separation, and clustering effects in HEAs.

The periodic variation in atomic intensity was visualized through high-resolution (HR) HAADF-STEM images, which leverage the contrast differences between neighbouring atomic columns based on their atomic number (Z-contrast). These findings reveal the orientation relationships between precipitates and the surrounding matrix, illustrating the regular arrangement of atoms in the superlattice structure. Specifically, the images in Figure 5.7a and b display various chemically ordered precipitates with a (Ni, Fe)3(Ta, W)-type D0a and D022 structure.

Fast Fourier transform (FFT) patterns obtained from selected areas of the δ phase (inset I in Figure 5.7a) and the FCC phase (inset II in Figure 5.7a) indicate a specific crystallographic orientation between the δ nano-lamellae and the FCC matrix, with the relationship given as [100] δ// [112] γ. This orientation aligns the atomic arrangement of the δ phase closely with that of the FCC matrix, which results in the formation of a coherent interface. The growth of the δ phase along this direction is constrained by minimal elastic distortion energy at the coherent interface, promoting the lamellar structure. The atom arrangement, with light atoms (Ni/Fe) shown as green spheres and heavy atoms (Ta/W) as red spheres, is depicted in the inset of Figure 5.7a confirm that Ni/Fe atoms predominantly occupy the central positions between the two rows of Ta/W atoms [51, 52].

Using the FFT images, we also characterized the crystal orientation relationship between the γ″ phase and the matrix as [010] γ″// [001] γ, shown in Figure 5.7b. The γ″ precipitates are evenly dispersed within the FCC matrix, exhibiting nearly completely coherent interfaces (Figure 5.7b). The lattice parameters for both the δ and γ″

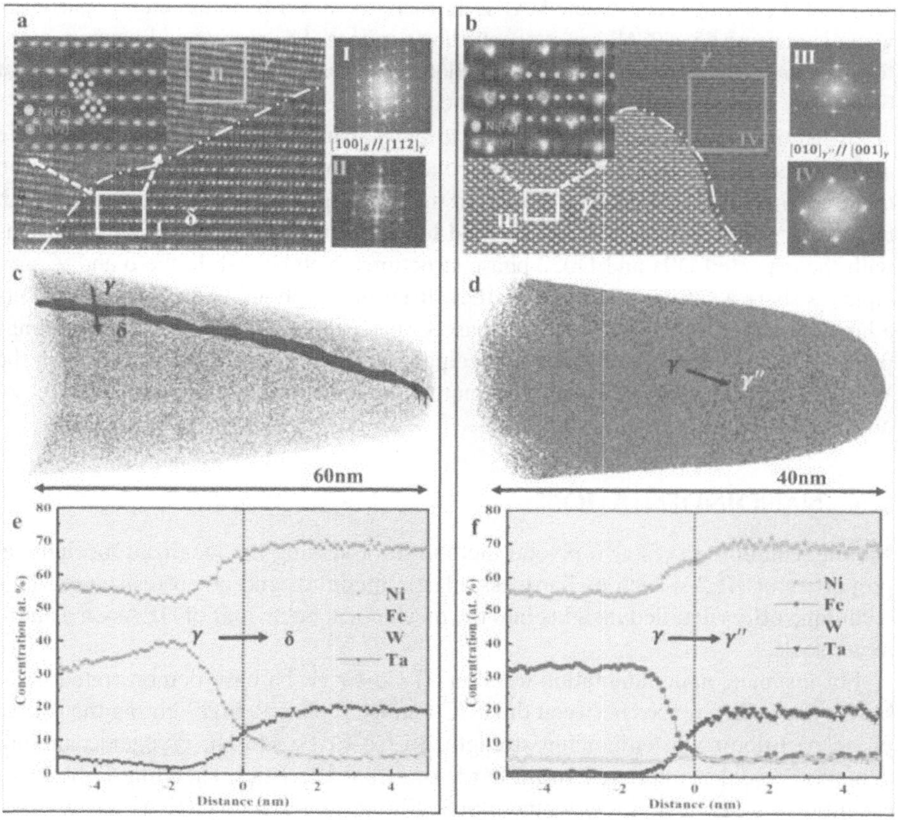

FIGURE 5.7 (a, b) Atomic-resolution HAADF-STEM images and corresponding FFT patterns of δ, γ″ and the FCC phase along different zone axes show the coherent relationship and specific coherent interfaces of the two different precipitations. I, II, III, and IV are the FFTs of the corresponding regions in (a) and (b), respectively. An accurate atomic placeholder map is also provided in the correlative illustrations (red balls represent heavy elements and green ones represent light elements). (c, d) 3D-APT atom maps show the elemental distribution of δ-lamellae and γ″-particles together with the 65 at.% Ni iso-concentration surfaces. (e, f) Proximity histograms show the elemental partitioning between the matrix and δ-lamellae and between the matrix and γ″-particles, respectively. The scale bars in (a) and (b) are 1 and 2 nm, respectively [51].

phases were measured directly from the HAADF-STEM images (Figures 5.7a and b), yielding values for δ as aδ=5.10 Å, bδ=4.24 Å, cδ=4.52 Å, and for γ″ as aγ″=3.62 Å, cγ″=7.31 Å. The axial ratio of the γ″ unit cell is very close to 2, suggesting that the ellipsoid shape of the γ″ phase can be attributed to high elastic distortion energy along the [001] direction, with additional details of atomic site occupations along the [001] γ″ // [001] FCC direction shown in Figure 5.7b [51–58].

To further investigate the chemical composition and elemental distribution of the two types of nano-precipitates, we utilized atom probe tomography (APT). Figures

5.7c and d present reconstructed APT datasets for the δ-lamellae and γ″-particles, visualized using 65 at.% Ni iso-concentration surfaces. The corresponding atom maps for the solute elements within the δ-lamellae and γ″-particles. These maps reveal that both types of nano-precipitates are enriched in Ni and Ta but depleted in Fe.

Proximity histograms for the δ-lamellae and γ″-particles, presented in Figures 5.7e and f, illustrate the variation in solute concentrations across the interface between the precipitates and the matrix. Based on these histograms, the composition of both the δ and γ″ precipitates was determined to be (Ni70Fe5)3(Ta20W5), which aligns with the expected D0a and D022 phase structures. Notably, while the δ and γ″ precipitates share a similar composition, they differ in morphology: the δ phase exhibits a lamellar structure, whereas the γ″ phase forms a spheroidal shape. These findings provide strong evidence for the nanoprecipitation of coherent δ and γ″ phases in the W-based refractory HEA, a phenomenon that, to our knowledge, has not been previously observed [51].

5.6 NANOINDENTATION

Nanoindentation serves as a pivotal method for assessing the localized mechanical properties of HEAs, such as hardness, elastic modulus, and creep resistance. This technique offers detailed insights into the mechanical behaviour of HEAs on a nanometer scale.

For instance, nanoindentation tests on Al-Co-Cr-Fe-Ni have demonstrated notable hardness differences between the FCC and BCC phases, highlighting the role of phase partitioning in enhancing strength. In Co-Cr-Fe-Mn-Ni, cryogenic nanoindentation revealed improved strain-hardening characteristics, shedding light on its exceptional toughness at low temperatures. Similarly, refractory HEAs like Hf-Nb-Ta-Ti-Zr display impressive indentation creep resistance, attributed to their slow diffusion properties.

However, nanoindentation results can be influenced by surface roughness and sample preparation methods. To address these challenges, refined polishing techniques and advanced modelling approaches, such as finite element analysis, are utilized. High-temperature nanoindentation has also been applied to study creep mechanisms in Nb-Mo-Ta-W, offering valuable data for applications at elevated temperatures.

5.7 CONCLUSION

Advanced characterization techniques—including SEM, EBSD, TEM, APT, and nanoindentation—are essential for unravelling the complex microstructures and mechanical behaviour of HEAs.

Each technique offers distinct insights, from surface morphology (SEM) to crystallographic orientation (EBSD) and atomic-scale composition (APT). However, to conclude whether certain phases exist or not, one must carefully select the appropriate technique based on specific requirements and applications. This selection aids in data interpretation, necessitating a multi-technique approach for a comprehensive understanding.

REFERENCES

1. Cantor, B., Chang, I. T. H., Knight, P., & Vincent, A. J. B. (2004). Microstructural development in equiatomic multicomponent alloys. *Materials Science and Engineering: A, 375–377*, 213–218.
2. Yeh, J. W. et al. (2004). Nanostructured high-entropy alloys with multiple principal elements: Novel alloy design concepts and outcomes. *Advanced Engineering Materials, 6(5)*, 299–303.
3. Otto, F. et al. (2013). The influences of temperature and microstructure on the tensile properties of a CoCrFeMnNi high-entropy alloy. *Acta Materialia, 61(15)*, 5743–5755.
4. Miracle, D. B., & Senkov, O. N. (2017). A critical review of high-entropy alloys and associated research. *Acta Materialia, 122*, 448–511.
5. Senkov, O. N. et al. (2010). Refractory high-entropy alloys. *Intermetallics, 18(9)*, 1758–1765.
6. Gludovatz, B. et al. (2014). Exceptional damage-tolerance of a medium-entropy alloy CrCoNi at cryogenic temperatures. *Science, 345(6201)*, 1153–1158.
7. Senkov, O. N. et al. (2011). Microstructure and properties of a refractory high-entropy alloy after thermomechanical processing. *Journal of Alloys and Compounds, 509(24)*, 6043–6048.
8. Zhang, Y. et al. (2012). Influence of chemical disorder on energy dissipation and defect evolution in high-entropy alloys. *Acta Materialia, 60(4)*, 1497–1507.
9. Tsai, M. H., & Yeh, J. W. (2014). High-entropy alloys: A critical review. *Materials Research Letters, 2(3)*, 107–123.
10. Hemphill, M. A. et al. (2012). Fatigue behavior of high-entropy alloys. *Acta Materialia, 60(16)*, 5723–5734.
11. Yeh, J. W., Lin, S. J., Chin, T. S. et al. (2004). Formation of simple crystal structures in Cu-Co-Ni-Cr-Al-Fe-Ti-V alloys with multiprincipal metallic elements. *Metallurgical and Materials Transactions A, 35*, 2533–2536. https://doi.org/10.1007/s11661-006-0234-4
12. Li, Z. et al. (2016). Metastability and grain boundary strengthening in high-entropy alloys. *Acta Materialia, 124*, 532–540.
13. George, E. P., Raabe, D., & Ritchie, R. O. (2019). High-entropy alloys: A new paradigm for metallic materials. *Nature Reviews Materials, 4(8)*, 515–534.
14. Senkov, O. N. et al. (2013). Mechanical properties of NbMoTaW and VNbMoTaW refractory high-entropy alloys. *Acta Materialia, 61(5)*, 1851–1864.
15. Feuerbacher, M., Heidelmann, M., & Thomas, C. (2015). Single-phase high-entropy alloys: An overview. *Journal of Alloys and Compounds, 646*, 60–64.
16. Tasan, C. C. et al. (2015). Composition-dependent deformation response in high-entropy alloys. *Acta Materialia, 86*, 1–12.
17. Ma, S. G., & Zhang, Y. (2012). Effect of valence electron concentration on stability of FCC or BCC phase in high-entropy alloys. *Journal of Alloys and Compounds, 536*, S270–S273.
18. Miracle, D. B. (2017). High-entropy alloys and their application. *Nature Materials, 16(1)*, 15–23.
19. Yeh, J. W. (2013). Alloy design strategies and future trends in high-entropy alloys. *Journal of Materials Research, 28(17)*, 2337–2344.
20. Guo, S., & Liu, C. T. (2011). Phase stability in high-entropy alloys: Formation of solid-solution phase or amorphous phase. *Progress in Natural Science: Materials International, 21(6)*, 433–446.
21. Wu, Z., Bei, H., Pharr, G. M., & George, E. P. (2014). Temperature dependence of the mechanical properties of equiatomic solid solution alloys with face-centered cubic crystal structures. *Acta Materialia, 81*, 428–441.

22. Laplanche, G. et al. (2015). Reasons for the superior mechanical properties of medium-entropy CrCoNi compared to high-entropy CrMnFeCoNi. *Acta Materialia*, *128*, 292–303.
23. Maiti, S., Steuren, P., & Senkov, O. (2016). Designing high-entropy alloys via thermodynamic and kinetic considerations. *Acta Materialia*, *121*, 130–150.
24. Juan, C. C. et al. (2018). Phase evolution in high-entropy alloys during processing. *Materials Chemistry and Physics*, *210*, 252–263.
25. Pandey, P., Kashyap, S., Palanisamy, D., Sharma, A., & Chattopadhyay, K. (2019). On the high temperature coarsening kinetics of γ' precipitates in a high strength $Co_{37.6}Ni_{35.4}Al_{9.9}Mo_{4.9}Cr_{5.9}Ta_{2.8}Ti_{3.5}$ fcc-based high entropy alloy. *Acta Materialia*, *177*, 82–95. https://doi.org/10.1016/j.actamat.2019.07.011
26. Kumar. S. (2021). On the development of Fe containing high entropy alloys with improved strength at high temperatures. Phd thesis. Indian Institute of Science Bangalore. http://hdl.handle.net/10603/433715
27. Yang, T., Xia, S., Liu, S., Wang, C., Liu, S., Zhang, Y., Xue, J., Yan, S., & Wang, Y. (2015). Effects of AL addition on microstructure and mechanical properties of AlxCoCrFeNi high-entropy alloy. *Materials Science and Engineering: A*, *648*, 15–22. https://doi.org/10.1016/j.msea.2015.09.034.
28. Dağ, İ. E., Avar, B., Panigrahi, M., Gündeş, A., Şimşek, T., & Rajendrachari, S. (2025). Evaluation of the microstructural, mechanical, and corrosion behavior of BN-, B4C-, and FeB-added distaloy AE composites produced by high-energy ball milling and hot-pressing. *Metallography, Microstructure, and Analysis*, *14*, 210–228.
29. George, E. P., Raabe, D., & Ritchie, R. O. (2019). High-entropy alloys. *Nature Reviews Materials*, *4*, 515–534.
30. Tong, Y., Velisa, G., Yang, T., Jin, K., Lu, C., Bei, H., ... & Zhang, F. X. (2017). Probing local lattice distortion in medium-and high-entropy alloys. *arXiv preprint arXiv:1707.07745*. https://doi.org/10.48550/arXiv.1707.07745
31. Owen, L. R., Pickering, E. J., Playford, H. Y., Stone, H. J., Tucker, M. G., & Jones, N. G. (2017). An assessment of the lattice strain in the CrMnFeCoNi high-entropy alloy. *Acta Materialia*, *122*, 11–18.
32. Aristeidakis, I. S., & Tzini, Maria-Ioanna T. (2016). High Entropy Alloys. PhD thesis. Department of Mechanical Engineering Laboratory of Materials, University of Thessaly.
33. Shi, P., Ren, W., & Zheng, T. et al. (2019). Enhanced strength–ductility synergy in ultrafine-grained eutectic high-entropy alloys by inheriting microstructural lamellae. *Nature Communications*, *10*, 489. https://doi.org/10.1038/s41467-019-08460-2
34. Schuh, B. et al. (2015). Mechanical properties, microstructure and thermal stability of a nanocrystalline CoCrFeMnNi high-entropy alloy after severe plastic deformation. *Acta Materialia*, *96*, 258–268.
35. Otto, F. et al. (2016). Decomposition of the single-phase high-entropy alloy CrMnFeCoNi after prolonged anneals at intermediate temperatures. *Acta Materialia*, *112*, 40–52.
36. He, F. et al. (2017). Phase separation of metastable CoCrFeNi high entropy alloy at intermediate temperatures. *Scripta Materialia*, *126*, 15–19.
37. Guo, S., Ng, C., & Liu, C. T. (2013). Sunflower-like solidification microstructure in a near-eutectic high-entropy alloy. *Materials Research Letters*, *1*, 228–232.
38. Gwalani, B. et al. (2018). Modifying transformation pathways in high entropy alloys or complex concentrated alloys via thermo-mechanical processing. *Acta Materialia*, *153*, 169–185.
39. Gludovatz, B., Hohenwarter, A. et al. (2014). Exceptional damage tolerance of a CrMnFeCoNi high-entropy alloy. *Science*, *345*(6201), 1153–1158.
40. Senkov, O. N., Miracle, D. B. et al. (2011). Refractory high-entropy alloys. *JOM*, *63*(7), 15–19.
41. Zou, Y., Ma, H., & Spolenak, R. (2015). Size-dependent plasticity in an NbMoTaW high-entropy alloy. *Acta Materialia*, *65*, 85–97.

42. Ming, K. et al. (2019). Grain boundary decohesion by nanoclustering Ni and Cr separately in CrMnFeCoNi high-entropy alloys. *Science Advances*, *5*, eaay0639. https://doi.org/10.1126/sciadv.aay0639
43. Laplanche, G., Berglund, S., Reinhart, C., Kostka, A., Fox, F., & George, E. P. (2018). Phase stability and kinetics of σ-phase precipitation in CrMnFeCoNi high-entropy alloys. *Acta Materialia*, *161*, 338–351.
44. Pickering, E. J., Muñoz-Moreno, R., Stone, H. J., & Jones, N. G. (2016). Precipitation in the equiatomic high-entropy alloy CrMnFeCoNi. *Scripta Materialia*, *113*, 106–109.
45. Cho, K., Fujioka, Y., Nagase, T., & Yasuda, H. Y. (2018). Grain refinement of non-equiatomic Cr-rich CoCrFeMnNi high-entropy alloys through combination of cold rolling and precipitation of σ phase. *Materials Science and Engineering: A*, *735*, 191–200.
46. Zhong, L., Wu, R., Freeman, A. J., & Olson, G. B. (1997). Effects of Mn additions on the P embrittlement of the Fe grain boundary. *Physical Review B*, *55*, 11133.
47. Wu, Q., He, F., & Li, J. *et al.* (2022). Phase-selective recrystallization makes eutectic high-entropy alloys ultra-ductile. *Nature Communications*, *13*, 4697. https://doi.org/10.1038/s41467-022-32444-4
48. Shukla, S., Wang, T., Cotton, S., & Mishra, R. S. (2018). Hierarchical microstructure for improved fatigue properties in a eutectic high entropy alloy. *Scripta Materialia*, *156*, 105–109.
49. Xiong, T. et al. (2021). Faceted Kurdjumov-Sachs interface-induced slip continuity in the eutectic high-entropy alloy, AlCoCrFeNi2.1. *Journal of Materials Science & Technology*, *65*, 216–227.
50. Wu, X., Yang, M., Yuan, F., Wu, G., & Zhu, Y. (2015). Heterogeneous lamella structure unites ultrafine-grain strength with coarse-grain ductility. *Proceedings of the National Academy of Sciences of the United States of America*, *112*, 14501–14505.
51. Lin, K., Hsu, C., & Lin, S. (2003). Precipitation mechanism of an MoNi type intermetallic phase in W–27.0at%Mo–35.6at%Ni–17.6at%Fe. *International Journal of Refractory Metals and Hard Materials*, *21*, 125–133.
52. Rajendrachari, S., Mahale, R. S., Adimule, V., Gulen, M., Mahesh, V., & Basavegowda, N. (2024). Fabrication of mechanically alloyed high entropy alloys (20-Fe-20Cr-20Ni-20Mn-20Ti) modified carbon paste electrode sensor for the cyclic voltammetric determination of methyl Orange. *Inorganic Chemistry Communications*, *164*, 112428. https://doi.org/10.1016/j.inoche.2024.112428.
53. Shashanka, R., & Chaira, D. (2014). Phase transformation and microstructure study of nano-structured austenitic and ferritic stainless steel powders prepared by planetary milling. *Powder Technology*, *259*, 125–136.
54. Shashanka, R., & Chaira, D. (2015). Development of nano-structured duplex and ferritic stainless steel by pulverisette planetary milling followed by pressureless sintering. *Materials Characterization*, *99*, 220–229.
55. Shashanka, R., & Chaira, D. (2015). Optimization of milling parameters for the synthesis of nanostructured duplex and ferritic stainless steel powders by high energy planetary milling. *Powder Technology*, *278*, 35–45.
56. Shashanka, R., Uzun, O., & Chaira, D. (2020). Synthesis of nano-structured duplex and ferritic stainless steel powders by dry milling and its comparison with wet milling. *Archives of Metallurgy and Materials*, *65*(1), 5–14.
57. Shashanka, R., Adimule, V. M., Jayaprakash, G. K., & Pandith, A. (2023). Electrochemical oxidation of methylene blue dye in wastewater using mechanically alloyed high entropy alloy modified carbon paste electrode using cyclic voltammetry. *Materials Research Express*, *10*, 054003. https://doi.org/10.1088/2053-1591/acd3d7
58. Gülen, M., Taş, R., Dünya, H., & Rajendrachari, S. (2025). Enlargement of the surface area of high-entropy alloys with acid treatment as positive electrode for high specific capacitance supercapacitors. *ChemistryOpen*, *14*(5), e202400312.

6 Preparation of High Entropy Alloys by the Ball Milling Method

Joan-Josep Suñol, Asma Wederni and Jason Daza

6.1 BALL MILLING OF HIGH ENTROPY ALLOYS: TECHNIQUE DESCRIPTION AND HEA CRITERIA

This chapter describes and summarizes scientific literature on the production and characterization of high entropy alloys by the ball milling technique. The influence of the processing conditions (milling time, milling media, milling energy, ball-to-powder ratio, addition of process control agent) on the final material is analyzed. The mechanical and functional properties depend largely on the microstructure. The particles obtained by ball milling are usually micrometric in size and their applicability can be direct (as powders) or as a first step in the manufacture of bulk specimens, either by consolidating the powder or as reinforcement of composite materials.

6.1.1 Ball Milling Technique

Ball milling (BM) involves a process that allows the production of alloys and compounds in powder form. It basically consists of introducing precursors (elemental or compound) in the form of powder or flakes into highly hard jars/containers where a mechanical synthesis is carried out by grinding. During the process, continuous fracture and welding (called cold welding) of the powder particles occur, facilitating diffusion and obtaining alloys. This technique has been used to produce multiple materials, the first being oxide dispersion strengthened alloys (OHS). Terms associated with ball milling are mechanical alloying and magnetochemistry.

Since the introduction of high entropy alloys (HEAs) as a new family of materials, production by means of various techniques has been analyzed. Among those used for their production in powder form are gas atomization and mechanical alloying. The size of the particles obtained by mechanical alloying is usually micrometric. If macroscopic pieces are to be developed, it is necessary to proceed to the consolidation or sintering of the particles. The particles can also be used as reinforcement for composite materials.

One of the characteristics of alloys produced by mechanical alloying is the high density of micro-deformations and micro-mechanical stresses (such as vacancies,

 DOI: 10.1201/9781003664093-8

dislocations, etc.). Thus, mechanical synthesis involves a high level of microstrains. This fact influences the functional response. For instance, in magnetic materials, it causes an increase in coercivity. To reduce microstrain, alloys are usually subjected to controlled annealing heat treatments. These heat treatments provoke structural relaxation exothermic processes and also help to increase the homogeneity of the alloys by promoting diffusion.

In general, the control of processing conditions in any material production technique influences its microstructure and, therefore, its properties. In mechanical alloying, the main processing conditions are the milling/grinding time, the energy associated with milling, the milling media, the addition of process control agents or the percentage of filling of the jars. Figure 6.1 shows a diagram of the main parameters linked to mechanical alloying.

During mechanical alloying, especially in alloys with the presence of ductile elements, layers of material can form which are deposited on the walls of the jars (or even on the surfaces of the balls). To reduce this effect, a process control agent (PCA) is introduced during milling. One of the objectives of this PCA is to favor the fracture process over the welding. The improvement of the fracture will lead to the production of smaller particles and their refinement. Normally, the PCA is a liquid, although solid materials area also used.

One of the main problems associated with ball milling is dust oxidation. The atmosphere during processing is usually inert to prevent said oxidation. Argon (or in some cases nitrogen) is commonly used to create the inert atmosphere. Nevertheless, handling after the end of milling often involves contact with the external atmosphere and, as the produced dust particles are micrometric, surface oxidation could be

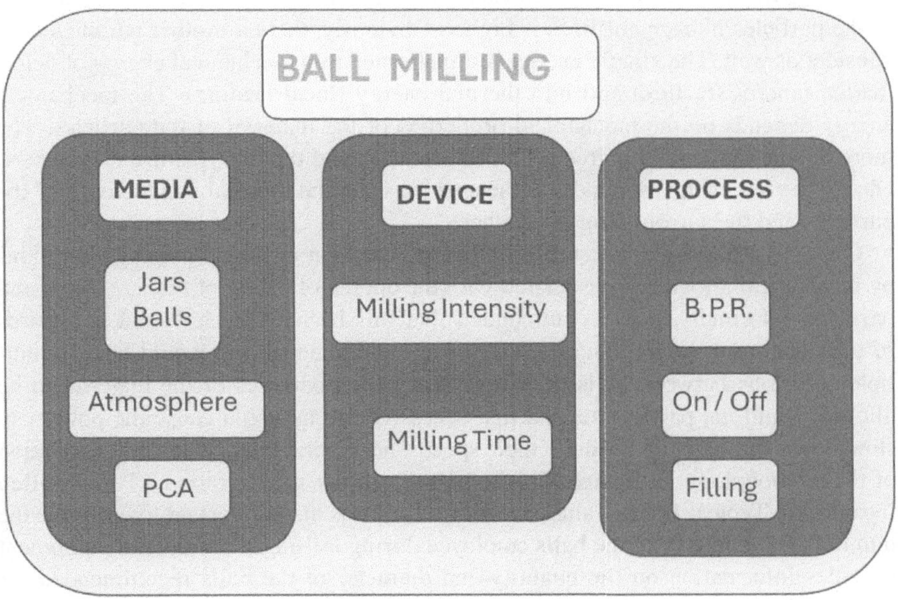

FIGURE 6.1 Schema of the main parameters linked to mechanical alloying technique.

detected, especially in metal particles. This oxide layer affects the properties of the final material. Another contamination comes from the milling media. The material from which the jars and balls are made wears out, adding new material and changing the composition. As the most commonly used materials for milling media are steels, the most common contamination is Fe. A third way of contamination is due to the addition of the process controlling agents. In many cases, they are organic compounds containing carbon, nitrogen and oxygen. An excessive percentage of PCA facilitates the formation of carbides or oxides if carbon or oxygen atoms are present in the PCA (hexane, cyclohexane, toluene).

One option to refine the particles size alternative to the addition of PCA is diminishing the milling temperature of the device. Thus, the welding and agglomeration of the powders is avoided. There are milling devices in which the milling is performed in liquid nitrogen (Mayer et al., 2022). In this work, cryomilling diminished the fraction of a minor intermetallic phase (Cr_5Si_3-type phase). As a consequence, an improvement of the soft magnetic behavior of the Al-Si-Cr-Mn-Fe-Ni-Cu HEA was found.

The interactions that occur during mechanical alloying are due to collisions between ball-particle(s)-ball or between ball-particle(s)-jar wall. The mechanisms associated with these interactions are percussion and abrasion. One of the main associated parameters is the energy of each interaction. This energy is of a kinetic nature and is due to the speed of the elements participating in the collision. The energy cannot be controlled directly, but there are parameters of the milling equipment/device that influence it. The main one is the frequency/intensity of rotation (or shaker motion) of the micro-mill. Increasing the frequency/intensity of milling increases the speed of jars, balls and particles. Increasing the device frequency/intensity causes more energy to be transmitted to the particles and also the frequency with which the impacts occur. Consequently, both the energy and the power transferred to the particles in each collision is higher. Obviously, there are other parameters to consider as well. The kinetic energy is transformed into mechanical energy of deformation (and/or fracture) and into thermal energy (local heating). The mechanical energy depends on the mechanical properties of the material of the particles. The more ductile they are, the greater the deformation and the less fracture they experience. Thermal energy depends on the specific heat and thermal conductivity of the particles and the surrounding atmosphere.

One of the parameters included in the optimization of the processing conditions by mechanical alloying is associated with the degree of filling of the jars (balls and particles). Normally, it is recommended that this be between half and two-thirds of the volume of the jar. The reason is that insufficient filling would lead to multiple collisions between balls or ball-wall in which particles of the material to be alloyed would not participate, and that excessive filling would cause the powder to slow down the balls and reduce their speed and kinetic energy. In the description of the methodology in the articles, the ball to powder weight ratio (B.P.R.) is often introduced. Typical B.P.R. values are 5:1 or 10:1. It is also important to optimize the number and geometry of the balls employed during milling. Commercial equipment provides information on the quantity and diameter of the balls recommended for each jar. An increase in B.P.R enhances the energy transfer, favoring the refinement of crystallite size and particle size.

In order to avoid an excessive heating of the particles during the milling process, an ON time is selected during which the equipment operates and an OFF time is selected during which it stops. Only the ON time is considered for the calculation of the effective milling time. It is also recommended to wait for the container to cool before opening it. Contact with oxygen at room temperature can facilitate oxidation or even combustion phenomena in metal particles.

In processing, small quantities of particles are usually extracted in intermediate extractions to study the evolution of the structure, microstructure and properties of the particles as a function of the milling time. This allows us to determine the minimum milling time required to produce particles with the required microstructure and functional properties. For industrial applicability, it is important that this time is as short as possible.

Another technological issue is the scaling from the laboratory to the industry. At laboratory level, a few grams are produced and to scale up to industrial level, the processing conditions described change.

6.1.2 HEA Criteria

There are more than a dozen parameters and criteria that are used in the classification and selection of HEAs with the aim of predicting their structure, whether it is a single phase or a combination. Among the most used criteria in the alloys produced by mechanical alloying, the following should be highlighted: a) atomic size difference, b) enthalpy of mixing, c) (configurational) entropy of mixing, d) valence electron concentration (VEC), e) electronegativity difference, f) the parameter Ω (combination of melting temperature, enthalpy of mixing and entropy of mixing) and g) the parameter Φ (also dependent on melting temperature, enthalpy and entropy of mixing). The main bibliography parameter is VEC.

By applying the different criteria, maps are constructed that allow to detect in which region (in which intervals of values of the criteria represented) a certain microstructure is obtained. One of the most used diagrams is the Darken-Gury plot, in which electronegativity is represented versus the difference in atomic size. Figure 6.2 shows two of these diagrams (Suryanarayana, 2022). In this specific case, the areas in which the BCC, FCC or amorphous phases are formed, among others, are detected.

There are other plots that represent different relations between the criteria previously mentioned. Another plot is that of the enthalpy of mixing as a function of the atomic ratio. It has been applied in metallic HEAs produced by powder metallurgy route, including ball milling results (Guo et al., 2013). It is a competence between the production of a solid solution, an amorphous phase or intermetallic compounds. The formation of solid solutions in HEAs is due to the high entropy effect. Figure 6.3 shows the representation of the parameter Φ in several HEAs by comparing with the parameter Φ_f (this parameter but the mixing enthalpy is provided by DFT) (Feng et al., 2016).

Ball milling is the preferred way to produce the fully powdered HEAs and spark plasma sintering (SPS) is the preferred consolidation method. Therefore, the most common combination is BM +SPS (Torralba et al., 2019).

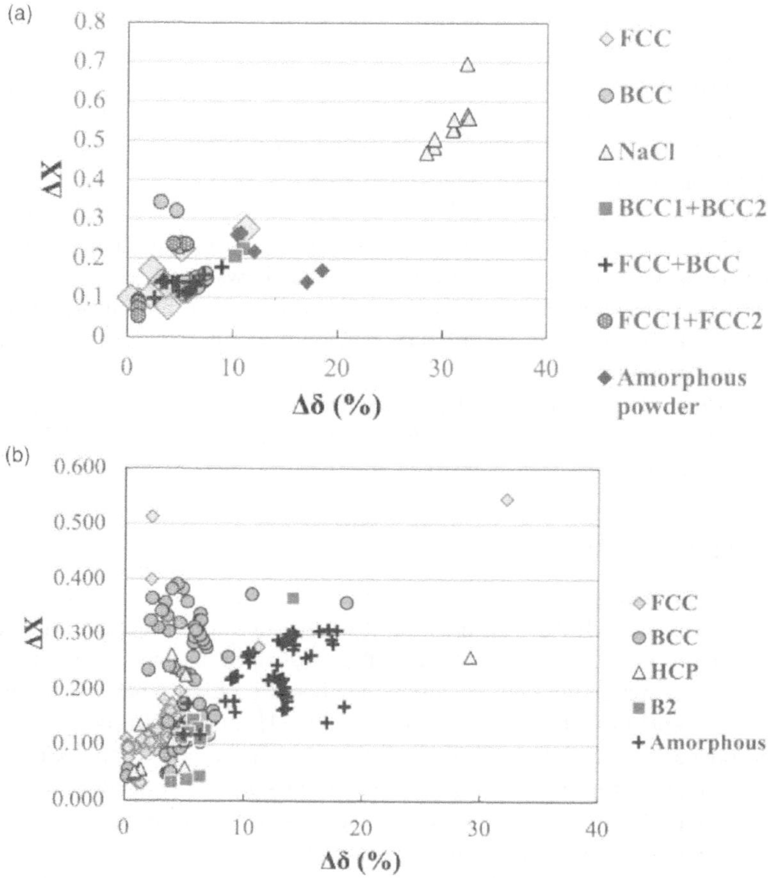

FIGURE 6.2 Electronegativity (ΔX) versus atomic size difference (Δδ) of HEAs. (a) Alloys processed by MA and (b) alloys processed by all the techniques (Suryanarayana 2022).

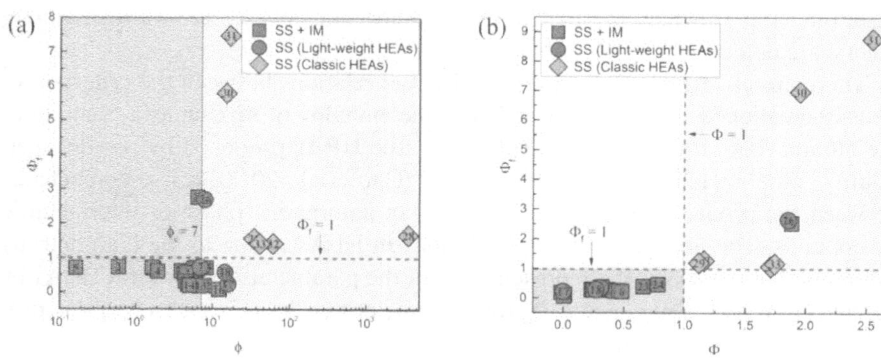

FIGURE 6.3 Comparison plots of (a) Φ_f versus φ and (b) Φ_f versus Φ (Feng et al., 2016). (Open access Creative Commons CC BY 4.0 license.)

Regarding the thermodynamic parameters, the mixing reaction leading to HEAs in the ball milling procedure is controlled by the entropy. The entropy increase causes a diminution of the Gibbs free energy, even at low temperatures. This phenomenon helps the development of supersaturated solid solutions and prevents the formation of equilibrium ordered phases. The extension of the solid solubility limits is typical of ball milled alloys. Ball milling allows the synthesis of compositions with low alloying affinity and a tendency to segregate.

Furthermore, in multicomponent elements, the diffusion sequence and the alloying rate is influenced by the brittleness of each element. In a $FeCoNiB_{0.5}Si_{0.5}$ HEA, diffusion sequence of the elements should be: Si $\rightarrow$ B $\rightarrow$ Co $\rightarrow$ Ni $\rightarrow$ Fe (Zaara et al., 2020). In this HEA, the Fe-Co favors the soft magnetic behavior, whereas the solid solution of Ni in the BCC Fe cubic structure leads to the increase of the magnetic permeability and the electrical resistivity.

The two traditional pathways to consolidate ball milled particles are pressing and sintering. These two pathways are sometimes combined. Likewise, in the last decades, the additive manufacturing techniques has been also introduced as a procedure to produce bulk materials. To facilitate the sintering process, it is necessary to control the size distribution, the average size and the morphology of the particles. Correct selection allows optimizing sinterability. Similarly, to favor the additive manufacturing process, it is also necessary to know and control the granulometry and morphometry of the powder particles obtained by ball milling. The characteristic to be optimized is flowability. In a study of ball milled HEAs (Adaan-Nyiak et al., 2023), it was found that the addition of PCA favors a smaller particle size, while increasing the milling time favors the reduction of the nanocrystalline size. One of the problems linked to the ball milled powders is the low spheroidicity of these powders. The general shape is that of corrugated powders with high surface/volume ratio. This irregular shape affects the flowability during the additive fabrication. One option is to apply the plasma spheroidization technique to the powders prior to the additive manufacturing procedure (Makhmutov et al., 2021). The additive manufacturing techniques applied in ball milled powders are laser powder bed fusion, selective laser melting and 3D-ink extrusion. An interesting aspect of materials consolidated by additive manufacturing is that the mechanical properties (especially those associated with the tensile response) are affected by the growth direction during consolidation/pressing.

Figure 6.4 shows a diagram of the different processing routes associated with the consolidation of particles synthesized by ball milling. As shown in the figure, pressing techniques can be followed by sintering processes or heat treatments (annealing). Warm pressing could also be introduced in pressing techniques. In the case of sintering, the technique most (commonly) used in the literature (spark plasma sintering) is highlighted, although other sintering techniques can be used, such as microwave sintering. Regarding additive manufacturing, this can be done directly using the particles produced by ball milling or, to ease their flowability, subjecting the particles to a laser spheroidization process.

One of the advantages of ball milling over other processing techniques is the extension of the solubility limit, probably due to the high density of mechanically induced crystallographic defects, defects that facilitate the formation of metastable

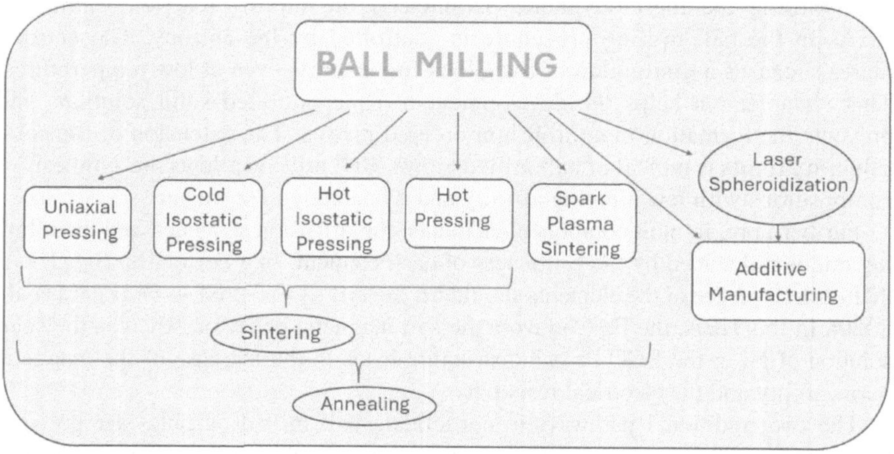

FIGURE 6.4 Schema of the pathways to produce bulk specimens with the ball milled powdered alloys.

and non-equilibrium phases. Another advantage of ball milling, compared to casting or melting techniques, is that it facilitates obtaining alloys in which elements with high vapor pressure, such as Li and Al, are present (Youssef et al., 2014). Likewise, the rate of structural refinement in ball milled HEAs is improved by a) the mechanical energy input and, b) the rate of work hardening among the material. Consequently, the ball milling method is a way to synthesize nanocrystalline with a uniform structure (Onawale et al., 2021). In Section 6.3., the positive effect of this phenomenon in the resulting mechanical properties of HEAs will be discussed.

6.2 MORPHOLOGY AND STRUCTURE OF POWDERS BEFORE AND AFTER PRESSING/SINTERING

One of the aspects to be considered when designing HEAs in which it is desired to verify their mechanical and functional properties is that these depend largely on the characteristics and properties of the powder particles obtained by ball milling and their microstructure; also on their microstructure and structure after pressing, sintering or additive manufacturing. This section is dedicated to the characterization of the morphometry and structure of HEAs before (ball milling) and after their consolidation (pressing, sintering, additive manufacturing).

Thus, one of the characteristics to control/check in the powders produced is its morphometry. For this reason, multiple articles show micrographs obtained by scanning electron microscopy (SEM) of the alloys obtained in powder form, depending on parameters such as composition, milling time or milling intensity. As an example, Figure 6.5 shows several micrographs of powder from a HEA (Al-Cr-Cu-Fe-Zn) (Cardoso et al., 2020).

Observation of a large number of powder particles from an HEA allows statistical studies to be carried out, determining the mean value or standard deviation of

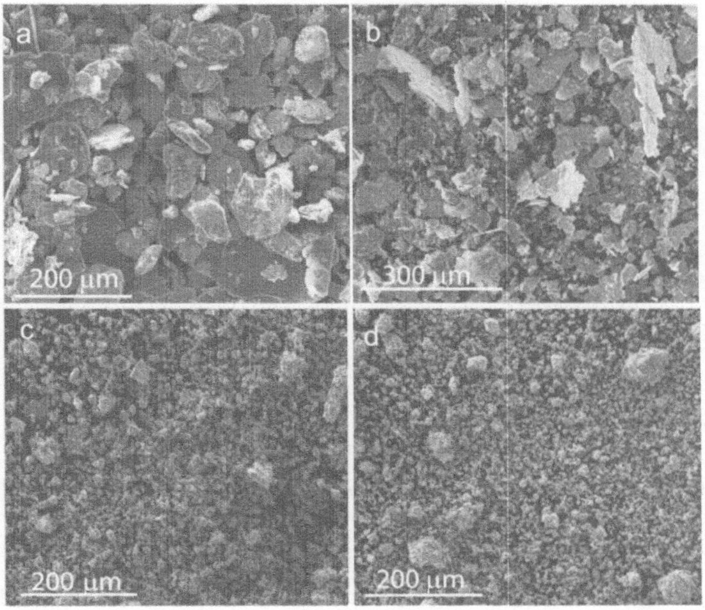

FIGURE 6.5 (a)–(d) SEM micrographs of Al-Cr-Cu-Fe-Zn multi-component HEA produced at different conditions (Cardoso et al., 2020).

the particle size, as well as the construction of histograms with the distribution of particle sizes. Software specific to electron microscopy equipment or free programs such as Fuji or ImageJ are usually used. Nevertheless, the application of laser as particle size distribution tester is highly recommended (Qi et al., 2019). In the mentioned study, firstly an increase of the particle size is found, secondly a decrease and finally the stabilization of the particle size. The decrease in particle size is associated with grinding in which cold welding predominates, the increase in particle size with the predominance of fracture and the stability in particle size with the balance between both processes.

The size of the powders is mostly micrometric and the crystalline structure is nanometric in size with a high density of crystallographic defects (microstrains) such as dislocations, vacancies or atoms in substitutional or interstitial positions. The density of crystallographic defects can be reduced by annealing at low temperature to favor the short-range diffusion necessary to cause structural relaxation (it involves processes with an activation energy between 0.2 and 2.0 eV/at).

To verify the nanocrystalline structure, various techniques can be used, the most common being X-ray diffraction (XRD) or transmission electron microscopy (TEM). Figure 6.6 (Niu et al., 2016) shows the nanocrystalline structure of HEA Co-Cr-Fe-Ni-Mn-Ti powder particles produced by mechanical alloying. The crystalline structure is a BCC cubic solid solution. The average particle size is close to 3 µm and the nanocrystals size mainly < 40 nm.

The nanocrystalline structure favors the increase in the percentage of atoms in the grain boundaries (between the nanocrystalline grains).

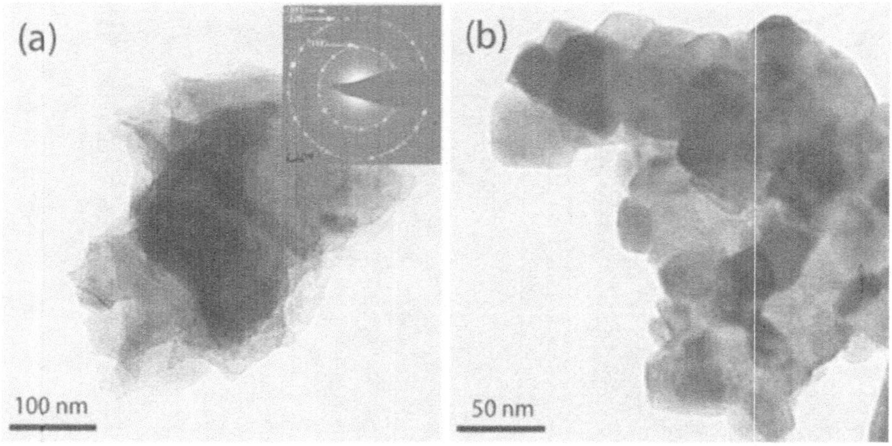

FIGURE 6.6 (a),(b)TEM images of nanocrystalline HEA Co-Cr-Fe-Ni-Mn-Ti (Niu et al., 2016).

Ball milling favors the contamination from the milling tools. For it, usually microanalysis (EDS, WDS, ICP) is performed to determine the contamination. Figure 6.7 shows the EDS microanalysis of a Fe-Ni-Co-Cr-Mn HEA (Mane et al., 2018). To determine whether the distribution of the different elements is homogeneous, several microanalyses are performed on various particles and statistical treatment is carried out. Another option is to generate maps of the areas where each element is present using EDS color mapping.

The consolidation of powders usually involves a thermal treatment of the alloys, favoring structural transformations, the appearance of new crystalline phases, the crystalline growth of the preexisting phase, the formation of oxides or compounds, or the formation of precipitates. Figure 6.8 shows the backscattered electron BSE-SEM

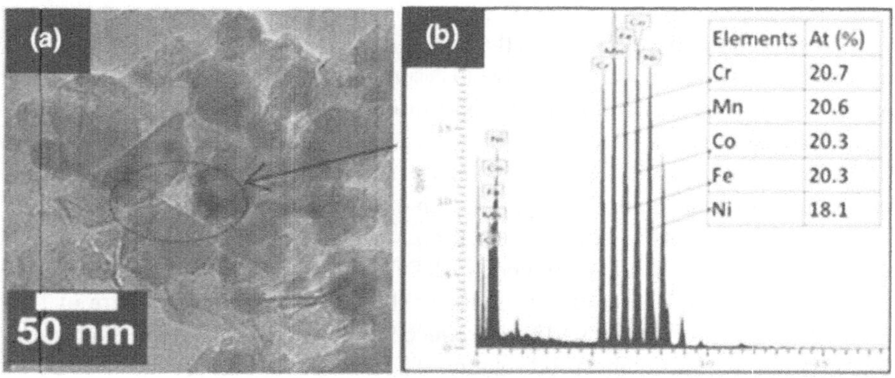

FIGURE 6.7 (a) TEM image and (b) EDS microanalysis of Fe-Ni-Co-Cr-Mn HEA (Mane et al., 2018).

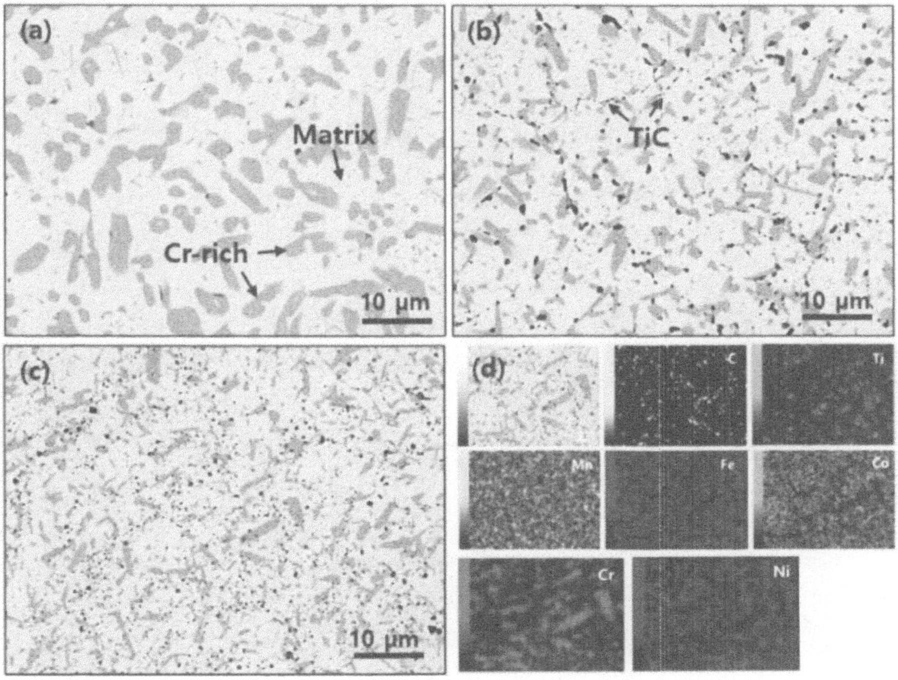

FIGURE 6.8 BSE-SEM cross-sectional microstructure of (a) Co-Cr-Fe-Mn-Ni HEA with the addition of TiC reinforcements (a) 0 wt-%, (b) 2 wt-% and (c) 4 wt-% composites fabricated by BM and SPS. (d) SEM-EDS maps of the TiC- 4 wt% reinforced (Nagarjuna et al., 2023).

images of a Co-Cr-Fe-Mn-Ni HEAs with the EDS mapping (Nagarjuna et al., 2023). In this work, a composite is formed by adding TiC. Thus, SEM images show the HEA matrix, the formation of Cr-rich precipitates and the added TiC precipitates.

The co-existence of different elements favor in-homogeneity. The larger atoms are usually located in the grain boundaries of the HEA nanocrystalline alloys whereas smaller atoms provoke the formation of voids of lower atomic density.

The analysis of the crystallographic structure of the HEA is carried out by XRD. As milling time increases, the formation of one or more solid solutions is typically observed, resulting from the incorporation of certain elements into the solution or the development of an amorphous phase. Also, milling leads to a decrease in the crystal size and an increase in microstrains.

Annealing of powders leads to a decrease in microstrains, an increase in crystallite size, and may facilitate the formation of new phases. Thus, XRD of annealed ball milled alloys provide information about the phase thermal stability and the formation of new phases. This knowledge is useful to design the thermal conditions of the sintering and consolidation processes. Figure 6.9 shows the XRD diffraction patterns of non-equiatomic Co-Cr-Fe-Ni-Mn-Ti HEA as milled and after annealing at different temperatures (Niu et al., 2016). In this work, after 36 hours of ball

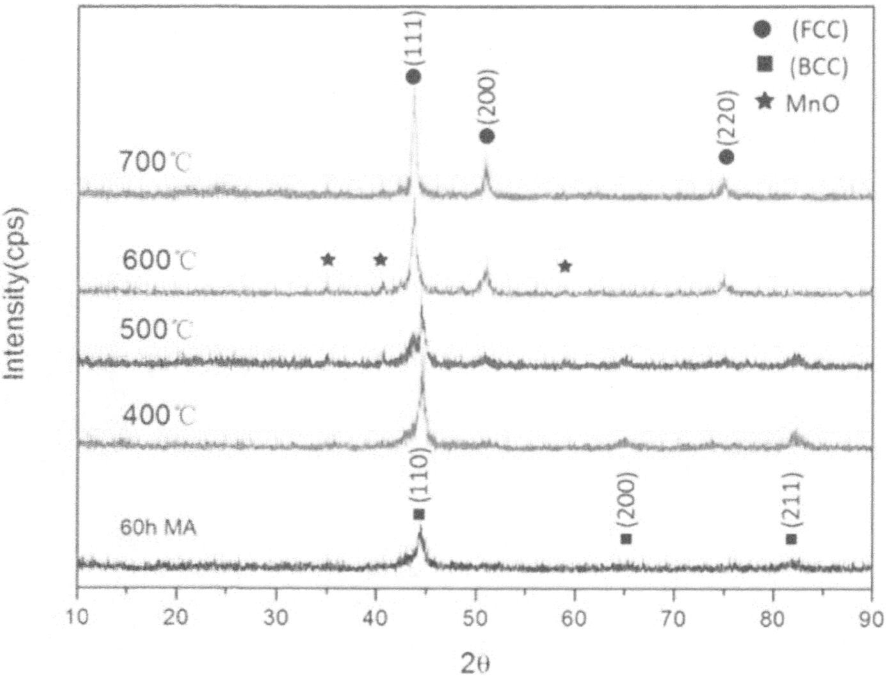

FIGURE 6.9 XRD pattern of Co-Cr-Fe-Ni-Mn-Ti HEA as milled and after annealing at different temperatures: 400, 500, 600 and 700°C (Niu et al., 2016).

milling only one phase remains. BCC-phase (solid solution) with a crystallite size lower than 20 nm of the milling time is increased up to 60 hours. The annealing at different temperatures induces the transformation of the BCC-phase into an FCC phase. It should be remarked that the liquidus point of the FCC-phase is around 1400°C. The undesired formation of a minor oxide (Mn rich) is also detected.

One of the distinctive characteristics and features of HEAs is the improvement of the lattice distortion. The severe lattice distortion is provoked by the coexistence of several elements with different atomic radii. Other parameters facilitating this distortion are the different bond energies and crystallographic microstructures/phases (Odetola et al., 2024).

Further consolidation of powders by pressing/sintering permits to attain high-density HEAs via controlled temperatures and pressures. Sintering bulk specimen's fabrication is associated to short intervals of time processing.

When the consolidation of powders is obtained, either through pressing or sintering, one of the parameters to optimize is the densification of the material. Normally, a degree of densification greater than 95% is expected, but sometimes it barely reaches 90%. A lower densification is associated with a decrease in the mechanical response and greater corrosion of the material. Open porosity favors corrosion by increasing the effective surface area in contact with the atmosphere. On the other hand, this porosity affects the comparison in the corrosion measurements published in the bibliography.

If there is no control of the atmosphere, corrosion is higher, for example, than if a flow of nitrogen has been used to introduce an inert gas against corrosion.

Pores may be either micrometric or nanometric in size. The Brunauer-Emmett-Teller (BET) method is one technique employed to assess the surface-to-volume ratio, allowing for the determination of the normalized surface area (m^2/g). This value increases with the average porous size and the porous density because it increases the total pore volume (cm^3/g).

A specific case occurs when the material being produced is foam. Thus, high densification is avoided. In the foams, the main interest is to control the pores distribution, size and structure. The homogeneity and uniformity in the foam material composition and in the size and distribution of pores is expected. Special interest is focused in nanopores. Nanoporous HEAs have been produced by chemically dealloying aluminum atoms in ball milled HEAs (Son et al., 2021). Regarding the composition influence, the results show that the high mixing enthalpy between the elements favors the de-alloying whereas low mixing enthalpy inhibit the de-alloying process. Obviously, open pores are those involved in de-alloying; in this work, by opening pathways for NaOH. The influence of the milling time on the mean ore diameter and specific surface area (BET measurements and calculus) has been analyzed before, showing the absorbed volume versus the relative pressure in BET experiments.

In addition to the consolidation processes for manufacturing massive parts, sintering can be used to produce layers (usually micrometric) for the production of composite coatings. It is necessary to optimize the thickness of the coating to make it as uniform as possible, as well as to control its compositional homogeneity, its microstructure and structure, and its mechanical properties, especially hardness, wear resistance and corrosion resistance. With respect to corrosion, it is necessary to obtain: (a) low weight gain, (b) optimal coating thickness, (c) fewer nodules formation and (d) low oxidation rate constant. In $Al_{70}(Cr\text{-}Fe\text{-}Ni)_{(30\text{-}x)}(SiC)_x$ composite coatings the effect of HEA coating on low carbon steel increase its corrosion resistance (Maulana et al., 2024). In this work, the best results correspond to 10% at.% of SiC, producing a larger thickness of Al_2O_3. Figure 6.10 shows schematically the steps of the coating process.

In order to determine the crystallographic structure of the bulk samples or coatings, XRD patterns are typically performed. This technique permits checking the composition influence in the final structure and the mechanical properties. In Co-Cr-Cu-Fe-Ni-Mo_x HEAs prepared by spark plasma sintering of ball milled precursors, the influence of the Mo addition ($x = 0$, 0.2, 0.4 and 0.8) has been analyzed (Yang et al., 2017). XRD analysis showed that the controlled addition of molybdenum changed the original FCC phase to a mixture of FCC + σ + μ. Moreover, the intensity of the (1 1 1) peak of the FCC phase decrease and the peak is shifted to lower angles. These analyses are complemented with field emission scanning electron microscopy. Figure 6.11 shows the diffraction patterns.

6.3 MECHANICAL AND FUNCTIONAL PROPERTIES

The mechanical and functional properties of HEA are discussed in several chapters of this handbook. In this section of the chapter, some examples of mechanical properties (mainly hardness, strength and stress) and functional properties (electrical,

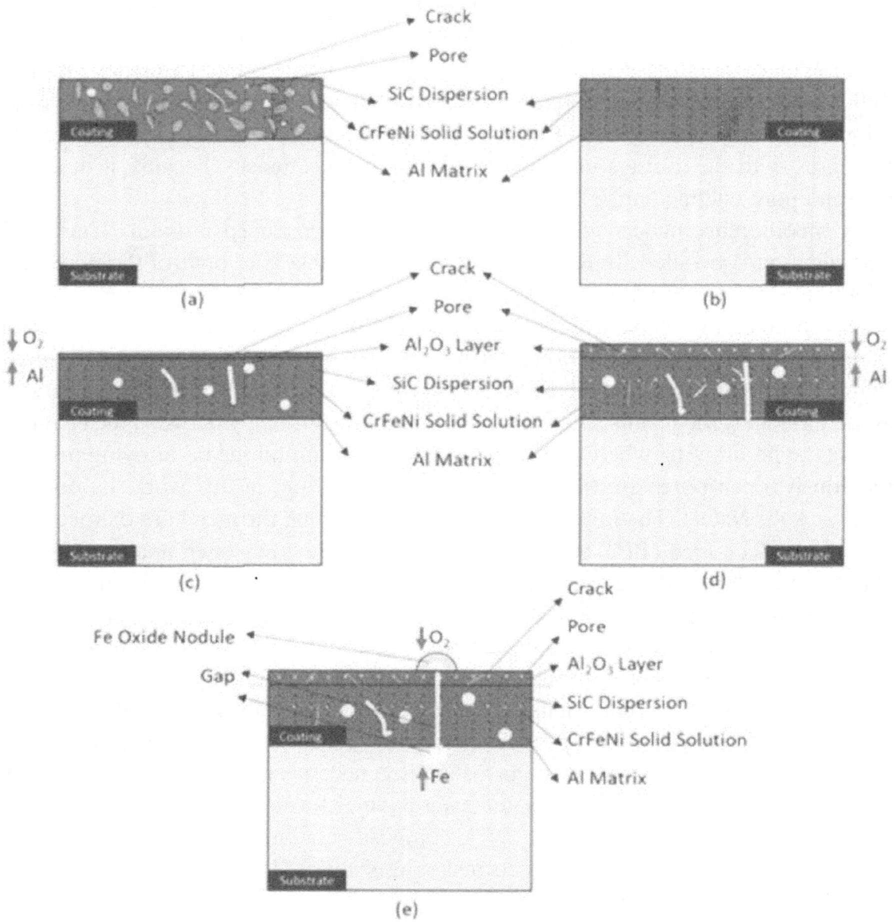

FIGURE 6.10 (a)–(e) Schema of the coating formation in $Al_{70}(Cr\text{-}Fe\text{-}Ni)_{(30\text{-}x)}(Si\ C)_x$ composite coatings (Maulana et al., 2024).

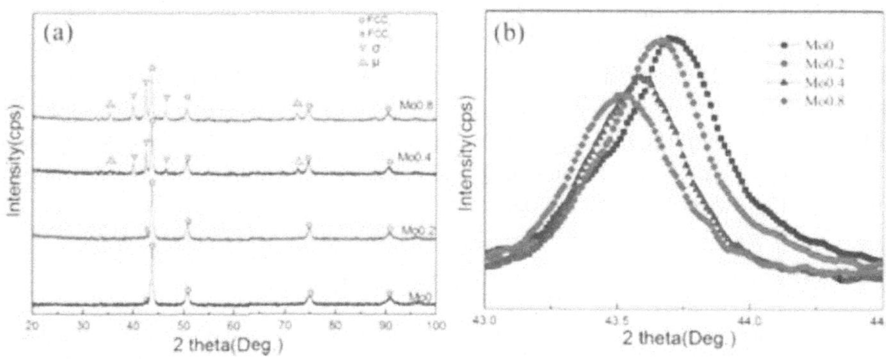

FIGURE 6.11 (a) XRD diffraction patterns of Co-Cr-Cu-Fe-Ni-Mo with molybdenum content (x = 0, 0.2, 0.4 and 0.8) and (b) inset in the FCC (1 1 1) peak 2 theta region (Yang et al., 2017).

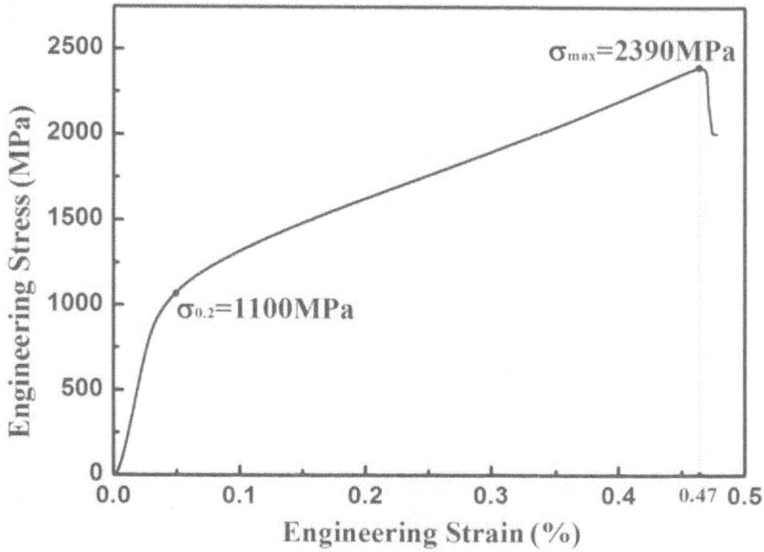

FIGURE 6.12 Engineering stress versus the engineering strain: HEA Cr-Mn-Fe-Co-Ni (Ni et al., 2018).

magnetic, electrochemical and hydrogen absorption) will be shown. The properties are related to the microstructure, which in turn is affected by the composition and processing conditions described in the first two sections of this chapter.

Optimal combination (compromise between both values) was found between strain and yield strength in Cr-Mn-Fe-Co-Ni bulk HEA produced by ball milling and subsequent spark plasma sintering. Figure 6.12 shows the stress versus strain curve (Ni et al., 2018). From the stress-strain curves, parameters such as the engineering yield strength (1100 MPa in the graph) and the engineering ultimate strength (2390 MPa in the graph) are determined.

The milling parameters (milling time and intensity) had a strong influence on the particle's morphology, structure and microstructure. It has been found that the final microstructure following sintering, as well as the mechanical and functional properties, are influenced by the ball milled powders characteristics.

Emamifar et al. found that the microstructure after spark plasma sintering of Al-Cr-Fe-Ni-Co-C HEA contained FCC, BCC ordered BCC and $C_{23}C_6$ carbide phases (Emamifar et al., 2019). Likewise, the increase of the milling time favors the strength increasing. This effect is due to the finer microstructure and the phases formation evolution. Figure 6.13 illustrates the evolution of the microhardness as milling time increases.

Obviously, the mechanical response depends on both the composition and the processing conditions. With prolonged milling times, the hardness usually increases due to hardening mechanisms present in the ball milled powders.

With respect to the composition, the addition of Mo in Co-Cr-Cu-Fe-Ni-Mo_x HEAs affects drastically the mechanical behavior, particularly in the stress-strain response (as shown in Figure 6.14), while also improving the wear resistance (Yang et al., 2017). The best set of results correspond to those of HEA with high Mo

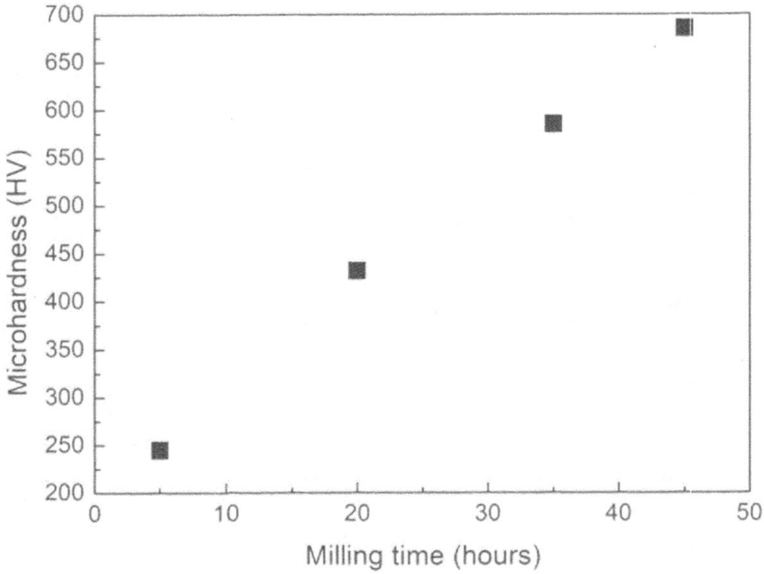

FIGURE 6.13 Microhardness evolution as a function of the milling time: HEA Al-Cr-Fe-Ni-Co-C (Emamifar et al., 2019).

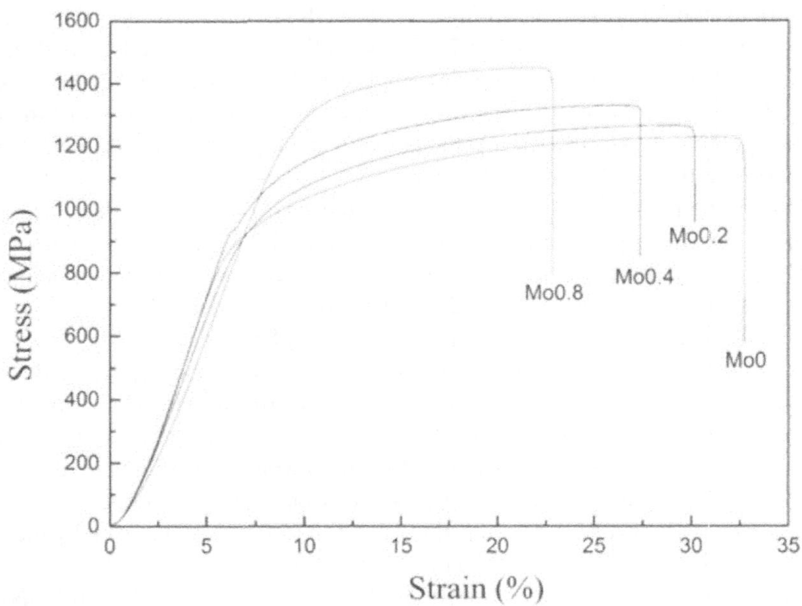

FIGURE 6.14 Stress-strain curves of Co-Cr-Cu-Fe-Ni-Mo_x HEAs (Yang et al., 2017).

content (x = 0.8): fracture strain 22.70%, fraction coefficient 0.65, hardness 530 HV and compressive strength 1448 MPa.

The main effects on the mechanical behavior of ball milled specimens are: a) hardening and b) strengthening. The primary mechanisms involved in both, hardening and strengthening are: a) dislocation, b) solid solution, c) precipitation and d) grain boundary.

The dislocation density (and the crystallographic defects density) increases with milling time due to the mechanically induced microstrains during the abrasion and percussion processes throughout ball milling. The continuous fracture and cold-welding of the particles cause a high density of dislocations. Annealing at selected temperatures is a pathway to reduce these microstrains. Dislocations provokes the lattice distortion.

Ball milling favors the extension of the solid solution limits with respect to the maximum solubility of the equilibrium phase diagrams. This effect was previously found in binary systems produced by ball milling. Likewise, in multi-component systems, such as the HEAs, this effect is nearly always found as a consequence of the coexistence of multiple elements in a chaotic ordered microstructure due to the high lattice distortion. The solid solution extension also increases the lattice distortion.

The ulterior annealing or sintering of the milled alloys facilitates the formation of precipitates. It is known that the thermal or thermomechanical treatment associated with pressing or sintering processes usually induces the formation of new phases. In the case of metallic HEAs, initially a specific crystalline phase can be obtained in the material in powder form, but subsequent treatment can induce the formation of new crystallographic phases (intermetallic compounds, oxides) in the form of precipitates or as a secondary phase. These precipitates hinder dislocation movement, resulting in increased hardness and strength. The ball milling originates the formation of structures far from the equilibrium. With respect to the crystallite size, ball milling induces the reduction of the crystal size (reduced to the nanometric range). The decrease of the crystallite size is associated with the increase of the atoms in the grain boundaries between the crystallites, being the percentage of atoms in the grain boundaries from the 5 to 50 atomic percentage.

There are other mechanisms as the Orowan strengthening due to the combination of two parameters: the interparticle mean free path for dislocation motion and the matrix distortion parameter.

Regarding the solid solution along the ball milling procedure, the addition of selected elements favors this phenomenon. It has been found that the addition of titanium and aluminum facilitate the formation of a BCC and FCC solid solution phase for Co-Cr-Fe-Ni-X and Co-Cu-Fe-Ni-X systems (X=Ti, Al). However, this effect is not found by adding Zn or Si (Joy et al., 2019). In this work, differential scanning calorimetry was applied to determine the transition temperatures and phase stability (Figure 6.15).

With respect to the precipitates influence, a similar effect is found by the controlled addition of secondary phases. The matrix is the HEA and the reinforcement a second phase favoring hardening, such as TiC. This effect is improved if the reinforcement is nanometric. The strengthening mechanisms involved are the grain boundary and the Orowan one.

Ball milling allows for the control of the microstructure with uniform dispersion of the nanoparticles reinforcement if compared to conventional methods.

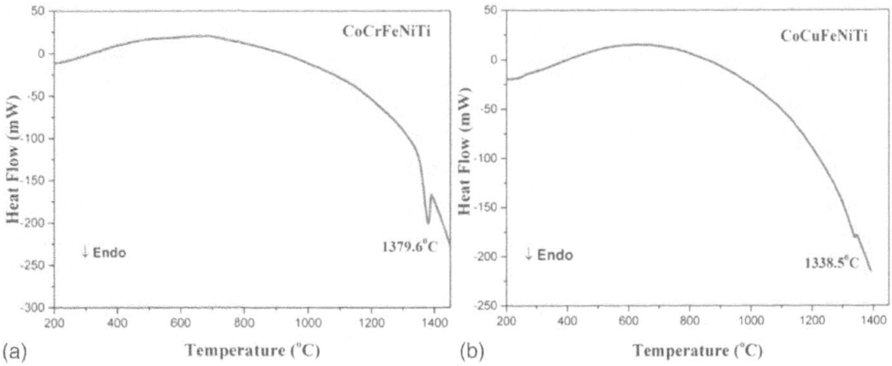

FIGURE 6.15 DSC scans of two HEAs: (a) Co-Cr-Fe-Ni-Ti and (b) Co-Cu-Fe-Ni-Ti (Joy et al., 2019).

Regarding the impact of ball milling on final properties, some research studies compare the mechanical response of HEAs consolidated from powder mixtures versus those produced through ball milling alloying. Ball milling provokes an improvement of the homogeneity of the samples, reduces the reactivity of the pure elements and facilitates the diffusivity of elements of the reinforcement in the HEA matrix. In Fe-Co-Ni-Cr-Al HEAs, reinforced uniformly with SiC, significant changes have been found in the hardness due to microstructural differences (Colombini et al., 2017). The crystalline phase after sintering is in both cases the same: BCC solid solution. The presence of metallic precursors in mixed samples favor the formation of stable fine precipitates: carbides and silicides. These precipitates, along with residual HEAs, contribute to the hardening of the material. Figure 6.16 shows the microhardness values.

The mechanical behavior of alloys is also influenced by temperature. Furthermore, the powders can be used to produce one- or two-dimensional specimens. For example, $Al_{0.1}$CoCrFeNi HEA microwires have been formed (Jiao et al., 2018), see Figure 6.17.

Dilatometry offers additional mechanical insights by measuring the thermal expansion coefficient as a function of temperature. For instance, the addition of TiC reinforcement leads to a reduction in the thermal expansion coefficient (Nagarjuna et al., 2023).

Dilatometric measurements are useful to determine the activation energies associated with the sintering process. A well-designed annealing treatment allows for the control of the densification rates. The sintering mechanisms of the HEA samples are not fully understood. The sintering of the ball milled powders causes energy and mass transfer. Scientific and technological interest is focused on the determination of the diffusion coefficients and the activation energies. The diffusion occurs in the volume (lattice diffusion), the grain boundary and the surface. As increasing the sintering temperature, the main mechanism becomes the volume diffusion. This effect was found in Co-Cr-Fe-Ni-Mn HEA ball milled and annealed (Mane et al., 2018). The activation energies (volume, grain boundary) increase in the annealed specimens.

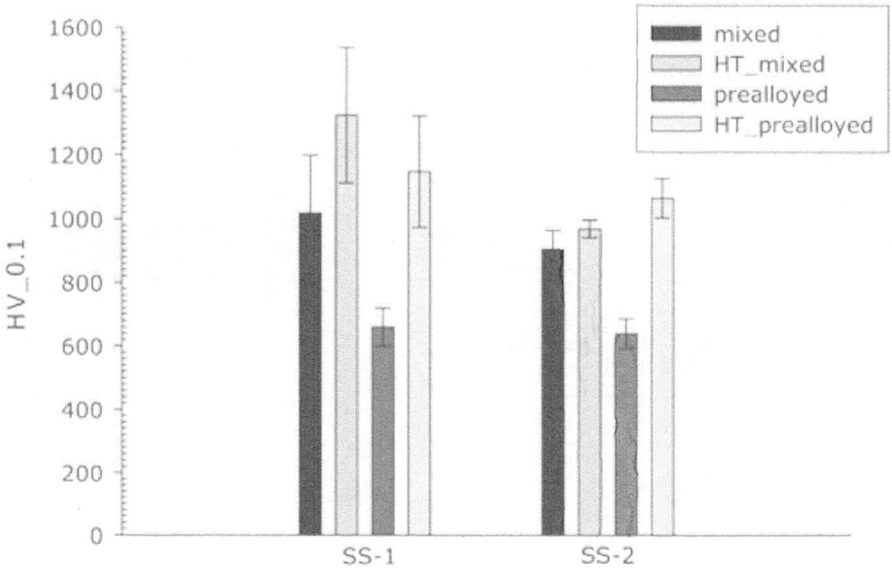

FIGURE 6.16 Vicker hardness of two SiC reinforced HEA solid solutions produced by mixing or ball milled powders sintered and after heat treatment (HT) (Colombini et al., 2017).

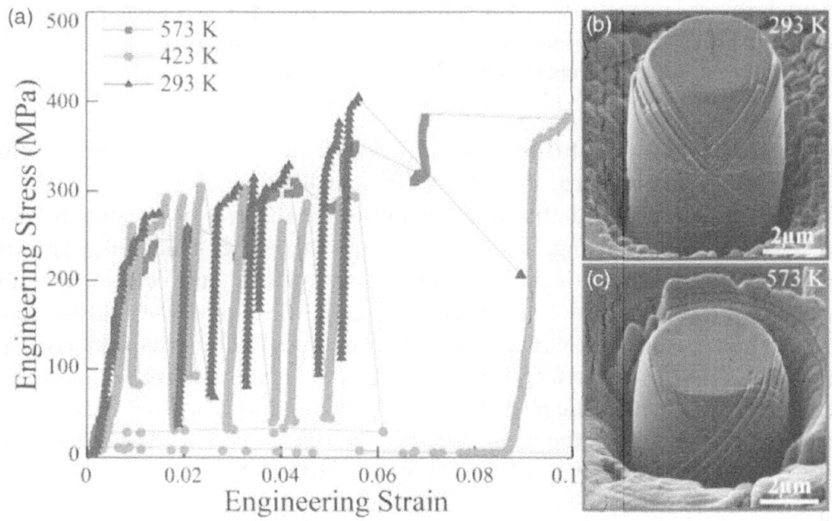

FIGURE 6.17 (a) Stress–strain curves for the $Al_{0.1}CoCrFeNi$ HEA tested at various temperatures. SEM micrographs of the deformed microcrystals at (b) 293 K and (c) 573 K (Jiao et al., 2018).

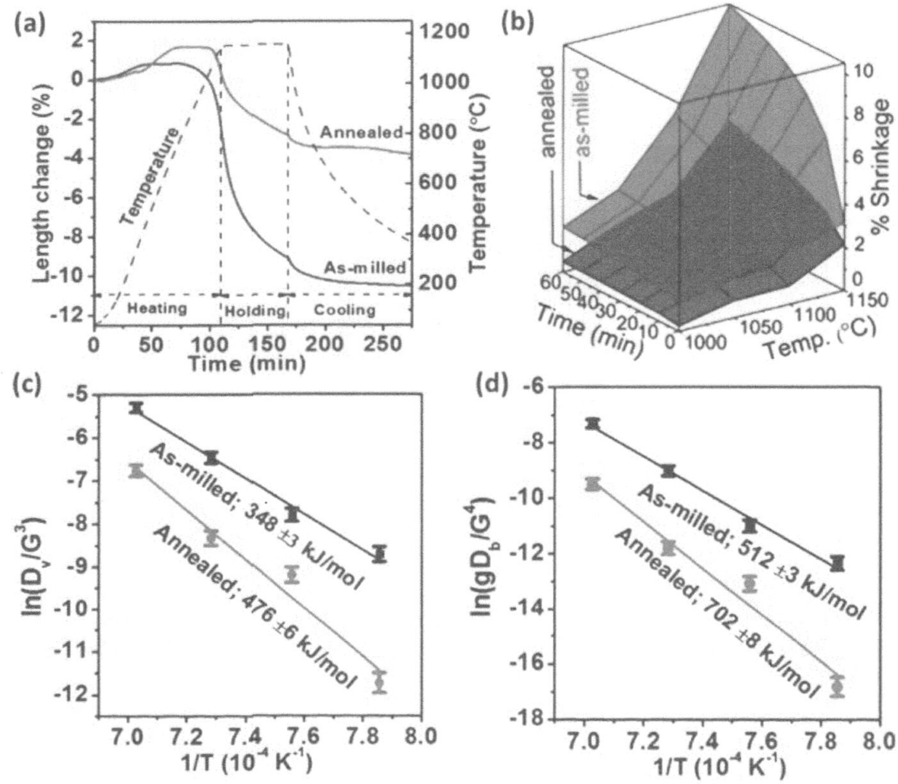

FIGURE 6.18 Co-Cr-Fe-Ni-Mn HEA ball milled and annealed. (a) Length change versus time abd (b) shrinkage % time maps in the sintering process, (c) volume activation energy and (d) grain boundary activation energy plots (Mane et al., 2018).

In Figure 6.18, D is the diffusion coefficient (sub index: v volume, b grain boundary), G is the mean grain size and g the grain boundary thickness.

Thermoelectricity: The lattice distortion effect in HEAs does not only influence hardening and strengthening. The electrical and thermal conductivities are influenced by the variations in local bond states associated to the electronic structure. The changes in the electronic characteristics/properties are due to: (a) the lattice distortion and (b) the complexity of nanostructures. Another mechanism associated with HEAs is the cocktail effect. This effect is based on the existence of unexpected and unpredictable synergistic responses, including physical properties in some materials as the HEAs. One of the applications of the functional HEA is in thermoelectricity. Thermoelectric materials need an optimized combination of properties: (a) electrical conductivity, (b) Seebeck coefficient and (c) thermal conductivity. These properties are all affected by several issues appearing in ball milled HEA samples: grain size, precipitates formation, structural complexity (Shi et al., 2022).

In thermoelectric materials the addition of a second phase as strengthening phase allows for the improvement of high-strength-conductive materials in ball milled compositions consolidated by spark plasma sintering TZNT HEA reinforced copper metal matrix composites (Chen et al., 2023). In this work, the controlled addition of hard particles as reinforcement in the matrix improved the mechanical response. Nevertheless, the electrical conductivity decreases. Connections have been found between the electrical conductivity, the Seebek coefficient and the power factor of complex GeTe-based HEA (Huang et al., 2022). It has been found that the high configurational entropy reduces the phase transition temperature close to room temperature and Mn-Zn-Cd co-alloying promotes the multivalent bands' convergence, inducing a high density-of-state effective mass.

From a materials perspective, the primary way to increase the efficiency of thermoelectric materials is through improving the figure of merit, ZT. This parameter determines the limiting conversion efficiency of a thermoelectric material. The powdered shape of ball milled HEAs exhibiting thermoelectric responses makes these materials strong candidates for creating thermoelectric specimens within porous media. The figure of merit is given in Figure 6.19 with other parameters as a function of the temperature.

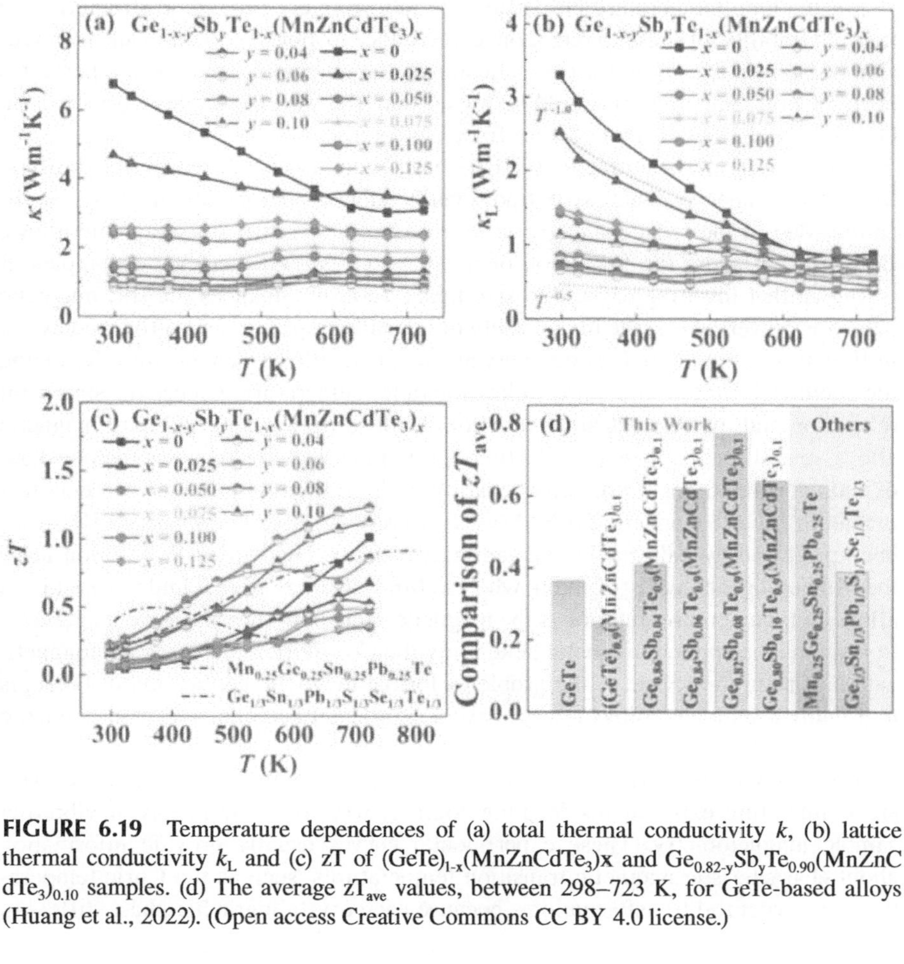

FIGURE 6.19 Temperature dependences of (a) total thermal conductivity k, (b) lattice thermal conductivity k_L and (c) zT of $(GeTe)_{1-x}(MnZnCdTe_3)x$ and $Ge_{0.82-y}Sb_yTe_{0.90}(MnZnCdTe_3)_{0.10}$ samples. (d) The average zT_{ave} values, between 298–723 K, for GeTe-based alloys (Huang et al., 2022). (Open access Creative Commons CC BY 4.0 license.)

Concerning the figure of merit, typical dependences are the composition and temperature. The fabrication of dense bulk thermoelectric specimens is usually done by combining ball milling and spark plasma sintering. This procedure has been applied in $BiSbTe_{1.5}Se_{1.5}$ HEAs doped with Ag (0.3–0.9 at.%). This alloy is monophasic and the density is > 96% (Raphel et al., 2024). The Ag addition enhanced the Seebeck coefficient. The combination of properties allows these HEAs to be candidates in the thermoelectric device fabrication due to minimal losses.

Electrochemistry: Another application of HEAs is in electrochemistry; for example, in electrodes. In this case, it is necessary to produce alloys with optimal corrosion resistance. Losses in the corrosion resistance are expected by increasing the temperature due to the formation of oxides at high temperature. Nevertheless, if oxide formation is inhibited, the increased sintered temperature is associated to an increase of the porosity, to the reduction of the lattice strain (and lattice distortion) and to the improvement of the corrosion resistance (Abhijith et al., 2024). The increased temperature of sintering enhances the diffusion of individual solute atoms. Some results included electrochemistry curves in Cr-Mn-Fe-Co-Ni HEAs are given (Wang and Kamada, 2024).

For the high-temperature properties, the following are needed: a) structural and thermal stability, b) strength retention, c) resistance to anneal softening and d) corrosion resistance. Likewise, for magnetic alloys, the magnetic transition temperatures (as the Curie temperature) should be higher than the working temperature interval. One family of HEAs is that of the refractory materials.

Magnetic properties: magnetic HEAs have been also produced by ball milling. Multi-component systems were traditionally fabricated for soft magnetic applications due to the reduced magnetic losses of some Fe-rich soft magnetic alloys. Ball milling favors the formation of nanocrystalline and amorphous samples. It is known that lowering crystallite size under 100 nm enhances the soft magnetic behavior. Nevertheless, the high density of crystallographic defects provoked by the milling process increase the coercivity and magnetically harden the alloy. Reducing the coercivity ulterior annealing at low-medium temperature induces the structural relaxation and the microstrains reduction. With respect to soft and hard magnetic alloys, one of the main issues affecting optimal magnetic response is the powders' oxidation. Sometimes a protective coating with an organic polymer prevents this oxidation. A soft magnetic Fe-Ni-Mn-Cu-X (X = Cr, Mo, Ti, W)-based HEA has been produced by ball milling (Jahani et al., 2023). The Ti and W addition favor the Mn homogeneous distribution, whereas Mn-rich areas were found in Cr and Mo alloys. A typical representation is the magnetic cycle of hysteresis of the magnetization versus the applied magnetic field. Only the Co addition increases the magnetization of saturation. One of the problems for the magnetic application of HEAs is that multi-component alloys always have a significant percentage of non-magnetic elements.

The magnetic properties also depend on temperature. Thus, magnetization versus temperature experiments (heating-cooling) were done by means of vibrating sample magnetometry. These experimental curves usually provide information about characteristic magnetic transition temperatures, such as the Curie temperature and structural transformations, because each crystallographic phase differs in

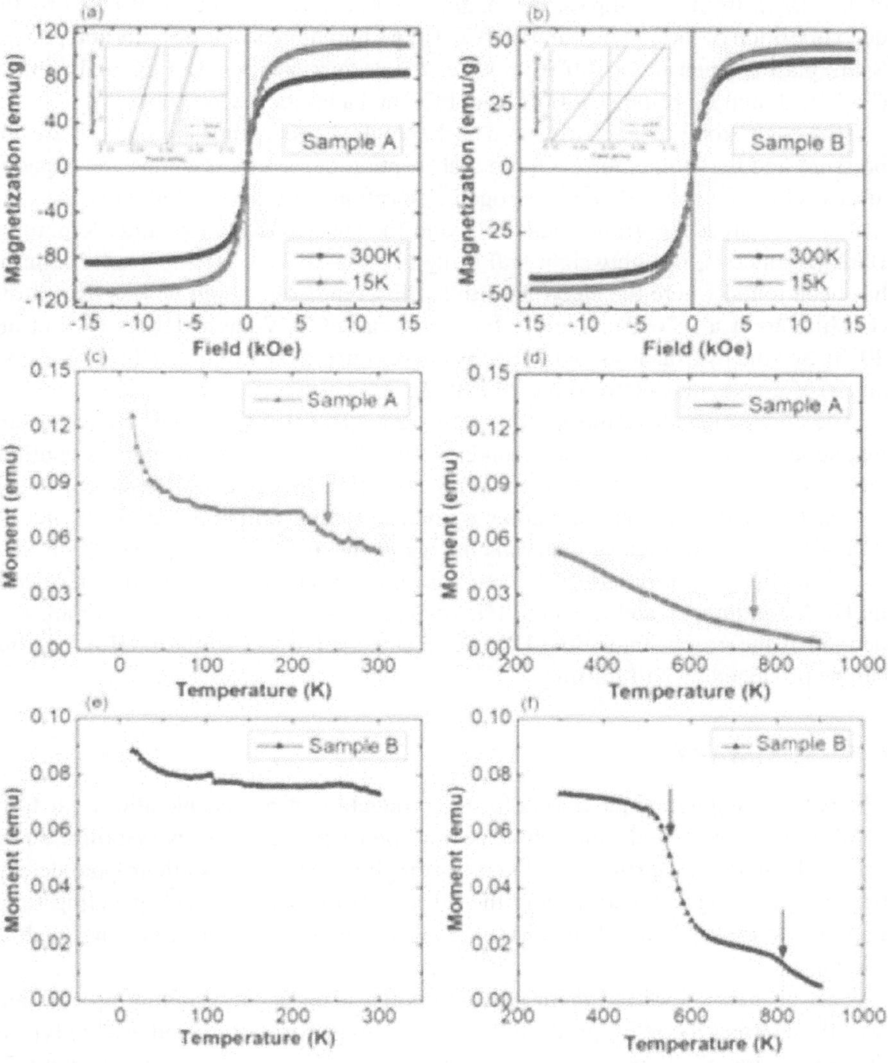

FIGURE 6.20 Fe-Co-Cr_x-Ni_2-Al HEAs with different Cr content (x = 0.5, 1.5), labeled as A and B, respectively. (a), (b) Magnetic hysteresis cycles of A and B, respectively, (c), (d) magnetization scans on heating, and (e), (f) cooling of A and B, respectively (Perumal et al., 2022).

the interatomic distances and, in the interaction between the magnetic moments of the atoms, modifying the magnetization.

Figure 6.20 shows the magnetic analysis of Fe-Co-Cr_x-Ni_2-Al alloys with different Cr content (x = 0.5, 1.5). The magnetization of saturation increases with the Cr content, whereas the coercivity is nearly constant (Perumal et al., 2022). See the magnetic hysteresis cycle of Fe-Co-Ni-B-Si HEAs as produced by ball milling (mechanical alloying), and ulterior annealing or spark plasma sintering at high temperature (Zaara et al., 2020). In these samples, the magnetic soft behavior is highly affected by the decomposition of the supersaturated FCC HEAs into more stable equilibrium phases. Spark plasma sintering at 700°C provokes the decomposition into FCC and a minor $(Fe,Ni)_{23}B_6$ and sintering at 1000°C into FCC and a minor $FeCo_2B$ phase.

Hydrogen storage: The two main technological processing parameters are the pressure and the temperature. Optimal absorption-desorption kinetics and capacity after cycling is expected. The hydrogen absorption activation is linked to the relative surface area, the grain boundaries and the surface oxidation state. Regarding HEA parameters, in lightweight ball milled Li-Mg-Al-Ti-V HEAs, a solid solution has been found where the electronegativity difference $\Delta\chi$ should be less than 0.285 (Hashimoto et al., 2023). Another family is that of Ti-V-based HEAs (Wu et al., 2023). See the hydrogen storage capacity curves during the cycling of Ti-V-Cr-Zr-Nb and Ti-V-Zr-Nb alloys (Bouzidi et al., 2022).

The applications here introduced are not an exhaustive list. Other properties are magnetic shape memory or superconductivity. The HEAs classification is continuously discussed. A classification could be: (a) 3D transition metal HEAs; (b) transition metals with larger atomic radius elements; (c) lightweight materials having Li, Al and/or Ti; (d) refractory materials having elements of the groups IVB, VB and/or VB; (e) lanthanide-based HEA having several lanthanide elements; (f) intermetallic HEA compounds and (g) other HEAs. Regarding specific issues, there are also defined as the groups interstitial HEAs and eutectic HEAs. All these HEA families can be produced by ball milling.

6.4 SUMMARY

The ball milling technique allows for the production of metastable alloys, far from equilibrium as extended solid solutions, amorphous phases or nanocrystalline structures. The lattice distortion is favored by a high density of crystallographic defects by increasing the microstrain and the density of dislocations with grinding time. Lattice distortion is one of the characteristics of multi-component systems such as HEAs.

This chapter shows the main parameters that control the grinding process and how these affect the HEA produced by ball milling. Also, discussed are the typical analyses of the alloys produced. When powdered alloys are obtained, it is common to consolidate them using pressing, sintering or additive manufacturing techniques.

On the other hand, examples of specific properties, whether mechanical or functional, are also shown: electrochemical, magnetic, hydrogen absorption, thermoelectric. The most studied in HEAs produced by ball milling are associated with the mechanical response: hardness, strength and wear resistance.

REFERENCES

Abhijith, N.V., Phutela, S., Kumar, D. & Kalyanasundaram, D. (2024). Transitional and refractory elements-based high entropy alloy developed through mechanical alloying and microwave sintering: tribological and corrosion responses. *Vacuum*, *222*, 113061. https://doi.org/10.1016/j.vacuum.2024.113061

Adaan-Nyiak, M.A., Alam, I. & Tiamiyu, A. (2023). Ball milling process variables optimization for high-entropy alloy development using design of experiment and genetic algorithm. *Powder Technology*, *427*, 118766. https://doi.org/10.1016/j.powtec.2023.118766

Bouzidi, A., Laversenne, L., Nassif, V., Elkaim, E. & Zlotea, C. (2022). Hydrogen storage properties of a new Ti-V-Cr-Zr-Nb high entropy alloy. *Hydrogen*, *3*(2), 270–284. https://doi.org/10.3390/hydrogen3020016

Cardoso, K.R., da Silva Izaias, B., de Souza Vieira, L. & Bepe, A.M. (2020). Mechanical alloying and spark plasma sintering of AlCrCuFeZn high entropy alloy. *Materials Science and Technology*, *36*(17), 1861–1869. https://doi:10.1080/02670836.2020.1839195

Chen, J., Xiang, T., Bao, W., Yu, B., Li, J., Wang, Y., Zhou, T., Du, P. & Xie, G. (2023). Novel strength-electrical conductivity synergy in Cu-based composites reinforced with TiZrNbTa high entropy alloy. *Materials Science and Engineering: A*, *878*, 145210. https://doi.org/10.1016/j.msea.2023.145210

Colombini, E., Lassinantti Gualtieri, M., Rosa, R., Tarterini, F., Zadra, M., Casagrande, A. & Veronesi, P. (2017). SPS-assisted synthesis of SIC_p reinforced high entropy alloys: reactivity of SIC and effects of pre-mechanical alloying and post-annealing treatment. *Powder Metallurgy*, *61*(1), 64–72. https://doi.org/10.1080/00325899.2017.1393162

Emamifar, A., Sadeghi, B., Cavaliere, P. & Ziaei, H. (2019). Microstructural evolution and mechanical properties of AlCrFeNiCoC high entropy alloy produced via spark plasma sintering. *Powder Metallurgy*, *62*(1), 61–70. https://doi.org/10.1080/00325899.2019.1576389

Feng, R., Gao, M.C., Lee, C., Mathes, M., Zuo, T., Chen, S., Hawk, J.A., Zhang, Y. & Liaw, P.K. (2016). Design of light-weight high-entropy alloys. *Entropy*, *18*, 333. https://doi.org/10.3390/e18090333

Guo, S., Hu, Q., Ng, C. & Liu, C.T. (2013). More than entropy in high-entropy alloys: Forming solid solutions or amorphous phase. *Intermetallics*, *41*, 96–103. https://doi.org/10.1016/j.intermet.2013.05.002

Hashimoto, H., Isobe, S., Hashimoto, N. & Oka, H. (2023). Synthesis of Li-Mg-Al-Ti based lightweight high entropy alloys by mechanical alloying and investigation of conditions for solid solution formation. *Journal of Alloys and Metallurgical Systems*, *4*, 100037. https://doi.org/10.1016/j.jalmes.2023.100037

Huang, Y., Zhi, S., Zhang, S., Yao, W., Ao, W., Zhang, C., Liu, F., Li, J. & Hu, L. (2022). Regulating the configurational entropy to improve the thermoelectric properties of $(GeTe)_{1-x}(MnZnCdTe_3)_x$ alloys. *Materials*, *15*, 6798. https://doi.org/10.3390/ma15196798

Jahani, N., Reihanian, M. & Gheisari, K. (2023). Alloying, phases and magnetic behaviour of mechanically alloyed FeNiMnCu-based high entropy alloys. *Materials Science and Technology*, *39*(14), 1745–1759. https://doi.org/10.1080/02670836.2023.2180902

Jiao, Q., Sim, G.D., Komarasamy, M., Mishra, R.S., Liaw, P.K. & El-Awady, J.A. (2018). Thermo-mechanical response of single-phase face-centered-cubic Al_xCoCrFeNi high-entropy alloy microcrystals. *Materials Research Letters*, *6*(5), 300–306. https://doi.org/10.1080/21663831.2018.1446468

Joy, J., Jadhav, M., Sahane, D., Davis, D. & Singh, S. (2019). Elemental effect on formation of solid solution phase in CoCrFeNiX and CoCuFeNiX (X=Ti, Zn, Si, Al) high entropy alloys. *Materials Science and Technology*, *35*(14), 1700–1707. https://doi.org/10.1080/02670836.2019.1639888

Makhmutov, T., Razumov, N., Kim, A., Ozerskoy, N., Mazeeva, A. & Popovich, A. (2021). Synthesis of $CoCrFeNiMnW_{0.25}$ high-entropy alloy powders by mechanical alloying and plasma spheroidization processes for additive manufacturing. *Metals and Materials International*, *27*, 50–54. https://doi.org/10.1007/s12540-020-00747-0

Mane, R.B., Y., R. & Panigrahi, B.B. (2018). Sintering mechanism of CoCrFeMnNi high-entropy alloy powders. *Powder Metallurgy*, *61*(2), 131–138. https://doi.org/10.1080/00325899.2018.1433268

Maulana, M.I., Aryanto, D., Prajatelistia, E., Nurrochman, A., Handika, G., Pratama, N.H. & Noviyanto, A. (2024). Oxidation resistance and hardness of high entropy alloy composite coating $Al_{70}(Cr\text{-}Fe\text{-}Ni)_{(30\text{-}X)}(SiC)_X$ prepared on AISI 1018 substrate by mechanical alloying. *Materials at High Temperatures*, *41*(4), 494–508. https://doi.org/10.1080/09603409.2024.2361196

Mayer, M., Ressel, G. & Svoboda, J. (2022). The effect of cryogenic mechanical alloying and milling duration on powder Particles' microstructure of an oxide dispersion strengthened FeCrMnNiCo high-entropy alloy. *Metall Mater Trans A*, *53*, 573–584. https://doi.org/10.1007/s11661-021-06532-x

Nagarjuna, C., Dewangan, S.K., Lee, K. & Ahn, B. (2023). Mechanical and thermal expansion behaviour of TiC-reinforced CoCrFeMnNi high entropy alloy prepared by mechanical alloying and spark plasma sintering. *Powder Metallurgy*, *66*(5), 613–622. https://doi:10.1080/00325899.2023.2219145

Ni, X., Dai, T., Lu, T., Pan, J., Li, M. & Dai, J. (2018). Microstructure and properties of CrMnFeCoNi high-entropy alloy prepared by mechanical alloying and spark plasma sintering. *Powder Metallurgy*, *62*(1), 38–43. https://doi.org/10.1080/00325899.2018.1554834

Niu, B., Ji, W., Li, N., Zhang, F. & Wu, Y. (2016). Alloying and thermal behaviour of $CoCrFeNiMn_{0.5}Ti_{0.5}$ high-entropy alloy synthesised by mechanical alloying. *Materials Science and Technology*, *32*(1), 94–98. https://doi.org/10.1179/1743284715Y.0000000124

Odetola, P.I., Babalola, B.J., Afolabi, A.E., Anamu, U.S., Olorundaisi, E., Umba, M.C., Phahlane, T., Ayodele, O.O. & Olubambi, P.A. (2024). Exploring high entropy alloys: A review on thermodynamic design and computational modeling strategies for advanced materials applications. *Heliyon*, e39660. https://doi.org/10.1016/j.heliyon.2024.e39660

Onawale, O.T., Cobbinah, P.V., Nzeukou, R.A. & Matizamhuka, W.R. (2021). Synthesis route, microstructural evolution, and mechanical property relationship of high-entropy alloys (HEAs): A review. *Materials*, *14*, 3065. https://doi.org/10.3390/ma14113065

Perumal, H.P., Jadhav, M., Abhinav, E.M., Sinha, J. & Singh, S. (2022). Enhanced magnetisation with increased chromium concentration in $FeCoCr_xNi_2Al$ high-entropy alloy. *Materials Science and Technology*, *38*(1), 12–18. https://doi.org/10.1080/02670836.2021.2021499

Qi, P.B., Liang, X.B., Tong, Y.B., Chen, Y.X. & Zhang, Z.B. (2019). Effect of milling time on preparation of NbMoTaW high entropy alloy powder by mechanical alloying. *Rare Metal Materials and Engineering*, *48*(8), 2623–2629. https://doi.org/10.12442/j.issn.1002-185X.20180616

Raphel, A., Vivekanandhan, P. & Kumaran, S. (2024). Tuning p-type semiconduction with ultra-low thermal conductivity in nanocrystalline n-type $BiSbTe_{1.5}Se_{1.5}$ high entropy alloy via Ag doping. *Applied Physics A*, *130*, 389. https://doi.org/10.1007/s00339-024-07553-0

Shi, Y., Shu, Q., Liaw, P.K., Wang, M., Teng, C.L., Zou, H., Wen, P., Xu, B., Wang, D. & Wang, J. (2022). Effect of annealing on mechanical and thermoelectric properties of a Al2CoCrFeNi high-entropy alloy. *Materials & Design*, *213*, 110313. https://doi.org/10.1016/j.matdes.2021.110313

Son, H., Nam, S. & Choi, H. (2021). Development of porous high-entropy alloys by mechanical alloying and chemical de-alloying. *Powder Metallurgy*, *64*(3), 211–218. https://doi.org/10.1080/00325899.2021.1901399

Suryanarayana, C. (2022). Mechanical alloying: a critical review. *Materials Research Letters*, *10*(10), 619–647. https://doi.org/10.1080/21663831.2022.2075243

Torralba, J.M., Alvaredo, P. & García-Junceda, A. (2019). High-entropy alloys fabricated via powder metallurgy. A critical review. *Powder Metallurgy*, *62*(2), 84–114. https://doi.org/10.1080/00325899.2019.1584454

Wang, R. & Kamada, S. (2024). Hardness and corrosion behavior of CrMnFeCoNi alloy fabricated by ball milling and spark plasma sintering. *Materials*, *17*(19), 4793. https://doi.org/10.3390/ma17194793

Wu, S., Li, W., Liu, X., Wang, Y., Cai, X. & Zhou, L. (2023). Evolution of the phase and hydrogen storage properties of new-type VTiCrFeMn high-entropy alloy prepared via mechanical alloying. *Integrated Ferroelectrics*, *235*(1), 17–28. https://doi.org/10.1080/10584587.2023.2192666

Yang, Q., Tang, Y., Wen, Y., Zhang, Q., Deng, D. & Nai, X. (2017). Microstructures and properties of CoCrCuFeNiMox high-entropy alloys fabricated by mechanical alloying and spark plasma sintering. *Powder Metallurgy*, *61*(2), 115–122. https://doi.org/10.1080/00325899.2017.1415248

Youssef, K.M., Zaddach, A.J., Niu, C., Irving, D.L. & Koch, C.C. (2014). A novel low-density, high-hardness, high-entropy alloy with close-packed single-phase nanocrystalline structures. *Materials Research Letters*, *3*(2), 95–99. https://doi.org/10.1080/21663831.2014.985855

Zaara, K., Chemingui, M., Le Gallet, S., Gaillard, Y., Escoda, L., Saurina, J., Suñol, J.J., Bernard, F., Khitouni, M. & Optasanu, V. (2020). High-entropy $FeCoNiB_{0.5}Si_{0.5}$ alloy synthesized by mechanical alloying and spark plasma sintering. *Crystals*, *10*(10), 929. https://doi.org/10.3390/cryst10100929

Section III

Properties of High Entropy Alloys

7 Mechanical and Structural Properties of High Entropy Alloys

D. Mishra, J. Singh and S.S. Nene

7.1 INTRODUCTION

High entropy alloys (HEAs) are increasingly recognized for their remarkable mechanical properties, rendering them appropriate for diverse applications in aerospace, automotive, and structural engineering [1–4]. HEAs are increasingly recognized for their exceptional mechanical properties, particularly their elevated yield and ultimate tensile strength. These enhancements can be largely attributed to solid solution strengthening and the ability to finely tailor their microstructures. This unique combination of features positions HEAs at the forefront of modern materials research, where they are being explored for a wide range of applications [4]. Moreover, many HEAs demonstrate improved ductility and toughness, which enhance their ability to absorb energy without fracturing [5]. They also exhibit superior fatigue and creep resistance, particularly at elevated temperatures [6, 7]. This combination of properties makes HEAs attractive for applications requiring robust performance under demanding conditions.

This chapter aims to provide a comprehensive review of the mechanical properties of HEAs, focusing on these critical aspects: tensile behavior, compressive properties, fatigue resistance, along with creep and superplasticity behavior. Each section examines the fundamental mechanisms that govern these properties, highlighting the impact of microstructural attributes and processing techniques on performance improvement. The tensile and compressive properties of HEAs are essential for understanding their load-bearing capacity in structural applications [8]. Fatigue resistance is vital for materials subjected to cyclic loading, while creep behavior is critical in high-temperature environments where materials experience prolonged yet lower stress [7]. The superplasticity of HEAs enables advanced forming methods, permitting the accurate fabrication of complex geometries [9]. Each section in this chapter is divided into transition metal (TM) based HEAs, recognized for their strength and endurance, and non-transition metal (NTM)-based HEAs, which emphasize high temperature characteristics for use in the automotive and aerospace sectors.

7.2 TENSILE PROPERTIES OF HEAs

The concept of HEA has revolutionized the worldwide drive for alloy design, marking a significant retreat from traditional metallurgical research practices [10]. The tensile attributes of HEAs are gaining considerable attention in material science due

DOI: 10.1201/9781003664093-10

to their remarkably improved performance under tension, surpassing that of conventional alloys [11, 12]. This section delves into the tensile properties of HEAs, particularly TM and NTM HEAs, emphasizing the unique behaviors and underlying deformation mechanisms that govern their tensile performance.

7.2.1 Tensile Properties of TM HEAs

HEAs containing transition metals, including chromium (Cr), manganese (Mn), iron (Fe), cobalt (Co), aluminum (Al), titanium (Ti), vanadium (V), and nickel (Ni) are typically classified in this group [13]. The Cantor alloy, named after Brian Cantor in 2004 [14], is the most often investigated HEA. It comprises five transition elements: Co, Cr, Fe, Mn, and Ni in equiatomic proportion. The production of an intermetallic phase was anticipated in Cantor alloy due to the high percentage of elements in the alloy; however, contrary to expectations, a single-phase solid solution was formed instead due to the high entropy effect [15]. The single-phase solid solution in the Cantor alloy contributes to a combination of high strength and good ductility. HEAs in the transition metal category can be viewed as advanced extensions of stainless steels and superalloys, achieved through progressive alloy design that imparts superior tensile properties [16]. An increasing amount of TM HEAs tensile data is becoming accessible, part of which is summarized in Table 7.1. This subsection discusses the tensile properties of TM HEAs in various conditions, including as-cast, thermomechanically processed (TMP), additively manufactured (AM), and processed through severe plastic deformation (SPD).

The microstructures of HEAs significantly influence their tensile properties; therefore, the fabrication route of HEAs is essential for the control of their microstructures [17]. Single-phase face centered cubic (FCC) structured CrMnFeCoNi HEA, fabricated through the casting process, behaves like a solid solution and results in moderate tensile strength and extended ductility. The equiaxed grained with a grain size of ~92.6 μm, single phase FCC microstructure of as-cast CrMnFeCoNi HEA is shown in Figure 7.1 (a) [18]. The statistics of electron back scattered diffraction (EBSD) results show that the as-cast alloy has a 90% fraction of high-angle grain boundaries, which are generally formed during casting. The tensile properties of the cast HEAs are significantly impacted by a variety of issues that arise during the casting process, including coarse dendrite structure, elemental segregation, porosity, and shrinkage cavity. Numerous TMP techniques, such as rolling, forging, and annealing, were implemented to resolve these issues. As discussed earlier, the plastic deformation in a material is primarily controlled by the evolution of microstructure, which can be obtained by altering alloy chemistry or processing methods. Therefore, to optimize the balance between strength and ductility in a material, microstructural engineering through TMP is one of the desirable approaches for a specific alloy composition. Dutta et al. [19] used the conventional TMP method to achieve a hierarchical microstructure to improve the tensile properties of $Fe_{44}Mn_{20}Cr_{15}Ni_{7.5}Co_6Si_{7.5}$. The as-cast alloy was homogenized, cold rolled up to 70%, and subsequently annealed at 900°C for 15 minutes. The microstructure of a TMP alloy is shown in Figure 7.1 (b), which contains complex γ-FCC dominant microstructure with an average grain size of ~9.7 μm. This alloy possesses a combination of fine and coarse grains (ranging

TABLE 7.1
Uniaxial Tensile Properties of TM HEAs, Phases Present, Processing Condition, Testing Temperature, YS (MPa), UTS (MPa), and Elongation (%)

S. No.	Alloy	Phases	Processing Condition	Testing Temperature	YS (MPa)	UTS (MPa)	Elongation (%)	Reference
1	$Fe_{20}Co_{20}Ni_{41}Al_{19}$	BCC + FCC	AC	RT	577	1103	18.7	[24]
2	$AlCoCrFeNi_{2.1}$	FCC + BCC	AM	RT	1109	1386	13.7	[25]
3	$Al_{0.3}FeCoNi$	BCC	AC	RT	221	486	41	[26]
	$Al_{0.3}Cr_{0.5}FeCoNi$				233	504	45	
	$Al_{0.3}Cr_{1}FeCoNi$				259	559	47	
	$Al_{0.3}Cr_{1.5}FeCoNi$	FCC + BCC			294	640	44	
	$Al_{0.3}Cr_{1.7}FeCoNi$				338	742	28	
	$Al_{0.3}Cr_{2}FeCoNi$				546	1078	11	
4	AlCoCrFeNi2.1	FCC + BCC	AC	RT	620	1115	20.4	[27]
				−90°C	675	1183	18.3	
				−150°C	697	1319	19.5	
				−196°C	857	1461	16.6	
5	AlCrCuNiFeCo	FCC1 + FCC2 + BCC	AC	RT	790	790	0.2	[28]
				600°C	542	551	0.4	
				700°C	350	360	4.7	
				800°C	161	180	12.1	
				900°C	88	100	30	
				1000°C	37	44	77	
		FCC1 + FCC2 + BCC. + σ	Forging	RT	1040	1170	1	[28]
				300°C	810	880	0.4	
				600°C	300	350	1.3	
				700°C	63	91	63	
				800°C	22	26	604	
				900°C	14	18	405	
				1000°C	9	22	864	

(Continued)

TABLE 7.1 *(Continued)*
Uniaxial Tensile Properties of TM HEAs, Phases Present, Processing Condition, Testing Temperature, YS (MPa), UTS (MPa), and Elongation (%)

S. No.	Alloy	Phases	Processing Condition	Testing Temperature	YS (MPa)	UTS (MPa)	Elongation (%)	Reference
6	$CoCr_3Fe_5Ni$	FCC + BCC	AC	RT	321	643	55	[29]
	$Al_{0.1}CoCr_3Fe_5Ni$				412	781	38.1	
	$Al_{0.2}CoCr_3Fe_5Ni$				568	853	28	
	$Al_{0.3}CoCr_3Fe_5Ni$				831	1018	16.1	
7	FeCoNiCrMn	FCC	AC	RT	221	505	64	[30]
	$(FeCoNiCrMn)_{96}Al_4$	FCC + BCC			214	500	60	
	$(FeCoNiCrMn)_{93}Al_7$				265	540	50	
	$(FeCoNiCrMn)_{92}Al_8$				308	648	38	
	$(FeCoNiCrMn)_{91}Al_9$				400	728	32	
	$(FeCoNiCrMn)_{90}Al_{10}$				680	992	21	
	$(FeCoNiCrMn)_{89}Al_{11}$				986	1171	11	
8	$Co_{25}Cr_{25}(FeNi)_{50}$	FCC	AC + HM (1200, 5 h) + Hot forging at 1200 (50% thickness reduction) + CR (40%)	RT	223	614	78	[31]
	$Co_{35}Cr_{25}(FeNi)_{40}$				258	723	96	
	$Co_{45}Cr_{25}(FeNi)_{30}$				373	844	76	
	$Co_{55}Cr_{25}(FeNi)_{20}$	FCC + HCP			400	747	20	
	$Co_{65}Cr_{25}(FeNi)_{10}$				506	860	13	
9	$Al_{0.5}CoCrFeNi$	FCC + BCC	AC	RT	312	587	24	[32]
			AC + HT (450)		335	652	40	
			AC + HT (650)		428	800	38	
			AC + HT (850)		600	980	19	
			AC + HT (1150)		365	720	33	
10	CrMnFeCoNi	FCC	CR	RT	1250	1327	19	[33]
			CR + AN (650, 1 h)		625	855	50	
			CR + AN (800, 1 h)		351	700	58	
			CR + AN (1000, 1 h)		194	600	63	

11	FeMnNiCoCr	FCC	AC + HM	RT	95	375	58	[34]
			CR + HM (1200, 2 h) + CR		700	760	17	
			CR + AN (900, 10 min)		250	645	58	
12	$FeCoNiCrTi_{0.2}$	FCC	HM (1100, 5 h) + CR (80%) + Recrystallization (1100, 1 h) + Aging (800, 1 h)	20°C	772	1242	38	[35]
				–196°C	945	1580	46	
13	$Al_{0.5}CoCrFeMnNi$	FCC + BCC	HM (1200, 6 h) + CR (78.6%) + AN (1000, 15 min)	RT	730	968	29	[36]
			HM (1200, 6 h) + CR (78.6%) + AN (1100, 10 min)		409	755	43	
		FCC	HM (1200, 6 h) + CR (78.6%) + AN (1200, 5 min)		278	619	60	
14	$Al_{0.3}CoCrFeNi$	FCC + L12	HM + CR (90%) + Recrystallization (1150, 2 min)	RT	123	396	68	[37]
			HM+CR (90%)+ Recrystallization (1150, 2 min) + AN (620, 50 h)		468	812	45	
			CR (90%) + AN (620, 50 h)		830	1044	34	
			CR (90%) + AN (550, 24 h)		1852	1852	4.4	
			CR (90%) + AN (550, 50 h)		1608	1651	6.6	
15	FeCoNiCr	FCC + unkonown	AC + HM (1200, 4 h)		171	447	70	[38]
	$(FeCoNiCr)_{94}Ti_2A_{14}$	FCC + unkonown	AC + HM (1200, 4 h)		223	515	70	
		$Ni_3(Ti,Al)$ + unknown	AC + HM (1200, 4 h) + CR (30%) + AN (1000, 2 h) + Aging (700, 18 h)		676	1100	43	
		$Ni_3(Ti,Al)$ + unknown	HM (1200, 4 h) + CR (70%) + Aging (650, 4 h)		1017	1285	21	

(Continued)

TABLE 7.1 *(Continued)*
Uniaxial Tensile Properties of TM HEAs, Phases Present, Processing Condition, Testing Temperature, YS (MPa), UTS (MPa), and Elongation (%)

S. No.	Alloy	Phases	Processing Condition	Testing Temperature	YS (MPa)	UTS (MPa)	Elongation (%)	Reference
16	NiFeCrCo	FCC	CR + AN 625	RT	540	786	49.3	[39]
			CR + AN 1100		197	582	82.4	
	NiFeCrCoMn	FCC	CR + AN 650		660	851	25	
			CR + AN 1100		135	497	55.4	
	$Ni_{18.5}Fe_{18.5}Cr_{18.5}Co_{26}Mn_{18.5}$		CR + AN 650		586	715	32.8	
			CR + AN 1100		134	414	73.5	
	$Ni_{14}Fe_{20}Cr_{26}Co_{20}Mn_{20}$	FCC + σ	CR + AN 650		1153	1187	1.8	
			CR + AN 1100		162	462	51.6	
17	CrMnFeCoNi	FCC	AC	25°C	241	571	75	[18]
				350°C	146	417	54	
				500°C	134	387	49	
				650°C	122	265	22	
				800°C	103	180	18	
				950°C	83	85	27	
			HM (1200, 2 h) + Forging (50% radial deformation)	25°C	311	603	60	
				350°C	323	468	23	
				500°C	307	451	30	
				650°C	294	379	17	
				800°C	185	200	31	
				950°C	89	89	46	

18	$Cr_{15}Fe_{20}Co_{35}Ni_{20}Mo_{10}$	FCC	AM + HM	RT	408	714	22	[40]
		FCC + HCP	AC + HM + HR + CR (70%) + AN (800, 1 h AC)		1311	1410	12	
		FCC + HCP	AC + HM + HR + CR (70%) + AN (850, 5min WQ)		1212	1360	15	
		FCC + HCP	AC + HM + HR + CR (70%) + AN (900, 5 min WQ)		1028	1249	18	
		FCC	AC + HM + HR + CR (70%) + AN (1000, 5 min WQ)		879	1194	25	
		FCC	AC + HM + HR + CR (70%) + AN (1000, 1 h AC)		799	1127	28	
		FCC	AC + HM + HR + CR (70%) + AN (1500, 1 h AC)		350	918	62	
19	$Al_{0.3}CoCrFeNi$	FCC + L12 + B2	AC + CR (90%) + SN (1150-60 min-WQ)	RT	159	410	65	[41]
			AC + CR (90%) + SN (1150, 5 min WQ)		220	550	60	
			AC + CR (90%) + SN (1150, 60 min WQ) + Aging (700, 50 h WQ)		215	520	43	
			AC + CR (90%) + SN (1150, 60 min WQ) + AG (550, 150 h WQ)		285	540	55	
			AC + CR (90%) + SN (1150, 2 min WQ)		263	589	60	
			AC + CR (90%) + SN (1150, 2 min WQ) + Aging (620, 50 h WQ)		490	840	45	

(Continued)

TABLE 7.1 *(Continued)*

Uniaxial Tensile Properties of TM HEAs, Phases Present, Processing Condition, Testing Temperature, YS (MPa), UTS (MPa), and Elongation (%)

S. No.	Alloy	Phases	Processing Condition	Testing Temperature	YS (MPa)	UTS (MPa)	Elongation (%)	Reference
20	$FeCoNiMn_{0.25}Al_{0.25}$	FCC	AC	RT	138	484	58	[42]
			AC + CR (70%)		623	1030	8	
			AC + CR (70%) + AN (900, 1 h WQ)		331	651	48	
21	$Fe_{40}Mn_{28}Ni_{28}Cr_4$	FCC	AC + HM (1000, 24 h)	RT	210	565	71	[43]
	$Fe_{40}Mn_{28}Ni_{20}Cr_{12}$		AC + HM (1000, 24 h)		245	545	66	
	$Fe_{40}Mn_{28}Ni_{14}Cr_{18}$		AC + HM (1000, 24 h)		310	605	54	
22	$(FeNiCrMn)_{95}Co_5$	FCC + σ	AC	RT	300	490	61	[44]
			AC + AN (850, 48 h)		315	468	12	
	$(FeNiCrMn)_{90}Co_{10}$	FCC	AC		295	536	65	
			AC + AN (850, 48 h)		303	475	32	
	$(FeNiCrMn)_{80}Co_{20}$		AC		290	516	69	
			AC + AN (850, 48 h)		290	500	62	
23	$Fe_{44}Mn_{20}Cr_{15}Ni_{7.5}Co_6Si_{7.5}$	FCC + SC	AC	RT	327	555	25	[19]
		FCC	AC + HT (1000 2.5 h WQ)		239	603	95	
		FCC	AC + HT + CR (70%) + AN (900, 15 min WQ)		402	873	110	
24	$Fe_{40}Mn_{20}Cr_{15}Co_{20}Si_5$	FCC + HCP	AC	RT	611	900	24	[45]
			AC + FSP		700	1286	46	
25	$Fe_{40}Mn_{20}Co_{20}Cr_{15}Si_5$	FCC + HCP	AC + FSP (S350)	RT	780	1132	29	[46]
			AC + FSP (S150)		725	1025	22	
			AC + FSP (D-pass)		662	1152	39	

26	$Fe_{39}Mn_{20}Co_{20}Cr_{15}Si_5Al_1$	FCC + HCP	AC + FSP (D-pass)	RT	703	1070	39	[22]
			AC + FSP (D-pass) + AN (650, 5 min)		682	1122	42	
			AC + FSP (D-pass) + AN (650, 15 min)		610	1110	50	
			AC + FSP (D-pass) + AN (650, 30 min)		680	1142	37	
			AC + FSP (D-pass) + AN (850, 5 min)		636	1053	27	
			AC + FSP (D-pass) + AN (850, 15 min)		542	1034	40	
			AC + FSP (D-pass) + AN (850, 30 min)		500	982	46	
27	$AlCoCrFeNi_{2.1}$	FCC + BCC	AM (PAAM)	RT	350	719	30	[20]
28	$Cu_{50}Mn_{30}Fe_{10}Co_{10}$	FCC1 + FCC2	AC	RT	400	630	36	[47]
29	$Fe_{38.5}Mn_{20}Co_{20}Cr_{15}Si_5Cu_{1.5}$	FCC	CR + HM (1100°C 1h)	RT	500	1140	27	[48]

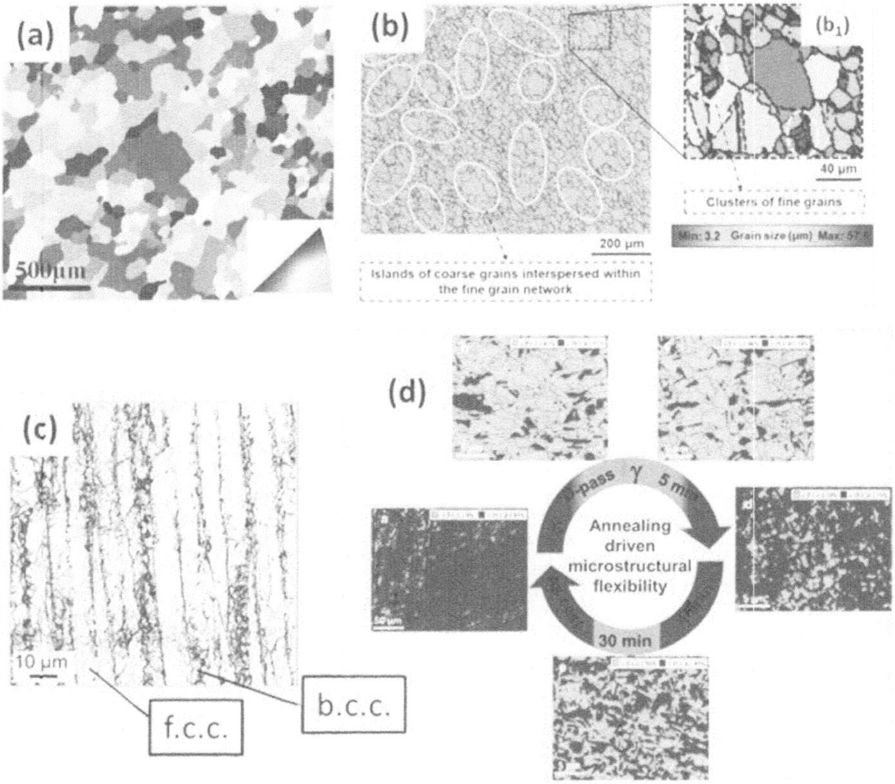

FIGURE 7.1 Initial microstructure of TM HEAs: (a) as-cast CrMnFeCoNi HEA [18], (b) rolled and annealed $Fe_{44}Mn_{20}Cr_{15}Ni_{7.5}Co_6Si_{7.5}$ HEA, (b_1) magnified region of fine grain clusters and coarse grains [19], (c) AM $AlCoCrFeNi_{2.1}$ HEA [20], (d) annealing-driven microstructural flexibility for FSP and annealed condition of $Fe_{39}Mn_{20}Co_{20}Cr_{15}Si_5Al_1$ HEA [22].

from ~3.2 μm to ~57.8 μm), and the clusters of fine grains around coarse grains resulted in a heterogeneous microstructure (shown in Figure 7.1 (b1)), resulting in good tensile strength and ductility.

In comparison to traditional manufacturing processes, additive manufacturing (AM) provides numerous benefits, such as the ability to produce complex or customized parts, the ability to design with flexibility, and the ability to experience ultrafast heating and cooling rates. Additionally, the microstructure of additively manufactured HEAs features fine grains and refined dendritic structure due to the rapid solidification. Dong et al. [20] devised a new powder-bed arc additive manufacturing (PAAM) approach to investigate AM processing to produce high-performance, crack-free HEAs. They used the PAAM approach to fabricate a pseudo-eutectic AlCoCrFeNi2.1 HEA consisting of FCC and body centered cubic (BCC) phases and optimized the number of deposition layers remelting. The middle region of build direction EBSD mapping (Figure 7.1 (c)) has columnar grains of FCC phase (~90 wt%) and fine dendritic grains of minor BCC phase (~10 wt%).

The tensile properties of HEAs are considerably enhanced by the refinement of their microstructure through SPD, which is an additional critical method for this purpose. SPD methods such as high-pressure torsion (HPT), equal channel angular pressing (ECAP), and friction stir processing (FSP) break down the microstructure into ultrafine grains, which increases the dislocation density, thereby enhancing yield strength (YS) and ultimate tensile strength (UTS) due to the Hall-Petch effect [21]. Additionally, SPD can improve ductility by promoting homogeneous deformation and reducing stress concentrations. The strength-ductility synergy achieved through SPD makes HEAs attractive for applications such as in the aerospace, automotive, and defense industries. In line with this, Nene et al. [22] incorporated FSP and subsequent annealing (temperature range of 650-850°C) in $Fe_{39}Mn_{20}Co_{20}Cr_{15}Si_5Al_1$ HEA, which resulted in an unexpected FCC (γ) $\rightarrow$ HCP (ε) transformation. This atypical phase evolution led to the formation of a refined, ε-dominated microstructure combined with a uniformly distributed fine γ phase (Figure 7.1 (d)).

Typical engineering stress-strain curves of as-cast CrMnFeCoNi HEA tested under tension at a wide temperature range (25-950°C) are shown in Figure 7.2 (a) [9]. This HEA shows substantial strain hardening from 25°C to 650°C, whereas, exhibits strain softening after 800°C, showing strong temperature sensitivity. For HEAs, each element might function as a solute atom and have the effect of sluggish diffusion. Consequently, a new mechanism was introduced, proposing that solute atoms do not need to propagate over extensive distances, but rather simply need to reorganize in the vicinity of the dislocation to capture the moving dislocation. In CrMnFeCoNi HEA, serrations in stress-strain curves are present in the range of 350-650°C, and the serration features vary with the change of temperature and strain. No serration occurs at room temperature as solute atoms find it difficult to overcome the energy barriers and rearrange themselves to lock the mobile dislocations. As the temperature rises, the atoms experience rapid vibrations, allowing them to overcome the energy barrier and rearrange themselves to immobilize moving dislocations, resulting in the formation of serrations. At a certain temperature, the formation of serration ceases to happen because the excessive thermal vibration of an atom is too great to effectively lock the dislocation. The key difference between CrMnFeCoNi HEA and conventional materials lies in the substantial enhancement of both strength and ductility as temperature decreases.

Figure 7.2 (b) shows room temperature tensile stress-strain curves in as-cast, homogenized, and TMP conditions of $Fe_{44}Mn_{20}Cr_{15}Ni_{7.5}Co_6Si_{7.5}$ HEA [10]. When compared with the as-cast and homogenized specimens, the TMP specimen exhibited excellent tensile behavior with a YS and UTS of ~400 $\pm$ 2 and ~860 $\pm$ 6 MPa, respectively, with ductility of (~106 $\pm$ 4%). The work hardening in the initial deformation stage results from hetero-deformation induced (HDI) strengthening as there is a mechanical incompatibility between small and large grains. This is then followed by extensive planar slip-induced hetero-deformation banding and deformation twinning in the later stages of tensile deformation, which results in uniform elongation of 82%. It is evident from the enhanced tensile properties after TMP that microsegregation and chemical inhomogeneity formed in an as-cast condition, which causes inferior tensile properties.

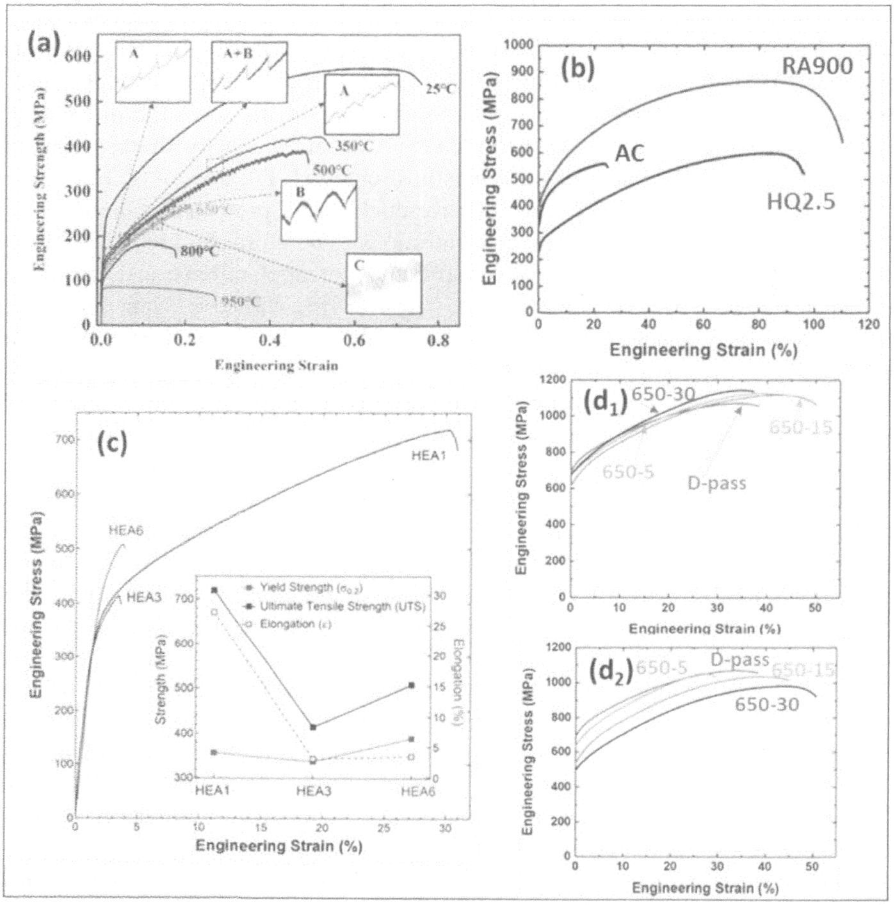

FIGURE 7.2 Tensile behavior of TM HEAs: (a) as-cast CrMnFeCoNi HEA [18], (b) as-cast, as-rolled, and annealed $Fe_{44}Mn_{20}Cr_{15}Ni_{7.5}Co_{6}Si_{7.5}$ HEA [19], (c) AM $AlCoCrFeNi_{2.1}$ HEA [20], and (d) FSP and annealed $Fe_{39}Mn_{20}Co_{20}Cr_{15}Si_{5}Al_{1}$ HEA [22].

HEAs produced using AM typically have improved tensile characteristics in comparison to their conventionally cast equivalents [23]. Tensile stress-strain curves of one-, three-, and six-time deposition remelted AlCoCrFeNi2.1 HEAs are shown in Figure 7.2 (b). The HEA1 (one-time deposition remelted) shows overall good comprehensive tensile properties with elongation to failure almost twice that of the casting counterpart [11]. The crack propagation behavior is linked to the columnar/dendritic microstructure with local misorientation localized near grain boundaries. Though grain boundary misorientation is insensitive to repeated remelting processing, tensile properties with an increasing number of remelting deteriorated. The EBSD analysis demonstrated that the accumulation and localization of thermally induced plasticity across the microstructure, as well as the appreciable growth of

BCC grains and the development of constrained dendritic morphology of BCC grains are the results of repeated remelting of deposited material. The observed degradation of ductility in the AlCoCrFeNi2.1 HEA after three- and six-time layer-remelting is believed to be caused by the localization of thermally induced plasticity in the vicinity of grain boundaries.

The untraditional behavior of the Al-HEA ($Fe_{39}Mn_{20}Co_{20}Cr_{15}Si_5Al_1$) is explained by the kinetic phase stabilization during FSP and the controlled transformation of the metastable matrix and twinning in the HCP (ε) phase during subsequent tensile deformation [13]. The controlled $\langle c+a \rangle$ slip and twinning in the ε phase, along with the transformation of a refined γ matrix, led to a significant increase in elongation of 52% and an improved ultimate tensile strength of 1.12 GPa during deformation (Figure 7.2 (d)). Although the ε (HCP) phase has fewer slip systems available for room temperature deformation, it still accommodates a significant amount of strain during extensive loading. Geometrically necessary dislocation (GND) measurements showed higher dislocation density values in ε (HCP) phase, which dictates the active involvement of this phase in plastic deformation. The Al-HEA demonstrates the ability to adjust its microstructure through FSP followed by annealing, which is achieved by self-tuning the matrix metastability via a combination of precipitation and grain growth events.

In conclusion, the tensile properties of TM HEAs are intricately connected to their microstructures, which exhibit substantial variations based on the specific processing techniques employed. As-cast HEAs typically exhibit poor tensile properties due to coarse dendritic structures, microsegregation, and residual stresses that lead to reduced strength and ductility. TMP HEAs experience improvements in their grain structures and dislocation density, leading to greater tensile strength and ductility. AM HEAs show improved tensile properties compared to as-cast alloys, attributed to fine-grained microstructures and reduced microsegregation, though their properties can be directionally dependent due to columnar grains aligned along the build direction. HEAs that have undergone SPD exhibit enhanced tensile performance due to the refinement of the grain structure and an increase in dislocation density. Each processing method has unique effects on the microstructure of HEAs, which in turn greatly affects their tensile behavior. Table 7.1 summarizes the tensile properties of recently worked HEAs, which are discussed briefly below.

Table 7.1 provides a comprehensive overview of the tensile properties of TM HEAs, showcasing their YS, UTS, and percentage elongation across various processing conditions. Diverse microstructural features such as single-phase solid solutions and multiphase mixtures result from differences in the type and concentration of elements present, whereas the extent of processing conditions affects the average grain size. The variations in processing methods and microstructures contribute to a wide range of tensile properties, highlighting the adaptability of these HEAs in tailoring strength and ductility for specific purposes. The table serves as a valuable resource for comparing how different manufacturing processes and microstructures influence the tensile behavior of TM HEAs, guiding their application in advanced engineering fields.

7.2.2 Tensile Properties of NTM HEAs

This HEA family mostly includes refractory HEAs, which consist of elements such as chromium (Cr), hafnium (Hf), molybdenum (Mo), niobium (Nb), tantalum (Ta), titanium (Ti), vanadium (V,) tungsten (W), and zirconium (Zr) [13]. Most NTM HEAs are designed with the motivation to develop new high-temperature structural materials as they possess elevated melting points [13]. NTM HEAs generally exhibit exceptional tensile strength but poor ductility at room temperature due to the presence of limited slip systems at lower temperatures [49]. In comparison with TM HEAs, the tensile properties of NTM HEAs are studied less frequently. In this subsection, tensile behavior of NTM HEAs is discussed in as-cast, thermomechanically processed, and severely plastic deformation conditions. Studies of tensile behavior in this family are not as mature as for TM HEAs. Table 7.2 summarizes the tensile properties of NTM HEAs.

Lightweight NTM HEAs exhibit limited formability when exposed to room temperature conditions. Poor formability is a significant restriction often addressed through post-fabrication processing and treatments, resulting in a considerable increase in cost. In response to this challenge, Balpande et al. [50] have developed a $Ti_{50}V_{16}Zr_{16}Nb_{10}A_{15}Mo_3$ HEA through suction casting, which can be used in the

TABLE 7.2
Uniaxial Tensile Properties of NTM HEAs: Phases Present, Processing Condition, Testing Temperature, YS (MPa), UTS (MPa), and Elongation (%)

S. No.	Alloy	Phases	Processing Condition	Testing Temperature	YS (MPa)	UTS (MPa)	Elongation (%)	Reference
1	TiZrHfNbTa	BCC	HM (1300, 5 h) + CR (70%) + AN (1200, 45 min)	−196°C	1547	1549	20	[55]
				4°C	875	885	22	
2	TiZrNbHfTa	BCC	AC + CR + HT (1100, 5 h)	RT	830	830	9	[53]
			AC + CR + HT (1100, 5 h) + HPT		1900	1900	8	
			AC + CR + HT (1100, 5 h) + HPT + AN (500, 1 h)		1520	1520	2	
			AC + CR + HT (1100, 5 h) + HPT + AN (800, 1 h)		795	795	5	
3	$Hf_{25}Nb_{25}Ti_{25}Zr_{25}$	BCC	AC	RT	879	969	15	[56]
4	HfNbTaTiZr	BCC	AC + CR (70%) + AN (GS:38 μm)	RT	958	974	20	[52]
			AC + CR (70%) + AN (GS:81 μm)		944	959	18	
			AC + CR (70%) + AN (GS:128 μm)		940	953	15	
5	$Ti_{50}Zr_{16}V_{16}Nb_{10}Al_5Mo_3$	BCC	AC	RT	1058	1241	35	[50]

as-cast condition for structural applications. An atypical equiaxed morphology mostly consisting of β-Ti (BCC) is observed in the as-cast state instead of the usual dendritic structure for this particular alloy as a result of alloy chemistry. The grain size is heterogeneously distributed (shown in Figure 7.3 (a)) with a limited presence of β-Zr phase (~3%).

Grain refinement is regarded as a method to enhance the tensile properties of metals and alloys. Moreover, controlling grain size in NTM HEAs is challenging compared to TM HEAs due to their slow diffusion rates [51]. Chien-Chang et al. [52] investigated the grain growth kinetics and tensile properties of HfNbTaTiZr HEA processed through cold rolling and subsequent annealing. This alloy exhibits a single-phase BCC structure, as shown in Figure 7.3 (b), which usually does not undergo a grain refining effect. In this alloy, the grain growth rate is controlled by

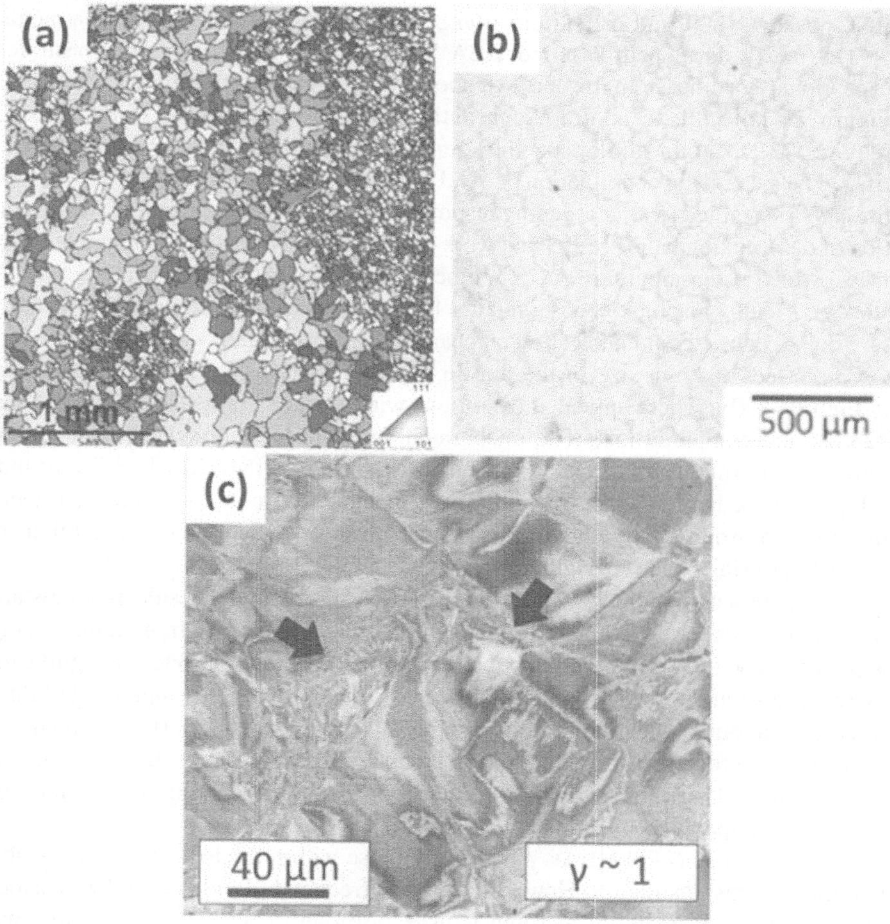

FIGURE 7.3 Initial microstructure of NTM HEAs: (a) IPF of as-cast $Ti_{50}V_{16}Zr_{16}Nb_{10}Al_5Mo_3$ HEA [43], (b) optical images of rolled and annealed HfNbTaTiZr HEA [45], and (c) backscatter image of the HPTed TiZrNbHfTa HEA [46].

the slowest components, i.e., Nb and Ta, as grain boundaries cannot migrate until the average composition is maintained. In other words, the species that move quickly must wait for the slower ones.

Refined structures increase the grain boundary area, which acts as a barrier to dislocation motion, leading to enhanced strength. However, due to poor ductility and very high strength, it is very difficult to process NTM HEAs through severe plastic deformation methods such as equal channel angular pressing (ECAP) and friction stir processing (FSP). However, these alloys can undergo high-pressure torsion (HPT) by the application of isostatic pressure during processing. Along similar lines, Scuh et al. [53] have processed TiZrNbHfTa by HPT to induce a nanocrystalline microstructure. HPT of a coarse-grained TiZrNbHfTa HEA leads to a substantial reduction in grain size (formation of substructures and localized deformation shown by arrows in Figure 7.3 (c)), reaching a minimum size of around 50 nm, while maintaining its single-phase nature, which decomposes into two different BCC phases (Nb-Ta and Zr-Hf rich) after being subjected to various heat treatments.

The newly developed Ti-based HEA $Ti_{50}V_{16}Zr_{16}Nb_{10}A_{15}Mo_3$ has demonstrated exceptional specific strength and extended ductility in its as-cast state (shown in Figure 7.4 (a)) [50]. In contrast, conventional alloys like Ti-6Al-4V and Ti-5553 require TMP to attain similar specific strength levels, albeit with only moderate ductility. The synergistic combination of YS, UTS, and ductility in this HEA is achieved through a sustained strain hardening response. This response is a result of the formation of deformation bands (DBs) and work hardening activity at the DB/matrix interfaces, which occur simultaneously in a heterogeneous-grained β microstructure. A steady softening response is observed in a heterogeneous equiaxed grain microstructure due to delayed void nucleation and early growth saturation.

The effect of grain size in the tensile behavior of HfNbTaTiZr HEA is shown in Figure 7.4 (b), where annealed conditions with average grain sizes of 38, 81, and 128 μm are compared [52]. The uncommon strength-ductility synergy with grain refinement in this HEA is related to Considere's criterion [54], which indicates that a higher work hardening rate would result in increased necking strain and necking strength. Additionally, the increased strength in smaller grain size conditions is attributed to Hall-Petch strengthening.

Further processing of NTM TiZrNbHfTa HEA through HPT results in a substantial enhancement in the strength of the material, with the tensile strength measuring approximately 1900 MPa. Perhaps more remarkable is that, despite the significant increase in strength, the specimens nevertheless maintain a rather high level of ductility, with a total elongation to failure of 7.9% (see Figure 7.4 (c)). Increased ductility in HEAs processed through HPT is attributed to a change in the deformation mechanism and potential alteration in the grain boundary condition (formation of DB), which impacts the mechanical characteristics [53].

To summarize, NTM HEAs, particularly those with refractory elements, exhibit remarkable tensile strength. However, they are frequently hindered by low plasticity and limited formability at room temperature due to their BCC structure and associated limited slip systems. Controlling grain size in these HEAs is challenging; however, HPT has shown promising results in refining grain size and a further improvement in tensile properties. Continued research into tensile behavior with

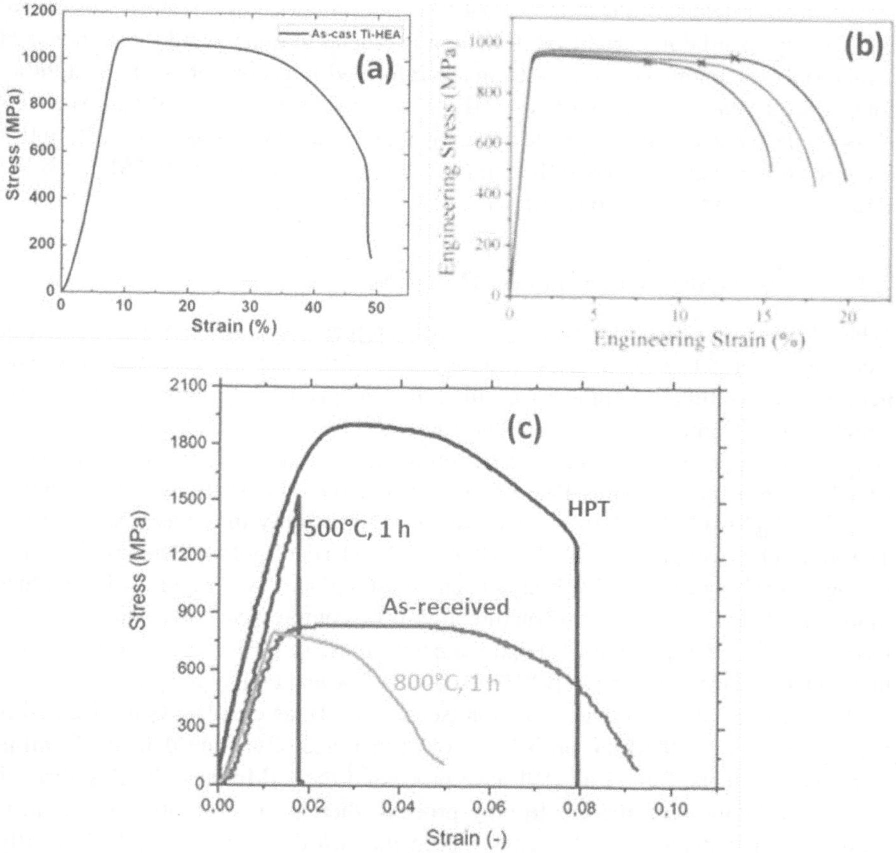

FIGURE 7.4 Tensile properties of NTM HEAs: (a) as-cast $Ti_{50}V_{16}Zr_{16}Nb_{10}Al_5Mo_3$ HEA [43], (b) rolled and annealed HfNbTaTiZr HEA [45], and (c) HPTed TiZrNbHfTa HEA [46].

different processing methods for NTM HEAs is crucial to fully harnessing their potential as advanced materials for challenging applications.

Table 7.2 summarizes the tensile properties of NTM HEAs, including YS, UTS, and percentage elongation across various processing conditions. These alloys exhibit diverse microstructural features, influenced by the type and concentration of refractory elements, which can include single-phase solid solutions and multiphase mixtures. This table highlights the variability in tensile properties due to different manufacturing methods and microstructures, emphasizing the potential for tailoring these materials for high-performance applications.

7.3 COMPRESSION PROPERTIES OF HEAs

Necking restricts uniform elongation during tension, rendering tension tests ineffective for examining plastic stress-strain relationships at elevated stresses, but higher strains can be achieved during compression loading. Moreover, compression testing reveals useful insights into failure modes, including buckling, yielding, and crushing, for any specific material [57]. Hence, it is crucial to understand the response of a material to compressive forces to make informed decisions about its suitability for particular engineering purposes. HEAs have recently been identified as a new class of materials that demonstrate advantageous compression properties [8]. In this section, authors have discussed the uniaxial compression behavior of TM and NTM HEAs fabricated through conventional and advanced processing routes.

7.3.1 Compression Properties of TM HEAs

Table 7.3 comprehensively summarizes the compression properties of TM-HEAs, including information on phases, processing conditions, compression testing temperature, YS, ultimate compressive strength, and percentage deformation for different alloys. Traditional alloys exhibit a low ability to withstand compression and adapt to strain because they have restricted methods of reinforcing and are prone to swift compression instability. Dutta et al. [58] emerged metastability engineering in $Fe_{44}Mn_{20}Cr_{15}Ni_{7.5}Co_6Si_{7.5}$ HEA to achieve super formability in the as-cast condition. The as-cast $Fe_{44}Mn_{20}Cr_{15}Ni_{7.5}Co_6Si_{7.5}$ HEA exhibited a dual-phase microstructure consisting of 94% γ-FCC and 6% β-simple cubic (SC) phases (see Figure 7.5 (a)) which transformed into a three-phase structure after 60% compressive deformation with the formation of ε-HCP phase (9.3%) and the nano-twins. At 90% compressive deformation γ-FCC reduced to 53.87%, β-SC increased to 7% and ε-HCP reduced to 5%.

As discussed in the previous section (section 7.2.1), as-cast HEAs are limited in applications due to the high probability of occurrence of casting defects. To minimize these defects, Sun et al. [59] have processed the CrMnFeCoNi alloy through hot forging. Following the hot forging process, there is a substantial reduction in grain size and a decrease in the scale of porosities within the material. Additionally, annealing twins emerged and the grains in the forged alloy no longer exhibited a conventional equiaxed structure, but instead displayed a characteristic deformed structure (shown in Figure 7.5 (b)).

Mechanical alloying (MA) combined with spark plasma sintering (SPS) enables rapid densification of as-milled powders and results in grain refinement, minimized porosities, homogeneous microstructure, phase stability, and overall load-bearing capacity in HEAs. Fu et al. [60] utilized MA followed by SPS techniques to incorporate these benefits into the CoNiFeCrAl0.6 Ti0.4 HEA, resulting in the formation of two FCC phases and a metastable BCC phase. The initial microstructure is depicted in Figure 7.5 (c).

Very few studies have been done to explore the compressive behavior of SPD HEAs. Weiss et al. [61] have used FSP on metastable $Fe_{42}Mn_{28}Co_{10}Cr_{15}Si_5$ HEA to achieve improved compressive strength and work hardening ability. The initial microstructure (after rolling followed by FSP and annealing) consists of γ-austenite and ε-martensite (see Figure 7.5 (d)), and the average grain size lies in the range of 8–90 μm in different processing methods.

$Fe_{44}Mn_{20}Cr_{15}Ni_{7.5}Co_6Si_{7.5}$ HEA showed maximum true stress in compression of 1.1 GPa in as-cast condition (see Figure 7.6 (a)). This exceptional strength is achieved through controlled hardening and sustained softening response during deformation.

TABLE 7.3
Uniaxial Compressive Properties: TM HEAs, Phases Present, Processing Conditions, Testing Temperature, YS (MPa), and Deformation (%)

S. No.	Alloy	Phases	Processing Condition	Testing Temperature	YS (MPa)	Deformation (%)	Reference
1	CrMnFeCoNi	FCC	Forged	−196°C	771	72	[59]
				−100°C	599	69	
				−50°C	584	69	
				25°C	458	70	
				650°C	392	69	
2	$CoNiFeCrAl_{0.6}Ti_{0.4}$	FCC1 + FCC2	AM	RT	2080	11.5	[60]
3	$Fe_{42}Mn_{28}Co_{10}Cr_{15}Si_5$	FCC	AC + CR + FSP150 (D-pass)		400	42	[61]
			AC + CR + FSP400		400	40	
			AC + CR + AN1100, 30 min		400	45	
			AC + CR + AN1100, 10 min		400	45	
4	$Ti_{0.5}CrFeCoNiCu_1$	FCC	AC	950°C	82	70	[62]
	$Ti_{0.5}CrFeCoNiAl_{0.25}Cu_{0.75}$				750	38.5	
	$Ti_{0.5}CrFeCoNiAl_{0.5}Cu_{0.5}$	FCC + BCC			1580	17.4	
	$Ti_{0.5}CrFeCoNiAl_{0.75}Cu_{0.25}$	BCC			1900	12	
5	CoCrFeNiTi	FCC				9	[63]
	$CoCrFeNiTiAl_{0.5}$	BCC + Laves phase				9.9	
	$CoCrFeNiTiAl_1$	BCC + Laves phase				6.4	
	$CoCrFeNiTiAl_{1.5}$	BCC1 + BCC2				9.8	
	$CoCrFeNiTiAl_2$	BCC 1 + BCC2				5.2	
6	CrFeCoNi	FCC			185	75	[64]
	$CrFeCoNiCu_{0.6}$	FCC1 + FCC2			360	75	

7	$Al_{0.7}CrFeMnNi_{1.3}$	BCC + B2	AC		1158	45	[65]
			AC + AN400		1164	46	
			AC + AN600		1160	46	
		BCC + B2 + σ	AC + AN700		1149	19	
			AC + AN800		1237	3	
			AC + AN900			2	
		BCC+ B2	AC + AN1000		1023	52	
8	$Fe_3CrCoNiAl_{0.6}$	FCC + BCC	AC		900	40	[66]
	$Fe_3CrCoNiAl_{0.7}$	FCC + BCC			1228	40	
	$Fe_3CrCoNiAl_{0.8}$	BCC + B2			1324	36	
	$Fe_3CrCoNiAl_{1.0}$	BCC + B2			1428	31	
	$Fe_3CrCoNiAl_{1.1}$	BCC+ B2			1420	29	
	$Fe_3CrCoNiAl_{1.2}$	BCC + B2			1464	20	
9	FeCoNiCuAl	FCC + BCC	AC	RT		9	[67]
	FeCoNiCuAlSi					10	
	FeCoNiCuAlCr					16	
	FeCoNiCuAlTi					11	
	FeCoNiCuAlZr	FCC +BCC + ZrFe3Al					
	FeCoNiCuAlNd						
10	$Co_{0.5}FeNiCrTi_{0.5}$	FCC + BCC			2650	10	[68]
11	$FeNiCrCo_{0.3}Al_{0.7}$	FCC + BCC			2033	8	[69]
12	AlCoCrFeNi	BCC	AC		1378	22	[70]
	$AlCoCrFeNiV_{0.2}$	BCC			1492	26	
	$AlCoCrFeNiV_{0.5}$	BCC			1597	5.1	
	$AlCoCrFeNiV_{0.8}$	BCC			1696	5.7	
	$AlCoCrFeNiV_{1.0}$	BCC			1726	0.7	

(Continued)

TABLE 7.3 *(Continued)*
Uniaxial Compressive Properties: TM HEAs, Phases Present, Processing Conditions, Testing Temperature, YS (MPa), and Deformation (%)

S. No.	Alloy	Phases	Processing Condition	Testing Temperature	YS (MPa)	Deformation (%)	Reference
13	TiVCrMnFeCoNiCu	FCC + BCC + σ			1312	0	[71]
	$Al_{11.1}$(TiVCrMnFeCoNiCu)$_{88.9}$	FCC + BCC			1862	0.95	
	Al_{20}(TiVCrMnFeCoNiCu)$_{80}$	BCC			1465	2.35	
	Al_{40}(TiVCrMnFeCoNiCu)$_{60}$	BCC + Al3Ti			1461	0	
14	AlCoCrFeNi	BCC			1110	-	[72]
	$AlCoCrFeNiSi_{0.2}$	BCC			1265	13.76	
	$AlCoCrFeNiSi_{0.4}$	BCC			1481	13.38	
	$AlCoCrFeNiSi_{0.6}$	BCC			1834	2.56	
	$AlCoCrFeNiSi_{0.8}$	BCC			2179	1.77	
	$AlCoCrFeNiSi_{1.0}$	BCC + σ			2411	1.77	
15	$Al_{0.5}CoCrFeCuNi$	FCC1 + FCC2	AC		487	40	[73]
		FCC1 + FCC2 + AlNi rich BCC	AC + AN600		600	40	
		FCC1 + FCC2 + AlNi rich BCC	AC + AN800		390	40	
		FCC1 + FCC2	AC + AN1000		268	40	
16	$CrFeNiTi_{0.25}$	Unknown	AC		1309	20.4	[74]
	$Al_{0.25}CrFeNiTi_{0.25}$	FCC + BCC			1390	33.4	
	$Al_{0.5}CrFeNiTi_{0.25}$	BCC			1879	40	
	$Al_{0.75}CrFeNiTi_{0.25}$	BCC			1726	20	
	$Al_{1.0}CrFeNiTi_{0.25}$	BCC			1444	7.2	

17	CoCrFeMnNi	FCC	AC + AN1000-24h	230	>75	[75]
	$CoCrFeMnNiV_{0.25}$	FCC + Tetragonal	AC + AN1000-24h	200	>75	
	$CoCrFeMnNiV_{0.5}$	FCC+ Tetragonal	AC + AN1000-24h	620	>75	
	$CoCrFeMnNiV_{0.75}$	FCC + Tetragonal	AC + AN1000-24h	740	7.8	
	$CoCrFeMnNiV_{1.0}$	FCC + Tetragonal	AC + AN1000-24h	1660	0.5	
18	CoCrCuFeNi	FCC	AC	230	0.56	[76]
	$CoCrCuFeNiTi_{0.5}$	FCC + Laves		700	0.8	
	$CoCrCuFeNiTi_{0.8}$	FCC + Laves		1042	0.89	
	$CoCrCuFeNiTi_{1.0}$	FCC + Laves		1272	1.6	
19	AlCoCrFeNi	BCC		1500	26.9	[77]
	$AlCoCrFeNiTi_{0.5}$	BCC		2260	23.3	
	$AlCoCrFeNiTi_{1.0}$	BCC		1860	8.8	
	$AlCoCrFeNiTi_{1.5}$	BCC + Laves phase		2220	5.3	
20	$Fe_{44}Mn_{20}Cr_{15}Ni_{7.5}Co_{6}Si_{7.5}$	FCC + SC		500	80	[58]
21	AlCrFeNi	BCC1 + BCC2		1406	28.5	[78]
	$AlCrFeNiMo_{0.2}$	BCC1 + BCC2		1487	28.7	
	$AlCrFeNiMo_{0.5}$	BCC1 + BCC2		1748	12.6	
	$AlCrFeNiMo_{0.8}$	BCC + σ		1512	0	
22	CoCrFeNi	FCC		190	>75	[79]
	CoCrFeNiV	FCC + Tetragonal		1435	2.5	
	CoCrFeNiMnV	FCC + Tetragonal		1660	0.5	

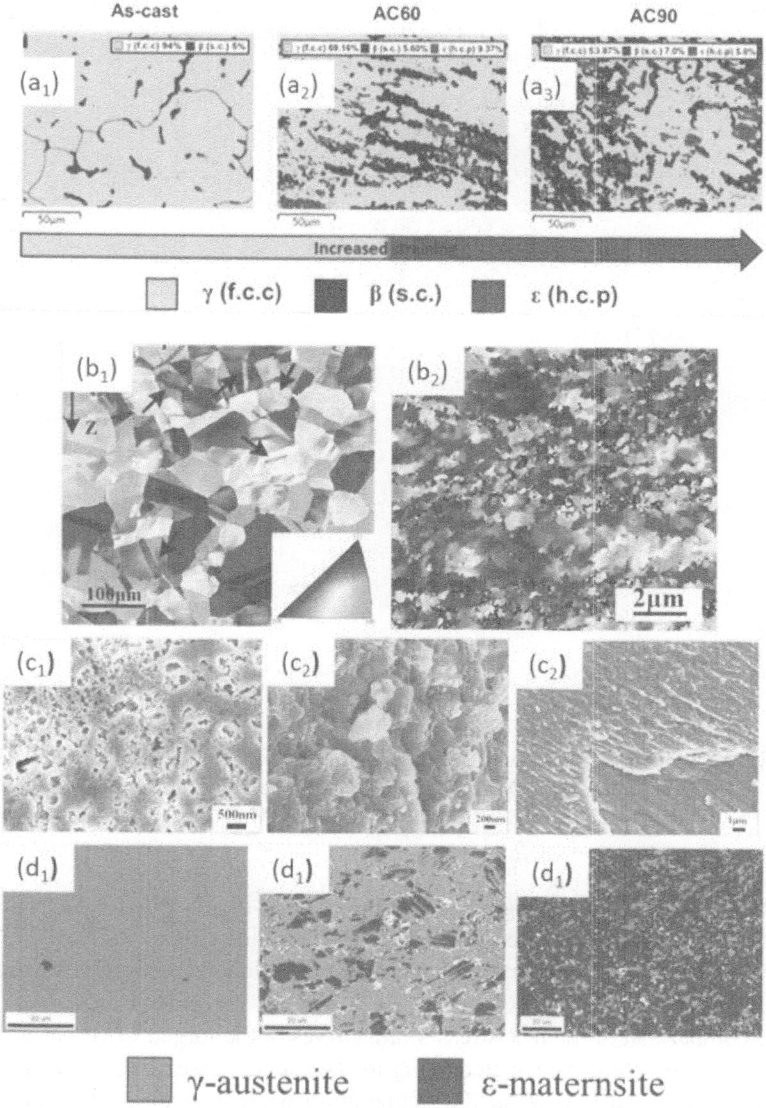

FIGURE 7.5 Microstructural evolution during compressive loading on TM HEAs, (a_1-a_3) as-cast, 60%, 905 compressed $Fe_{44}Mn_{20}Cr_{15}Ni_{7.5}Co_6Si_{7.5}$ HEA [54], (b_1-b_2) forged and compressed CrMnFeCoNi HEA [55], (C1-C3) as-SPS and after compression fracture $CoNiFeCrAl_{0.6}Ti_{0.4}$ HEA [56], (d1-d3) FSP and after 105 and 30% compressed $Fe_{38.5}Mn_{20}Co_{20}Cr_{15}Si_5Cu_{1.5}$ HEA [57].

The controlled hardening phenomenon, which occurs up to a strain of 60%, is caused by the combination of transformation-induced plasticity (TRIP) in the γ-phase and nano-twinning in the untransformed γ-phase. The sustained softening is facilitated by extensive strain accommodation of both the γ and β phases, with the γ phase undergoing dynamic recovery and the β phase undergoing shearing during deformation.

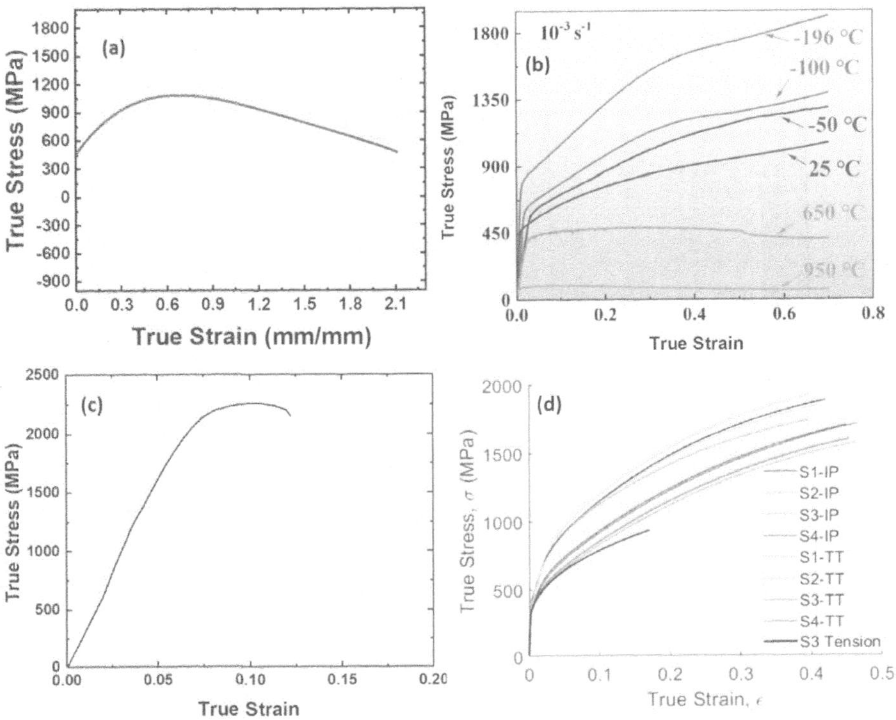

FIGURE 7.6 Uniaxial compressive behavior of TM HEAs, (a) as-cast $Fe_{44}Mn_{20}Cr_{15}Ni_{7.5}Co_6Si_{7.5}$ HEA [55], (b) forged CrMnFeCoNi HEA at temperature range (–196 to 950°C) [56], (c) as-SPS $CoNiFeCrAl_{0.6}Ti_{0.4}$ HEA [57], and (d) FSP $Fe_{38.5}Mn_{20}Co_{20}Cr_{15}Si_5Cu_{1.5}$ HEA [58].

The combination of hardening and softening in $Fe_{44}Mn_{20}Cr_{15}Ni_{7.5}Co_6Si_{7.5}$ HEA also allows it to be super formable with a strength-formability index (SFI) of 2717 MPa at a true strain of 2.2, which makes it a promising option for cost-effective structural applications [54].

The comparison of the true stress-strain curves of the forged CrMnFeCoNi HEA at various temperatures is depicted in Figure 7.6 (b). It is observed that the YS exhibits a noticeable increase when the temperature decreases during the quasi-static compression tests. It is important to emphasize that the YS reaches up to 771 MPa at the cryogenic temperature of –196°C and the specimen does not experience crushing, indicating exceptional flexibility. During compression at low temperatures, the atoms have a reduced ability to undergo thermal activation, resulting in increased resistance to the motion of dislocations, emerging in the increase in YS of forged CrMnFeCoNi at cryogenic temperature [55].

The true stress-strain curve of bulk $CoNiFeCrAl_{0.6}Ti_{0.4}$ HEAs prepared through SPS is presented in Figure 7.6 (c), where the YS is 2.08 GPa. The presence of Al and Ti with higher atomic radii in the lattice sites leads to a significant increase in lattice distortion energy, ultimately amplifying the effect of solid solution strengthening. Moreover, the presence of nanoscale twins, acting as strong barriers for dislocation motion, can be considered equivalent to grain boundaries with respect to the Hall-Petch effect. Hence, the exceptional compressive properties are the result of nanoscale twins and solid solution strengthening, along with optimal concentrations of all constituent elements [56].

The probability of achieving a fine-grain structure after FSP is very high for $Fe_{38.5}Mn_{20}Co_{20}Cr_{15}Si_5Cu_{1.5}$ HEA. Moreover, processing parameters also affect its microstructure as follows: dual-pass FSP at 150 RPM resulted in more refined grains than FSP at 400 RPM, thereby showing compressive strength of nearly 2 GPa in the former processing condition. At 150 RPM FSP, the material exhibited rapid hardening as the smallest initial grain size complemented with the dynamic Hall-Petch effect, unlike at 400 RPM, which provides higher heat input, thereby promoting recovery and recrystallization. The compressive true stress-strain curve of FSP followed by annealing is shown in Figure 7.6 (d), where S1 and S2 are FSP samples processed at 150 RPM and 400 RPM, respectively [57].

Compressive properties of as-cast and rolled HEAs are generally influenced by the inherent phase composition, whereas AM HEAs show unique microstructures due to rapid solidification, resulting in fine grains and low porosity, however, challenges such as residual stresses and anisotropic behavior of HEAs fabricated through AM can affect formability under compressive load. On the other hand, FSP of TM HEAs improves the grain structure and equalizes the distribution of phases, resulting in increased compressive strength and work-hardening capability. The process of dynamic recrystallization and texture development that occurs during FSP leads to an improved ability to withstand plastic strain and resistance to failure under compression, particularly in metastable HEAs that are susceptible to martensitic transformation. Table 7.3 presents a detailed summary of the compressive characteristics of TM HEAs. It displays the YS and percentage deformation during uniaxial compressive loading of HEAs processed through different methods.

7.3.2 Compression Properties of NTM HEAs

Alloys of this category have garnered attention as promising candidates for high-temperature structural applications. Due to their ability to maintain excellent compression properties at temperatures between 1000–1200°C, these materials are well suited for aerospace-grade applications [13]. Additionally, researchers have directed their attention toward minimizing the density of NTM-HEAs, which is a crucial necessity for aerospace materials. Shen et al. [80] developed and examined the compressive behavior of lightweight $TiZrTa_xNbMo$ HEAs that display a dendritic microstructure primarily consisting of a BCC1 phase and a secondary BCC2 phase. As Ta content decreases, the volume fraction of the BCC1 phase reduces, causing the dendritic arm structure to shift from dense to more dispersed (shown in Figure 7.7 (a)). This change in Ta concentration alters the stability and morphology of the phases, impacting the alloy's solidification behavior and mechanical properties.

Senkov et al. [81] modified earlier-developed HfNbTaTiZr and $CrMo_{0.5}NbTa_{0.5}TiZr$ HEAs by the addition of Al and produced lightweight $Al_{0.4}$Hf0.6NbTaTiZr and $AlMo_{0.5}NbTa_{0.5}TiZr$ HEAs through hot isostatic pressing (HIP) and subsequent annealing. The $AlMo_{0.5}NbTa_{0.5}TiZr$ HEA consists of dual BCC phases, whereas

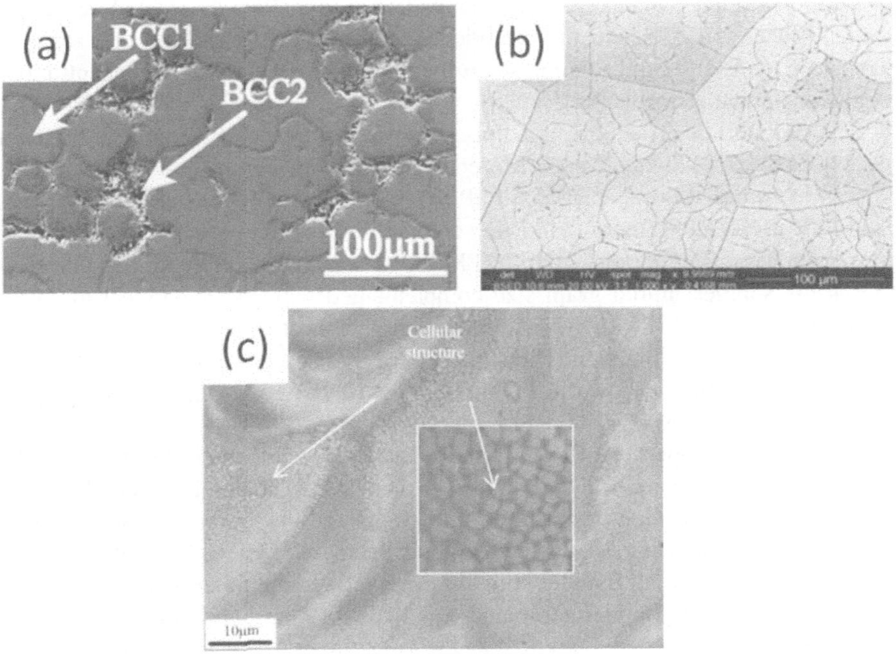

FIGURE 7.7 Initial microstructures of NTM HEAs, (a) as-cast $TiZrTa_{1.0}NbMo$ HEA [80], (b) HIPed $AlMo_{0.5}NbTa_{0.5}TiZr$ HEA [81], and (c) LPBF $Ti_{1.5}Nb_1Ta_{0.5}Zr_1Mo_{0.5}$ HEA [82].

the $Al_{0.4}Hf_{0.6}NbTaTiZr$ HEA has a single-phase BCC structure (Figure 7.7 (b)). Replacing Cr with Al reduced the density in $CrMo_{0.5}NbTa_{0.5}TiZr$, whereas partial substitution of Hf with Al in HfNbTaTiZr eliminated the formation of a brittle Laves phase. AM of NTM HEAs is challenging due to the wide range of melting points of constituent elements; hence, information on AM of NTM HEAs is limited. Liu et al. [82] fabricated $Ti_{1.5}Nb_1Ta_{0.5}Zr_1Mo_{0.5}$ by AM (laser powder bed fusion, LPBF), which is characterized by distinctive cellular structures (shown in Figure 7.7 (c)), and explored its compressive behavior.

A reduction in Ta content in TiZrTaxNbMo HEAs leads to a significant improvement in compressive plasticity. The $Ta_{0.7}$ alloy exhibits the highest plasticity, reaching 34.8%, while maintaining YS of 1253 MPa (shown in Figure 7.8 (a)). This enhancement in plasticity, alongside high-strength retention, highlights the favorable mechanical properties of the alloy with optimized alloy composition. During the stress-loading process, the formation of deformation bands is facilitated in the brittle dendritic zone. As the deformation bands stretched further, they encountered resistance from the interdendritic area, which exacerbated plasticity [80]. $AlMo_{0.5}NbTa_{0.5}TiZr$ and $Al_{0.4}Hf_{0.6}NbTaTiZr$ HIPed and annealed NTM HEAs have shown enhanced compressive properties at varied temperatures (Figure 7.8 (b1 and b2)) in the range of (23°C–1200°C), where compressive strength decreased with increasing temperature. Elongation of the grains was observed in both HEAs in the direction of plastic flow during the compressive loading [81].

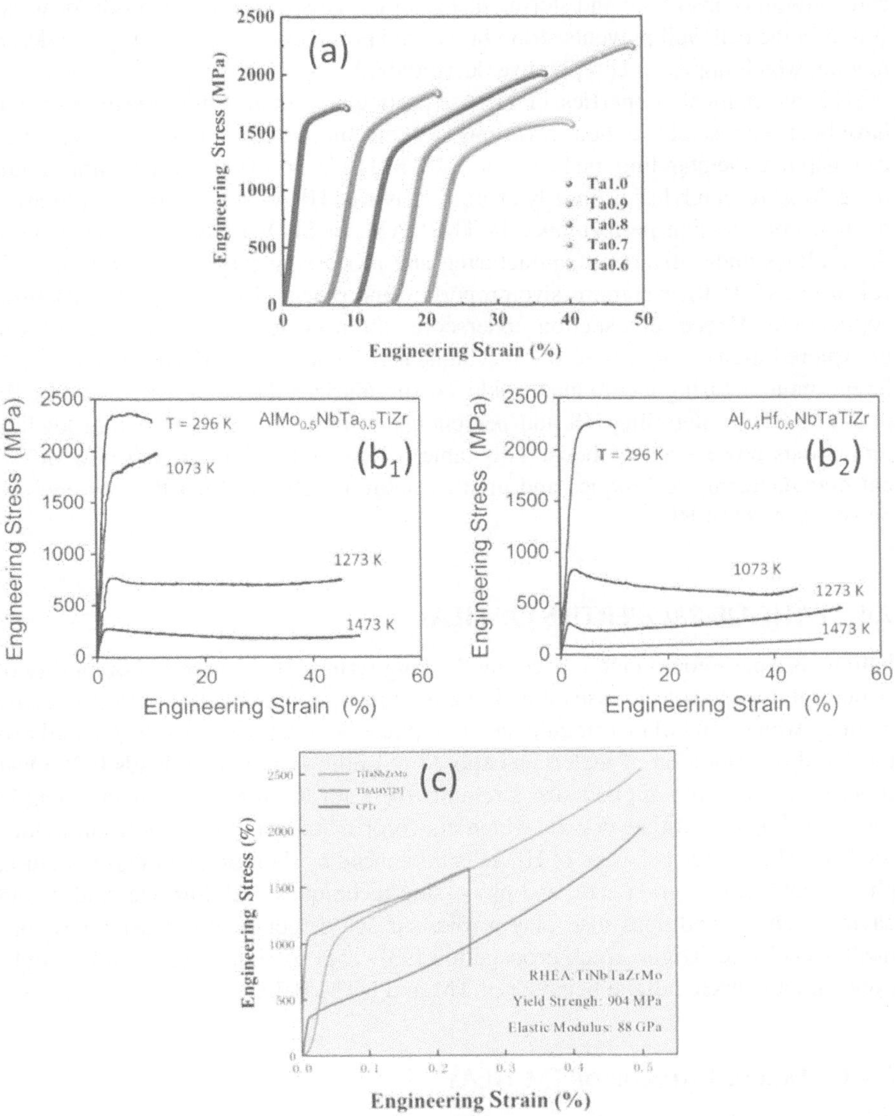

FIGURE 7.8 Uniaxial compressive behavior of NTM HEA, (a) as-cast $TiZrTa_{1.0}NbMo$ HEA [80], (b_1-b_2) HIPed $AlMo_{0.5}NbTa_{0.5}TiZr$ and $AlHf_{0.6}NbTaTiZr$ HEAs [81], and (c) LPBF $Ti_{1.5}Nb_1Ta_{0.5}Zr_1Mo_{0.5}$ HEA [82].

AM has been proven to be an effective approach for achieving excellent compressive properties of $Ti_{1.5}Nb_1Ta_{0.5}Zr_1Mo_{0.5}$ HEA (YS; 1000 MPa, shown in Figure 7.8 (c)). The superior compressive behavior of this HEA is due to the combined effect of hierarchical heterogeneous microstructure, dislocations, grain boundaries, and cellular structures. In the initial stage of deformation, the cell walls impede dislocation movements, which increases the YS; moreover, when the dislocation density increases, cell walls start absorbing and storing dislocations. This characteristic of dislocations stored in the cell wall prevents strain bursts and promotes homogeneous grain deformation, which improves compressive ductility [82].

The mechanical properties of HEAs, particularly their compressive behavior, have been extensively studied in various systems; moreover, there remains a significant gap in understanding the behavior of NTM HEAs in different processing conditions. Most research has primarily examined as-cast HEAs, with limited exploration of their compression performance in TMP, AM, or SPD conditions. Investigating these alloys under diverse manufacturing and processing routes could reveal critical insights into their compressive properties and potential for advanced structural applications. Hence, this section underscores the need for further studies in these unexplored areas to optimize the mechanical behavior of NTM HEAs across different manufacturing techniques. Table 7.4 summarizes the compressive properties of NTM HEAs, detailing YS and percentage deformation under uniaxial loading for various processing methods. This table is essential for comparing how different manufacturing techniques and microstructural features affect the compressive behavior of NTM HEAs.

7.4 FATIGUE PROPERTIES OF HEAs

Fatigue is a paramount factor affecting the long-term performance and durability of structural materials in industrial and engineering applications. In contrast to static loading, which can fail in a singular stress application, fatigue entails the gradual and localized deterioration of materials exposed to cyclic or repetitive loads [92]. Over time, even when the applied stress remains beneath the material's yield strength, cyclic loading can initiate cracks, which may propagate and ultimately result in failure [57]. The fatigue behavior of HEAs is influenced by their unique microstructure, phase composition, grain size, and processing techniques [92]. Surface quality and environmental conditions also play a role, surface defects and corrosive environments accelerate fatigue crack propagation [93]. This section delves into the high-cycle and low-cycle fatigue behavior of TM and NTM HEAs.

7.4.1 Fatigue Behavior of TM HEAs

The majority of research on HEAs concentrates on the high-cycle (stress-controlled) fatigue (HCF) characteristics of single and multiphase alloys, while low-cycle (strain-controlled) fatigue (LCF) has been minimally investigated [7]. The HCF life is determined by the Basquin equation (eq 7.1), in which predetermined stress amplitudes are applied and the cycles to failure are recorded [94].

$$\Delta\sigma_a = \sigma'_f (N_f)^{-b} \tag{7.1}$$

Where $\Delta\sigma_a$ is the applied stress amplitude, σ'_f is the fatigue-strength coefficient, N_f is the number of cycles to failure, and b is the fatigue strength exponent.

TABLE 7.4
Uniaxial Compressive Properties: NTM HEAs, Phases Present, Processing Conditions, Testing Temperature, YS (MPa), and Deformation (%)

S. No.	Alloy	Phases	Processing Condition	Testing Temperature	YS (MPa)	Deformation (%)	Reference
1	TiNbCrTa	BCC1 + BCC2 + Laves phase	AC	RT	1400	10	[83]
2	$TiZrTa_{1.0}NbMo$	BCC1 + BCC2	AC	RT	1536	9	[80]
	$TiZrTa_{0.9}NbMo$	BCC1 + BCC2		RT	1402	17	
	$TiZrTa_{0.8}NbMo$	BCC1 + BCC2		RT	1317	26	
	$TiZrTa_{0.7}NbMo$	BCC1 BCC2		RT	1253	34	
	$TiZrTa_{0.6}NbMo$	BCC1 + BCC2		RT	1168	20	
3	WTaVCrTi	BCC + TiO0.5	AC	RT	1628	8.5	[84]
		BCC + TiO0.6		400°C	1370	12.2	
		BCC + TiO0.7		600°C	1215	12.4	
		BCC + TiO0.8		800°C	1055	14.4	
4	ZrNbAlVTi	BCC B2 + Al3Z4 + Al2Ti + Zr2Al	Induction Levitation melting	RT	1045	25	[85]
5	$Nb_{25}Mo_{25}Ta_{25}W_{25}$	BCC	AC	RT	1058	2.5	[86]
				600°C	561	>25	
				800°C°C	552	>25	
				1000°C	548	>25	
				1200°C	506	>25	
				1400°C	421	>25	
				1600°C	405	>25	

(Continued)

TABLE 7.4 *(Continued)*
Uniaxial Compressive Properties: NTM HEAs, Phases Present, Processing Conditions, Testing Temperature, YS (MPa), and Deformation (%)

S. No.	Alloy	Phases	Processing Condition	Testing Temperature	YS (MPa)	Deformation (%)	Reference
	$V_{20}Nb_{20}Mo_{20}Ta_{20}W_{20}$	BCC	AC	RT	1246	1.7	
				600°C	862	13	
				800°C	846	17	
				1000°C	842	19	
				1200°C	735	7.5	
				1400°C	656		
				1600°C	477		
6	$Al_{0.4}Hf_{0.6}NbTaTiZr$	BCC	AC + (HIP 1200°C) + AN 1200°C, 24 h	RT	2000	10	[81]
				800°C	1597	11	
				1000°C	745	>50	
				1200°C	250	>50	
	$AlMo_{0.5}NbTa_{0.5}TiZr$	BCC + B2	AC + (HIP 1400°C) + AN 1400°C	RT	1841	10	
				800°C	796	>50	
				1000°C	298	>50	
				1200°C	89	>50	

7	$Al_{0.3}NbTa_{0.8}Ti_{1.4}V_{0.2}Zr_{1.3}$	BCC	AC, HIP 1200°C, AN 1200°C, 24 h	RT	1965	5	[87]
				800°C	678	>50	
				1000°C	166	>50	
	$Al_{0.5}NbTa_{0.8}Ti_{1.5}V_{0.2}Zr$	BCC1 + BCC2	AC, HIP 1200°C, AN 1200°C, 24 h	RT	2035	4.5	
				800°C	796	>50	
				1000°C	220	>50	
	$Al_{0.3}NbTaTi_{1.4}Zr_{1.3}$	BCC1 + BCC2	AC, HIP 1200°C, AN 1200°C, 24 h	RT	1965	5	
				800°C	362	>50	
				1000°C	236	>50	
	$AlNb_{1.5}TaTi_{1.5}Zr_{0.5}$	BCC	AC, HIP 1200°C, AN 1200°C, 24 h	RT	1280	3.5	
				800°C	728	30	
				1000°C	403	>50	
8	AlNbTiV	BCC	AC, AN 1200°C, 24 h	RT	1020	5	[88]
				600°C	810	12	
				800°C	685		
				1000°C	158		
9	HfMoNbTiZr	BCC	AC	RT	1719	10.12	[89]
			AC + AN 1100°C, 10 h	RT	1575	9.08	
				800°C	825	>60	
				900°C	728	>60	
				1000°C	635	>60	
				1100°C	397	>60	
				1200°C	187	>60	
10	HfNbTiZr	BCC	AC + AN 1300°C, 6 h	RT	879	14.3	[56]

(Continued)

TABLE 7.4 *(Continued)*
Uniaxial Compressive Properties: NTM HEAs, Phases Present, Processing Conditions, Testing Temperature, YS (MPa), and Deformation (%)

S. No.	Alloy	Phases	Processing Condition	Testing Temperature	YS (MPa)	Deformation (%)	Reference
11	NbTiVZr	BCC	AC + HIP 1200°C + AN 1200, 24 h	RT	1105	>50	[90]
				600°C	834	>50	
				800°C	187	>50	
				1000°C	58	>50	
	$NbTiV_2Zr$	BCC	AC + HIP 1200°C + AN 1200°C, 24 h	RT	918	>50	
				600°C	571	>50	
				800°C	240	>50	
				1000°C	72	>50	
12	$Ti_{1.5}Nb_1Ta_{0.5}Zr_1Mo_{0.5}$	BCC	AM	RT	904	50%	[82]
13	NbMoTaTi	BCC	SLM	RT	1201	23	[91]
	NbMoTaNi			RT	1350	11	
	$NbMoTaTi_{0.5}Ni_{0.5}$			RT	1750	15	
				600°C	1279	28	
				800°C	756	28	
				1000°C	554	11	

The low-cycle fatigue life is determined via the Coffin-Manson equation (eq 7.2), wherein predetermined strain amplitudes are applied to the specimen, and the cycles to failure are recorded [94].

$$\Delta\epsilon_p = \epsilon_f' \, (N_f)^c \tag{7.2}$$

Where $\Delta\epsilon_p$ is the applied strain amplitude, ϵ_f' is the fatigue-ductility coefficient, N_f is the number of cycles to failure, and c is the fatigue-ductility exponent.

The most prominent study of fatigue has been done on the single-phase FCC equiatomic CoCrFeMnNi HEA as it possesses high strength-ductility synergy and outstanding fracture toughness. Studies show that the fatigue behavior of the CoCrFeMnNi HEA improves when the grain size is reduced to the lower side of the micron scale [95]. According to Lee and coworkers [93], dislocation twins increase the yield strength and additionally improve the fatigue properties of CoCrFeMnNi HEA, but the formation of numerous deformation twins results in micro-voids and eventually swift the formation of fatigue cracks. Hu et al. [96] have studied the fatigue behavior of CoCrFeMnNi HEA in as-cast conditions to investigate the fatigue crack initiation mechanism. The as-cast alloy showed FCC phase, arranged in columnar grains with an average grain size of 250 μm (see Figure 7.9 (a)). It also contains oxides of Cr-Mn and casting defects, such as gas pores and shrinkage porosity, which are detrimental to fatigue properties. As discussed in earlier sections, composition optimization and processing conditions are two ways of enhancing the mechanical properties of HEAs. In similar lines, Liu et al. [97] have explored $Al_{0.3}CoCrFeNi$ HEA with optimized Al content to achieve BCC and σ phases in FCC matrix further processed through cold rolling and subsequent annealing to pursue an ultrafine grain structure. The initial microstructure (see Figure 7.9 (b)) displays fine BCC (14%) and σ (5%), formed simultaneously at the grain interior and boundary during annealing. Li et al. [98] in their study observed that crack initiation was postponed because of BCC Precipitates obstructed the path of persistent slip bands in a heat-treated eutectic $AlCoCrFeNi_{2.1}$ HEA. These studies underline the critical role of phase distribution and processing techniques in improving the fatigue behavior of HEAs. Fatigue behavior and deformation mechanism of AM (selective laser melting) CoCrFeMnNi HEA have been studied by Kim et al. [99]. The AM alloy has an FCC single-phase heterogeneous grain structure and dislocation network induced sub-structure with dispersed nanosized oxides (see Figure 7.9 (c)).

Previous research on the fatigue behavior of transformation-induced plasticity (TRIP) steels indicated that the damage accumulation commences only when austenite transforms into martensite, influencing the void nucleation mechanism along with it [100]. Along similar lines, $\gamma \rightarrow \varepsilon$ transformation at the fracture tip is advantageous for inhibiting crack propagation due to the significant strain mismatch at the γ/ε interfaces, which includes a localized work hardening effect. Liu et al. [101] utilized controlled transformation and twinning-assisted deformation mechanisms while studying the fatigue behavior of FSP $Fe_{42}Mn_{28}Cr_{15}Co_{10}Si_5$ (metastable) HEA, where Si acted as ϵ phase stabilizer and assisted in grain refinement during FSP (see Figure 7.9 (d)). Multiple passes of FSP effectively induced a fine microstructural evolution, leading to a reduction in grain size and concurrent redistribution of phases.

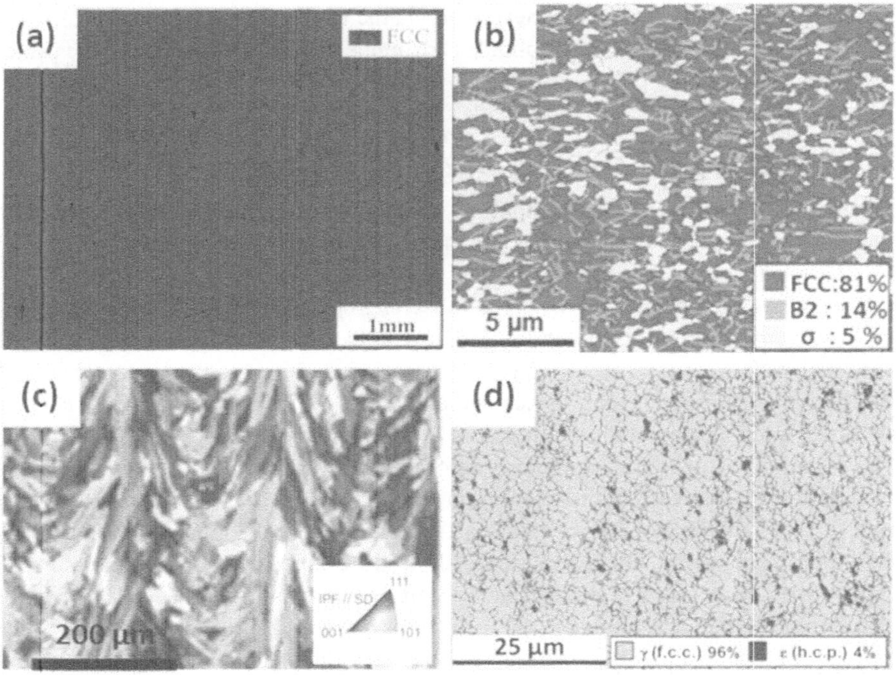

FIGURE 7.9 Initial microstructure of TM HEAs: (a) as-cast CoCrFeNiMn HEA [96], (b) TMP Al0.3CoCrFeNi HEA [97], (c) AM CoCrFeMnNi HEA [99], and (d) FSP Fe42Mn28Cr15Co10Si5 HEA [101].

The as-cast CoCrFeMnNi HEA has shown uniaxial tensile YS and UTS as 234 MPa and 603 MPa, respectively. The stress amplitude-number of cycles to failure (S-N) curve is depicted in Figure 7.10 (a); the alloy exhibits a long fatigue life at low-stress amplitudes (< 240 MPa), whereas it shows a shorter fatigue life at higher stress amplitudes (> 240 MPa) [96]. The cracks were initiated by the casting defects and crystallographic facets during fatigue testing. Additionally, segregations are observed near the shrinkage pores, which cause stress concentrations due to heterogeneity in the microstructure. Some defects aided the formation of crystallographic facets, originating from the matrix; the fatigue micro-cracks tend to initiate near a triple junction grain boundary. Commonly, {1 1 1} slip planes are preferential for fatigue crack initiation for this alloy but this study proved that the {1 1 0} plane may also activate it.

$Al_{0.3}$CoCrFeNi HEA exhibits remarkable tensile properties (YS: 900 MPa, UTS: 1074 MPa) in the TMP state, with grain boundary strengthening and composite strengthening from hard phases being significant factors. Fatigue specimens were tested up to the stress level of 680 MPa, which is ~76% of UTS (S-N curve is given in Figure 7.10 (b)). Nevertheless, the alloy remained within a high cycle fatigue regime (fatigue life between 10^4 and 10^7) [96]. For the $Al_{0.3}$CoCrFeNi HEA, the formation of nanotwins resulted in superior fatigue resistance with a fatigue endurance limited to a UTS ratio of 0.43. According to microstructural analysis, fatigue cracks started to form around the tougher phases during cyclic deformation, resulting in specimen failure.

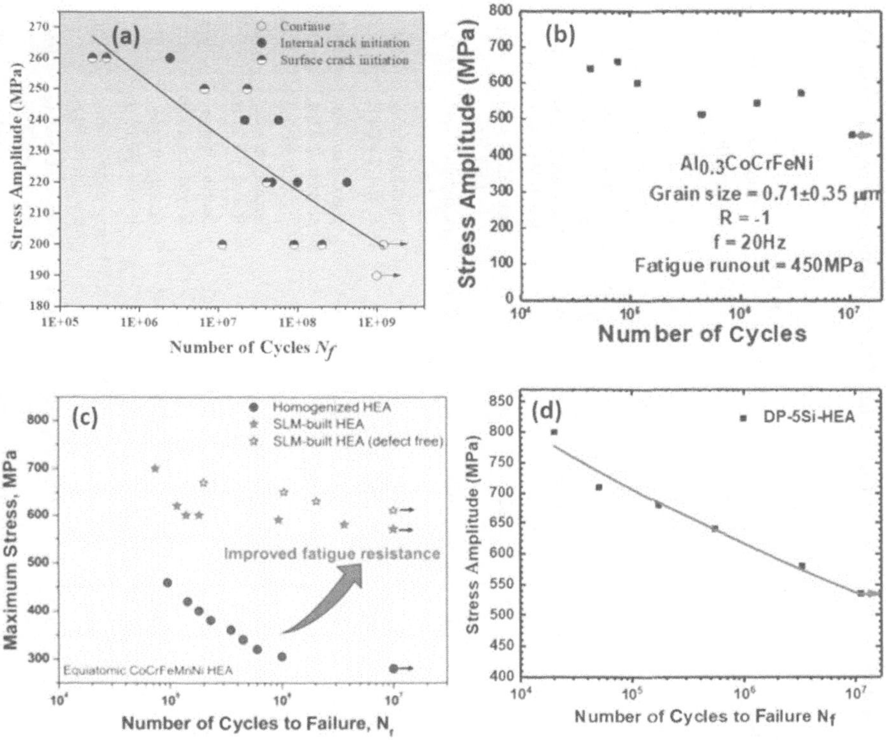

FIGURE 7.10 Stress amplitude vs number of cycles plots of TM HEA: (a) as-cast CoCrFeNiMn HEA [96], (b) TMP Al0.3CoCrFeNi HEA [97], (c) AM CoCrFeMnNi HEA [99], (d) FSP Fe42Mn28Cr15Co10Si5 HEA [101].

SLM-built CoCrFeMnNi HEA contained defects such as pores and un-melted powder, still, it showed a superior fatigue limit of 570 MPa (S-N curve is shown in Figure 7.10 (c)) with a fatigue ratio of 0.60 [99]. Fatigue cracks originated at the surface due to the formation of persistent slip bands in defect-free specimens. The exceptionally high-cycle fatigue resistance of the SLM-fabricated HEA is ascribed to its distinctive microstructural characteristics, particularly the elevated density of in-situ formed nanoscale oxides. Moreover, the deformation twinning generated by fatigue under cyclic loading significantly enhanced fatigue resistance.

Fine-grain structure achieved by FSP on the $Fe_{42}Mn_{28}Cr_{15}Co_{10}Si_5$ HEA aided enhanced YS (957 MPa) and UTS (1158 MPa) due to the optimization of interfacial interactions. This grain-refined HEA exhibited an atypical fatigue response due to the controlled metastability of the γ matrix. The significant $\gamma \rightarrow \varepsilon$ transformation within the crack tip plastic zone augmented work hardening, hence retarding the crack propagation. The simultaneous enhancement of fatigue strength (535 MPa) and fatigue ratio (0.46) of FSP $Fe_{42}Mn_{28}Cr_{15}Co_{10}Si_5$ HEA (shown in Figure 7.10 (d)) makes it a robust contender for cyclic loading applications [101].

TM HEAs exhibit remarkable mechanical properties, including excellent fatigue resistance under both LCF and HCF conditions. Tables 7.5 and 7.6 give explicit

TABLE 7.5
High-Cycle Fatigue Properties of TM HEAs: Alloy Composition, Phases Present, Processing Condition, Testing Temperature, Uniaxial YS, UTS, Fatigue Strength, Fatigue Ratio, and Fatigue Test Condition

S. No.	Alloy	Phases	Processing Condition	Testing Temperature	σ_y (Uniaxial YS)	σ_{UTS}	σ_{af}	Fatigue Ratio (σ_{af}/σ_{UTS})	Fatigue Test Condition	Reference
1	CoCrFeMnNi	FCC	AC + HM + Hot forged + CR + AN	RT	300	676	190	0.28	push-pull, R = −1, f = 30 Hz	[102]
2	CoCrFeMnNi	FCC	AC + HM + HR + AN	RT	851	903	550	0.3	bending fatigue, R = −1, f = 50 Hz	[103]
				−196°C	770	848	470	0.5		
3	$Al_{0.3}CoCrFeNi$	FCC + B2 + σ	AC + HM + CR + AN	RT	900	1074	450	0.43	bending fatigue, R = −1, f = 20 Hz	[96]
6	$Al_{0.5}CoCrCuFeNi$	FCC1 + FCC2	AC + HM + CR	RT	1284	1344	383±71	0.29	four-point bending, R = 0.1, f = 10 Hz	[104]
7	CoCrFeMnNi	FCC	AC + HR + HM	RT	293.1	625.6	280	0.45	push-pull, R = 0.1, f = 2 Hz	[99]
11	CrMnFeCoNi	FCC	AC + HM + CR + AN	RT	262	650	170	0.26	push-pull, R = −1, f = 30 Hz	[105]
12	$Al_{0.7}CoCrFeNi$	FCC+ L12	AC + HR + CR + HM	RT	800	1000			bending fatigue, R = −1, f = 20 Hz	[106]
			AC + CR + HM + AN	RT	1000	1400				
13	$Al_{0.3}CoCrFeNi$	FCC + B2 + σ	AC + AN + CR + AN	RT	700	900	281	0.3	uniaxial tension-tension, R = 0.1, f = 40	[100]
14	CoCrFeMnNi	FCC	AC	RT	234	604			bending fatigue, R = −1, f = 20 Hz	[96]
15	$Fe_{38.5}Mn_{20}Co_{20}Cr_{15}Si_5Cu_{1.5}$	FCC + HCP	AC + FSP	RT	800	1126	700	0.62	bending fatigue, R = −1, f = 20 Hz	[107]
16	$Fe_{42}Mn_{28}Cr_{15}Co_{10}Si_5$	FCC + HCP	AC + FSP	RT	950	1150	535	0.46	bending fatigue, R = −1, f = 20 Hz	[101]
17	$Al_{0.5}CoCrCuFeNi$	FCC1 + FCC2	AC + AN + CR	RT	1284	1344	540–945	0.402–0.703	four-point bending, R = 0.1, f = 10 Hz	[108]
18	CoCrFeNi	FCC	LPBF	RT	535	601	450	0.74	four-point bending, R = 0.1, f = 20 Hz	[109]
19	CoCrFeMnNi	FCC	SLM	RT	774	923	570	0.62	four-point bending, R = 0.1, f = 20 Hz	[99]

TABLE 7.6
LCF Properties of TM HEAs: Alloy Composition, Phases Present, Processing Condition, Testing Temperature, Strain Amplitude, and Fatigue Life Cycle

S. No.	Alloy	Phases	Processing Condition	Testing Temperature	Strain Amplitude (%)	Fatigue Life (Cycle)	Reference
4	CoCrFeMnNi	FCC	AC + HM + Rotary swaging (cold worked rods) + AN	RT	0.3, 0.5, and 0.7	~5×10^4	[100]
5	CoCrNiFeMn	FCC	AC + Hot extrusion		0.2	~5×10^3	[110]
					0.4	~10^4	
					0.6	~5×10^4	
			AC + Hot extrusion + ECAP		0.2	~10^3	
					0.4	~10^4	
					0.6	~10^5	
8	CoCrFeNiMn	FCC	AC + HR + HM (650)	RT	0.4	~2×10^4	[111]
			AC + HR + HM (650)		0.7	~3×10^3	
			AC + HR + HM (700)		0.4	~1.5×10^4	
			AC + HR + HM (700)		0.7	~3×10^3	
			AC + HR + HM (1000)		0.4	~1.2×10^4	
			AC + HR + HM (1000)		0.7	~4×10^3	
9	CoCrFeMnNi	FCC	AC + HM + Rotary swaging + AN	550°C	0.2	~7×10^4	[112]
					0.3	~6×10^4	
					0.4	~8×10^3	
					0.5	~7×10^3	
					0.75	~10^3	
					0.8	~9×10^2	
10	CoCrFeMnNi	FCC	AC + HM + HR + HT	RT	0.3	~10^4	[113]
					0.4		
					0.55		
					0.7		
					0.85		

information on LCF and HCF properties and reveal the distinct responses of these materials under different loading conditions. In the LCF table, properties such as fatigue life and strain amplitude are highlighted, emphasizing how HEAs sustain significant plastic deformation over a limited number of cycles. In contrast, the HCF table focuses on parameters like stress amplitude and fatigue strength at higher numbers of cycles, showing the alloy's capacity to withstand smaller deformations over extended periods.

7.4.2 Fatigue Properties of NTM HEAs

Tough NTM HEAs have high-temperature strength and the potential to substitute Ni-based superalloys in the aerospace sector; regrettably, the fatigue characteristics of NTM HEAs are infrequently examined and inadequately comprehended in comparison to other categories of HEAs, notwithstanding limited efforts [91].

The fatigue behavior of NTM equivalents has remained largely unexplored, particularly in comparison to TM HEAs [114]. The study done by Guennec and coworkers [115] seeks to address this gap, as it is the inaugural examination of the fatigue properties of the BCC NTM HfNbTaTiZr HEA after 50% thickness reduction through cold rolling and subsequent annealing at 1200°C for 5 hours. Figure 7.11 (a) displays typical inverted pole figure (IPF) maps derived from a plane perpendicular to the rolling direction (RD), illustrating an equiaxed grain microstructure where a noticeable grain size gradient is detected. The same alloy is studied by S. Chen et al. in different rolling (70% thickness reduction) and annealing conditions and the initial material possesses equiaxed grains of random orientation, with an average grain size of ~45 μm (see Figure 7.11 (b)) [114].

Four-point bending fatigue tests conducted at room temperature (maximum stress vs. number of cycles to failure are shown in Figure 7.12 (a)) and reveal several key findings for 50% cold rolled HfNbTaTiZr HEA. Firstly, maximum stress levels exceeding the yield stress (953 MPa) are necessary to induce fatigue failure, leading to the adoption of an elastoplastic model for stress assessment [115]. Secondly, fatigue crack propagation exhibits transgranular crystallographic features from initiation to rupture, alongside an atypical torn pattern, while the final rupture surface displays shallow dimple patterns akin to those seen under tensile loading. The initiation of fatigue cracks consistently occurs at the specimen surface, characterized by intergranular

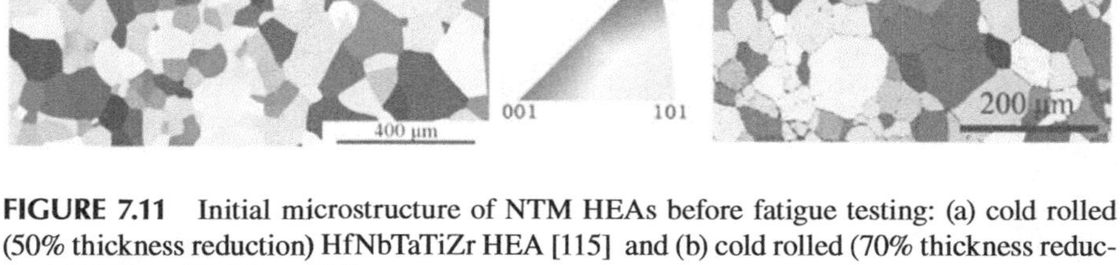

FIGURE 7.11 Initial microstructure of NTM HEAs before fatigue testing: (a) cold rolled (50% thickness reduction) HfNbTaTiZr HEA [115] and (b) cold rolled (70% thickness reduction) HfNbTaTiZr HEA [114].

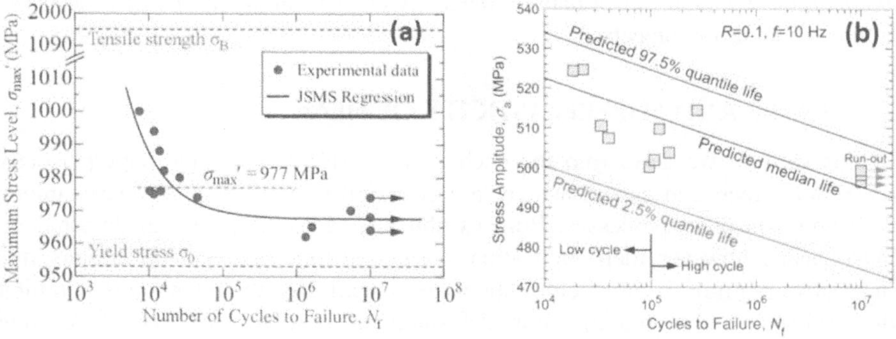

FIGURE 7.12 Fatigue properties of NTM HEAs: (a) cold rolled (50% thickness reduction) HfNbTaTiZr HEA [115] and (b) cold rolled (70% thickness reduction) HfNbTaTiZr HEA [114].

nucleation and a fine crystallographic structure, which significantly enhances the material's fatigue strength, particularly at grain boundaries aligned perpendicular to the loading direction. The 70% cold rolled and annealed HfNbTaTiZr HEA exhibits YS of ~1113 MPa and UTS of ~1139 MPa, and during the four-point bending test shows fatigue strength and fatigue ratio as 497 MPa and 0.44, respectively (see Figure 7.12 (b)) [114]. This HCF resistance is due to high strength and the reason behind delayed cracking is retarding the formation of localized persistent slip bands.

Table 7.7 summarizes the HCF behavior of NTM HEAs, including alloy composition phases present, processing conditions, uniaxial tensile properties (YS (σ_y) and UTS (σ_{UTS})), fatigue stress amplitude (σ_{af}), fatigue ratio (σ_{af}/σ_{UTS}), and fatigue test condition. Though NTM HEAs have shown remarkable fatigue properties, the

TABLE 7.7
HCF Behavior of NTM HEAs: Alloy Composition, Phases Present, Processing Condition, Testing Temperature, YS (σ_y), UTS (σ_{UTS}), Fatigue Stress Amplitude (σ_{af}), Fatigue Ratio (σ_{af}/σ_{UTS}), and Fatigue Test Condition

S. No.	Alloy	Phases	Processing Condition	Testing Temperature	σ_y (Uniaxial YS)	σ_{UTS}	σ_{af}	Fatigue Ratio (σ_{af}/σ_{UTS})	Fatigue Test Condition	Reference
1	HfNbTaTiZr	BCC	AC + CR + AN (1200°C, 10 min)	RT	1113	1139	497	0.44	four-point bending, R = 0.1, f = 10 Hz	[114]
2	HfNbTaTiZr	BCC	AC + CR + AN (1100°C, 5 h)	RT	953	1095	433	0.114	four-point bending, R = 0.1, f = 10 Hz	[115]
3	HfNbTaTiZr	BCC	AC + CR + AN (1100°C, 5 h)	RT	953	1095	470	0.43	four-point bending, R = 0.1, f = 10 Hz	[116]

availability of a small set of data on the study of their fatigue properties shows the future scope of the subject.

7.5 CREEP AND SUPERPLASTICITY OF HEAs

In this section, we delve into the enchanting world of creep and superplasticity in HEAs. Creep and superplasticity are two well-researched phenomena among diffusion-controlled processes. Both exhibit analogous fundamental deformation mechanisms; however, their applications are contrasting. In creep applications, it is imperative to minimize the deformation rate; specifically, the intention is to inhibit the mechanisms of high-temperature deformation. Conversely, superplastic forming is employed for unitized structures, and the aim is to improve the kinetics of high-temperature deformation.

7.5.1 CREEP IN HEAs

Creep deformation is a critical consideration for materials employed at elevated temperatures (often greater than 0.5 T_m) under constant load or stress (below yield stress), which constitutes time-dependent plastic deformation [7, 117]. In previous studies, it has been reported that HEAs can improve diffusion resistance due to the different local cohesive energy at any lattice site and the lattice distortion [118, 119]. Considering this, researchers have commenced investigations into the creep behavior of HEAs to clarify their microstructural evolution, stress-strain response, and phase stability under extended high-temperature loading. The deformation of materials at elevated temperatures is typically analyzed using the following expression [120]:

$$\dot{\varepsilon} = \frac{ADGb}{kT}\left(\frac{b}{d}\right)^p\left(\frac{\sigma}{G}\right)^n \tag{7.3}$$

Where, $\dot{\varepsilon}$ is the strain rate, A is the dimensionless constant, D is the diffusion coefficient (m^2/sec), G is the shear modulus (GPa), b is the magnitude of the Burger vector, k is Boltzmann's constant, d is the grain size (μm), σ is the stress (MPa), p is the inverse grain size exponent and n is the stress exponent. A constitutive connection for high-temperature deformation is articulated in Norton's law [120–122], which correlates the deformation rate with temperature and deformation activation energy.

$$\dot{\varepsilon} = \mathrm{A}\sigma^n \exp\left(-\frac{Q}{RT}\right)$$

where R is the gas constant and Q is the activation energy.

The creep behavior in HEAs is profoundly affected by activation energy (Q, kJ/mol) and the stress exponent (n). Q denotes the minimal energy necessary for atomic mobility and deformation, which can fluctuate owing to the intricate microstructure of HEAs, influencing their creep resistance [8]. The stress exponent measures the relationship between creep rate and applied stress, with specific values reflecting the primary deformation mechanisms: n = 1 corresponds to diffusion creep, n = 2 to grain boundary sliding, and n ≥ 3 to dislocation creep [8]. Comprehensive data on the creep behavior of HEAs is shown in Table 7.8, encompassing processing conditions;

TABLE 7.8
Creep Behavior of HEAs: Phases Present, Processing Condition, Testing Method, Testing Temperature, Activation Energy (Q), Stress Exponent (n), and Deformation Mechanism

S. No.	Alloy	Phases	Processing Condition	Testing Method	Testing Temperature	Q (KJ/mol)	n	Deformation Mechanism	Reference
1	Al0.15CoCrFeNi	FCC	AC	Stress relaxation creep	650–700°C	385	5.56	Cross-slip	[120]
	Al0.60CoCrFeNi	FCC + BCC			580–700°C	334	8.46		
2	CoCrFeNiMn	FCC	Mechanical alloying + SPS	Uniaxial compression creep	700–800°C	210–580	1.8–13.2	Diffusion creep and Harper-Dorn creep	[130]
3	AlTiVNbZr0.25	B2 + Zr5Al3	AC + HM (1200°C, 24 h)	Tensile creep	800°C		2	Dislocation dominated creep	[131]
4	CrMnFeCoNi	FCC	AC + HR + AN (1050°C, 1 h)	Tensile creep	650–700°C	303	4.13–5.94	Dislocation dominated creep	[132]
	CrFeCoNi	FCC			650–725°C	295	5.36–6.75		
5	Al0.5CoCrCuFeNi	FCC	AC	Nanoindentation creep	RT		121–129	Dislocation glide-climb	[133]
					200°C		84–97		
6	Al20Ce20La20Ni20Y20		AC	Tensile creep	30–210°C			Difusion	[134]
7	Al0.5CoCrFeMnNi	BCC	AC + HM (1100°C, 6 h)	Compression creep	750–1050°C	270	3	Solute drag creep, dislocation climb creep	[135]
	Al1CoCrFeMnNi				900–1100°C	251	7		
8	(Hf-Ta-Zr-Nb)C		SPS	Uniaxial compression creep	1400–1600°C	212	2.34–2.89	Grain-boundary sliding and dislocation glide/climb	[128]
9	FeCoNiCrMn	FCC	AC + CR + Aging	Tensile creep	750–850°C	330	5	Dislocation climb control	[119]
10	CoCrFeMnNi	FCC	AC + AN (1000°C, 6 h)	Tensile creep	500–700°C	270–313	3	Dislocation viscous glide	[127]
11	CoCrFeMnNi	FCC	AC + HPT	Sperical nanoindentation creep	RT		3	Sluggish diffusion	[129]

(Continued)

TABLE 7.8 *(Continued)*

Creep Behavior of HEAs: Phases Present, Processing Condition, Testing Method, Testing Temperature, Activation Energy (Q), Stress Exponent (n), and Deformation Mechanism

12	CoCrFeMnNi	FCC	AC + HPT	Nanoindentation creep	RT –300°C			Grain boundary diffusion	[136]
13	CoCrFeMnNi	FCC	SLM	Nanoindentation	RT			Diffusion creep and grain boundary creep	[137]
14	AlxCrMnFeCoNi	FCC + BCC	AC + HM (1150°C, 12 h) + CR + Recrystallized (1100°C, 2 h)	Tensile creep	600–700°C	126	2.8–5.8	Grain boundary sliding	[138]
15	CoCrFeMnNi	FCC	AC + HM + CR + AN (900°C)	Tensile creep	535–650°C			Viscous glide creep	[139]
16	CoCrFeNiMn	FCC	AC + HM (1100°C, 24 h) + CR + Severe cross rolling + AN (900°C, 1 h)	Tensile creep	500–600°C	268–380	5–14	Lattice diffusion controlled dislocation climb	[140]
17	HfTaTiVZr TaTiVWZr	BCC	AC	Nanoindentation creep	25–300°C		20–140	Dislocation creep	[141]
18	CrMnFeCoNi	FCC	AC (single crystal)	Tensile creep	700–1200°C		4.5–7.3	Grain size effect, grain boundary sliding, diffusion	[142]
19	CoCrFeNiMn	FCC	AC + HM + HR	Tensile creep	600°C	394	6.2	Dislocation creep	[143]
20	FeCoNiCrMn	FCC	AC + CR + HM (750°C, 1 h)	Tensile creep	750–850°C		3–5	Dislocation climb	[119]

type of testing; and temperature, Q, n, and deformation mechanisms. This table will elucidate insight into the effect of processing conditions, Q and n, on the deformation mechanism during creep testing of various HEAs.

The microstructure characteristics, such as grain size, existence of secondary phases, and phase distribution of HEAs, profoundly affect their creep behavior. Fine-grained materials typically demonstrate superior creep resistance attributed to grain boundary strengthening, whereas coarser grains may promote dislocation mobility, resulting in elevated creep rates [123–126]. The existence of several phases in HEAs can promote stability and obstruct dislocation motion; hence, increasing high-temperature performance [123]. From a microstructural perspective, the interaction between the solvent and solutes in HEAs becomes indistinct, as elemental species of varying sizes infiltrate the crystal lattice, resulting in a significantly distorted deformation unit. Consequently, HEAs demonstrate unique and significantly more complex deformation characteristics, even though they possess basic crystal structures akin to normal metals.

Song et al. [127] have explored the nanoindentation creep behavior of CoCrFeMnNi HEA, which consists of a dendritic structure (with Mn, Ni-deficient in dendrite and Mn, Ni-rich in inter-dendrite) in the as-cast condition, whereas a homogeneous equiaxed grain structure after heat treatment at 1200°C for 6 hours (see Figure 7.13 (a)). The grain size of as-annealed CoCrFeMnNi is around 100 μm, sufficiently big to facilitate dislocation creep. Typically, FCC structured alloys, characterized by less self-diffusivity and increased atomic stacking density relative to BCC alloys are anticipated to demonstrate enhanced heat resistance. Moreover, He et al. [119] showed that TMP of FCC FeCoNiCrMn HEA (see the microstructure in Figure 7.13 (b)) promotes a preferred crystallographic orientation, markedly affecting slip systems during creep and improving creep performance, particularly at high temperatures.

HEAs improve diffusion resistance through variations in local cohesive energy and lattice distortion. This characteristic suggests that SLM-fabricated HEAs may exhibit unique creep behavior. Xu et al. [127] fabricated CoCrFeMnNi HEA through selective laser melting and found that the rapid solidification and steep temperature gradients in the molten pool during the SLM process resulted in a microstructure consisting of cellular subgrains with grain boundary angles of less than 5° (see Figure 7.13 (c)) and examined its creep behavior. Recent studies demonstrate that minimizing the grain size of HEAs to the nanocrystalline scale can improve their mechanical properties [128]. D. H. Lee et al. [129] have incorporated HPT to significantly reduce the grain size of as-cast CrMnCoFeNi HEA from ~ 46 μm to ~ 33 nm at the periphery of the HPT disk after two turns (see Figure 7.13 (d)).

In Figure 7.14 (a), creep curves $\epsilon \sim t$ of annealed CoCrFeMnNi HEA at 180 MPa for different temperatures ranging from 500–700°C are shown, where the fracture time increased significantly with decreasing temperature, but the fracture strain increased with temperature [126]. The creep curves, analogous to those of traditional alloys and nanocrystalline Cu, can be categorized into three stages: primary, secondary (stationary state), and tertiary. However, the tertiary stage is not observed at 500°C and 550°C due to the limited creep time. Here in the steady state creep, dislocation climb, coupled with solute Mn diffusion, predominates at elevated temperatures, whereas dislocation viscous glide, influenced by solute Mn, prevails at reduced temperatures.

FIGURE 7.13 Initial microstructure before creep deformation of HEAs, (a) as-csat CoCrFeMnNi HEAs [126], (b) TMP FeCoNiCrMn HEA [129], (c) AM CoCrFeMnNi HEA [127], and (d) HPT CoCrFeMnNi HEA [128].

The steady-state tensile flow characteristics of the rolled and annealed FeCoNiCrMn HEA were examined at temperatures between 750 and 850°C (see Figure 7.14 (b_1 and b_2)) [129]. The correlation between steady-state stress and strain rate is effectively characterized by the conventional power-law equation, revealing two distinct deformation areas contingent upon the strain rate. In the high strain rate regime ($> 2 \times 10^5$ s^{-1}, referred to as Region II), the stress exponent is approximately 5, and the activation energy is around 330 kJ/mol, indicating a dislocation-climb mechanism, with the slowest diffusing species, nickel, governing the rate process. Conversely, in the low strain rate regime ($< 2 \times 10^{-5}$ s^{-1}, Region I), the stress exponent is below 3 and the activation energy is approximately 280 kJ/mol, indicating that the mechanism involves the drag of gliding dislocations, with the deformation rate governed by the diffusion of a constituent element acting as the solute atom. Microstructural analyses indicate that, despite the alloy's initial simple FCC structure, diverse Mn- and Cr-enriched precipitates developed under stress, particularly in samples subjected to elevated temperatures and reduced strain rates, i.e., prolonged thermal exposure.

The nanoindentation creep results of SLM fabricated CoCrFeMnNi HEA, fitted with various P_{max} values, show excellent agreement ($R^2 > 0.99$), as illustrated

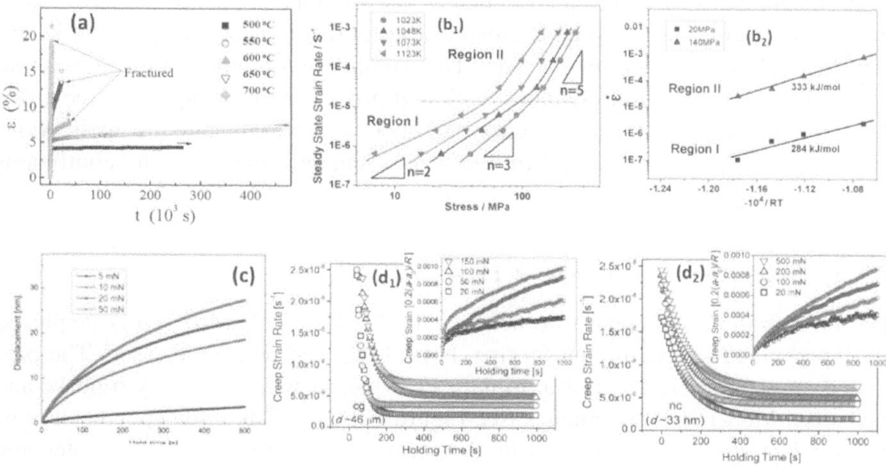

FIGURE 7.14 Creep behavior of HEAs, (a) creep curves of as-cast CoCrFeMnNi HEA at 180 MPa and different temperatures [126], (b_1) steady-state strain rate as a function of the true stress on a double logarithmic plot which shows two deformation regions, (b_2) steady-state strain rate versus the inverse of temperature on an Arrhenius plot which shows the activation energy for Region I and II for TMP FeCoNiCrMn HEA [129], (c) fit of creep displacements during the holding stage obtained at various P_{max} [127], and (d) strain vs. holding time for coarse-grained and HPT fine-grained CoCrFeMnNi HEA [128].

in Figure 7.14 (c) [127]. The initial creep displacements at different P_{max} values are aligned for comparison, revealing larger creep displacements at higher P_{max}. The curves exhibit two distinct stages: transient and steady-state creep, lacking a third (accelerated) stage due to the absence of catastrophic failure at low stress. At P_{max} = 20 mN, the initial strain rate is high (10^{-3} to $10^{-4}\,s^{-1}$) during transient creep, gradually stabilizing to ~$10^{-5}\,s^{-1}$ in the steady-state stage as holding time increases. It is suggested that dislocation motion primarily drives the creep deformation.

The creep strain (ε_{creep}) vs. holding time (t_{hold}) curves for as-cast (cg) and after HPT (nc) CrMnCoFeNi HEA are shown in Figure 7.14 (d_1) and (d_2), which are somewhere analogous to the early stages of the typical high-temperature creep curves, consisting of both transient and steady-state creep regimes [128]. The increase in ε_{creep} is highly correlated with peak load (P_{max}), suggesting that the observed creep behavior is not a mere artifact of thermal drift unrelated to load. Moreover, the creep stress exponent, n, derived from the quasi-steady-state strain rate and stress, along with the activation volume V*, indicates a transition in the primary time-dependent deformation mechanism from dislocation-mediated processes in coarse-grained HEAs to grain boundary diffusion and associated phenomena in nanocrystalline HEAs. A theoretical assessment of the creep rate in nanocrystalline HEAs and its comparison with that in typical nanocrystalline (Ni) indicates that the creep resistance of nanocrystalline HEAs is markedly improved due to reduced diffusion rates in HEAs.

Table 7.8 provides a detailed review of the creep behavior demonstrated by various HEAs.

It encompasses essential factors such as HEA compositions, processing conditions, temperature ranges, type of creep testing, activation energy (Q), stress exponent (n), and deformation mechanism. This compilation highlights the substantial impact of microstructural characteristics and processing techniques on the creep behavior of HEAs, offering insights into their appropriateness for applications requiring outstanding resistance to time-dependent deformation under continuous loading.

7.5.2 Superplastic Behavior of HEAs

The investigation of superplastic behavior in $Al_{0.2}CoCrFeNiMo_{0.5}$ HEA employed TMP techniques including hot forging, hot rolling, and homogenization. The processed microstructure features a primary FCC matrix with secondary σ phase and nanosized Mo/Cr-rich precipitates (see Figure 7.15 (a)). This study specifically targets the identification of superplastic deformation mechanisms through systematic analysis of flow behavior across diverse temperature regimes and strain rates.

The high-temperature mechanical properties of HEAs have been explored in various studies. A recent review highlighted key findings related to superplastic behavior, including notable results from a CoCrFeNiMn HEA that demonstrated superplastic-like flow [144]. Superplasticity typically necessitates a grain size of less than approximately 10 μm, which can be effectively achieved through SPD techniques. At present, fine-grained HEAs are manufactured by techniques such as HPT, CR, and FSP. HPT has successfully produced the CoCrFeNiMn HEA with an average grain size of 10 nm, attaining an exceptional elongation of 600%. The restricted

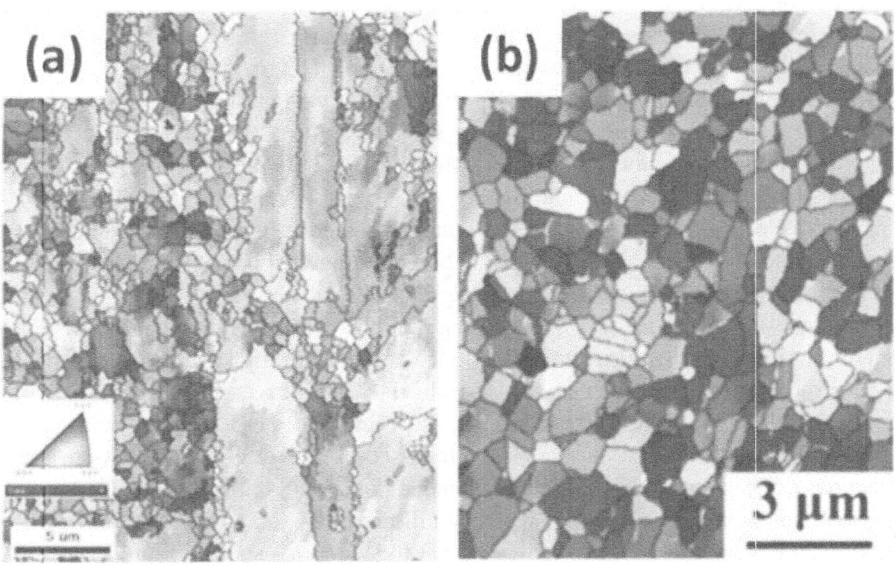

FIGURE 7.15 Initial microstructure of superplastic HEAs, (a) TMP $Al_{0.2}CoCrFeNiMo_{0.5}$ HEA [145] and (b) FSP CoCrFeNiCu HEA [146].

size of HPT samples limits their utility in superplastic formation. Conversely, fine-grained CoCrFeNiMn HEAs produced by CR demonstrate a maximum elongation of approximately 300%, and their SP properties necessitate improvement. FSP, a brief yet intense plastic deformation technique, produces a fine/ultrafine grain structure distinguished by a significant presence of high-angle grain boundaries (HAGBs) and a reduced dislocation density, unlike the microstructures observed in CR and HPT. Li et al. [145] incorporated the FSP on CoCrFeNiMn HEA, which is characterized by an equiaxed ultrafine grain structure and a significant presence of high-angle grain boundaries, demonstrating commendable stability. The grain size of 689 nm was preserved upon annealing at 675°C (see Figure 7.15 (b)).

The influence of temperature and strain rate on the deformation characteristics of TMP $Al_{0.2}CoCrFeNiMo_{0.5}$ HEA was examined, and it attained a tensile elongation of roughly 275% at 900°C with an initial strain rate of $1 \times 10^{-3}\ s^{-1}$ (see Figure 7.16 (a)) [146]. Extensive dislocation activity inside the grains, a strain rate sensitivity of 0.32, and texture rotation suggest that dislocation glide-controlled deformation is likely the predominant mechanism responsible for this significant elongation. The primary failure mode is defined by cavity nucleation at the secondary phase-matrix interface, which subsequently results in the interconnection of cavities.

The FSP CoCrFeNiMn HEA demonstrated a maximum elongation of 870% at 675°C and a strain rate of $3 \times 10^{-4}\ s^{-1}$ (see Figure 7.16 (b)) [145]. The superior superplasticity was primarily ascribed to the equiaxed ultrafine grain structure (489 nm), the elevated HAGB ratio (88%), the twin boundary (TB) ratio (10%), the hard chromium-rich phase, and the sluggish diffusion impact of the CoCrFeNiMn HEA. During superplastic deformation, a high density of dislocations and high-angle grain boundaries serve as diffusion pathways for chromium, hence expediting the precipitation of chromium and the production of chromium-rich phases near grain boundaries. As deformation intensifies, two distinct Cr-rich phases coalesce at the grain boundaries. Grain boundary strengthening (GBS) was primarily facilitated by the spread of Cr and the movement of dislocations.

Table 7.9 includes the summary of HEAs, which showed superplastic behavior under different processing conditions. This detailed data emphasizes the variety of

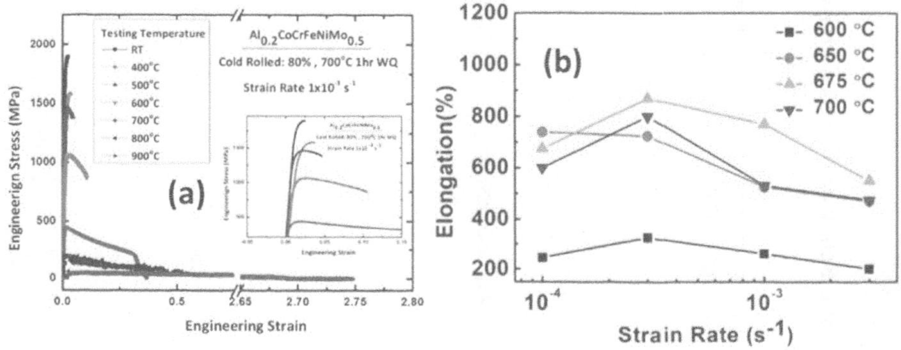

FIGURE 7.16 Superplastic behavior of HEAs: (a) TMP Al0.2CoCrFeNiMo0.5 HEA [144] and (b) FSP CoCrFeNiCu HEA [145].

TABLE 7.9

Superplastic Behavior of HEAs: Phases Present, Processing Condition, Testing Temperature, Strain Rate, Ductility (%)

S. No.	Alloy	Phases	Processing Condition	Testing Temperature	Strain Rate (s^{-1})	Ductility (%)	Reference
1	AlCoCrCuFeNi	FCC + BCC	AC + HM (960°C, 50 h) + Hot forging	700°C	10^{-3}	63	[147]
				800°C	10^{-3}	604	
				900°C	10^{-3}	405	
				1000°C	10^{-4}	753	
				1000°C	10^{-3}	864	
				1000°C	10^{-2}	858	
				1000°C	10^{-1}	442	
2	$Al_{0.2}CoCrFeNiMo_{0.5}$	FCC + σ	AC + HM + Hot forging + HR (1100°C) + HM + CR + HT	RT	10^{-3}	5	[148]
				400°C	10^{-3}	10	
				500°C	10^{-3}	13	
				600°C	10^{-3}	15	
				700°C	10^{-3}	40	
				800°C	10^{-3}	65	
				900°C	10^{-4}	190	
				900°C	10^{-3}	275	
				900°C	10^{-2}	210	
3	CoCrFeNiMn	FCC	AC + HF + HM (1000°C) + HPT	600°C	10^{-2}	460	[149]
				700°C	10^{-2}	830	
				800°C	10^{-2}	570	
				700°C	10^{-3}	650	
				700°C	10^{-1}	630	

4	CoCrFeNiCu	FCC	AC + FSP	900°C	3×10^{-4}	225	[145]
				900°C	1×10^{-4}	500	
				900°C	3×10^{-3}	225	
				900°C	1×10^{-3}	275	
				925°C	1×10^{-4}	300	
				925°C	3×10^{-3}	400	
				925°C	1×10^{-2}	425	
				925°C	3×10^{-2}	375	
				950°C	1×10^{-4}	300	
				950°C	3×10^{-3}	610	
				950°C	1×10^{-2}	300	
				950°C	3×10^{-2}	225	
5	CoCrFeNiMn	FCC	AC + FSP	600°C	10^{-4}	250	[150]
				600°C	3×10^{-4}	300	
				600°C	10^{-3}	250	
				600°C	3×10^{-3}	200	
				650°C	10^{-4}	750	
				650°C	3×10^{-4}	700	
				650°C	10^{-3}	500	
				650°C	3×10^{-3}	475	
				675°C	10^{-4}	675	
				675°C	3×10^{-4}	870	
				675°C	10^{-3}	770	
				675°C	3×10^{-3}	550	
				700°C	10^{-4}	600	
				700°C	3×10^{-4}	800	
				700°C	10^{-3}	500	
				700°C	3×10^{-3}	550	

(Continued)

TABLE 7.9 *(Continued)*

Superplastic Behavior of HEAs: Phases Present, Processing Condition, Testing Temperature, Strain Rate, Ductility (%)

S. No.	Alloy	Phases	Processing Condition	Testing Temperature	Strain Rate (s^{-1})	Ductility (%)	Reference
6	$Al_{0.5}CoCrFeMnNi$	FCC	AC + HM + CR + AN (1200°C, 2 h) + HPT	650°C	5×10^{-1}	450	[151]
				650°C	10^{-1}	575	
				650°C	5×10^{-2}	550	
				650°C	10^{-2}	650	
				650°C	5×10^{-3}	500	
				650°C	10^{-3}	575	
				650°C	5×10^{-4}	450	
				700°C	5×10^{-1}	800	
				700°C	10^{-1}	900	
				700°C	5×10^{-2}	750	
				700°C	10^{-2}	700	
				700°C	5×10^{-3}	600	
				700°C	10^{-3}	615	
				700°C	5×10^{-4}	525	
				800°C	5×10^{-1}	950	
				800°C	10^{-1}	1125	
				800°C	5×10^{-2}	1000	
				800°C	10^{-2}	950	
				800°C	5×10^{-3}	1050	
				800°C	10^{-3}	675	
				800°C	5×10^{-4}	575	

7	$AlCoCrFeNi_{2.1}$	FCC + BCC	AC + CR + AN (800°C, 1 h)	700°C	10^{-1}	80	[152]
				800°C	10^{-1}	330	
				900°C	10^{-1}	960	
				900°C	10^{-2}	840	
				900°C	10^{-3}	860	
				900°C	3×10^{-4}	590	
8	AlCoCrCuFeNi	FCC	AC + HM (960, 50 h) + Multidirectional forging	800°C	10^{-4}	325	[153]
				800°C	10^{-3}	160	
				800°C	10^{-2}	60	
				900°C	10^{-4}	490	
				900°C	10^{-3}	585	
				900°C	10^{-2}	350	
				1000°C	10^{-4}	850	
				1000°C	10^{-3}	1240	
				1000°C	10^{-2}	600	
9	CoCrFeNiMn	FCC	AC + HM (1000°C, 16 h) + HPT	500°C	10^{-3}	150	[144]
				600°C	10^{-3}	525	
				600°C	10^{-2}	400	
				600°C	10^{-1}	340	
				700°C	10^{-3}	625	
				700°C	10^{-2}	500	
				700°C	10^{-1}	440	
				800°C	10^{-3}	400	
				800°C	10^{-2}	360	
				800°C	10^{-1}	320	

(Continued)

TABLE 7.9 *(Continued)*
Superplastic Behavior of HEAs: Phases Present, Processing Condition, Testing Temperature, Strain Rate, Ductility (%)

S. No.	Alloy	Phases	Processing Condition	Testing Temperature	Strain Rate (s^{-1})	Ductility (%)	Reference
10	$V_{10}Cr_{15}Mn_5Fe_{35}Co_{10}Ni_{25}$	FCC	AC + HM (1100°C, 6 h) + CR + AN (900°C, 10 min) + HPT	600°C	1×10^{-2}	130	[148]
				600°C	5×10^{-3}	225	
				600°C	1×10^{-3}	325	
				600°C	5×10^{-4}	390	
				700°C	1×10^{-2}	490	
				700°C	5×10^{-3}	480	
				700°C	3.3×10^{-3}	770	
				700°C	1×10^{-3}	650	
				700°C	5×10^{-4}	410	
				800°C	1×10^{-2}	580	
				800°C	5×10^{-3}	580	
				800°C	3.3×10^{-3}	700	
				800°C	1×10^{-3}	630	
				800°C	5×10^{-4}	615	
11	$Fe_{42}Mn_{28}Co_{10}Cr_{15}Si_5$	FCC + HCP.	AC + HR + HM (1200°C, 1 h) + FSP	500°C	10^{-3}	45	[154]
				600°C	10^{-3}	55	
				700°C	10^{-3}	550	
				700°C	5×10^{-3}	215	
				700°C	10^{-4}	520	
				700°C	10^{-2}	220	

12	FeCoCrNiMn	FCC	AC + HM (1150°C, 6 h) + CR + AN (300°C–1100)	500°C	5×10^{-4}	30	[155]
				600°C	5×10^{-4}	20	
				650°C	5×10^{-4}	90	
				700°C	5×10^{-4}	270	
				700°C	10^{-3}	150	
				700°C	10^{-2}	100	
				700°C	10^{-1}	75	
				800°C	5×10^{-4}	335	
				800°C	10^{-3}	175	
				800°C	10^{-2}	120	
				800°C	10^{-1}	85	
13	CoCrFeMnNi	FCC	AC + HM (1100°C, 6 h) + CR + AN (800°C, 1 h) + WR	750°C	10^{-4}	325	[156]
				750°C	10^{-3}	290	
				750°C	10^{-1}	160	

HEAs and the impact of various fabrication methods on their superplastic characteristics, illustrating how differences in microstructure and processing conditions affect their performance under superplastic flow examination.

7.6 SUMMARY

The complex reciprocation between the composition, microstructure, and mechanical behavior of HEAs has prompted researchers to delve deeply into the underlying mechanisms governing their extraordinary properties. Understanding the microstructural evolution of these alloys and their impact on mechanical properties has become critical as they depart from the conventional material-design approach. Furthermore, in HEAs, multiple strengthening mechanisms, such as solid solution, precipitation, work hardening, stacking fault, and grain boundary strengthening, work concurrently. In this context, this chapter provides in-depth knowledge of the captivating realm of HEAs, presenting TM HEAs, and NTM HEAs along with their active deformation mechanisms. Each section contains tables that give comprehensive information about the different compositions, their crystal structures, processing, and deformation mechanisms.

The chapter probes into the insights of the effects of different processing conditions on the microstructures and mechanical properties of TM and NTM HEAs. During TMP, metals and alloys undergo dynamic recovery, dynamic and static recrystallization which aid in tailoring microstructure and crystallographic texture. In addition to conventional processing methods like rolling, forging, and extrusion, HEAs can be microstructurally engineered through SPD techniques such as FSP and HPT. These approaches enable the development of submicron and nanocrystalline grain structures, which enhance mechanical properties and yield superior strength and toughness. Moreover, AM provides the flexibility to produce complex and intricate shapes in HEAs, which can be challenging to achieve using traditional casting and forming methods.

List of Abbreviations Used in Tables

AC As-cast
AM Additive manufacturing
AN Annealed
CR Cold rolling
D-pass Dual pass
ECAP Equal channel angular pressing
FSP Friction stir processing
GS Grain size
HIP Hot isostatic pressing
HM Homogenized
HPT High-pressure torsion
HR Hot rolling
HT Heat treated
LPBF Laser powder bed fusion
PAAM Powder-bed arc additive manufacturing

RT Room temperature
SLM Selective laser melting
SN Solutionizing
SPS Spark plasma sintering
WQ Water quenched
WR Warm rolling

REFERENCES

1. D. Miracle, J. Miller, O. Senkov, C. Woodward, M. Uchic and J. Tiley, "Exploration and development of high entropy alloys for structural applications," *Entropy*, vol. 16, no. 1, pp. 494–525, Jan. 2014. doi: 10.3390/e16010494.
2. S. Praveen and H. S. Kim, "High-entropy alloys: potential candidates for high-temperature applications – An overview," *Adv Eng Mater*, vol. 20, no. 1, Jan. 2018. doi: 10.1002/adem.201700645.
3. M. C. Gao, J.-W. Yeh, P. K. Liaw and Y. Zhang, Eds., *High-Entropy Alloys*. Cham: Springer International Publishing, 2016. doi: 10.1007/978-3-319-27013-5.
4. V. E. Gromov, Y. A. Shlyarova, S. V. Konovalov, S. V. Vorob'ev and O. A. Peregudov, "Application of high-entropy alloys," *Steel Transl*, vol. 51, no. 10, pp. 700–704, Oct. 2021. doi: 10.3103/S096709122110003X.
5. S. S. Nene *et al.*, "Corrosion-resistant high entropy alloy with high strength and ductility," *Scr Mater*, vol. 166, pp. 168–172, Jun. 2019. doi: 10.1016/j.scriptamat.2019.03.028.
6. W. Li, S. Chen and P. K. Liaw, "Discovery and design of fatigue-resistant high-entropy alloys," *Scr Mater*, vol. 187, pp. 68–75, Oct. 2020. doi: 10.1016/j.scriptamat.2020.05.047.
7. W. Li, G. Wang, S. Wu and P. K. Liaw, "Creep, fatigue, and fracture behavior of high-entropy alloys," *J Mater Res*, vol. 33, no. 19, pp. 3011–3034, Oct. 2018. doi: 10.1557/jmr.2018.191.
8. E. P. George, W. A. Curtin and C. C. Tasan, "High entropy alloys: A focused review of mechanical properties and deformation mechanisms," *Acta Mater*, vol. 188, pp. 435–474, Apr. 2020. doi: 10.1016/j.actamat.2019.12.015.
9. R. Motallebi, Z. Savaedi and H. Mirzadeh, "Superplasticity of high-entropy alloys: A review," *Arch Civil Mech Eng*, vol. 22, no. 1, p. 20, Dec. 2021. doi: 10.1007/s43452-021-00344-x.
10. S. S. Nene, S. Sinha, D. K. Yadav and A. Dutta, "Metallurgical aspects of high entropy alloys," *J Alloys Compd*, vol. 1005, p. 175849, Nov. 2024. doi: 10.1016/j.jallcom.2024.175849.
11. Z. Li, S. Zhao, R. O. Ritchie and M. A. Meyers, "Mechanical properties of high-entropy alloys with emphasis on face-centered cubic alloys," *Prog Mater Sci*, vol. 102, pp. 296–345, May 2019. doi: 10.1016/j.pmatsci.2018.12.003.
12. W. Li, D. Xie, D. Li, Y. Zhang, Y. Gao and P. K. Liaw, "Mechanical behavior of high-entropy alloys," *Prog Mater Sci*, vol. 118, p. 100777, May 2021. doi: 10.1016/j.pmatsci.2021.100777.
13. D. B. Miracle and O. N. Senkov, "A critical review of high entropy alloys and related concepts," *Acta Mater*, vol. 122, pp. 448–511, Jan. 2017. doi: 10.1016/j.actamat.2016.08.081.
14. B. Cantor, I. T. H. Chang, P. Knight and A. J. B. Vincent, "Microstructural development in equiatomic multicomponent alloys," *Mater Sci Eng A*, vol. 375–377, pp. 213–218, Jul. 2004. doi: 10.1016/j.msea.2003.10.257.
15. W. Chen, A. Hilhorst, G. Bokas, S. Gorsse, P. J. Jacques and G. Hautier, "A map of single-phase high-entropy alloys," *Nat Commun*, vol. 14, no. 1, p. 2856, May 2023. doi: 10.1038/s41467-023-38423-7.

16. S. S. Nene, "Microstructure and Mechanical Properties of High Entropy Alloys," in *High Entropy Alloys*, Singapore: Springer Nature Singapore, 2024, pp. 99–175. doi: 10.1007/978-981-99-7173-2_4.
17. Y. Zhang *et al.*, "Microstructures and properties of high-entropy alloys," *Prog Mater Sci*, vol. 61, pp. 1–93, Apr. 2014. doi: 10.1016/j.pmatsci.2013.10.001.
18. J. Sun *et al.*, "High temperature tensile properties of as-cast and forged CrMnFeCoNi high entropy alloy," *Mater Sci Eng A*, vol. 850, p. 143570, Aug. 2022. doi: 10.1016/j.msea.2022.143570.
19. A. Dutta *et al.*, "Excellent work hardening ability in a novel compositionally complex alloy by hierarchical microstructuring," *Appl Mater Today*, vol. 39, p. 102300, Aug. 2024. doi: 10.1016/j.apmt.2024.102300.
20. B. Dong *et al.*, "On the development of pseudo-eutectic AlCoCrFeNi2.1 high entropy alloy using Powder-bed Arc Additive Manufacturing (PAAM) process," *Mater Sci Eng A*, vol. 802, p. 140639, Jan. 2021. doi: 10.1016/j.msea.2020.140639.
21. F. Ghader, K. H. Seop and K. H. Torabzadeh, *Severe Plastic Deformation: Methods, Processing and Properties*, 1st ed. Chennai: Elsevier, 2018.
22. S. S. Nene *et al.*, "Unexpected strength–ductility response in an annealed, metastable, high-entropy alloy," *Appl Mater Today*, vol. 13, pp. 198–206, Dec. 2018. doi: 10.1016/j.apmt.2018.09.002.
23. P. Niu, R. Li, Z. Fan, T. Yuan and Z. Zhang, "Additive manufacturing of TRIP-assisted dual-phases Fe50Mn30Co10Cr10 high-entropy alloy: Microstructure evolution, mechanical properties and deformation mechanisms," *Mater Sci Eng A*, vol. 814, p. 141264, May 2021. doi: 10.1016/j.msea.2021.141264.
24. X. Jin, Y. Zhou, L. Zhang, X. Du and B. Li, "A novel Fe20Co20Ni41Al19 eutectic high entropy alloy with excellent tensile properties," *Mater Lett*, vol. 216, pp. 144–146, Apr. 2018. doi: 10.1016/j.matlet.2018.01.017.
25. S. Wang, Y. Li, D. Zhang, Y. Yang, S. Marwana Manladan and Z. Luo, "Microstructure and mechanical properties of high strength AlCoCrFeNi2.1 eutectic high entropy alloy prepared by selective laser melting (SLM)," *Mater Lett*, vol. 310, Mar. 2022. doi: 10.1016/j.matlet.2021.131511.
26. X. Yan *et al.*, "Al0.3CrxFeCoNi high-entropy alloys with high corrosion resistance and good mechanical properties," *J Alloys Compd*, vol. 860, Apr. 2021. doi: 10.1016/j.jallcom.2020.158436.
27. T. Bhattacharjee *et al.*, "Effect of low temperature on tensile properties of AlCoCrFeNi2.1 eutectic high entropy alloy," *Mater Chem Phys*, vol. 210, pp. 207–212, May 2018. doi: 10.1016/j.matchemphys.2017.06.023.
28. A. V. Kuznetsov, D. G. Shaysultanov, N. D. Stepanov, G. A. Salishchev and O. N. Senkov, "Tensile properties of an AlCrCuNiFeCo high-entropy alloy in as-cast and wrought conditions," *Mater Sci Eng A*, vol. 533, pp. 107–118, 2012. doi: 10.1016/j.msea.2011.11.045.
29. J. Zhang, H. Qiu, H. Zhu and Z. Xie, "Effect of Al additions on the microstructures and tensile properties of AlxCoCr3Fe5Ni high entropy alloys," *Mater Charact*, vol. 175, May 2021. doi: 10.1016/j.matchar.2021.111091.
30. J. Y. He *et al.*, "Effects of Al addition on structural evolution and tensile properties of the FeCoNiCrMn high-entropy alloy system," *Acta Mater*, vol. 62, no. 1, pp. 105–113, 2014. doi: 10.1016/j.actamat.2013.09.037.
31. D. Wei *et al.*, "Novel co-rich high entropy alloys with superior tensile properties," *Mater Res Lett*, vol. 7, no. 2, pp. 82–88, Feb. 2019. doi: 10.1080/21663831.2018.1553803.
32. S. Z. Niu, H. C. Kou, J. Wang and J. S. Li, "Improved tensile properties of Al0.5CoCrFeNi high-entropy alloy by tailoring microstructures," *Rare Metals*, vol. 40, no. 9, Sep. 2021. doi: 10.1007/s12598-016-0860-y.

33. J. W. Bae *et al.*, "Trade-off between tensile property and formability by partial recrystallization of CrMnFeCoNi high-entropy alloy," *Mater Sci Eng A*, vol. 703, pp. 324–330, Aug. 2017. doi: 10.1016/j.msea.2017.07.079.
34. M. J. Yao, K. G. Pradeep, C. C. Tasan and D. Raabe, "A novel, single phase, non-equiatomic FeMnNiCoCr high-entropy alloy with exceptional phase stability and tensile ductility," *Scr Mater*, vol. 72–73, pp. 5–8, Feb. 2014. doi: 10.1016/j.scriptamat.2013.09.030.
35. Y. Tong *et al.*, "Outstanding tensile properties of a precipitation-strengthened FeCoNiCrTi0.2 high-entropy alloy at room and cryogenic temperatures," *Acta Mater*, vol. 165, pp. 228–240, Feb. 2019. doi: 10.1016/j.actamat.2018.11.049.
36. M. Park, J. Moon, J. W. Bae, J. Jung, S. Lee and H. S. Kim, "Effect of annealing heat treatment on microstructural evolution and tensile behavior of Al0.5CoCrFeMnNi high-entropy alloy," *Mater Sci Eng A*, vol. 728, pp. 251–258, Jun. 2018. doi: 10.1016/j.msea.2018.05.041.
37. B. Gwalani, S. Gorsse, D. Choudhuri, Y. Zheng, R. S. Mishra and R. Banerjee, "Tensile yield strength of a single bulk Al0.3CoCrFeNi high entropy alloy can be tuned from 160 MPa to 1800 MPa," *Scr Mater*, vol. 162, pp. 18–23, Mar. 2019. doi: 10.1016/j.scriptamat.2018.10.023.
38. J. Y. He *et al.*, "A precipitation-hardened high-entropy alloy with outstanding tensile properties," *Acta Mater*, vol. 102, pp. 187–196, Jan. 2016. doi: 10.1016/j.actamat.2015.08.076.
39. A. J. Zaddach, R. O. Scattergood and C. C. Koch, "Tensile properties of low-stacking fault energy high-entropy alloys," *Mater Sci Eng A*, vol. 636, pp. 373–378, Jun. 2015. doi: 10.1016/j.msea.2015.03.109.
40. K. Ming, X. Bi and J. Wang, "Precipitation strengthening of ductile Cr 15 Fe 20 Co 35 Ni 20 Mo 10 alloys," *Scr Mater*, vol. 137, pp. 88–93, Aug. 2017. doi: 10.1016/j.scriptamat.2017.05.019.
41. B. Gwalani, V. Soni, M. Lee, S. Mantri, Y. Ren and R. Banerjee, "Optimizing the coupled effects of Hall-Petch and precipitation strengthening in a Al 0.3 CoCrFeNi high entropy alloy," *Mater Des*, vol. 121, pp. 254–260, May 2017. doi: 10.1016/j.matdes.2017.02.072.
42. P. Li, A. Wang and C. T. Liu, "A ductile high entropy alloy with attractive magnetic properties," *J Alloys Compd*, vol. 694, pp. 55–60, Feb. 2017. doi: 10.1016/j.jallcom.2016.09.186.
43. N. D. Stepanov, D. G. Shaysultanov, M. A. Tikhonovsky and G. A. Salishchev, "Tensile properties of the Cr–Fe–Ni–Mn non-equiatomic multicomponent alloys with different Cr contents," *Mater Des*, vol. 87, pp. 60–65, Dec. 2015. doi: 10.1016/j.matdes.2015.08.007.
44. Z. G. Zhu, K. H. Ma, X. Yang and C. H. Shek, "Annealing effect on the phase stability and mechanical properties of (FeNiCrMn)(100–)Co high entropy alloys," *J Alloys Compd*, vol. 695, pp. 2945–2950, Feb. 2017. doi: 10.1016/j.jallcom.2016.11.376.
45. M. Frank *et al.*, "Correlating work hardening with co-activation of stacking fault strengthening and transformation in a high entropy alloy using in-situ neutron diffraction," *Sci Rep*, vol. 10, no. 1, p. 22263, Dec. 2020. doi: 10.1038/s41598-020-79492-8.
46. S. Sinha, S. S. Nene, M. Frank, K. Liu, R. A. Lebensohn and R. S. Mishra, "Deformation mechanisms and ductile fracture characteristics of a friction stir processed transformative high entropy alloy," *Acta Mater*, vol. 184, pp. 164–178, Feb. 2020. doi: 10.1016/j.actamat.2019.11.056.
47. D. Mishra *et al.*, "Phase separation induced heterostructure promotes strength-ductility synergy in Cu-rich compositionally complex alloy in as-cast state," *Mater Today Commun*, vol. 44, p. 112020, Mar. 2025, doi: 10.1016/j.mtcomm.2025.112020.

48. R. S. Haridas *et al.*, "Work hardening in metastable high entropy alloys: a modified five-parameter model," *J. Mater. Res. Technol.* vol. 18, pp. 3358–3372, May 2022, doi: 10.1016/j.jmrt.2022.04.016.
49. F. Maresca and W. A. Curtin, "Theory of screw dislocation strengthening in random BCC alloys from dilute to 'High-Entropy' alloys," *Acta Mater*, vol. 182, pp. 144–162, Jan. 2020. doi: 10.1016/j.actamat.2019.10.007.
50. A. R. Balpande *et al.*, "Excellent specific strength-ductility synergy in novel complex concentrated alloy after suction casting," *Mater Des*, vol. 242, p. 112988, Jun. 2024. doi: 10.1016/j.matdes.2024.112988.
51. S. Chen *et al.*, "Grain growth and Hall-Petch relationship in a refractory HfNbTaZrTi high-entropy alloy," *J Alloys Compd*, vol. 795, pp. 19–26, Jul. 2019. doi: 10.1016/j.jallcom.2019.04.291.
52. C.-C. Juan *et al.*, "Simultaneously increasing the strength and ductility of a refractory high-entropy alloy via grain refining," *Mater Lett*, vol. 184, pp. 200–203, Dec. 2016. doi: 10.1016/j.matlet.2016.08.060.
53. B. Schuh *et al.*, "Thermodynamic instability of a nanocrystalline, single-phase TiZrNbHfTa alloy and its impact on the mechanical properties," *Acta Mater*, vol. 142, pp. 201–212, Jan. 2018. doi: 10.1016/j.actamat.2017.09.035.
54. G. E. Dieter, *Mechanical Metallurgy*. McGraw-Hill, 1998.
55. S. Wang, M. Wu, D. Shu, G. Zhu, D. Wang and B. Sun, "Mechanical instability and tensile properties of TiZrHfNbTa high entropy alloy at cryogenic temperatures," *Acta Mater*, vol. 201, pp. 517–527, Dec. 2020. doi: 10.1016/j.actamat.2020.10.044.
56. Y. D. Wu *et al.*, "A refractory Hf25Nb25Ti25Zr25 high-entropy alloy with excellent structural stability and tensile properties," *Mater Lett*, vol. 130, pp. 277–280, Sep. 2014. doi: 10.1016/j.matlet.2014.05.134.
57. W. F. Hosford, *Mechanical Behavior of Materials*, 2nd ed. New York: Cambridge University Press, 2010.
58. A. Dutta, S. Tung, S. K. Gupta, M. Tsai and S. S. Nene, "Room-temperature superformability in novel as-cast high-entropy alloy during compressive loading," *Adv Eng Mater*, vol. 25, no. 8, Apr. 2023. doi: 10.1002/adem.202201347.
59. J. Sun, W. Zhao, P. Yan, X. Xia, L. Jiao and X. Wang, "Effect of temperature and strain rate on quasi-static and dynamic compressive behavior of forged CrMnFeCoNi high entropy alloy," *Mater Sci Eng A*, vol. 870, p. 144846, Apr. 2023. doi: 10.1016/j.msea.2023.144846.
60. Z. Fu, W. Chen, S. Fang, D. Zhang, H. Xiao and D. Zhu, "Alloying behavior and deformation twinning in a CoNiFeCrAl0.6Ti0.4 high entropy alloy processed by spark plasma sintering," *J Alloys Compd*, vol. 553, pp. 316–323, Mar. 2013. doi: 10.1016/j.jallcom.2012.11.146.
61. J. Weiss, D. J. Savage, S. C. Vogel, B. A. McWilliams, R. S. Mishra and M. Knezevic, "Evolution of microstructure and strength of a high entropy alloy undergoing the strain-induced martensitic transformation," *Mater Sci Eng: A*, vol. 887, p. 145754, Nov. 2023. doi: 10.1016/j.msea.2023.145754.
62. F. J. Wang, Y. Zhang and G. L. Chen, "Atomic packing efficiency and phase transition in a high entropy alloy," *J Alloys Compd*, vol. 478, no. 1–2, pp. 321–324, Jun. 2009. doi: 10.1016/j.jallcom.2008.11.059.
63. K. B. Zhang *et al.*, "Microstructure and mechanical properties of CoCrFeNiTiAl high-entropy alloys," *Mater Sci Eng A*, vol. 508, no. 1–2, pp. 214–219, May 2009. doi: 10.1016/j.msea.2008.12.053.
64. Q. Wang, X. Wu, Z. Bai, L. Gu and L. Zeng, "Microstructure and compressive deformation behavior of as-cast CrFeCoNiCux high entropy alloys," *J Alloys Compd*, vol. 1004, p. 175880, Nov. 2024. doi: 10.1016/j.jallcom.2024.175880.
65. T. Wang, Y. Wu and H. Jiang, "Effect of matrix reversible transition on mechanical and tribological properties of a novel co-free high-entropy alloy," *J Alloys Compd*, vol. 1003, p. 175556, Oct. 2024. doi: 10.1016/j.jallcom.2024.175556.

66. S. Xu *et al.*, "Multiscale-phase-driven strength-ductility synergy in Fe3Cr2CoNiAlx high entropy alloys," *Intermetallics*, vol. 156, p. 107865, May 2023. doi: 10.1016/j.intermet.2023.107865.
67. Y. X. Zhuang, W. J. Liu, Z. Y. Chen, H. D. Xue and J. C. He, "Effect of elemental interaction on microstructure and mechanical properties of FeCoNiCuAl alloys," *Mater Sci Eng A*, vol. 556, pp. 395–399, Oct. 2012. doi: 10.1016/j.msea.2012.07.003.
68. Z. Fu, W. Chen, H. Xiao, L. Zhou, D. Zhu and S. Yang, "Fabrication and properties of nanocrystalline Co0.5FeNiCrTi0.5 high entropy alloy by MA–SPS technique," *Mater Des*, vol. 44, pp. 535–539, Feb. 2013. doi: 10.1016/j.matdes.2012.08.048.
69. W. Chen, Z. Fu, S. Fang, H. Xiao and D. Zhu, "Alloying behavior, microstructure and mechanical properties in a FeNiCrCo0.3Al0.7 high entropy alloy," *Mater Des*, vol. 51, pp. 854–860, Oct. 2013. doi: 10.1016/j.matdes.2013.04.061.
70. Y. Dong, K. Zhou, Y. Lu, X. Gao, T. Wang and T. Li, "Effect of vanadium addition on the microstructure and properties of AlCoCrFeNi high entropy alloy," *Mater Des*, vol. 57, pp. 67–72, May 2014. doi: 10.1016/j.matdes.2013.12.048.
71. Y. J. Zhou, Y. Zhang, Y. L. Wang and G. L. Chen, "Microstructure and compressive properties of multicomponent Alx(TiVCrMnFeCoNiCu)100−x high-entropy alloys," *Mater Sci Eng A*, vol. 454–455, pp. 260–265, Apr. 2007. doi: 10.1016/j.msea.2006.11.049.
72. J. M. Zhu, H. M. Fu, H. F. Zhang, A. M. Wang, H. Li and Z. Q. Hu, "Synthesis and properties of multiprincipal component AlCoCrFeNiSix alloys," *Mater Sci Eng A*, vol. 527, no. 27–28, pp. 7210–7214, Oct. 2010. doi: 10.1016/j.msea.2010.07.049.
73. H. F. Sheng, M. Gong and L. M. Peng, "Microstructural characterization and mechanical properties of an Al0.5CoCrFeCuNi high-entropy alloy in as-cast and heat-treated/quenched conditions," *Materials Science and Engineering: A*, vol. 567, pp. 14–20, Apr. 2013. doi: 10.1016/j.msea.2013.01.006.
74. S. Liu, M. C. Gao, P. K. Liaw and Y. Zhang, "Microstructures and mechanical properties of Al CrFeNiTi0.25 alloys," *J Alloys Compd*, vol. 619, pp. 610–615, Jan. 2015. doi: 10.1016/j.jallcom.2014.09.073.
75. N. D. Stepanov *et al.*, "Effect of V content on microstructure and mechanical properties of the CoCrFeMnNiVx high entropy alloys," *J Alloys Compd*, vol. 628, pp. 170–185, Apr. 2015. doi: 10.1016/j.jallcom.2014.12.157.
76. X. F. Wang, Y. Zhang, Y. Qiao and G. L. Chen, "Novel microstructure and properties of multicomponent CoCrCuFeNiTix alloys," *Intermetallics*, vol. 15, no. 3, pp. 357–362, Mar. 2007. doi: 10.1016/j.intermet.2006.08.005.
77. Y. J. Zhou, Y. Zhang, Y. L. Wang and G. L. Chen, "Solid solution alloys of AlCoCrFeNiTix with excellent room-temperature mechanical properties," *Appl Phys Lett*, vol. 90, no. 18, Apr. 2007. doi: 10.1063/1.2734517.
78. Y. Dong, Y. Lu, J. Kong, J. Zhang and T. Li, "Microstructure and mechanical properties of multi-component AlCrFeNiMox high-entropy alloys," *J Alloys Compd*, vol. 573, pp. 96–101, Oct. 2013. doi: 10.1016/j.jallcom.2013.03.253.
79. G. A. Salishchev *et al.*, "Effect of Mn and V on structure and mechanical properties of high-entropy alloys based on CoCrFeNi system," *J Alloys Compd*, vol. 591, pp. 11–21, Apr. 2014. doi: 10.1016/j.jallcom.2013.12.210.
80. X. Shen *et al.*, "Microstructural evolution and mechanical behavior of novel TiZrTaxNbMo refractory high-entropy alloys," *J Alloys Compd*, vol. 990, p. 174459, Jun. 2024. doi: 10.1016/j.jallcom.2024.174459.
81. O. N. Senkov, S. V. Senkova and C. Woodward, "Effect of aluminum on the microstructure and properties of two refractory high-entropy alloys," *Acta Mater*, vol. 68, pp. 214–228, Apr. 2014. doi: 10.1016/j.actamat.2014.01.029.
82. C. Liu, Y. Wang, Y. Zhang, L.-C. Zhang and L. Wang, "Deformation mechanisms of additively manufactured TiNbTaZrMo refractory high-entropy alloy: The role of cellular structure," *Int J Plast*, vol. 173, p. 103884, Feb. 2024. doi: 10.1016/j.ijplas.2024.103884.

83. A. Gupta, G. Shankar, S. Pawar, S.-H. Choi and S. Suwas, "Microstructural stability and mechanical properties of the as-cast and heat-treated newly developed TiNbCrTa refractory complex concentrated alloy," *J Appl Phys*, vol. 136, no. 5, Aug. 2024. doi: 10.1063/5.0206425.
84. X.-Y. Ding, H.-Y. Zheng, P.-P. Zhang, L.-M. Luo, Y.-C. Wu and J.-H. Yao, "Microstructure and mechanical properties of WTaVCrTi refractory high-entropy alloy by vacuum levitation melting for fusion applications," *J Mater Eng Perform*, vol. 32, no. 17, pp. 7869–7878, Sep. 2023. doi: 10.1007/s11665-022-07688-2.
85. P. Chakraborty *et al.*, "Design and development of low density, high strength ZrNbAlVTi high entropy alloy for high temperature applications," *Int J Refract Metals Hard Mater*, vol. 113, p. 106222, Jun. 2023. doi: 10.1016/j.ijrmhm.2023.106222.
86. O. N. Senkov, G. B. Wilks, J. M. Scott and D. B. Miracle, "Mechanical properties of Nb25Mo25Ta25W25 and V20Nb20Mo20Ta20W20 refractory high entropy alloys," *Intermetallics*, vol. 19, no. 5, pp. 698–706, May 2011. doi: 10.1016/j.intermet.2011.01.004.
87. O. N. Senkov, C. Woodward and D. B. Miracle, "Microstructure and properties of aluminum-containing refractory high-entropy alloys," *JOM*, vol. 66, no. 10, pp. 2030–2042, Oct. 2014. doi: 10.1007/s11837-014-1066-0.
88. N. D. Stepanov, D. G. Shaysultanov, G. A. Salishchev and M. A. Tikhonovsky, "Structure and mechanical properties of a light-weight AlNbTiV high entropy alloy," *Mater Lett*, vol. 142, pp. 153–155, Mar. 2015. doi: 10.1016/j.matlet.2014.11.162.
89. N. N. Guo *et al.*, "Microstructure and mechanical properties of refractory MoNbHfZrTi high-entropy alloy," *Mater Des*, vol. 81, pp. 87–94, Sep. 2015. doi: 10.1016/j.matdes.2015.05.019.
90. O. N. Senkov, S. V. Senkova, D. B. Miracle and C. Woodward, "Mechanical properties of low-density, refractory multi-principal element alloys of the Cr–Nb–Ti–V–Zr system," *Mater Sci Eng A*, vol. 565, pp. 51–62, Mar. 2013. doi: 10.1016/j.msea.2012.12.018.
91. H. Zhang *et al.*, "High-strength NbMoTaX refractory high-entropy alloy with low stacking fault energy eutectic phase via laser additive manufacturing," *Mater Des*, vol. 201, p. 109462, Mar. 2021. doi: 10.1016/j.matdes.2021.109462.
92. P. Chen *et al.*, "Fatigue behavior of high-entropy alloys: A review," *Sci China Technol Sci*, vol. 61, no. 2, pp. 168–178, Feb. 2018. doi: 10.1007/s11431-017-9137-4.
93. S. A. McKelvey and A. Fatemi, "Surface finish effect on fatigue behavior of forged steel," *Int J Fatigue*, vol. 36, no. 1, pp. 130–145, Mar. 2012. doi: 10.1016/j.ijfatigue.2011.08.008.
94. S. Kwofie and H. Chandler, "Fatigue life prediction under conditions where cyclic creep–fatigue interaction occurs," *Int J Fatigue*, vol. 29, no. 12, pp. 2117–2124, Dec. 2007. doi: 10.1016/j.ijfatigue.2007.01.022.
95. K. Lu *et al.*, "Deformation mechanisms of CoCrFeMnNi high-entropy alloy under low-cycle-fatigue loading," *Acta Mater*, vol. 215, p. 117089, Aug. 2021. doi: 10.1016/j.actamat.2021.117089.
96. J. Hu, K. Yang, Q. Wang, Q. C. Zhao, Y. H. Jiang and Y. J. Liu, "Ultra-long life fatigue behavior of a high-entropy alloy," *Int J Fatigue*, vol. 178, p. 108013, Jan. 2024. doi: 10.1016/j.ijfatigue.2023.108013.
97. K. Liu, M. Komarasamy, B. Gwalani, S. Shukla and R. S. Mishra, "Fatigue behavior of ultrafine grained triplex Al0.3CoCrFeNi high entropy alloy," *Scr Mater*, vol. 158, pp. 116–120, Jan. 2019. doi: 10.1016/j.scriptamat.2018.08.048.
98. Y. Li *et al.*, "Uncovering the critical factor governing the fatigue crack propagation under the coexistence of multi-interfaces in AlCoCrFeNi2.1 eutectic high-entropy alloy," *Eng Fract Mech*, vol. 258, p. 108132, Dec. 2021. doi: 10.1016/j.engfracmech.2021.108132.
99. Y.-K. Kim, M.-S. Baek, S. Yang and K.-A. Lee, "In-situ formed oxide enables extraordinary high-cycle fatigue resistance in additively manufactured CoCrFeMnNi high-entropy alloy," *Addit Manuf*, vol. 38, p. 101832, Feb. 2021. doi: 10.1016/j.addma.2020.101832.

100. C. Song *et al.*, "Effect of multiphase microstructure on fatigue crack propagation behavior in TRIP-assisted steels," *Int J Fatigue*, vol. 133, p. 105425, Apr. 2020. doi: 10.1016/j.ijfatigue.2019.105425.
101. K. Liu, S. S. Nene, M. Frank, S. Sinha and R. S. Mishra, "Metastability-assisted fatigue behavior in a friction stir processed dual-phase high entropy alloy," *Mater Res Lett*, vol. 6, no. 11, pp. 613–619, Nov. 2018. doi: 10.1080/21663831.2018.1523240.
102. Y. Z. Tian, S. J. Sun, H. R. Lin and Z. F. Zhang, "Fatigue behavior of CoCrFeMnNi high-entropy alloy under fully reversed cyclic deformation," *J Mater Sci Technol*, vol. 35, no. 3, pp. 334–340, Mar. 2019. doi: 10.1016/j.jmst.2018.09.068.
103. G. T. Lee *et al.*, "Effect of microstructural features on the high-cycle fatigue behavior of CoCrFeMnNi high-entropy alloys deformed at room and cryogenic temperatures," *Metal Mater Int*, vol. 27, no. 4, pp. 593–602, Apr. 2021. doi: 10.1007/s12540-020-00786-7.
104. Z. Tang, T. Yuan, C.-W. Tsai, J.-W. Yeh, C. D. Lundin and P. K. Liaw, "Fatigue behavior of a wrought Al0.5CoCrCuFeNi two-phase high-entropy alloy," *Acta Mater*, vol. 99, pp. 247–258, Oct. 2015. doi: 10.1016/j.actamat.2015.07.004.
105. M. Z. Ghomsheh *et al.*, "High cycle fatigue deformation mechanisms of a single phase CrMnFeCoNi high entropy alloy," *Mater Sci Eng A*, vol. 777, p. 139034, Mar. 2020. doi: 10.1016/j.msea.2020.139034.
106. K. Liu, B. Gwalani, M. Komarasamy, S. Shukla, T. Wang and R. S. Mishra, "Effect of nano-sized precipitates on the fatigue property of a lamellar structured high entropy alloy," *Mater Sci Eng A*, vol. 760, pp. 225–230, Jul. 2019. doi: 10.1016/j.msea.2019.06.012.
107. K. Liu, S. S. Nene, M. Frank, S. Sinha and R. S. Mishra, "Extremely high fatigue resistance in an ultrafine grained high entropy alloy," *Appl Mater Today*, vol. 15, pp. 525–530, Jun. 2019. doi: 10.1016/j.apmt.2019.04.001.
108. M. A. Hemphill *et al.*, "Fatigue behavior of Al0.5CoCrCuFeNi high entropy alloys," *Acta Mater*, vol. 60, no. 16, pp. 5723–5734, Sep. 2012. doi: 10.1016/j.actamat.2012.06.046.
109. Y. Chen, B. Li, B. Chen and F. Xuan, "High-cycle fatigue induced twinning in CoCrFeNi high-entropy alloy processed by laser powder bed fusion additive manufacturing," *Addit Manuf*, vol. 61, p. 103319, Jan. 2023. doi: 10.1016/j.addma.2022.103319.
110. S. Picak *et al.*, "On the low-cycle fatigue response of CoCrNiFeMn high entropy alloy with ultra-fine grain structure," *Acta Mater*, vol. 205, p. 116540, Feb. 2021. doi: 10.1016/j.actamat.2020.116540.
111. H. Shahmir, P. Saeedpour, M. S. Mehranpour, A. Shams and C. S. Lee, "Low cycle fatigue properties of CoCrFeNiMn high-entropy alloy with heterogeneous microstructure," *Mater Sci Eng A*, vol. 912, p. 147004, Oct. 2024. doi: 10.1016/j.msea.2024.147004.
112. K. Lu *et al.*, "High-temperature low cycle fatigue behavior of an equiatomic CoCrFeMnNi high-entropy alloy," *Mater Sci Eng A*, vol. 791, p. 139781, Jul. 2020. doi: 10.1016/j.msea.2020.139781.
113. S. A. A. Shams *et al.*, "Low-cycle fatigue properties of CoCrFeMnNi high-entropy alloy compared with its conventional counterparts," *Mater Sci Eng A*, vol. 792, p. 139661, Aug. 2020. doi: 10.1016/j.msea.2020.139661.
114. S. Chen *et al.*, "Stress-controlled fatigue of HfNbTaTiZr high-entropy alloy and associated deformation and fracture mechanisms," *J Mater Sci Technol*, vol. 114, pp. 191–205, Jul. 2022. doi: 10.1016/j.jmst.2021.10.026.
115. B. Guennec *et al.*, "Four-point bending fatigue behavior of an equimolar BCC HfNbTaTiZr high-entropy alloy: Macroscopic and microscopic viewpoints," *Materialia*, vol. 4, pp. 348–360, Dec. 2018. doi: 10.1016/j.mtla.2018.09.040.
116. B. Guennec *et al.*, "Analysis of the fatigue crack growth mechanisms in equimolar body centered cubic HfNbTaTiZr high-entropy alloy: Discussions on its singularities and consequences on the crack propagation rate properties," *Intermetallics*, vol. 110, p. 106459, Jul. 2019. doi: 10.1016/j.intermet.2019.04.002.

117. M. E. Kassner, *Fundamentals of Creep in Metals and Alloys*. Los Angeles: Butterworth-Heinemann, 2015.
118. K.-Y. Tsai, M.-H. Tsai and J.-W. Yeh, "Sluggish diffusion in Co–Cr–Fe–Mn–Ni high-entropy alloys," *Acta Mater*, vol. 61, no. 13, pp. 4887–4897, Aug. 2013. doi: 10.1016/j.actamat.2013.04.058.
119. S.-Y. Chang, C.-E. Li, Y.-C. Huang, H.-F. Hsu, J.-W. Yeh and S.-J. Lin, "Structural and thermodynamic factors of suppressed interdiffusion kinetics in multi-component high-entropy materials," *Sci Rep*, vol. 4, no. 1, p. 4162, Feb. 2014. doi: 10.1038/srep04162.
120. J. Y. He, C. Zhu, D. Q. Zhou, W. H. Liu, T. G. Nieh and Z. P. Lu, "Steady state flow of the FeCoNiCrMn high entropy alloy at elevated temperatures," *Intermetallics*, vol. 55, pp. 9–14, Dec. 2014. doi: 10.1016/j.intermet.2014.06.015.
121. T. Cao, J. Shang, J. Zhao, C. Cheng, R. Wang and H. Wang, "The influence of Al elements on the structure and the creep behavior of Al x CoCrFeNi high entropy alloys," *Mater Lett*, vol. 164, pp. 344–347, Feb. 2016. doi: 10.1016/j.matlet.2015.11.016.
122. Z. Tobolová and J. Čadek, "An interpretation of steady state creep," *Philosophical Magazine*, vol. 26, no. 6, pp. 1419–1428, Dec. 1972. doi: 10.1080/14786437208220352.
123. F. Liu *et al.*, "Creep behaviors of fine-grained Ni-base powder metallurgy superalloys at elevated temperatures," *J Alloys Compd*, vol. 867, p. 158865, Jun. 2021. doi: 10.1016/j.jallcom.2021.158865.
124. T. Zhen, M. W. Barsoum, S. R. Kalidindi, M. Radovic, Z. M. Sun and T. El-Raghy, "Compressive creep of fine and coarse-grained T3SiC2 in air in the 1100–1300°C temperature range," *Acta Mater*, vol. 53, no. 19, pp. 4963–4973, Nov. 2005. doi: 10.1016/j.actamat.2005.05.037.
125. T. G. Langdon, "A Lifetime of Research in Creep, Superplasticity, and Ultrafine-Grained Materials," *Adv Eng Mater*, vol. 22, no. 1, Jan. 2020. doi: 10.1002/adem.201900442.
126. Y.-L. Tsai, S.-F. Wang, H.-Y. Bor and Y.-F. Hsu, "Effects of alloy elements on microstructure and creep properties of fine-grained nickel-based superalloys at moderate temperatures," *Mater Sci Eng A*, vol. 571, pp. 155–160, Jun. 2013. doi: 10.1016/j.msea.2013.02.002.
127. C. Song, G. Li, G. Li, G. Zhang and B. Cai, "Tensile creep behavior and mechanism of CoCrFeMnNi high entropy alloy," *Micron*, vol. 150, p. 103144, Nov. 2021. doi: 10.1016/j.micron.2021.103144.
128. X. Han *et al.*, "Improved creep resistance of high entropy transition metal carbides," *J Eur Ceram Soc*, vol. 40, no. 7, pp. 2709–2715, Jul. 2020. doi: 10.1016/j.jeurceramsoc.2019.12.036.
129. D.-H. Lee *et al.*, "Spherical nanoindentation creep behavior of nanocrystalline and coarse-grained CoCrFeMnNi high-entropy alloys," *Acta Mater*, vol. 109, pp. 314–322, May 2016. doi: 10.1016/j.actamat.2016.02.049.
130. F. Dobeš, H. Hadraba, Z. Chlup, A. Dlouhý, M. Vilémová and J. Matějíček, "Compressive creep behavior of an oxide-dispersion-strengthened CoCrFeMnNi high-entropy alloy," *Mater Sci Eng A*, vol. 732, pp. 99–104, Aug. 2018. doi: 10.1016/j.msea.2018.06.108.
131. P. Kral *et al.*, "Creep behavior of an AlTiVNbZr0.25 high entropy alloy at 1073 K," *Mater Sci Eng A*, vol. 783, p. 139291, May 2020. doi: 10.1016/j.msea.2020.139291.
132. M.-G. Jo *et al.*, "High temperature tensile and creep properties of CrMnFeCoNi and CrFeCoNi high-entropy alloys," *Mater Sci Eng A*, vol. 838, p. 142748, Mar. 2022. doi: 10.1016/j.msea.2022.142748.
133. S. Chen *et al.*, "Nanoscale serration and creep characteristics of Al0.5CoCrCuFeNi high-entropy alloys," *J Alloys Compd*, vol. 752, pp. 464–475, Jul. 2018. doi: 10.1016/j.jallcom.2018.04.137.
134. J. Wu *et al.*, "Structure related potential-upsurge during tensile creep of high entropy Al20Ce20La20Ni20Y20 metallic glass," *J Alloys Compd*, vol. 827, p. 154298, Jun. 2020. doi: 10.1016/j.jallcom.2020.154298.

135. H. T. Jeong, H. K. Park, H. S. Kang and W. J. Kim, "Operation of solute-drag creep in an AlCoCrFeMnNi high-entropy alloy and enhanced hot workability," *J Alloys Compd*, vol. 824, p. 153829, May 2020. doi: 10.1016/j.jallcom.2020.153829.
136. D.-H. Lee *et al.*, "Activation energy for plastic flow in nanocrystalline CoCrFeMnNi high-entropy alloy: A high temperature nanoindentation study," *Scr Mater*, vol. 156, pp. 129–133, Nov. 2018. doi: 10.1016/j.scriptamat.2018.07.014.
137. Z. Xu *et al.*, "Microstructure and nanoindentation creep behavior of CoCrFeMnNi high-entropy alloy fabricated by selective laser melting," *Addit Manuf*, vol. 28, pp. 766–771, Aug. 2019. doi: 10.1016/j.addma.2019.06.012.
138. C. M. Cao, J. Xu, W. Tong, Y. X. Hao, P. Gu and L. M. Peng, "Creep behaviour and microstructural evolution of Al_xCrMnFeCoNi high-entropy alloys," *Mater Sci Technol*, vol. 35, no. 10, pp. 1283–1290, Jul. 2019. doi: 10.1080/02670836.2019.1619306.
139. Y. Bin Kang, S. H. Shim, K. H. Lee and S. I. Hong, "Dislocation creep behavior of CoCrFeMnNi high entropy alloy at intermediate temperatures," *Mater Res Lett*, vol. 6, no. 12, pp. 689–695, Dec. 2018. doi: 10.1080/21663831.2018.1543731.
140. C. Cao *et al.*, "Intermediate-Temperature Creep Deformation and Microstructural Evolution of an Equiatomic FCC-Structured CoCrFeNiMn High-Entropy Alloy," *Entropy*, vol. 20, no. 12, p. 960, Dec. 2018. doi: 10.3390/e20120960.
141. M. Sadeghilaridjani, S. Muskeri, M. Pole and S. Mukherjee, "High-Temperature Nano-Indentation Creep of Reduced Activity High Entropy Alloys Based on 4-5-6 Elemental Palette," *Entropy*, vol. 22, no. 2, p. 230, Feb. 2020. doi: 10.3390/e22020230.
142. C. Gadelmeier, S. Haas, T. Lienig, A. Manzoni, M. Feuerbacher and U. Glatzel, "Temperature dependent solid solution strengthening in the high entropy alloy CrMnFeCoNi in single crystalline state," *Metals (Basel)*, vol. 10, no. 11, p. 1412, Oct. 2020. doi: 10.3390/met10111412.
143. K. A. Rozman, M. Detrois, T. Liu, M. C. Gao, P. D. Jablonski and J. A. Hawk, "Long-Term Creep Behavior of a CoCrFeNiMn High-Entropy Alloy," *J Mater Eng Perform*, vol. 29, no. 9, pp. 5822–5839, Sep. 2020. doi: 10.1007/s11665-020-05103-2.
144. H. Shahmir, J. He, Z. Lu, M. Kawasaki and T. G. Langdon, "Evidence for superplasticity in a CoCrFeNiMn high-entropy alloy processed by high-pressure torsion," *Mater Sci Eng A*, vol. 685, pp. 342–348, Feb. 2017. doi: 10.1016/j.msea.2017.01.016.
145. N. Li *et al.*, "Achieving superior superplasticity in CoCrFeNiCu high entropy alloy via friction stir processing with an improved convex tool," *Mater Sci Eng A*, vol. 873, p. 145034, May 2023. doi: 10.1016/j.msea.2023.145034.
146. Y. Palguna and R. Korla, "Superplastic-like flow behavior in Al0.2CoCrFeNiMo0.5 high entropy alloy," *Mater Lett*, vol. 358, p. 135789, Mar. 2024. doi: 10.1016/j.matlet.2023.135789.
147. A. V. Kuznetsov, D. G. Shaisultanov, N. Stepanov, G. A. Salishchev and O. N. Senkov, "Superplasticity of AlCoCrCuFeNi High Entropy Alloy," *Mater Sci Forum*, vol. 735, pp. 146–151, Dec. 2012. doi: 10.4028/www.scientific.net/MSF.735.146.
148. N. Thi-Cam Nguyen *et al.*, "Superplasticity of V10Cr15Mn5Fe35Co10Ni25 high-entropy alloy processed using high-pressure torsion," *Mater Sci Eng A*, vol. 764, p. 138198, Sep. 2019. doi: 10.1016/j.msea.2019.138198.
149. H. Shahmir, M. Nili-Ahmadabadi, A. Shafiee and T. G. Langdon, "Effect of a minor titanium addition on the superplastic properties of a CoCrFeNiMn high-entropy alloy processed by high-pressure torsion," *Mater Sci Eng A*, vol. 718, pp. 468–476, Mar. 2018. doi: 10.1016/j.msea.2018.02.002.
150. N. Li *et al.*, "Deformation mechanism and microstructure evolution during superplastic deformation of friction stir processed CoCrFeNiMn high entropy alloy," *Mater Charact*, vol. 212, p. 113970, Jun. 2024. doi: 10.1016/j.matchar.2024.113970.
151. N. T.-C. Nguyen, P. Asghari-Rad, H. Park and H. S. Kim, "Differential superplasticity in a multi-phase multi-principal element alloy by initial annealing," *J Mater Sci*, vol. 57, no. 38, pp. 18154–18167, Oct. 2022. doi: 10.1007/s10853-022-07616-8.

152. S. R. Reddy, X. Li, S. Guo, P. P. Bhattacharjee and A. H. Chokshi, "High Strain Rate Superplastic Flow and Fracture Characteristics of a Fine-Grained Eutectic High Entropy Alloy," *Metall Mater Trans A*, vol. 55, no. 1, pp. 173–182, Jan. 2024. doi: 10.1007/s11661-023-07240-4.
153. D. G. Shaysultanov, N. D. Stepanov, A. V. Kuznetsov, G. A. Salishchev and O. N. Senkov, "Phase Composition and Superplastic Behavior of a Wrought AlCoCrCuFeNi High-Entropy Alloy," *JOM*, vol. 65, no. 12, pp. 1815–1828, Dec. 2013. doi: 10.1007/s11837-013-0754-5.
154. S. S. Nene, K. Liu, S. Sinha, M. Frank, S. Williams and R. S. Mishra, "Superplasticity in fine grained dual phase high entropy alloy," *Materialia*, vol. 9, p. 100521, 2020. doi: 10.1016/j.mtla.2019.100521.
155. Z. Li *et al.*, "Effect of annealing temperature on microstructure and mechanical properties of a severe cold-rolled FeCoCrNiMn high-entropy alloy," *Metall Mater Trans A*, vol. 50, no. 7, pp. 3223–3237, Jul. 2019. doi: 10.1007/s11661-019-05231-y.
156. S. R. Reddy, S. Bapari, P. P. Bhattacharjee and A. H. Chokshi, "Superplastic-like flow in a fine-grained equiatomic CoCrFeMnNi high-entropy alloy," *Mater Res Lett*, vol. 5, no. 6, pp. 408–414, Nov. 2017. doi: 10.1080/21663831.2017.1305460.

8 Thermal Stability of Nanocrystalline High Entropy Alloys (HEAs) Synthesized by Mechanical Alloying (MA)

Koushik Sikdar, Debdas Roy and Rahul Mitra

High entropy alloys (HEAs) are one of the latest developed structural materials. Introduced in 2002, the unprecedented physical and mechanical properties and their combinations have drawn enormous research interest. Furthermore, the reduced microstructure in a nanometer length scale resulted in a significant improvement in strength. The mechanical alloying (MA) route was successfully adapted for synthesizing these alloy groups, overcoming the difficulties of the conventional ingot metallurgy route. This development opened a new paradigm with widened composition space and the flexibility of microstructure tailoring. However, the phase stability of the supersaturated solid solution obtained by MA, and the issues related to the grain coarsening during the consolidation of the mechanically alloyed nano and/or ultrafine alloy powder need to be addressed. This chapter emphasizes the thermal stability of nanostructured HEAs prepared via the MA route. The chapter begins with the fundamental concepts of HEAs, followed by the phase selection criteria, processing route, and phase and mechanical property stability has been included in mostly chronological order. Along with this, prescribed methodologies and major developments in this field have been illustrated.

8.1 INTRODUCTION

There is an ever-increasing demand for increasing the operating temperature in power generation and the automotive industry. For such applications, materials need a high-strength and stable microstructure to withstand this temperature for an extended period without plastic deformation. High entropy alloys (HEAs) are one of the newly developed classes of material that have garnered enormous academic research interest because of their unique compositions, microstructures, and properties, unattainable with the existing terminal compositions. The concept of HEAs was

DOI: 10.1201/9781003664093-11

first published in the year 2004 [1, 2]. The "*composition-based*" definition of HEAs based on one of the earliest publications is "*those composed of five or more principal elements in equimolar ratios*", where the "*principal elements with the concentration of each element being between 35 and 5 at.%.*" [2]. Based on the configurational entropy (S) of these multicomponent alloys, the "*entropy-based definition*" has been proposed as $S \geq 1.61R$, where R is the universal gas constant [3]. However, Senkov et al. [4, 5] have considered that $S = 1.39R$ can be considered as HEAs. With an agreement Miracle et al. [3] have concluded that alloy systems having $S \geq 1.36R$ might be considered as HEAs. Considering the indistinctness in the definitions of HEAs, these alloys were generalized as multi-principal element alloys, multi-component alloys, concentrated solid solution alloys, complex concentrated alloys, or metal buffets. Note that the term "HEA" is predominant, and the composition space is not restricted to its original "*rigid*" definition. For instance, equiatomic ternary alloys are often also referred to as HEAs [6]. There are four core effects associated with HEAs, represented as:

High entropy effect
Sluggish diffusion
Severe lattice distortion
Cocktail effect

8.1.1 High Entropy Effect

If the kinetic constraint is omitted, then the feasibility of phase formation is thermodynamically controlled by the Gibbs free energy given by,

$$\Delta G_{mix} = \Delta H_{mix} - T\Delta S_{mix} \tag{8.1}$$

where ΔG_{mix} is the Gibbs free energy of mixing, ΔH_{mix} is the enthalpy of mixing, T is the absolute temperature, and ΔS_{mix} is the entropy of mixing. The ΔS_{mix} can be calculated as:

$$\Delta S_{mix} = -R\sum x_i \ln(x_i) \tag{8.2}$$

where x_i is the mole fraction of constituting elements. For the equiatomic composition, it reduces to:

$$\Delta S_{mix} = R\ln(n) \tag{8.3}$$

where n is the number of elements present in the alloy. All entropy sources such as configurational, vibrational, electronic, and magnetic contributions, are included in the term ΔS_{mix} [7]. At a particular temperature, the phase selection of any alloy has been determined by the relative contributions of these two competing terms (ΔH_{mix} and ΔS_{mix}). The phases with higher entropy have lower Gibbs free energy and tend to be stable. The "*entropy*" term in HEAs mainly refers to configurational entropy.

Based on the total configurational molar entropy (ΔS_{conf}), the alloy groups are classified as:

Low entropy: $\Delta S_{conf} < 0.69R$
Medium entropy: $0.69R < \Delta S_{conf} < 1.61R$
High entropy: $\Delta S_{conf} > 1.61R$

The high entropy effect is essential for the formation of a solid solution (SS) rather than the brittle intermetallic phases [3]. Another important aspect of the HEA is that the number of phases is less than that found from Gibbs criteria. According to the Gibbs phase rule, the number of phases (P) under constant pressure and equilibrium conditions can be predicted as:

$$P = C + 1 - F \tag{8.4}$$

where C is the number of components and F is the degrees of freedom of the system. For example, in the equiatomic AlCoCrCuFeNi HEA, several Body-Centered Cubic (BCC)/ Face-centered Cubic (FCC) phases were observed, depending on the processing route; however, the total number of phases is always less than the Gibbs condition [8]. Gorsse et al. [9] have used the Computer Coupling of Phase Diagrams and Thermochemistry (CALPHAD) technique to reevaluate the role of increased configurational entropy for SS formation, as shown in Figure 8.1. It indicates that as the number of alloying elements increases, the likelihood of single-phase formation reduces. Nevertheless, the addition of alloying elements raises the system's configurational entropy, potentially challenging Boltzmann's hypothesis and suggesting multiple factor influence phase formation.

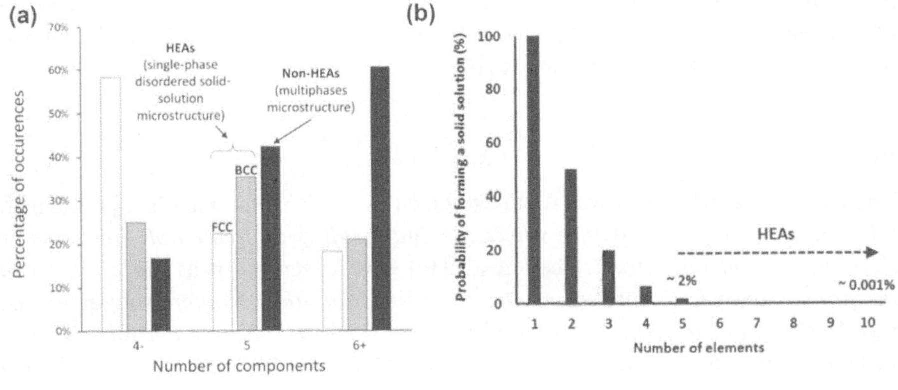

FIGURE 8.1 (a) Occurrence of HEAs and non-HEAs as a function of the number of elements and (b) probability of forming a single phase as a function of the number of alloying elements [9].

8.1.2 Sluggish Diffusion Effect

It means that the diffusion in HEAs is much slower than the conventional alloys. This is necessary for retaining high-temperature strength, structural stability, and formation of nanocrystalline (NC) microstructure. By measuring the Ni tracer diffusion, Vaidya et al. [10] have found that in CoCrFeNi and CoCrFeMnNi, the diffusion rate decreases with an increase in the number of components. Tsai et al. [11] have investigated the diffusion in CoCrFeMnNi in a quasi-binary approach and proposed that the presence of different surrounding atoms increases the activation barrier, thereby slowing down diffusion. Contrary to this, Dabrowa et al. [12] have claimed that "*the sluggish diffusion in HEAs is just an artifact resulting from the study on a specific system (Mn-containing Co-Cr-Fe-Mn-Ni) and unjustified generalization*". Mehta et al. [13] have also presented evidence that goes against the idea of "sluggish diffusion", based on the potential energy fluctuation and configurational entropy of mixing.

8.1.3 Severe Lattice Distortion

Lattice distortion results from different-sized atoms that make the crystal structure. The displacement of each lattice site depends on the atom occupying that site and the types of atoms. This distortion is much more severe than the conventional alloys. However, a study on the VNbMoTaW HEAs elicits that the lattice strain is not more than 5% of the lattice parameter, comparable to both dilute and concentrated alloys [14]. Chen et al. [14] have explained that when the lattice distortion is very high, ordered phases are formed. The instability induced by high lattice distortion cannot be counterbalanced by high configurational entropy. In addition, Song et al. [15] have also demonstrated that the local lattice distortion is higher in refractory HEAs than the 3*d* HEAs. In quaternary and quinary BCC alloys, stability is contributed from both the lattice distortion energy and the mixing entropy, whereas in all FCC alloys, the mixing entropy is dominant. Severe lattice distortion and a high density of dislocations in the TiZrNbTa alloy result in increased oxygen atom diffusion and oxidation rates in the TiZr-rich areas [16].

8.1.4 Cocktail Effect

The term "*cocktail*" effect was first introduced by Prof. S. Ranganathan [17], which refers to a "*synergistic mixture where the end result is unpredictable and greater than the sum of the parts*". Chen et al. [14] have described it as "*in addition to the properties of individual elements, the interaction among composing elements, which may result in some exceptional properties, should also be considered*". For instance, they cite the example of the AlCoCrCuFeNi alloy system, where the Cu precipitation causes the formation of the $L1_2$ ($(Ni,Cu)_3Al$) phase. In addition to this, an increase in the Al content causes increases in hardness due to the formation of a hard BCC phase. In addition to these four core effects, Yang et al. [18] have pointed out the local atomic configuration (LAC) as a core effect, which primarily stems from the fundamental variances in local chemical configuration at the nanoscale.

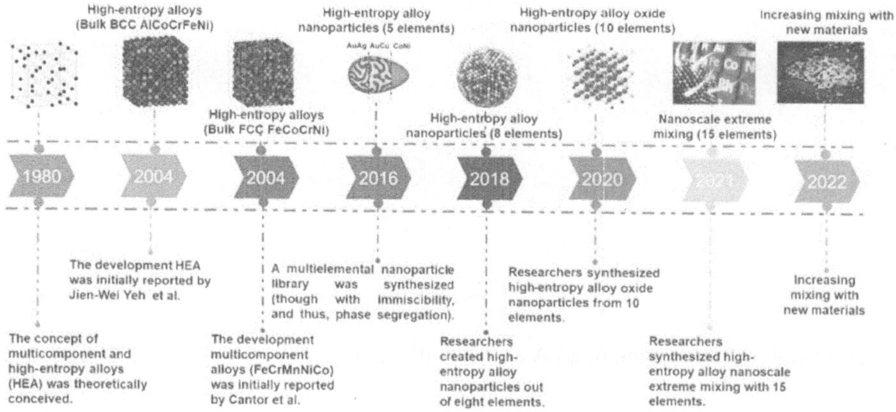

FIGURE 8.2 A history of the evolution of HEAs [55].

A timeline graph showing the HEA's property enhancement has been presented in Figure 8.2. The historical origin, development, design philosophy, fabrication methods, mechanical properties (high-temperature and/or dynamic), deformation mechanism, and microstructure-property correlation have been thoroughly reviewed by several researchers [3, 6, 14, 19–60]. Diao et al. [50] categorized HEA systems into four families, based on their mechanical responses, shown in Figure 8.3. The classification proposed by Krishna et al. [55] in 2024 is the latest one, presented in Figure 8.4.

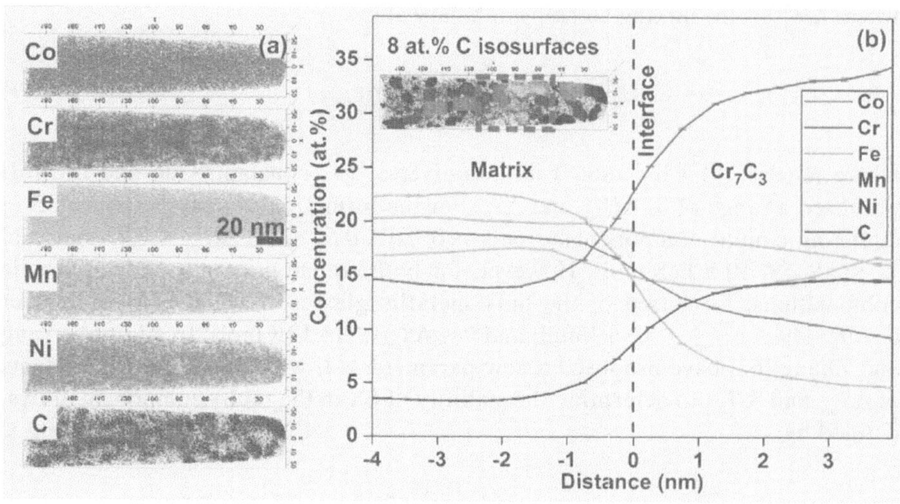

FIGURE 8.3 The classifications of HEAs [50].

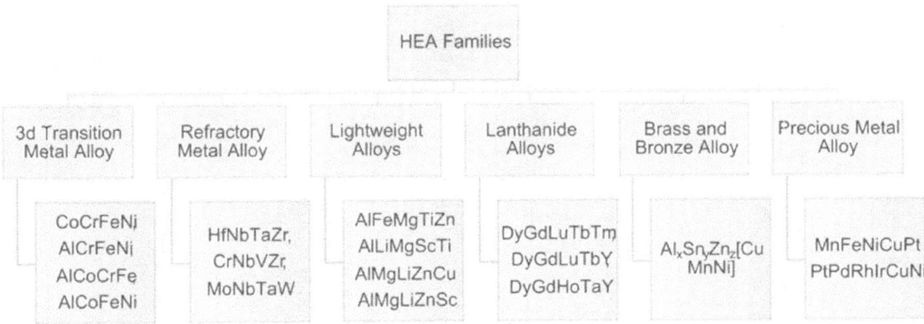

FIGURE 8.4 Classification of HEA alloy families [55].

8.2 PHASE FORMATION CRITERIA IN HEAs

Three major constraints characterize the behavior of the constituent elements in multi-component alloys: the atomic size difference (δ), the mixing enthalpy (ΔH_{mix}), and the mixing entropy (ΔS_{mix}), estimated as [61]:

$$\delta = 100\sqrt{\sum_{i=1}^{n} c_i \left(1 - r_i / \bar{r}\right)^2} \tag{8.5}$$

where $\bar{r} = \sum_{i=1}^{n} c_i r_i$, c_i and r_i are the atomic percentage and atomic radius of the ith element.

$$\Delta H_{mix} = \sum_{i=1, i \neq j}^{n} 4\Delta H_{ij}^{mix} c_i c_j \tag{8.6}$$

where ΔH_{ij}^{mix} is the mixing enthalpy of binary alloys.

$$\Delta S_{mix} = -R\sum_{i=1}^{n} c_i \ln c_i \tag{8.7}$$

where R (= 8.314 J K^{-1} mol^{-1}) is the universal gas constant. Guo and Liu [61] proposed a range of δ, ΔH_{mix} and ΔS_{mix} for the formation of a single-phase SS in equiatomic multi-component systems as $0 \leq \delta \leq 8.5$, $-22 \leq \Delta H_{mix} \leq 7$ kJ/mol, and $11 \leq \Delta S_{mix} \leq 19.5$ J/(K·mol). However, for both equiatomic and non-equiatomic compositions, formation of the bulk metallic glasses (BMGs) is favorable for $\delta > 9$, $-49 \leq \Delta H_{mix} \leq 5.5$ kJ/mol, and $7 \leq \Delta S_{mix} \leq 16$ J/(K·mol). In addition, Yang and Zhang [62] have proposed a new parameter, Ω, which combines the effects of ΔS_{mix} and ΔH_{mix} to determine the stability of SS in the multi-component alloys, defined by:

$$\Omega = \frac{T_m \Delta S_{mix}}{\left|\Delta H_{mix}\right|} \tag{8.8}$$

The melting temperature of the multi-component alloy, T_m, is calculated using the rule of mixtures:

$$T_m = \sum_{i=1}^{n} c_i \left(T_m\right)_i \tag{8.9}$$

where $\left(T_m\right)_i$ is the melting point of the ith component of the alloy. They claimed that the high entropy (HE) stabilized the SS phase for $\Omega \geq 1.1$ and $\delta \leq 6.6\%$. Guo et al. [63] have also pointed out that a critical atomic size polydispersity ~ 6.5 (%) determines the phase selection. Below this, the SS, and above this, the amorphous phase formation, have been found dominant. The amorphous phase formation is also favored by a high negative mixing enthalpy ($\Delta H_{mix} < -12$ kJ/mol). In addition to these basic parameters, as an extension of the classical Hume-Rothery's rule to form solid solution phases, the electronegativity difference, $\Delta\chi$, was also considered. It is defined by [61]:

$$\Delta\chi = \sqrt{\sum_{i=1}^{n} c_i \left(\chi_i - \bar{\chi}\right)^2} \tag{8.10}$$

where $\bar{\chi} = \sum_{i=1}^{n} c_i \chi_i$, χ_i is the Pauling electronegativity for the ith element. The valence electron concentration (VEC) has been found useful in determining the phase stability of HEAs [64]. VEC is defined by:

$$VEC = \sum_{i=1}^{n} c_i \left(VEC\right)_i \tag{8.11}$$

where $\left(VEC\right)_i$ is the VEC of the ith element. When VEC ≤ 6.87, the HEA has a BCC structure; if VEC ≥ 8, the HEA has an FCC structure; and for 6.87 < VEC < 8, the HEA has FCC and BCC mixed structure (cf. Figure 8.5). It is worthwhile to mention that the VEC is different from the electron atom ratio (e/a). The average number of itinerant electrons per atom in that VEC counts the total electrons, including the *d*-shell electrons accommodated in the valence band. Due to the difficulty in defining the e/a values for transition metallic elements, the use of VEC is more convenient for predicting the phase in HEAs. Poletti and Battezzati [65] evaluated the phase stability of HEAs using both e/a and VEC. They considered the electronegativity of an element based on "*Allen*", rather than the "*Pauling*". The investigation shows that the FCC phases are stable at VEC > 7.5 and 1.6 < e/a < 1.8, whereas the BCC phases are found to be stable at VEC < 7.5 and 1.8 < e/a < 2.3. Ye et al. [66] have demonstrated that the influence of root-mean-square (RMS) residual strain significantly influences the phase selection. They found that, in the case of glass-forming HEAs and bulk metallic glasses, amorphization occurs when the RMS residual strain increases above ~10%. In crystalline HEAs, when the RMS residual strain approaches ~5%, a transition from the single-phase SS to a multi-phase structure takes place. Figure 8.6 shows the factors influencing the phase formation of HEAs. A comprehensive explanation of the phase formation in HEAs can be found in Ref. [7]. Praveen et al. [67] have suggested that the these criteria of SS formation are the necessary

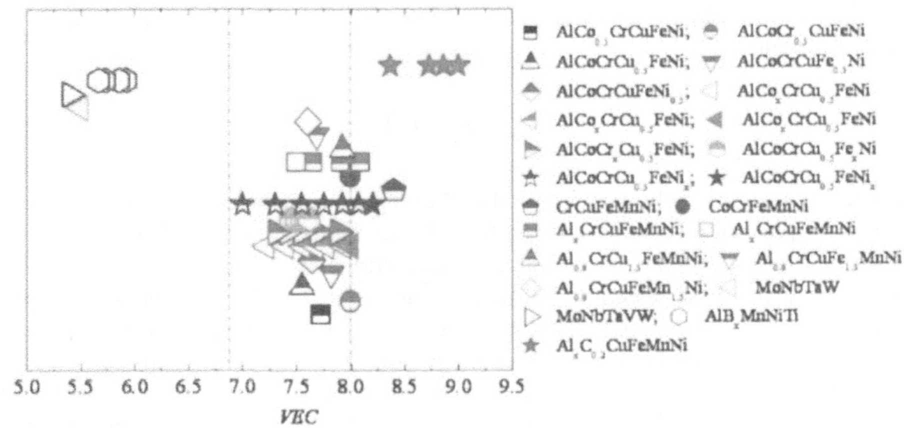

FIGURE 8.5 Relationship between VEC and the phase stability for FCC and BCC solid solutions in various HEAs. Note on the legends: Fully closed symbols for sole FCC phases, fully open symbols for sole bcc phases, and top-half closed symbols for mixed FCC and BCC phases [64].

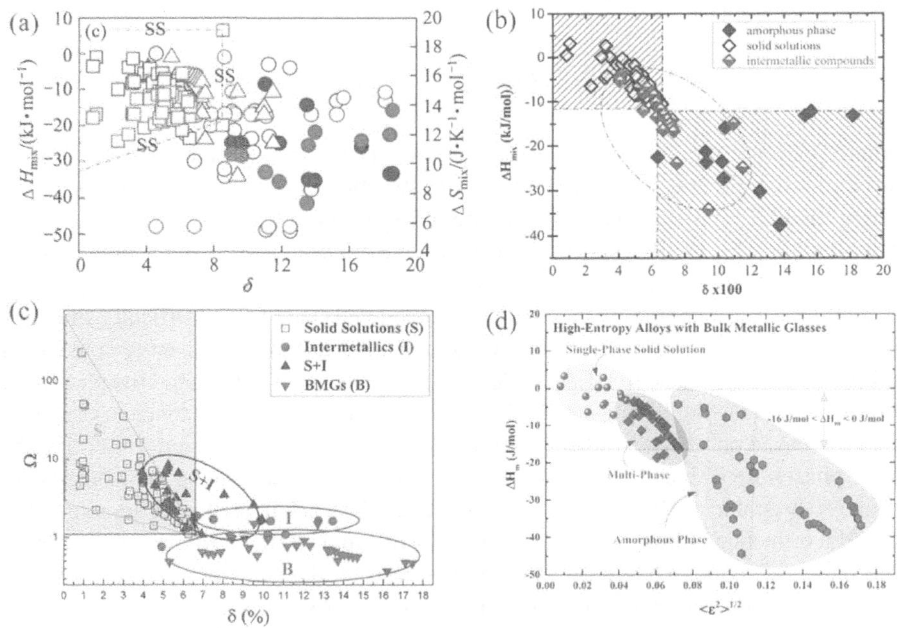

FIGURE 8.6 Relationship between the structure of HEAs and parameters, (a) superimposed effect of ΔH_{mix}, δ and ΔS_{mix}, (b) superimposed effect of ΔH_{mix}, δ, (c) superimposed effect of Ω and δ, and (d) superimposed effect of ΔH_m and $< \varepsilon^2 >^{1/2}$ ($< \varepsilon^2 >^{1/2}$ is the residual strain, which is related to δ) [32].

condition, but it is not sufficient enough to predict the formation of single-phase SS. To support this, they cited the examples of NiCoCuFe, NiCoCrFe, NiCoCrCuFe, and AlCoCrCuFe. All satisfied the condition of ΔH_{mix} ΔS_{mix} and δ, but showed a BCC+FCC phase mixture. In addition, these empirical rules were primarily established for the HEAs synthesized by the conventional melting route and, therefore, may not predict the anticipated nonequilibrium processes like mechanical alloying (MA). Mridha et al. [68] have investigated phase selection in the $Cu_xZn_yTi_{20}Fe_{20}Cr_{20}$ system (x/y = 1/0, 3/1, 1; and x + y = 40) HEAs. They reported that even though the VEC criterion (> 8) predicts the formation of the single-phase FCC structure, all compositions were a mixture of FCC and BCC phases, resulting from the nonequilibrium processing route.

8.3 NANOCRYSTALLINE HEAs BY THE MECHANICAL ALLOYING (MA) ROUTE

Owing to the increased Hall-Petch strengthening contribution, the nano and/or ultrafine-grained metals and alloys show superior mechanical behavior compared to their coarse-grained counterpart. Nanocrystalline (NC) CoCrFeMnNi processed by high-pressure torsion (HPT) possesses a very high tensile strength of 1.9 GPa [69]. Fu et al. [70] have found that the bulk NC $Co_{25}Ni_{25}Fe_{25}Al_{7.5}Cu_{17.5}$ prepared by MA and spark plasma sintering (SPS) exhibits a yield strength and hardness of ~1795 MPa and 454 Hv, as compared to 192 MPa and 129 Hv obtained by the conventional ingot metallurgy route. The NC AlCoCrCuFeNi synthesized by the combination of MA and SPS attains a hardness level of 6.5 GPa [71]. There are several top-down and bottom-up approaches for preparing NC metals and alloys. These methods, along with their advantages and limitations, are presented in Table 8.1.

The majority of the earlier developments of HEAs were concentrated on conventional melting-casting routes. In this process, the possibility of synthesis of a new alloy is primarily dictated by the phase diagram and/or mutual solubility of the elements. Constructing and/or the availability of an equilibrium diagram containing five or more elements is difficult. In addition, the segregation of the alloying elements may require an extended homogenization time. Mechanical alloying (MA) is a nonequilibrium processing technique and has a potential of departure from equilibrium equivalent to irradiation/ion implantation (30 kJ/mol.) [72]. It was invented by Benjamin and coworkers to produce oxide dispersion-strengthened (ODS) alloys. However, with time, this technique has been successfully adapted for producing several materials like SS alloys, quasicrystalline phases, intermetallics, ordered compounds, high-pressure phases, amorphous materials, and nanomaterials, shown in Figure 8.7. The high energy throughput of this nonequilibrium process has been found effective in extending the solid solubility limit in immiscible systems. Varalakshmi et al. [73] first reported the synthesis of NC AlFeTiCrZnCu HEAs by the MA route. Since then, a new process paradigm was opened up, as MA is one for the robust nonequilibrium processing route that can develop various phases and composition overcoming the solubility-induced barrier in controlled grain scale, in scalable quantity [74]. The number of publications on HEAs by

TABLE 8.1
Advantages and Disadvantages of Five Major Classes of NC Material Synthesis Techniques

Method	Advantages	Limitations
Mechanical alloying (Ball milling and cryo-milling)	• Scalable. • No melting required. • Allows alloying of normally immiscible elements.	• Rapid grain growth upon consolidation of powders. • Void formation on compaction and sintering. • High energy input. • High risk of contamination. • Large variation in particle size.
Thermal spraying (Cold-spray, plasma-spray, flame-spray, etc...)	• High deposition rate. • Near-net-shape coating technology. • Scalable. • No thickness limitations. • High deposition rate	• Voids. • Constrained geometry. • High equipment cost. • High energy input. • Nonuniform grain-size distribution.
Physical vapor deposition (Gas condensation, DC/RF sputtering, vacuum arc deposition)	• Wide range of materials can be synthesized. • Good control of cluster sizes. • Continuous process. • Low risk of contamination. • Low thermal load to substrate. • High coating density and adhesion. • Excellent uniformity.	• Void formation upon compaction and sintering of vapor deposited. • Powders. • High cost (need ultra-high vacuum). • Difficult to scale up (low yield). • Requires vacuum environment. • Limited to thin films.
Severe plastic deformation (High-pressure torsion, equal channel angular pressing)	• Straightforward relationship between strain and grain size. • Simple grain-refinement mechanism.	• Small-scale. • Constrained geometry. • Poor efficiency. • High waste.
Electrodeposition	• Low cost. • Low energy requirements. • Scalable. • Can obtain uniform grain sizes. • Few size or shape limitations. • Low initial capital investment.	• Difficult to co-deposit multiple elements. • Constrained to certain elements. • Significant internal stresses can be built up in the material.

Adapted from ref. [29].

the MA route is presented in Figure 8.8. HEAs synthesized by the MA route tend to stabilize supersaturated SS phases [22]. Excellent chemical homogeneity with the nano or ultrafine microstructure is a characteristic of mechanically alloyed (MAed) products [67]. The majority of HEA synthesis by MA involves planetary ball mills; very few reported using the SPEX mills and shaker rod mills [22]. After synthesis, the powders are consolidated into a usable shape. The relative

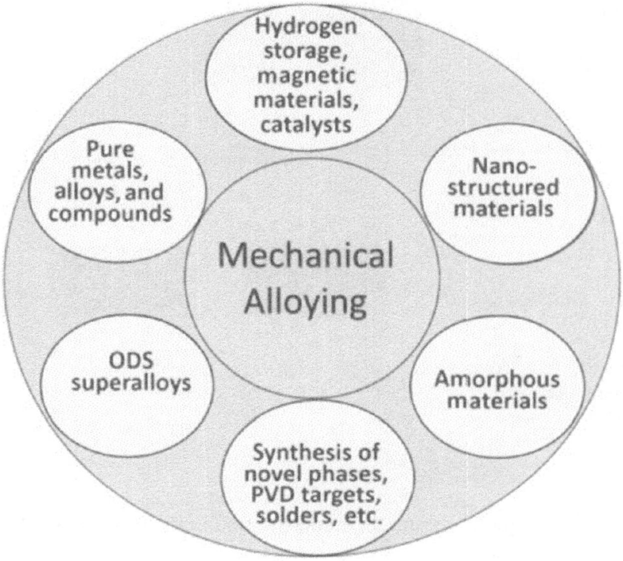

FIGURE 8.7 An outline of the application areas of mechanical alloying [45].

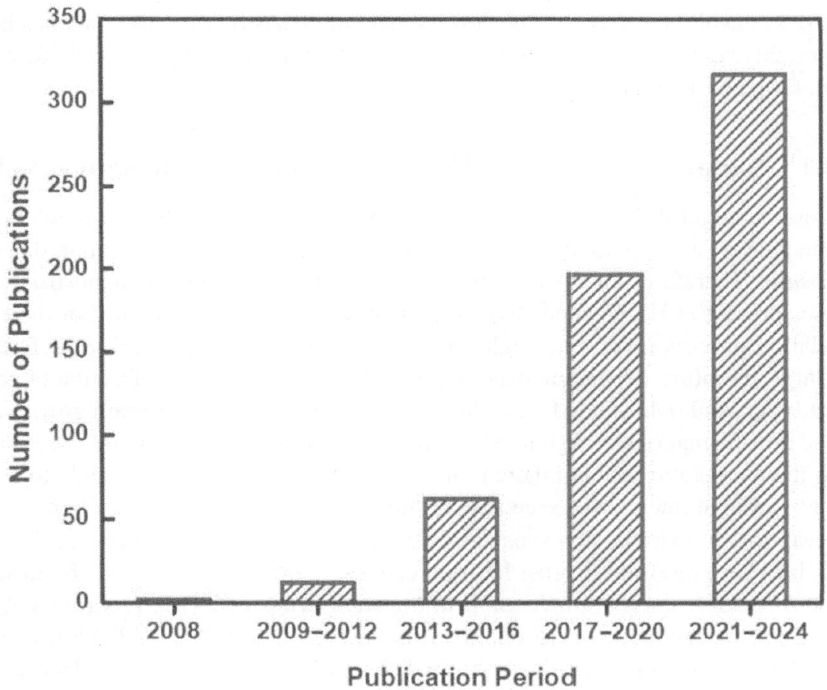

FIGURE 8.8 Number of publications on HEAs by the mechanical alloying route. (Source: Google Scholar.)

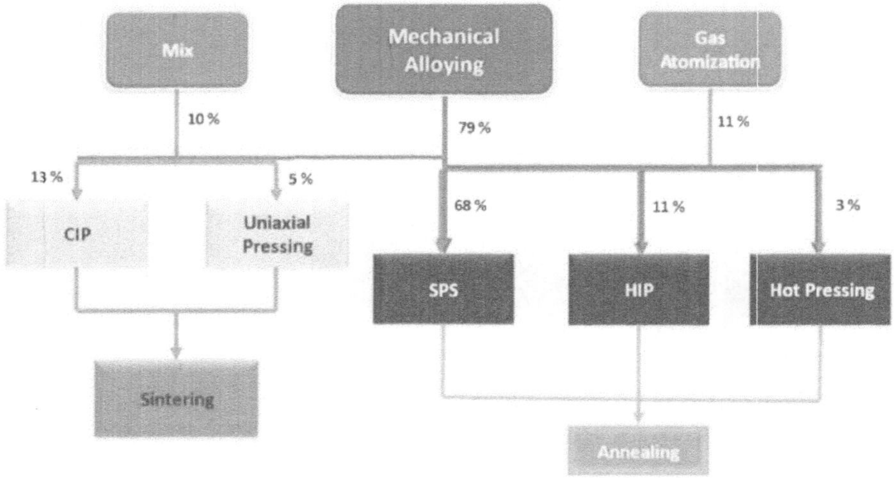

FIGURE 8.9 Different methods and their contribution in the processing of HEAs [44].

contributions from various powder consolidation processes have been pictorially represented in Figure 8.9. It can be observed that the majority of the effort has been channeled toward the development of bulk compact by combining MA followed by the SPS consolidation route. The development of HEAs by the MA route has been thoroughly reviewed by Koch [39], Vaidya et al. [22], Sharma et al. [60], Torralba et al. [30], and Kumar et al. [44].

8.3.1 Thermal Stability of NC High Entropy Alloys Synthesized by MA

Being a nonequilibrium synthesis route, MA may lead to the formation of metastable phases, which may decompose into stable equilibrium phases upon thermal exposure. Therefore, an investigation of the thermal stability of nanostructured phases present in HEAs needs to be explored. In addition to this, most of the consolidation process involves a high operating temperature (cf. Figure 8.9). Due to the large curvature-dependent driving force, the increased volume fraction of grain boundaries, and reduced diffusion distances, the driving force for grain growth and phase transformations is high in NC metals and alloys, and the kinetics accelerates with the temperature. This inherent metastability makes them susceptible to grain growth, even at low to moderate temperatures. For example, NC pure Cu prepared by evaporation exhibits abnormal grain growth at room temperatures [75]. Tekin et al. [76] have reported that the NC microstructure of CoCrFeNi HEA obtained by the MA route is "*not thermally stable at high temperatures*". They reported that the average grain size increases to 1.45 μm after annealing at 1100°C (1373 K), accompanied by ~54% drop in hardness (from 477 ± 20 HV to 220 ± 14 HV). Hence, the grain growth during consolidation should be controlled; otherwise, it may result in diminishing its attractive properties (mainly strength and hardness). There are several approaches for designing a thermally stable NC alloy. The conventional paradigm incorporates two fundamental mechanisms, thermodynamic mechanism and kinetic mechanism, and their synergy. The kinetic mechanism relies upon second-phase drag, grain boundary pinning by fine intermetallic particles, solute drag, chemical ordering, and grain size stabilization [77]. On the other hand, the thermodynamic mechanism involves the minimization of overall system energy by preferential grain boundary (GB) alloying [78]. Vaidya et al. [22, 79] have pointed out that there are two aspects of thermal stability of NC HEAs synthesized by MA that need to be addressed: (i) stability of the phases formed and (ii) the retention of NC microstructure. In addition, they also pointed out that existing literature lacks in "*systematic study of thermal stability of nanocrystalline HEAs*". Garlapati et al. [80] have also mentioned that "*studies on the thermal stability of nanocrystalline HEAs are very seldom*". Our work has reviewed the thermal stability of NC HEAs obtained by the MA route regarding their mechanical property stability, phase, and microstructural stability.

8.3.2 Initial Studies on HEAs Prepared by MA Route

Initially, Varalakshmi et al. [73] have studied the phase formation and stability in AlFeTiCrZnCu HEAs starting from the binary Al-Fe to the senary composition. After MA, it possesses a BCC crystal structure and crystallite size of 9 nm. The alloy was then consolidated by sintering at 800°C (1073 K) for 1 hour. The sintered compact had a hardness of 2 GPa, which was accompanied by an increase in the crystallite size to 20 nm. A similar single-phase formation was also reported for CuNiCoZnAlTi and NiFeCrCoMnW alloy systems [73]. In a subsequent study, they consolidated this alloy by vacuum hot press (VHP) and hot isostatic pressing (HIP) at 800°C (1073 K) to confirm the structural stability [81]. They found that the single-phase BCC decomposed into two BCC phases with a small volume fraction of FCC phase. To identify the phase transition temperature, they conducted the differential scanning calorimetry (DSC) of the as-milled powder. However, no predominant endothermic or exothermic peak was observed up to 800°C (1073 K). Based on this, they concluded that no phase transition occurred during consolidation, rather the second BCC and the FCC phase might form during the MA itself. However, very close lattice parameters of two BCC phases (d_{BCC1} = 0.209 nm, and d_{BCC2} = 0.205 nm, respectively) caused a significant overlapping of their X-ray diffraction (XRD) peaks, and therefore could not be detected after MA. In addition, the peak broadening due to the NC microstructure formation made the FCC reflection very difficult to detect. The hardness and compressive strength of the bulk alloy prepared by VHP were found to be 9.50 GPa and 2.19 GPa, respectively. In the HIPed HEA, these values were 10.04 GPa and 2.83 GPa, respectively.

After MA, the CuNiCoZnAlTi showed BCC (a = 0.29 nm) along with a small fraction of FCC phase [82]. It has been observed that the single-phase (BCC) SS is stable up to 800°C (1073 K). The bulk alloy prepared by VHP at 1073 K, under 30 MPa pressure for 0.5 hour attained 99.95% densification. The bulk alloy has one BCC (a = 0.283 nm) and two FCC phases (a = 0.420 nm and 0.357 nm, respectively). The backscattered scanning electron (BSE-SEM) micrograph revealed a gray and white contrast in the microstructure. The gray phase was found enriched in Al, Co, and Ti, and the white phase contained a higher fraction of Cu and Zn. High hardness (7.55 GPa) and compressive strength (2.36 GPa) were attributed to the retention of 53 nm crystallite size, SS hardening, and the coexistence of FCC and BCC phases. The phase evolution and thermal stability of CuNiCoZnAlTi with varying Cu content 0, 8.33, 33.33, and 49.98 at.% was studied [83]. It was found that up to the 8.33 at.% Cu, the BCC phase was observed as dominant, with a small fraction of the FCC phase. As the Cu content increased beyond this, the FCC phase was found as the dominant one. The $Cu_{33.33}$ NiCoZnAlTi, and $Cu_{49.98}$ NiCoZnAlTi were composed of a single-phase FCC and two FCC phases, respectively. Phase stability was examined by annealing at 800°C (1073 K) for 1 hour. Lattice parameter analysis elicits that a second FCC phase (FCC2, a = 0.351 nm) precipitated out during heat treatment. In the $Cu_{33.33}$ NiCoZnAlTi alloy, along with the major FCC phase (a = 0.365 nm), two other FCC phases (a = 0.432 nm and a = 0.350 nm, respectively) and a BCC phase (a = 0.298 nm) have formed. The alloy composition containing the highest concentration of Cu (49.98 at.%) showed the formation of two more FCC phases (a = 0.428 nm, and a = 0.355 nm, respectively). Unlike the as-milled state, the BCC phase was reported to be the major phase up to 8.33 at.% Cu, and beyond the FCC phase, became predominant after annealing. The high mixing entropy was attributed to the forming of the SS over intermetallic phases.

8.3.3 Studies on Alloy Groups Containing 3D Transition Metals

The popular HEAs synthesized using *3d* transition elements, say CoCrFeMnNi, CoCrFeNi, CoFeNi, CoFeMnNi, and CoCrCuFeNi, all showed the FCC structure when processed by the conventional melting-casting route, where segregation and clustering of Cu are observed in Cu-containing alloys [84]. A large fraction of HEAs developed so far consists of *3d*-transition elements [3]. Phase evolution in equiatomic CoCrCuFeNi, CoCuFeNi, CoCrCuNi, and CoCrFeNi alloys was studied by Praveen et al. [85]. It was observed that after MA, XRD patterns of Cr-containing compositions showed a mixture of BCC and FCC phases, whereas in the absence of Cr, formation of the single-phase FCC SS was observed. Ni was found as the host lattice for FCC and Cr for the BCC phase. Phase decomposition was assessed by annealing the milled powder at 1000°C (1273 K) for 15 minutes. In the presence of Cu, the FCC phase decomposed into two FCC phases, and the Cr-containing composition showed the formation of the σ phase. Development of the Cu-enriched FCC phase was due to its high positive enthalpy with Fe, Cr, and Co. They proposed that the σ phase was similar to the Cr-Fe, Cr-Co, or Cr-Fe-Co intermetallic compound, as observed in the Cr-Fe/Cr-Co binary phase diagrams or in the Cr-Fe-Co ternary phase diagram. The SPS was carried out for single-phase FCC-CoCrFeNi, and two-phase FCC phases of CoCuFeNi. In addition, the ternary CoFeNi and Mn (FCC stabilizer) substituted CoFeMnNi were also processed by SPS. FCC was found to be dominant for all compositions in both milled and SPSed states. Cr-added alloy showed the transformation of the BCC phase to the σ phase. To trace the transformation temperature, CoFeNiCr was annealed up to 900°C (1173 K). The XRD spectra of annealed powder illustrated that beyond 600°C (873 K), the BCC phase transformed to the σ phase. This transformation was possibly responsible for the superior densification in CoFeNiCr.

Consequently, it showed the maximum relative density and hardness of 95% and 570 ± 10 HV, respectively.

The influence of a very small fraction of Mo and Nb additions in one phase formation and retaining the high-temperature properties were studied by Praveen et al. [86]. Phase formation was examined during MA, and both phase stability and hardness stability were assessed in the SPSed pellet. Unlike the base CoCrFeNi, a major FCC and minor BCC phase were observed for Mo- and Nb-added alloys after the completion of MA. However, a partial dissolution of Mo was observed even after extending the milling time to 30 hours in $CoCrFeNiMo_{0.2}$ and $CoCrFeNiMo_{0.14}Nb_{0.14}$. Contrary to this, complete dissolution of Nb was observed for both compositions. They claimed that it might be due to the higher melting point, higher diffusivity, and lower activation energy of Nb. After SPS, in $CoCrFeNiMo_{0.2}$, a new phase (μ) was formed along with the major FCC phase and σ phase. They suggested that the μ phase is the (Mo, Cr)-rich phase. The phase stability of $CoCrFeNiMo_{0.2}$ was assessed by annealing it up to 900°C (1173 K). The corresponding XRD patterns showed that peaks corresponding to the BCC phase and undissolved Mo disappeared as the annealing temperature reached above 500°C (773 K), and the new carbide phase (Ni_3Mo_3C) formed. With the increase in the annealing temperature, this carbide dissolved, and the μ phase and σ phase formed below 900°C (1173 K). The niobium carbide (NbC) was also formed, along with the major FCC and σ phase. It was concluded that the carbon was coming from the unsolicited contamination from the milling media. The XRD pattern of the annealed $CoCrFeNiMo_{0.2}$ pellet exhibited no further decomposition and only a 15% drop in hardness when annealed at 800°C (1073 K) for 1 day. With the addition of Mo, the SPSed compact hardness of CoCrFeNi increased from 570 HV to 620 HV, and the maximum hardness (710 HV) was observed for the combined addition of Mo and Nb. After sintering, a major FCC phase with crystallite size in the range of 60–70 nm was observed for all compositions.

Pradeep et al. [87] studied phase stability in bulk NC AlCrCuFeNiZn HEAs prepared by MA followed by hot compaction at 600°C (873K) under 650 MPa pressure for 15 minutes. The 20-hour MAed alloy comprised a major BCC (a = 287 pm) with 12 vol. % FCC (a = 367 pm) phase. The bulk compact showed an increase in FCC fraction to 40 vol. %. Transmission electron microscopy (TEM) investigation revealed that the crystallite size of the milled alloy (10 ± 2 nm) grew to 20 ± 8 nm after consolidation. The chemical composition and elemental distribution by the atom probe tomography (APT) of the BCC phase revealed alternating Al-Ni and Fe-Cr phases with similar structure and lattice parameters, which were separated by the α-brass type interphase, with a composition of 63 at.% Cu and 32 at.% Zn. They proposed that segregated α-brass type phases along the grain boundaries stabilized the nanostructure. The high hardness of (870 ± 10 HV) of the bulk alloy was contributed from the NC microstructure and individual hardness of phases (i.e., Al-Ni and Fe-Cr) that constituted the matrix.

In another study, the phase stability of Y_2O_3 dispersed AlCoCrFe HEA powder was studied by annealing it from 500°C (773 K) to 800°C (1073 K), with 100°C (373 K) intervals for 15 minutes in the Ar atmosphere [88]. It was reported that during annealing, the Al, together with Co and Fe, resulted in the formation of an

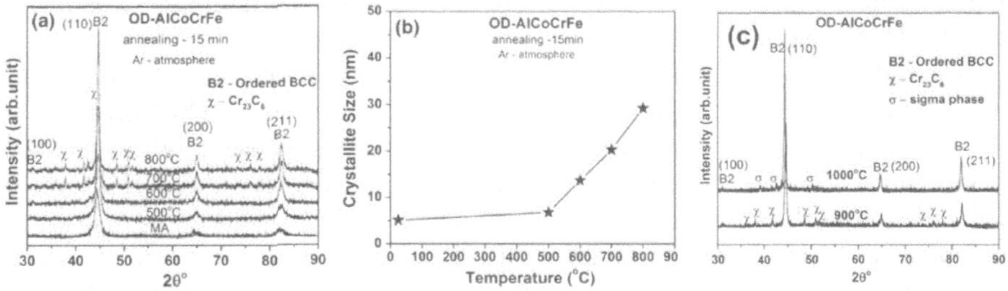

FIGURE 8.10 (a) Phase evolution and (b) crystallite size variation of oxide-dispersed AlCoCrFe HEA with annealing temperature. (c) Phase evolution of oxide-dispersed AlCoCrFe HEA after sintering [88].

ordered B2 phase, along with the chromium carbide ($Cr_{23}C_6$). Very little crystallite growth was observed during annealing (Figure 8.10b). The phase constituent was found to be similar in the 900°C (1173 K) SPSed compact; however, the σ-phase instead of the chromium carbide was observed in the 1000°C (1273 K) sintered compact (Figure 8.10c). Dispersion of oxide increased the hardness of the SPSed compact from 1050 ± 20 HV_1 to 1070 ± 20 HV_1.

CoCrFeNiMnAl showed a predominant BCC phase after MA, and it transformed to the FCC when the annealing temperature exceeded 500°C (773 K). SPS at 800°C (1073 K) resulted in the formation of a bimodal microstructure. The mean grain size of BCC and FCC were < 50 nm and 200 nm, respectively. It showed a hardness and compressive strength of 662 HV and 2142 MPa. Very high grain growth resistance was observed for bulk NC CoCrFeNi consolidated by the SPS for an annealing duration of 600 hours at a very high homologous temperature of 0.68 T_m (1173 K) [89]. The FCC phase remained the major phase for both MAed and SPSed conditions. After SPS, the BCC phase disappeared, and secondary phases, Cr_7C_3 and Cr_2O_3, were formed. The formation of carbide and oxide were attributed to the carbon and/or oxygen contamination of the elemental powder blend from the process-controlling agent or the milling medium during the MA. The composite microstructure obtained after SPS was composed of FCC grains, Cr-rich carbide and Cr-rich oxide; showed very high resistance to grain growth; and synergistically contributed by the Zener pinning effect from the fine dispersion of oxide and the presence of a two-phase mixture (cf. Figure 8.11).

From the aforementioned discussion, it is clear that the unsolicited carbon and oxygen contamination during milling leads to the formation of a composite microstructure contributing toward superior microstructural stability. In the subsequent work, the grain growth kinetics and Hall-Petch analysis of SPSed CoCrFeNi was conducted [90]. The MAed powder was annealed up to 1200°C (1473 K). The XRD analysis revealed concurrence of the BCC phase dissolution with Cr_7C_3 evolution for a sintering temperature of more than 500°C (773 K). The carbides dissolved beyond 1000°C (1273 K), and the XRD spectrum showed peaks corresponding to the FCC phase and Cr_2O_3. Considering the stability of the carbide phase up to 1000°C (1273 K), the authors assessed the thermal stability of the SPSed pellet up to 900°C (1173 K), as shown in Figure 8.12. They also reported that although the formation of Cr_7C_3

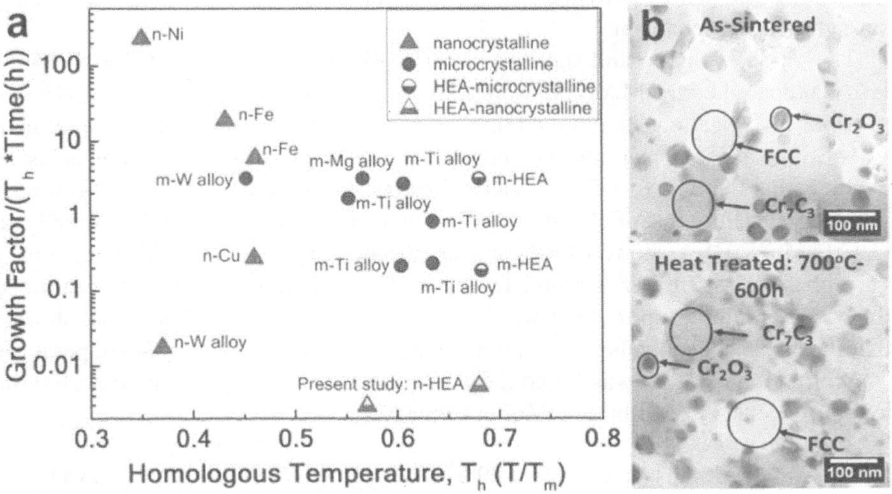

FIGURE 8.11 (a) Normalized grain growth factor Vs homologous temperature summarizing on grain growth (n indicates nanocrystalline materials, m indicates microcrystalline materials). (b) The scanning transmission electron microscopy (STEM) image of as-sintered compact and specimen that was heat treated at 700°C for 600 hours; the precipitates remain nearly unchanged [89].

requires higher C activity than $Cr_{23}C_6$, the presence of Co (graphitizing element) and the activity of C might be higher, which stabilizes the Cr_7C_3. In addition, the TGA revealed the volatilization of Cr_2O_3 above 1000°C (1273 K). Therefore, the Cr_2O_3 peaks in the XRD patterns of powders heated above 1000°C (1273 K) might have formed during cooling. The grain growth exponent (n ~4) was almost similar to the microcrystalline alloy (n ~3), indicating pronounced resistance to grain growth.

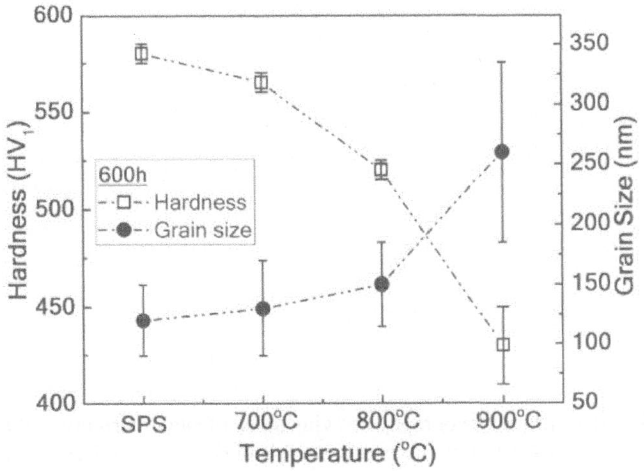

FIGURE 8.12 Hardness and grain size variation of CoCrFeNi at different heat treatment conditions [90].

The Hall-Petch analysis elicited that the grain boundary strengthening was the key strengthening mechanism.

The phase evolution and stability in NC CoFeNi, CoCrFeNi, and CoCrFeMnNi HEAs prepared by the MA route were also investigated by Vaidya et al. [79]. Unlike previous investigations, they found that the 15-hour MA resulted in the formation of FCC SS for all compositions. The phase stability of as-milled alloys was assessed by annealing them up to 1373 K. XRD investigation showed retention of the single-phase SS structure in CoFeNi after annealing. Contrary to this, both CoCrFeNi and CoCrFeMnNi HEA showed formation of carbide (Cr_7C_3), along with the primary FCC phase. They further consolidated these HEAs by SPS (1173 K), and examined their phase stability up to 1373 K. The XRD patterns of the SPSed pellet showed that the phase constituents were similar to the annealed powder at that temperature. The formation of Cr_7C_3 was found thermodynamically favorable as it possessed the lowest free energy of formation (ΔG in kJ/mol) as:

$$\Delta G_{Cr7C3}(-158.3) < \Delta G_{Mn23C6}(-89.5) < \Delta G_{Cr23C6}(-67.3) < \Delta G_{Cr3C2}(-53.1)$$
$$< \Delta G_{Ni3C}(-26.3) < \Delta G_{Fe3C}(2.2) < \Delta G_{Co2C}$$

They also reported that due to its superior stability, the volume fraction of Cr_7C_3 did not change significantly during annealing of the SPSed pellets. The phase stability of CoCrFeNi and CoCrFeMnNi were ascribed to the extended FCC SS present in the phase diagram of the multi-component Co-Cr-Fe-Mn-Ni system [91], which remained unaffected by the loss of Cr and Mn from the matrix phase. In addition, to trace the origin of Cr_7C_3, they conducted the APT investigation (Figure 8.13) of

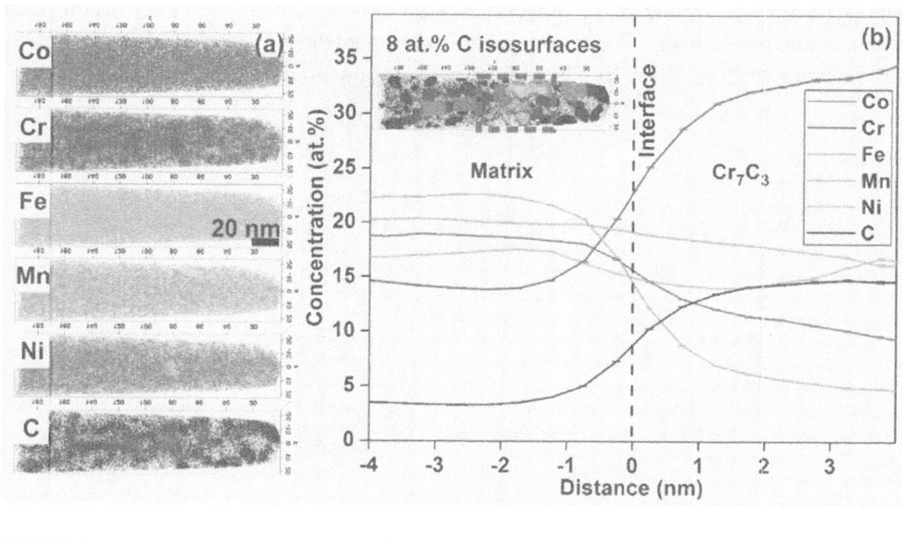

FIGURE 8.13 Atom probe tomography showing (a) elemental distribution of MA (15 hours) CoCrFeMnNi HEA powder and (b) proximity histogram obtained from the highlighted region (dotted box) of the 8 at.% C isoconcentration surfaces revealing the presence of chromium carbide (Cr_7C_3) [79].

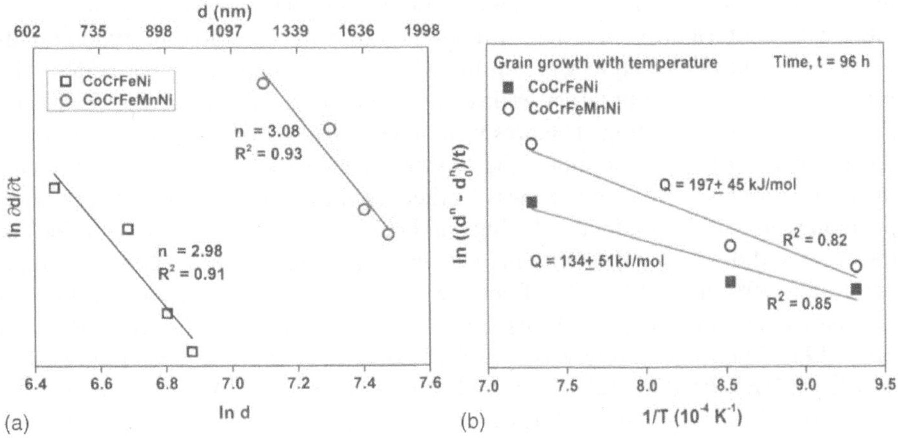

FIGURE 8.14 Parameters of grain growth kinetics for FCC phase in CoCrFeNi and CoCrFeMnNi HEAs: (a) grain growth exponent, *n*, and (b) activation energy, *Q* [92].

as-milled CoCrFeMnNi powder, which indicates the formation of carbide takes place during milling. They also analyzed the grain growth kinetics of SPSed CoCrFeNi and CoCrFeMnNi HEAs by heat treating both as a function of temperature (1073, 1173, and 1373 K for 96 hours) and time (24, 48, 72, and 96 hours at 1373 K) [92]. The value of grain growth exponent (n) = 3 was found for both HEAs (cf. Figure 8.14). This showed that the grain growth of two-phase materials was controlled primarily by long-range diffusion. The activation energy for grain growth (Q) for CoCrFeNi and CoCrFeMnNi was found as 134 kJ/mol and 194 kJ/mol, respectively. They concluded that the rate-controlling step for grain growth was the coarsening of carbide particles residing at GBs. It involved the diffusion of Cr along grains, GBs, and the matrix/particle interfaces. The Hall-Petch analysis elicits that the FCC-carbide phase boundaries provided the maximum contribution to strengthening in SPSed HEA.

Nanoscale oxide dispersion (ODS) was found to be effective in improving the mechanical property stability of HEAs. Rogal et al. [93] have demonstrated that the addition of 5 wt. % nano-Al_2O_3 (α-Al_2O_3, size 5–35 nm) to CoCrFeMnNi improved its thermal stability up to 1100°C (1373 K). The alloy was synthesized by MA and consolidated by HIP at 1173 K for 15 minutes under 150 bar pressure. The HIPed CoCrFeMnNi HEA consisted of FCC SS (a = 3.598 nm), dispersion of α-Al_2O_3, and a small fraction (2%) of $M_{23}C_6$. The presence of nanoparticles impedes coarsening during thermal exposure and retains the average grain size of ~108 nm against the average size of grains of ~168 nm obtained without Al_2O_3 addition. The addition of nano-Al_2O_3 improved the yield strength (YS) from 1180 MPa to 1600 MPa, with compromised plasticity (from 34.5% to 6.5%).

In another study, Hadraba et al. [94] prepared CoCrFeNiMn-based ODS HEA by MA and SPS at 1150°C (1423 K). Results indicated that oxide dispersion improved high temperatures' mechanical properties. The ODS HEA exhibits an increment in both ultimate tensile strength and YS to 30% at room temperature and about 70%

at 1073 K. Improvement in the room temperature mechanical property was correlated with the formation of a finer microstructure, which resulted in grain boundary strengthening as well as the pinning of dislocation movement on oxides. The formation of twins within larger grains was observed as an additional deformation mechanism during loading. The presence of nano-scale oxide dispersion inhibited the grain growth, and therefore, the twining deformation mechanism was suppressed, and plastic deformation was limited even at elevated temperatures. This phenomenon could explain a widening gap between the strength characteristics of the HEA and the ODS HEA with increasing temperature (from 30% at room temperature to 70% at 1073 K). The effect of SPS temperature for attaining the highest mechanical property in Y_2O_3–Ti dispersed CoCrCuFeNi was evaluated by Rao and Sinha [95]. The MAed alloy was consolidated at different SPS temperatures, i.e., 900°C (1173K), 1000°C (1273K), and 1100°C (1373 K). The maximum YS (903 ± 10 MPa) and hardness (468 ± 5 HV) were observed for the 1373 K sintered alloy. The Orowan mechanism was found to be the most dominant strengthening mechanism, contributing ~70% of total strength, followed by the Hall-Petch strengthening (~23%) and dispersion strengthening (~7%). The mechanical property stability of NC CoCrFeNi HEA by adding 1 and 4 wt. % nanosized Y_2O_3 (designated as CCFN-1Y_2O_3 and CCFN-4Y_2O_3) was evaluated by Tekin et al. [76]. The thermal stability of the MAed HEA was examined by annealing treatments with temperatures up to 1100°C (1373 K). The results have shown FCC SS remained stable, even after annealing at 1100°C (Figure 8.15). The undoped HEA underwent rapid coarsening after exposure to 1100°C and reached the ultrafine domain (avg. grain size ~1.45 μm). This resulted in a hardness drop from 477 ± 20 HV (as-milled hardness) to 220 ± 14 HV after annealing at 1100°C. In contrast, the mean grain size of CCFN-4Y_2O_3 remained in the pure NC domain (~72 nm) after annealing at 1100°C. In addition, the hardness of CCFN-4Y_2O_3 HEA after annealing at 1100 °C (441 ± 23 HV) closely matched the hardness of as-milled CCFN HEA (477 ± 20 HV). The grain growth and hardness drop are presented in Figure 8.16. They attributed the superior thermal stability to the Zener pinning of the grain boundaries by the nanosized Y_2O_3 phases.

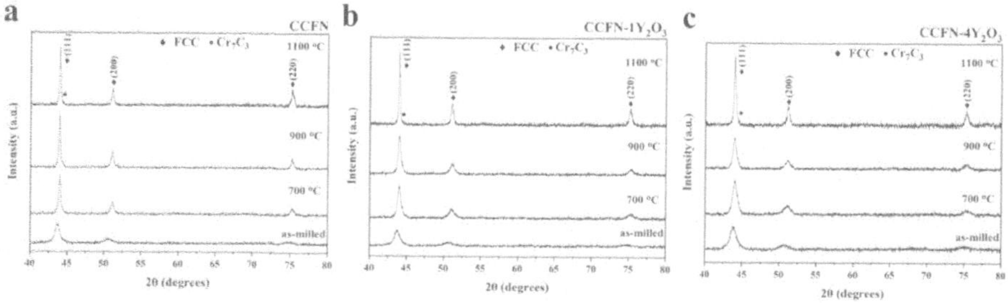

FIGURE 8.15 XRD scans of as-milled and annealed HEAs: (a) CCFN, (b) CCFN-1Y_2O_3, and (c) CCFN-4Y_2O_3 [76].

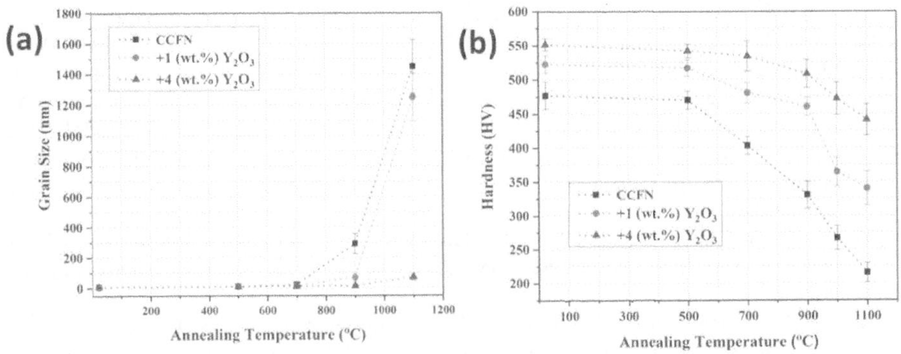

FIGURE 8.16 (a) Estimated grain sizes from XRD, Focused Ion Beam (FIB), or TEM versus annealing temperature for the investigated CCFN-xY_2O_3 (x: 0, 1, and 4 wt. %) HEAs. Room-temperature grain sizes are the as-milled values. (b) Hardness values of CCFN-xY_2O_3 (x: 0, 1, and 4 wt. %) HEAs as a function of annealing temperature. Room-temperature hardness values are the as-milled values [76].

8.4 SUMMARY, CURRENT CHALLENGES, AND FUTURE OPPORTUNITIES

The majority of the work focused on the phase stability. Little effort was channeled for the temperature-dependent mechanical property alteration. Since MA is a non-equilibrium process operated mostly at ambient condition, it tends to stabilize the metastable state (the super-saturated SS), which tends to decompose during consolidation. Therefore, the synthesis of a single-phase alloy is difficult in the bulk state. In addition, the role of the carbide-forming elements is very crucial. Selecting the proper alloying element and/or proper milling-heat treatment cycle is required for preparation of the bulk NC alloy. Furthermore, a bulk alloy with an NC microstructure with fine-scale dispersion of carbides may prove beneficial for elevated temperature applications. A BCC/FCC phase mixture in proper volume fraction may lead to superior mechanical properties. The decomposition-induced fine-scale intermetallic phases lead to improvement in mechanical properties. High hardness can also improve the tribological response of the alloy.

REFERENCES

1. B. Cantor, I.T.H. Chang, P. Knight, A.J.B. Vincent, Microstructural development in equiatomic multicomponent alloys, Mater. Sci. Eng. A. 375–377 (2004) 213–218. https://doi.org/10.1016/j.msea.2003.10.257.
2. J.W. Yeh, S.K. Chen, S.J. Lin, J.Y. Gan, T.S. Chin, T.T. Shun, C.H. Tsau, S.Y. Chang, Nanostructured high-entropy alloys with multiple principal elements: Novel alloy design concepts and outcomes, Adv. Eng. Mater. 6 (2004) 299–303. https://doi.org/10.1002/adem.200300567.

3. D.B. Miracle, O.N. Senkov, A critical review of high entropy alloys and related concepts, Acta Mater. 122 (2017) 448–511. https://doi.org/10.1016/j.actamat.2016.08.081.
4. O.N. Senkov, G.B. Wilks, J.M. Scott, D.B. Miracle, Mechanical properties of Nb25Mo25Ta25W25 and V20Nb20Mo20Ta20W20 refractory high entropy alloys, Intermetallics. 19 (2011) 698–706. https://doi.org/10.1016/j.intermet.2011.01.004.
5. O.N. Senkov, G.B. Wilks, D.B. Miracle, C.P. Chuang, P.K. Liaw, Refractory high-entropy alloys, Intermetallics. 18 (2010) 1758–1765. https://doi.org/10.1016/j.intermet.2010.05.014.
6. W. Li, D. Xie, D. Li, Y. Zhang, Y. Gao, P.K. Liaw, Mechanical behavior of high-entropy alloys, Prog. Mater. Sci. 118 (2021) 100777. https://doi.org/10.1016/j.pmatsci.2021.100777.
7. M.C. Gao, P.K. Liaw, J.W. Yeh, Y. Zhang, High-entropy Alloys: Fundamentals and Applications. (2016) 1–516. https://doi.org/10.1007/978-3-319-27013-5.
8. R. Sriharitha, B.S. Murty, R.S. Kottada, Phase formation in mechanically alloyed Al_xCoCrCuFeNi (x = 0.45, 1, 2.5, 5 mol) high entropy alloys, Intermetallics. 32 (2013) 119–126. https://doi.org/10.1016/j.intermet.2012.08.015.
9. S. Gorsse, F. Tancret, Current and emerging practices of CALPHAD toward the development of high entropy alloys and complex concentrated alloys, J. Mater. Res. 33 (2018) 2899–2923. https://doi.org/10.1557/jmr.2018.152.
10. M. Vaidya, K.G. Pradeep, B.S. Murty, G. Wilde, S.V. Divinski, Bulk tracer diffusion in CoCrFeNi and CoCrFeMnNi high entropy alloys, Acta Mater. 146 (2018) 211–224. https://doi.org/10.1016/j.actamat.2017.12.052.
11. K.Y. Tsai, M.H. Tsai, J.W. Yeh, Sluggish diffusion in Co-Cr-Fe-Mn-Ni high-entropy alloys, Acta Mater. 61 (2013) 4887–4897. https://doi.org/10.1016/j.actamat.2013.04.058.
12. J. Dąbrowa, M. Zajusz, W. Kucza, G. Cieślak, K. Berent, T. Czeppe, T. Kulik, M. Danielewski, Demystifying the sluggish diffusion effect in high entropy alloys, J. Alloys Compd. 783 (2019) 193–207. https://doi.org/10.1016/j.jallcom.2018.12.300.
13. A. Mehta, Y. Sohn, Investigation of sluggish diffusion in FCC Al0.25CoCrFeNi high-entropy alloy, Mater. Res. Lett 9 (2021) 239–246. https://doi.org/10.1080/21663831.2021.1878475.
14. J. Chen, X. Zhou, W. Wang, B. Liu, Y. Lv, W. Yang, D. Xu, Y. Liu, A review on fundamental of high entropy alloys with promising high–temperature properties, J. Alloys Compd. 760 (2018) 15–30. https://doi.org/10.1016/j.jallcom.2018.05.067.
15. H. Song, F. Tian, Q.M. Hu, L. Vitos, Y. Wang, J. Shen, N. Chen, Local lattice distortion in high-entropy alloys, Phys. Rev. Mater. 1 (2017) 1–8. https://doi.org/10.1103/PhysRevMaterials.1.023404.
16. R. Wang, Y. Tang, S. Li, Y. Ai, Y. Li, B. Xiao, L. Zhu, X. Liu, S. Bai, Effect of lattice distortion on the diffusion behavior of high-entropy alloys, J. Alloys Compd. 825 (2020). https://doi.org/10.1016/j.jallcom.2020.154099.
17. S. Ranganathan, Alloyed pleasures: Multimetallic cocktails, Curr. Sci 85 (2003) 1404–1406.
18. Y.C. Yang, C. Liu, C.Y. Lin, Z. Xia, Core effect of local atomic configuration and design principles in AlxCoCrFeNi high-entropy alloys, Scr. Mater. 178 (2020) 181–186. https://doi.org/10.1016/j.scriptamat.2019.11.016.
19. Y. Zou, Nanomechanical studies of high-entropy alloys, J. Mater. Res 33 (2018) 3035–3054. https://doi.org/10.1557/jmr.2018.155.
20. O. Maulik, D. Kumar, S. Kumar, S.K. Dewangan, V. Kumar, Structure and properties of lightweight high entropy alloys: A brief review, Mater. Res. Express 5 (2018) 52001. https://doi.org/10.1088/2053-1591/aabbca.
21. Z. Lyu, C. Lee, S.Y. Wang, X. Fan, J.W. Yeh, P.K. Liaw, Effects of constituent elements and fabrication methods on mechanical behavior of high-entropy alloys: A review, Metall. Mater. Trans. A Phys. Metall. Mater. Sci. 50 (2019). https://doi.org/10.1007/s11661-018-4970-z.

22. M. Vaidya, G.M. Muralikrishna, B.S. Murty, High-entropy alloys by mechanical alloying: A review, J. Mater. Res 34 (2019) 664–686. https://doi.org/10.1557/jmr.2019.37.
23. F. Alijani, M. Reihanian, K. Gheisari, Study on phase formation in magnetic FeCoNiMnV high entropy alloy produced by mechanical alloying, J. Alloys Compd. 773 (2019) 623–630. https://doi.org/10.1016/j.jallcom.2018.09.204.
24. E.P. George, D. Raabe, R.O. Ritchie, High-entropy alloys, Nat. Rev. Mater. 4 (2019) 515–534. https://doi.org/10.1038/s41578-019-0121-4.
25. J.M. Torralba, P. Alvaredo, A. García-Junceda, High-entropy alloys fabricated via powder metallurgy. A critical review, Powder Metall. 62 (2019) 84–114. https://doi.org/10.1080/00325899.2019.1584454.
26. A.O. Soto, A.S. Salgado, E.B. Niño, Thermodynamic analysis of high entropy alloys and their mechanical behavior in high and low-temperature conditions with a microstructural approach - A review, Intermetallics. 124 (2020). https://doi.org/10.1016/j.intermet.2020.106850.
27. E.P. George, W.A. Curtin, C.C. Tasan, High entropy alloys: A focused review of mechanical properties and deformation mechanisms, Acta Mater. 188 (2020) 435–474. https://doi.org/10.1016/j.actamat.2019.12.015.
28. Y. Zhang, R. Li, New advances in high-entropy alloys, Entropy 22(10)(2020) 1–7.
29. M.J.R. Haché, C. Cheng, Y. Zou, Nanostructured high-entropy materials, J. Mater. Res 35 (2020) 1051–1075. https://doi.org/10.1557/jmr.2020.33.
30. J.M. Torralba, P. Alvaredo, A. García-Junceda, Powder metallurgy and high-entropy alloys: Update on new opportunities, Powder Metall. 63 (2020) 227–236. https://doi.org/10.1080/00325899.2020.1807713.
31. C. Han, Q. Fang, Y. Shi, S.B. Tor, C.K. Chua, K. Zhou, Recent advances on high-entropy alloys for 3D printing, Adv. Mater 32 (2020) 1–41. https://doi.org/10.1002/adma.201903855.
32. J. Chen, X. Jiang, H. Sun, Z. Shao, Y. Fang, R. Shu, Phase transformation and strengthening mechanisms of nanostructured high-entropy alloys, Nanotechnol. Rev. 10 (2021) 1116–1139. https://doi.org/10.1515/ntrev-2021-0071.
33. M. Tokarewicz, M. Grądzka-Dahlke, Review of recent research on alcocrfeni high-entropy alloy, Metals (Basel). 11 (2021). https://doi.org/10.3390/met11081302.
34. Z.U. Arif, M.Y. Khalid, E. ur Rehman, S. Ullah, M. Atif, A. Tariq, A review on laser cladding of high-entropy alloys, their recent trends and potential applications, J. Manuf. Process. 68 (2021) 225–273. https://doi.org/10.1016/j.jmapro.2021.06.041.
35. T.-X. Li, J.-W. Miao, E.-Y. Guo, H. Huang, J. Wang, Y.-P. Lu, T.-M. Wang, Z.-Q. Cao, T.-J. Li, Tungsten-containing high-entropy alloys: A focused review of manufacturing routes, phase selection, mechanical properties, and irradiation resistance properties, Tungsten. 3 (2021) 181–196. https://doi.org/10.1007/s42864-021-00081-x.
36. X. Wang, W. Guo, Y. Fu, High-entropy alloys: Emerging materials for advanced functional applications, J. Mater. Chem. A 9 (2021) 663–701. https://doi.org/10.1039/D0TA09601F.
37. I. Hussain, C. Lamiel, M. Ahmad, Y. Chen, S. Shuang, M.S. Javed, Y. Yang, K. Zhang, High entropy alloys as electrode material for supercapacitors: A review, J. Energy Storage. 44 (2021) 103405. https://doi.org/10.1016/j.est.2021.103405.
38. M. Srikanth, A.R. Annamalai, A. Muthuchamy, C.P. Jen, A review of the latest developments in the field of refractory high-entropy alloys, Crystals. 11 (2021) 1–15. https://doi.org/10.3390/cryst11060612.
39. C.C. Koch, Nanocrystalline high-entropy alloys, J. Mater. Res 32 (2017) 3435–3444. https://doi.org/10.1557/jmr.2017.341.
40. X. Xie, N. Li, W. Liu, S. Huang, X. He, Q. Yu, H. Xiong, E. Wang, X. Hou, Research Progress of Refractory High Entropy Alloys: A Review, Chinese J. Mech. Eng. 35 (2022). https://doi.org/10.1186/s10033-022-00814-0.

41. Z.U. Arif, M.Y. Khalid, A. Al Rashid, E. ur Rehman, M. Atif, Laser deposition of high-entropy alloys: A comprehensive review, Opt. Laser Technol. 145 (2022) 107447. https://doi.org/10.1016/j.optlastec.2021.107447.
42. J. Zhou, Y. Cheng, Y. Chen, X. Liang, Composition design and preparation process of refractory high-entropy alloys: A review, Int. J. Refract. Met. Hard Mater. 105 (2022) 105836. https://doi.org/10.1016/j.ijrmhm.2022.105836.
43. P. Sathiyamoorthi, H.S. Kim, High-entropy alloys with heterogeneous microstructure: Processing and mechanical properties, Prog. Mater. Sci. 123 (2022) 100709. https://doi.org/10.1016/j.pmatsci.2020.100709.
44. A. Kumar, A. Singh, A. Suhane, Mechanically alloyed high entropy alloys: Existing challenges and opportunities, J. Mater. Res. Technol. 17 (2022) 2431–2456. https://doi.org/10.1016/j.jmrt.2022.01.141.
45. S.K. Dewangan, A. Mangish, S. Kumar, A. Sharma, B. Ahn, V. Kumar, A review on high-temperature applicability: A milestone for high entropy alloys, Eng. Sci. Technol. an Int. J 35 (2022) 101211. https://doi.org/10.1016/j.jestch.2022.101211.
46. S. Rajendrachari, An overview of high-entropy alloys prepared by mechanical alloying followed by the characterization of their microstructure and various properties, Alloys. 1 (2022) 116–132. https://doi.org/10.3390/alloys1020008.
47. X.-J. Hua, P. Hu, H.-R. Xing, J.-Y. Han, S.-W. Ge, S.-L. Li, C.-J. He, K.-S. Wang, C.-J. Cui, Development and property tuning of refractory high-entropy alloys: A review, Acta Metall. Sin. 35 (2022) 1231–1265. https://doi.org/10.1007/s40195-022-01382-x.
48. Y. Tang, R. Wang, B. Xiao, Z. Zhang, S. Li, J. Qiao, S. Bai, Y. Zhang, P.K. Liaw, A review on the dynamic-mechanical behaviors of high-entropy alloys, Prog. Mater. Sci. 135 (2023) 101090. https://doi.org/10.1016/j.pmatsci.2023.101090.
49. C. Zhu, L. Xu, M. Liu, M. Guo, S. Wei, A review on improving mechanical properties of high entropy alloy: Interstitial atom doping, J. Mater. Res. Technol. 24 (2023) 7832–7851. https://doi.org/10.1016/j.jmrt.2023.05.002.
50. H.Y. Diao, R. Feng, K.A. Dahmen, P.K. Liaw, Fundamental deformation behavior in high-entropy alloys: An overview, Curr. Opin. Solid State Mater. Sci 21 (2017) 252–266. https://doi.org/10.1016/j.cossms.2017.08.003.
51. D. Kumar, Recent advances in tribology of high entropy alloys: A critical review, Prog. Mater. Sci. 136 (2023) 101106. https://doi.org/10.1016/j.pmatsci.2023.101106.
52. Z. Cheng, J. Sun, X. Gao, Y. Wang, J. Cui, T. Wang, H. Chang, Irradiation effects in high-entropy alloys and their applications, J. Alloys Compd. 930 (2023) 166768. https://doi.org/10.1016/j.jallcom.2022.166768.
53. W. Xiong, A.X.Y. Guo, S. Zhan, C.T. Liu, S.C. Cao, Refractory high-entropy alloys: A focused review of preparation methods and properties, J. Mater. Sci. Technol. 142 (2023) 196–215. https://doi.org/10.1016/j.jmst.2022.08.046.
54. P. Martin, C. Aguilar, J.M. Cabrera, A review on mechanical alloying and spark plasma sintering of refractory high-entropy alloys: Challenges, microstructures, and mechanical behavior, J. Mater. Res. Technol. 30 (2024) 1900–1928. https://doi.org/10.1016/j.jmrt.2024.03.205.
55. S.A. Krishna, N. Noble, N. Radhika, B. Saleh, A comprehensive review on advances in high entropy alloys: Fabrication and surface modification methods, properties, applications, and future prospects, J. Manuf. Process. 109 (2024) 583–606. https://doi.org/10.1016/j.jmapro.2023.12.039.
56. Y. Zhang, T.T. Zuo, Z. Tang, M.C. Gao, K.A. Dahmen, P.K. Liaw, Z.P. Lu, Microstructures and properties of high-entropy alloys, Prog. Mater. Sci. 61 (2014) 1–93. https://doi.org/10.1016/j.pmatsci.2013.10.001.
57. M.-H. Tsai, J.-W. Yeh, High-entropy alloys: A critical review, Mater. Res. Lett. 2 (2014) 107–123. https://doi.org/10.1080/21663831.2014.912690.

58. K.K. Alaneme, M.O. Bodunrin, S.R. Oke, Processing, alloy composition and phase transition effect on the mechanical and corrosion properties of high entropy alloys: A review, J. Mater. Res. Technol. 5 (2016) 384–393. https://doi.org/10.1016/j.jmrt.2016.03.004.
59. W. Li, P. Liu, P.K. Liaw, Microstructures and properties of high-entropy alloy films and coatings: A review, Mater. Res. Lett. 6 (2018) 199–229. https://doi.org/10.1080/21663831.2018.1434248.
60. A.S. Sharma, S. Yadav, K. Biswas, B. Basu, High-entropy alloys and metallic nanocomposites: Processing challenges, microstructure development and property enhancement, Mater. Sci. Eng. R Reports. 131 (2018) 1–42. https://doi.org/10.1016/j.mser.2018.04.003.
61. S. Guo, C.T. Liu, Phase stability in high entropy alloys: Formation of solid-solution phase or amorphous phase, Prog. Nat. Sci. Mater. Int. 21 (2011) 433–446. https://doi.org/10.1016/S1002-0071(12)60080-X.
62. X. Yang, Y. Zhang, Prediction of high-entropy stabilized solid-solution in multicomponent alloys, Mater. Chem. Phys. 132 (2012) 233–238. https://doi.org/10.1016/j.matchemphys.2011.11.021.
63. S. Guo, Q. Hu, C. Ng, C.T. Liu, More than entropy in high-entropy alloys: Forming solid solutions or amorphous phase, Intermetallics. 41 (2013) 96–103. https://doi.org/10.1016/j.intermet.2013.05.002.
64. S. Guo, C. Ng, J. Lu, C.T. Liu, Effect of valence electron concentration on stability of fcc or bcc phase in high entropy alloys, J. Appl. Phys. 109 (2011). https://doi.org/10.1063/1.3587228.
65. M.G. Poletti, L. Battezzati, Electronic and thermodynamic criteria for the occurrence of high entropy alloys in metallic systems, Acta Mater. 75 (2014) 297–306. https://doi.org/10.1016/j.actamat.2014.04.033.
66. Y.F. Ye, C.T. Liu, Y. Yang, A geometric model for intrinsic residual strain and phase stability in high entropy alloys, Acta Mater. 94 (2015) 152–161. https://doi.org/10.1016/j.actamat.2015.04.051.
67. S. Praveen, B.S. Murty, R.S. Kottada, Alloying behavior in multi-component AlCoCrCuFe and NiCoCrCuFe high entropy alloys, Mater. Sci. Eng. A 534 (2012) 83–89. https://doi.org/10.1016/j.msea.2011.11.044.
68. S. Mridha, S. Samal, P.Y. Khan, K. Biswas Govind, Processing and consolidation of nanocrystalline Cu-Zn-Ti-Fe-Cr high-entropy alloys via mechanical alloying, Metall. Mater. Trans. A Phys. Metall. Mater. Sci. 44 (2013) 4532–4541. https://doi.org/10.1007/s11661-013-1824-6.
69. B. Schuh, F. Mendez-Martin, B. Völker, E.P. George, H. Clemens, R. Pippan, A. Hohenwarter, Mechanical properties, microstructure and thermal stability of a nanocrystalline CoCrFeMnNi high-entropy alloy after severe plastic deformation, Acta Mater. 96 (2015) 258–268. https://doi.org/10.1016/j.actamat.2015.06.025.
70. Z. Fu, W. Chen, H. Wen, D. Zhang, Z. Chen, B. Zheng, Y. Zhou, E.J. Lavernia, Microstructure and strengthening mechanisms in an FCC structured single-phase nanocrystalline Co25Ni25Fe25Al7.5Cu17.5 high-entropy alloy, Acta Mater. 107 (2016) 59–71. https://doi.org/10.1016/j.actamat.2016.01.050.
71. R. Sriharitha, B.S. Murty, R.S. Kottada, Alloying, thermal stability and strengthening in spark plasma sintered AlxCoCrCuFeNi high entropy alloys, J. Alloys Compd. 583 (2014) 419–426. https://doi.org/10.1016/j.jallcom.2013.08.176.
72. C. Suryanarayana, Mechanical alloying and milling, Mech. Alloy. Milling. (2004) 1–472. https://doi.org/10.4150/kpmi.2006.13.5.371.
73. S. Varalakshmi, M. Kamaraj, B.S. Murty, Synthesis and characterization of nanocrystalline AlFeTiCrZnCu high entropy solid solution by mechanical alloying, J. Alloys Compd. 460 (2008) 253–257. https://doi.org/10.1016/j.jallcom.2007.05.104.

74. C. Suryanarayana, Mechanical alloying and milling, Prog. Mater. Sci. 46 (2001) 1–184. https://doi.org/10.1016/S0079-6425(99)00010-9.
75. V.Y. Gertsman, R. Birringer, On the room-temperature grain growth in nanocrystalline copper, Scr. Metall. Mater. 30 (1994) 577–581. https://doi.org/10.1016/0956-716X(94)90432-4.
76. M. Tekin, G. Polat, Y.E. Kalay, H. Kotan, Grain size stabilization of oxide dispersion strengthened CoCrFeNi-Y2O3 high entropy alloys synthesized by mechanical alloying, J. Alloys Compd. 887 (2021) 161363. https://doi.org/10.1016/j.jallcom.2021.161363.
77. C.C. Koch, R.O. Scattergood, M. Saber, H. Kotan, High temperature stabilization of nanocrystalline grain size: Thermodynamic versus kinetic strategies, J. Mater. Res 28 (2013) 1785–1791. https://doi.org/10.1557/jmr.2012.429.
78. C.C. Koch, R.O. Scattergood, K.A. Darling, J.E. Semones, Stabilization of nanocrystalline grain sizes by solute additions, J. Mater. Sci. 43 (2008) 7264–7272. https://doi.org/10.1007/s10853-008-2870-0.
79. M. Vaidya, A. Karati, A. Marshal, K.G. Pradeep, B.S. Murty, Phase evolution and stability of nanocrystalline CoCrFeNi and CoCrFeMnNi high entropy alloys, J. Alloys Compd. 770 (2019) 1004–1015. https://doi.org/10.1016/j.jallcom.2018.08.200.
80. M.M. Garlapati, M. Vaidya, A. Karati, S. Mishra, R. Bhattacharya, B.S. Murty, Influence of Al content on thermal stability of nanocrystalline AlxCoCrFeNi high entropy alloys at low and intermediate temperatures, Adv. Powder Technol. 31 (2020) 1985–1993. https://doi.org/10.1016/j.apt.2020.02.032.
81. S. Varalakshmi, G. Appa Rao, M. Kamaraj, B.S. Murty, Hot consolidation and mechanical properties of nanocrystalline equiatomic AlFeTiCrZnCu high entropy alloy after mechanical alloying, J. Mater. Sci. 45 (2010) 5158–5163. https://doi.org/10.1007/s10853-010-4246-5.
82. S. Varalakshmi, M. Kamaraj, B.S. Murty, Processing and properties of nanocrystalline CuNiCoZnAlTi high entropy alloys by mechanical alloying, Mater. Sci. Eng. A 527 (2010) 1027–1030. https://doi.org/10.1016/j.msea.2009.09.019.
83. S. Varalakshmi, M. Kamaraj, B.S. Murty, Formation and stability of equiatomic and nonequiatomic nanocrystalline CuNiCoZnAlTi high-entropy alloys by mechanical alloying, Metall. Mater. Trans. A Phys. Metall. Mater. Sci. 41 (2010) 2703–2709. https://doi.org/10.1007/s11661-010-0344-x.
84. B.S. Murty, J.W. Yeh, S. Ranganathan, High Entropy Alloys, Butterworth-Heinemann Ltd (Verlag), (2014).
85. S. Praveen, B.S. Murty, R.S. Kottada, Phase evolution and densification behavior of nanocrystalline multicomponent high entropy alloys during spark plasma sintering, JOM. 65 (2013) 1797–1804. https://doi.org/10.1007/s11837-013-0759-0.
86. S. Praveen, B.S. Murty, R.S. Kottada, Effect of molybdenum and niobium on the phase formation and hardness of nanocrystalline CoCrFeNi high entropy alloys, J. Nanosci. Nanotechnol. 14 (2014) 8106–8109. https://doi.org/10.1166/jnn.2014.9441.
87. K.G. Pradeep, N. Wanderka, P. Choi, J. Banhart, B.S. Murty, D. Raabe, Atomic-scale compositional characterization of a nanocrystalline AlCrCuFeNiZn high-entropy alloy using atom probe tomography, Acta Mater. 61 (2013) 4696–4706. https://doi.org/10.1016/j.actamat.2013.04.059.
88. S. Praveen, A. Anupam, T. Sirasani, B.S. Murty, R.S. Kottada, Characterization of oxide dispersed AlCoCrFe high entropy alloy synthesized by mechanical alloying and spark plasma sintering, Trans. Indian Inst. Met. 66 (2013) 369–373. https://doi.org/10.1007/s12666-013-0268-4.
89. S. Praveen, J. Basu, S. Kashyap, R.S. Kottada, Exceptional resistance to grain growth in nanocrystalline CoCrFeNi high entropy alloy at high homologous temperatures, J. Alloys Compd. 662 (2016) 361–367. https://doi.org/10.1016/j.jallcom.2015.12.020.

90. P. Sathiyamoorthi, J. Basu, S. Kashyap, K.G. Pradeep, R.S. Kottada, Thermal stability and grain boundary strengthening in ultrafine-grained CoCrFeNi high entropy alloy composite, Mater. Des. 134 (2017) 426–433. https://doi.org/10.1016/j.matdes.2017.08.053.
91. G. Bracq, M. Laurent-Brocq, L. Perrière, R. Pirès, J.M. Joubert, I. Guillot, The fcc solid solution stability in the Co-Cr-Fe-Mn-Ni multi-component system, Acta Mater. 128 (2017) 327–336. https://doi.org/10.1016/j.actamat.2017.02.017.
92. M. Vaidya, A. Anupam, J.V. Bharadwaj, C. Srivastava, B.S. Murty, Grain growth kinetics in CoCrFeNi and CoCrFeMnNi high entropy alloys processed by spark plasma sintering, J. Alloys Compd. 791 (2019) 1114–1121. https://doi.org/10.1016/j.jallcom.2019.03.341.
93. Ł. Rogal, D. Kalita, L. Litynska-Dobrzynska, CoCrFeMnNi high entropy alloy matrix nanocomposite with addition of Al2O3, Intermetallics. 86 (2017) 104–109. https://doi.org/10.1016/j.intermet.2017.03.019.
94. H. Hadraba, Z. Chlup, A. Dlouhy, F. Dobes, P. Roupcova, M. Vilemova, J. Matejicek, Oxide dispersion strengthened CoCrFeNiMn high-entropy alloy, Mater. Sci. Eng. A 689 (2017) 252–256. https://doi.org/10.1016/j.msea.2017.02.068.
95. K.R. Rao, S.K. Sinha, Effect of sintering temperature on microstructural and mechanical properties of SPS processed CoCrCuFeNi based ODS high entropy alloy, Mater. Chem. Phys. 256 (2020) 123709. https://doi.org/10.1016/j.matchemphys.2020.123709.

9 Corrosion and Radiation Resistance Properties of High Entropy Alloys

Trinath Talapaneni, Avala Lava Kumar and Vatsala Chaturvedi

High entropy alloys (HEAs) are a novel class of materials characterized by their multi-principal element compositions, which provide unique microstructures and remarkable mechanical, thermal, corrosion, and radiation resistance properties. This chapter comprehensively analyzes HEAs, focusing on their classification, processing methods, and phase formation while emphasizing their potential in demanding environments. The chapter explores the core effects of HEA composition, such as their influence on microstructure and material properties like hardness, wear resistance, fatigue strength, and thermal stability. The corrosion resistance of HEAs is examined through the lens of alloying elements and their impact in various environments, particularly resistance to chloride-induced and acidic corrosion. The corrosion testing methods, including potentiodynamic polarization and electrochemical impedance spectroscopy, are also thoroughly discussed. The radiation resistance of HEAs is another critical aspect, highlighted by their atomic arrangement, compositional complexity, and radiation-induced mechanical changes. Experimental techniques, including ion irradiation, positron annihilation spectroscopy, and atom probe tomography, are evaluated to understand HEAs' response to radiation. The synergistic effects of corrosion and radiation resistance, particularly in nuclear, aerospace, and defense applications, are analyzed to evaluate the impact of irradiation on corrosion behavior. Finally, this chapter explores the challenges and future directions in developing corrosion and radiation-resistant HEAs, emerging trends, and future research directions.

9.1 INTRODUCTION

High entropy alloys (HEAs) are multi-component alloys typically consisting of five or more principal elements in approximately equal atomic percentages [1]. The atomic concentration of each component generally falls between 5% and 35% [2, 3]. Initially proposed by Professors Jien-Wei Yeh and later termed "multi-component alloys" by Brian Cantor, these alloys are also referred to as multi-principal element alloys, equimolar alloys, and substitutional alloys [4–6]. HEAs often form simple solid solutions with high mixing entropy, usually adopting face-centered cubic (FCC), body-centered cubic (BCC), hexagonal close-packed (HCP) structures, or

DOI: 10.1201/9781003664093-12

combinations of these [7]. Positioned at the center of a multi-component phase diagram, HEAs maximize configuration entropy in the solution phase. The random arrangement of various elements creates a unique, locally disordered chemical environment, contributing to their distinct corrosion-resistant properties [8]. This resistance is primarily attributed to developing protective passive layers facilitated by alloying elements like Cr, Ni, and Mo, which act as barriers, preventing further corrosion [9]. Additionally, HEAs offer exceptional properties, including high strength-ductility combinations, enhanced fatigue resistance, high hardness, wear resistance, creep resistance, superior fracture toughness, and excellent thermal stability. These attributes make them valuable for demanding industrial applications, particularly in extreme environments like nuclear reactors, turbines, and aerospace sectors [10–17].

9.1.1 Overview of High Entropy Alloys (HEAs)

HEAs are a unique class of compositionally complex alloys, distinguished by several principal alloying elements in substantial concentrations. The primary goal in developing HEAs is to reduce the tendency for ordering and segregation, leveraging the high mixing entropy that arises from incorporating several principal elements [18]. HEAs are typically characterized as alloys containing a minimum of five principal components, with each element contributing between 5% and 35% of the atomic concentration, with the additional criterion that the configurational entropy is equal to or greater than 1.5R. This classification, based on configurational entropy, has led to the identification of other alloy categories: medium-entropy alloys, with entropy values between R and 1.5R, and low-entropy alloys, which include traditional alloys with configurational entropy below R.

9.1.1.1 Classification of HEAs

HEAs are classified into three types: single-phase HEA, multi-phase HEA, or complex engineered HEA. Single-phase HEAs are alloys that form a single random solid solution (RSS) as their entire microstructure. This RSS can adopt a FCC, BCC, or HCP crystal structure, with the constituent elements distributed in a fully disordered manner. This type of HEA is precious for investigating the material's intrinsic kinetic phenomena and mechanical properties, as the absence of competing phases or mechanisms provides a clearer understanding of the behavior-driven solely by the RSS phase. Multi-phase HEAs meet the widely accepted definition by having a microstructure consisting of RSS phases and ordered crystalline structures. These phases can include various combinations, such as multiple RSS phases, a BCC RSS combined with a B2-ordered structure, or an FCC RSS paired with an L12-ordered structure. Complex-engineered HEAs are alloys designed around a RSS composition, utilizing various techniques to improve thermal stability and mechanical performance. These techniques involve modifications to the bulk composition, such as intermetallic and precipitate formation, minor alloying with substitutional or interstitial elements (less than 5 at.%), carbide reinforcement, and oxide dispersion strengthening. Despite variations in composition, all complex-engineered HEAs rely on the RSS's influence to enhance the overall material properties [19].

9.1.1.2 Processing Methods

Various processing methods have been employed to synthesize HEAs, broadly categorized into three states: liquid, solid, and gaseous. The liquid-state methods include melting and casting with uniform and non-uniform cooling rates, such as arc melting, vacuum induction melting, laser-enhanced net shaping, and laser cladding (LC) [20]. In the solid state, powder metallurgy techniques like mechanical alloying (MA) followed by consolidation are commonly used. HEAs can also be synthesized from the vapor or gaseous state using techniques such as sputtering, pulsed laser deposition (PLD), molecular beam epitaxy (MBE), and atomic layer deposition (ALD) [21]. Also, new preparation methods, including 3D printing, magnetron sputtering, and directional solidification, are used to study HEAs. The techniques mentioned can fabricate HEAs in various forms, including three-dimensional bulk, two-dimensional film-like structures, and one-dimensional fibrous shapes.

9.1.1.3 Phase Formation

Despite comprising more than five elements, HEAs typically solidify into relatively simple phases, such as FCC, BCC, or HCP structures. However, extensive research has revealed that these alloys can also include ordered intermetallic phases, amorphous structures, and nanocrystalline precipitates [22–24].

A single-phase solid solution in the HEA refers to a microstructure where several principal elements are randomly dispersed within a single crystalline lattice, such as BCC, FCC, and HCP structures. In this structure, the constituent elements do not segregate into separate phases or compounds but instead form a uniform, disordered arrangement. The high configurational entropy from this random distribution stabilizes the solid solution, leading to improved mechanical and thermal properties. This simple solid-solution phase is a crucial characteristic of HEAs, contributing to their unique combination of strength, ductility, and thermal stability.

Intermediate phases and other complex structures in HEAs: Intermediate phases and other complex structures in HEAs arise when the alloy's composition and processing conditions form phases beyond simple solid solutions. These phases may include intermetallic compounds, amorphous regions, or nanocrystalline precipitates, depending on factors like atomic size difference, mixing enthalpy, and valence electron concentration. The coexistence of these phases, often alongside solid-solution structures, can significantly influence HEAs' mechanical and thermal properties, enhancing hardness, strength, or corrosion resistance through complex interactions between phases [25, 26].

Nanocrystalline and amorphous phases in HEAs: Nanocrystalline and amorphous phases in HEAs form due to unique effects like lattice distortion and sluggish diffusion, which inhibit conventional grain growth. Nanocrystalline phases contain excellent grains, while amorphous phases lack a defined crystal structure. These phases can develop during casting or heat treatment, contributing to enhanced mechanical properties such as increased strength, hardness, and wear resistance. The presence of these phases in HEAs offers a balance of strength and ductility, as well as improved thermal stability and corrosion resistance, making them suitable for advanced engineering applications [6, 23, 27].

9.1.1.4 Benefits of HEAs

The primary benefits of HEAs include:

1. Exceptional Mechanical Properties: HEAs typically display a distinctive high strength and excellent ductility balance, making them well suited for challenging applications.
2. High-Temperature Stability: These alloys maintain their structural integrity and performance at elevated temperatures, which is crucial for applications in extreme environments.
3. Corrosion Resistance: The complex microstructures and protective oxide layers formed by certain alloying elements enhance resistance to various corrosive conditions.
4. Fatigue Resistance: HEAs typically demonstrate improved resistance to fatigue and crack propagation, increasing their longevity in service.
5. Irradiation: HEAs demonstrate enhanced irradiation resistance, making them suitable for applications in radiation-intensive environments, such as nuclear reactors.
6. Thermoelectric Property: HEAs exhibit promising thermoelectric properties, making them potential candidates for efficient energy conversion applications.

9.1.2 Importance of Corrosion and Radiation Resistance of HEAs

HEAs corrosion and radiation resistance are critical attributes that enhance their suitability for demanding applications, particularly in harsh environments like nuclear reactors and aerospace systems.

1. **Corrosion Resistance:** HEAs often exhibit superior corrosion resistance due to their unique microstructures and alloying elements forming protective oxide layers. Compared to traditional alloys, this makes them more resilient to corrosive environments, such as high-temperature steam or acidic solutions. Their ability to maintain structural integrity in aggressive conditions is essential for extending the life span of components in industries such as chemical processing and maritime applications.
2. **Radiation Resistance:** HEAs also show promise in radiation resistance, making them ideal candidates for applications in nuclear environments. Their complex microstructures can help mitigate damage from radiation by accommodating defects and preventing the formation of harmful phases that typically occur in conventional materials. This resistance to radiation-induced degradation is vital for maintaining performance and safety in nuclear reactors and other radiation-rich environments.

The combination of corrosion and radiation resistance in HEAs contributes to their potential for use in extreme service conditions, where traditional materials may fail, thereby enhancing the reliability and longevity of critical components.

9.2 ESSENTIALS OF HEAs

9.2.1 Composition and Microstructure

HEAs are advanced materials characterized by their unique composition and microstructure. Unlike conventional alloys, typically based on one or two principal elements, HEAs consist of five or more principal elements in near-equiatomic or significant concentrations. This lets HEAs achieve unique properties.

Composition: In most HEAs, the elements are present in nearly equal atomic ratios (e.g., 20% each in a five-element alloy). This equal distribution is critical for achieving high configurational entropy, stabilizing the alloy's structure. Deviations from equal ratios are possible, but the overall goal is maintaining high configurational entropy for stabilizing simple solid-solution phases like BCC, FCC, and HCP rather than complex intermetallic ones. The elements used in HEAs are typically metals, including transition metals (Fe, Co, Cr, Ni, Mn, Cu), refractory metals (Ti, Zr, Hf, Ta, Mo, W, Nb), and lightweight elements (Al, Mg, Si, etc.). Ni, Co, and Cr often enhance strength and corrosion resistance. Al and Cr can improve oxidation resistance, especially at high temperatures. Al, Ti, or Mg can reduce the density of HEAs.

9.2.1.1 Core Effects of Composition in HEAs [6, 19, 28]

HEAs' composition leads to several core effects that distinguish them from conventional alloys. These effects result from the unique mixing of elements in nearly equal proportions, contributing to their superior mechanical, thermal, and chemical properties. The four core effects of HEAs are:

1. **High Entropy Effect:** The entropy effect in HEAs arises from mixing multiple principal elements in nearly equal atomic proportions, leading to significantly higher configurational entropy than in conventional alloys. This increased entropy stabilizes simple solid-solution phases such as BCC, FCC, and HCP structures, even in complex multi-element systems, rather than forming brittle intermetallic compounds. As a result, HEAs tend to exhibit enhanced phase stability and thermal resistance, particularly at elevated temperatures, making them more stable and durable for high-performance applications.
2. **Lattice Distortion:** Different atomic sizes in HEAs lead to significant lattice distortion because the atoms do not fit perfectly in the crystal lattice. The varying atomic radii cause distortions in the structure, increasing internal strain and resistance to dislocation motion. This distortion enhances mechanical properties such as strength, hardness, and toughness by making it more difficult for the material to deform under stress. The lattice distortion also contributes to improved wear resistance and can help maintain the alloy's mechanical integrity at elevated temperatures.
3. **Sluggish Diffusion:** The sluggish diffusion effect in HEAs arises from multiple elements with varying atomic sizes and bonding characteristics, hindering atoms' movement within the lattice. This diverse atomic environment creates energy barriers that slow down diffusion processes, making it difficult for atoms to migrate through the material. As a result, HEAs exhibit excellent resistance to creep and maintain their microstructural

stability over long periods, especially at high temperatures. This sluggish diffusion also contributes to retaining desirable properties like strength and hardness during extended service in harsh environments, making HEAs ideal for high-temperature and high-stress applications.

4. **Cocktail Effect:** The cocktail effect in HEAs refers to the synergistic interaction of multiple principal elements, resulting in unique and enhanced properties that surpass those of individual components or traditional alloys. By combining various elements in near-equal atomic proportions, HEAs benefit from the complex interactions between different atoms, which can improve mechanical, thermal, and chemical properties. This synergy often leads to enhanced strength, toughness, wear resistance, and corrosion resistance, as the alloy's performance is not dictated by any single element but rather by the collective influence of all constituents. The cocktail effect allows for tailoring HEAs for specific high-performance aerospace, automotive, and energy applications.

9.2.1.2 Microstructures in HEAs

HEAs typically form a simple solid-solution microstructure characterized by a random arrangement of constituent elements within FCC, BCC, or HCP lattices. The microstructure of HEAs often includes fine grains, which can be achieved through controlled processing techniques like rapid solidification or thermo-mechanical processing. Fine-grained structures enhance strength via the Hall-Petch effect, where smaller grains contribute to increased resistance to dislocation motion. In some HEAs, additional phases such as ordered intermetallic, amorphous regions, or nanocrystalline precipitates may form, depending on processing conditions and the specific elements used. These phases can enhance specific properties, such as strength and wear resistance. The microstructure of HEAs can be fine-tuned by controlling processing parameters such as cooling rates, heat treatment, and thermo-mechanical processing. This allows for customizing grain sizes, phase composition, and precipitate distribution to optimize mechanical properties for specific applications.

9.2.2 High Entropy Effect and Its Influence on Material Properties

HEAs are a captivating category of materials defined by their unique compositions, typically consisting of five or more principal elements in nearly equal atomic ratios. This diverse mix results in high configurational entropy, significantly impacting their material properties. Over 300 HEAs have been developed, marking a new frontier in metallic materials. HEAs exhibit various desirable characteristics, including high hardness, exceptional wear resistance, good fatigue resistance, excellent high-temperature strength, strong thermal stability, and generally effective oxidation and corrosion resistance. These attributes highlight their promising potential for a variety of applications.

9.2.2.1 Hardness

The hardness of HEAs has been assessed across various scales, using methods such as Vickers microhardness testing and instrumented micro- and nano-indentation techniques. Depending on the alloy systems and related processing methods, HEAs exhibit a wide hardness range, from 150 to 1200 HV. Alloys with a BCC structure generally demonstrate higher hardness than those with an FCC phase. Wu et al. [29] revealed that incorporating Al into Co-Cr-Cu-Fe-Ni alloys increases hardness non-linearly. Up to 6% Al content, the hardness increase is gradual due to the BCC phase; beyond that, it rises sharply at 26% Al due to the formation of a dual-phase (FCC and BCC) microstructure. The hardness of the $Al_{0.4}$-$Co_{0.5}$-V_x-Fe-Ni HEA increases as the vanadium (V) content is raised from 0 to 0.6 [30]. The hardness of $(Fe\text{-}Mn\text{-}Ni\text{-}Co)_{100-x}$-$Mo_x$ HEAs was observed to increase with the addition of molybdenum (Mo), as the Mo content increased from x = 0 to x = 10 in increments of 5 atomic wt% [31]. The enhancement in hardness of the HEAs can be attributed to solid solution strengthening, grain boundary strengthening, and precipitation strengthening mechanisms. It was noticed that increasing the Nb content from x = 0 to x = 1 with an increment of 0.25 atomic wt. % in $Mo_{0.25}$-$V_{0.25}$-$Ti_{1.5}$-$Zr_{0.5}$-Nb_x HEA resulted in a decline in hardness [32]. Huang et al. [33] added Sc and Y to Ti-Zr-Hf HEA improved hardness, with Y contributing more significantly than Sc. Furthermore, the study revealed that combining both Sc and Y enhanced the hardness more effectively than when either element was added individually. Guo et al. [34] have shown that adding silicon (Si) to Nb-Ta-W-Mo-Si_x HEAs increases hardness as the Si content varies from x = 0 to 0.75. This improvement is attributed to forming a BCC structure and silicide phases. Similarly, Huang et al. [35], in their study on Fe-Co-Cr-Ni-Si_x HEAs, found that varying Si content from 0 to 1 (in increments of 0.5 atomic wt. %) resulted in a 600% increase in hardness. This significant enhancement was due to the formation of a NiSi phase and increased lattice distortion caused by the smaller atomic size of Si. Cong Ye et al. [36] found that adding 0.8 wt. % of Ti to Cu-Co-Fe-Ni-Ti_x HEA show superior hardness of HV 461.6. Chen et al. [37] discovered that as the Fe content in Al-Co-Cr-Fe_x-Ni (x = 0.2 to 2 wt. %) HEA increased, the hardness decreased from HV637.2 to HV460.2. This reduction in hardness was attributed to the disappearance of the Cr_3Ni_2 phases and a decrease in lattice distortion. Churyumov et al. [38] investigated the effect of Nb addition (x = 4 and 10) on the mechanical properties of heat-treated Fe-Co-Ni-Cu-Cr HEAs, comparing it to a Nb-free alloy. They found that at 1173 K, the hardness of the $(Fe\text{-}Co\text{-}Ni\text{-}Cu\text{-}Cr)_{96}Nb_4$ alloy increased due to the precipitation of dispersed particles from the supersaturated solid solution. In contrast, the hardness of the Nb-free alloy and the $(Fe\text{-}Co\text{-}Ni\text{-}Cu\text{-}Cr)_{90}Nb_{10}$ alloy decreased, attributed to a reduction in initial dislocation density and the growth of dispersed eutectic particles. Zhu et al. [39] found that the V addition above 1% resulted in the formation of FCC solid solution and intermetallic sigma phase, which enhanced the hardness of Co-Fe-Ni-Mn-Vx (x = 0.25 to 1.50 wt. %) from 213 HV to 716 HV. Pauzi et al. [40] have studied the effect of Zr addition on the hardness of Fe-Cr-Ni-Mn-Co-Zr_x and $Al_{0.5}$-Fe-Cr-Ni-Mn-Co-Zr_x HEA and found that hardness value increased with the addition of Zr (x = 0 to 0.3) in both the alloys and also it is observed that 0.5% Al with 0.3% Zr alloy has shown highest hardness value of 519 HV due to BCC as significant phase. Chandrakar et al. [41] investigated the impact of silicon addition on Al-Co-Cr-Cu-Fe-Ni-Si_x HEAs (x = 0 to 0.9, with increments of 0.3%) and found that hardness increased with the addition of Si. This improvement was attributed to the evolution of BCC and sigma phases and solid solution strengthening. Liu et al.

[42] observed increased Fe content from 0.5 to 3.5 wt. % showed a decrement of hardness from 312 HV to 218 HV in an Al-Co-$Cr_{0.5}$-Fe_x-$Ni_{2.5}$ HEA due to a change of phase from BCC to FCC with Fe content.

9.2.2.2 Wear Resistance

The wear behavior of HEAs has been evaluated across various systems under both abrasive and adhesive conditions. Tong et al. [43] investigated the wear behavior of Al_x-Co-Cr-Cu-Fe-Ni HEAs (x = 0.5, 1.0, 2.0). They found that the $Al_{0.5}$-Co-Cr-Cu-Fe-Ni alloy displayed higher wear resistance despite having lower hardness due to its significant work hardening ability. Hsu et al. [44] observed the wear mechanism of Cu-Co-Ni-Cr-$Al_{0.5}$-Fe-B_x HEA and found that boron addition (x = 0 to 1.0) results in the formation of boride, which enhances the wear resistance property of HEA. Chuang et al. [45] discovered that the wear resistance of $Co_{1.5}$-Cr-Fe-$Ni_{1.5}$-Ti and $Al_{0.2}$-$Co_{1.5}$-Cr-Fe-$Ni_{1.5}$-Ti alloys is at least twice as effective as traditional wear-resistant steels with comparable hardness. These HEAs' superior oxidation and thermal softening resistance are suggested as the primary factors contributing to their exceptional wear performance. Wu et al. [29] investigated the effect of Al addition on the wear property of Al_x-Co-Cr-Cu-Fe-Ni (x = 0.5, 1.0, 2.0) HEA. The high aluminum content (x = 2.0) significantly enhances wear resistance due to its increased hardness. This prevents plastic deformation and delamination and promotes oxidative wear, where forming an oxide film further improves wear resistance. Chen et al. [46] explored the wear behavior of $Al_{0.5}$-Co-Cr-Cu-Fe-Ni-Ti_x HEA (x = 0 to 2.0 with 0.2 variations). They observed a significant improvement in the wear resistance of the alloy as the Ti content increased from 0.6 to 1.0, reaching its peak at x = 1.0. However, further increases in Ti content resulted in a gradual decline in wear resistance. The initial improvement is attributed to the development of a Co-Cr-like phase. At the same time, the subsequent decline is linked to the appearance of lateral cracks on the worn surfaces, indicating more brittle deformation behavior, which corresponds with the reduction in wear resistance. Hsu et al. [47] studied the wear mechanism of Al-Co-Cr-Fe_x-$Mo_{0.5}$-Ni HEA with varying Fe (0.6, 1.0, 1.5, and 2.0) content and observed that at Fe1.0, the wear resistance reaches its maximum. In contrast, the wear resistance of the Fe2.0 alloy is significantly lower than that of Fe1.5. This decrease is attributed to the markedly higher oxidation rate of the Fe2.0 alloy than Fe1.5. Chen et al. [48] studied the effect of V on the wear resistance of $Al_{0.5}$-Co-Cr-Cu-Fe-Ni-V_x (x = 0 to 2.0), and it is observed that wear resistance improves by approximately 20% as the V content increases from x = 0.6 to 1.2, then stabilizes beyond 1.2. This enhancement is associated with the phase transformation from FCC to BCC and the formation of the Sigma phase. Moazzen et al. [49] examined the influence of Zr on the wear resistance of Ni-Co-Cr-Fe-Zr_x HEA. They observed that adding Zr enhanced wear resistance and facilitated the transition from an abrasive to an adhesive wear mechanism. Xiaotao et al. [50] investigated the influence of boron on the wear properties of $Al_{0.5}$-Co-Cr-Cu-Fe-Ni-Bx (x = 0–1.0) HEA, fabricated through arc melting. Their findings revealed that adding more than 0.4% boron significantly improves wear resistance and alters the wear phenomena from debonding to oxidative wear. This enhancement is attributed to the presence of an FCC matrix and the formation of hard borides within the alloy. Gorban et al. [51]

examined the wear resistance of a Co-Cr-Fe-Mn-Ni (counterpart) and Ti_{30}-Zr_{25}-Hf_{15}-Nb_{20}-Ta_{10} (finger) friction pair over a temperature range of 77–873 K. Their findings showed that wear on the finger reduced as the temperature increased compared to the counterpart. This improvement is attributed to the formation of oxides on the finger's surface at higher temperatures, which increased its hardness to 18 GPa, leading to enhanced wear resistance. Yadav et al. [52] investigated the effect of soft dispersoids, such as Pb and Bi, on the wear behavior in equiatomic Cu-Cr-Fe-Ti-Zn and Al-Cr-Fe-Mn-V HEAs. They found that increasing the Pb and Bi content in HEAs Reduced the wear rate by 21% and 25%, respectively. The improvement in wear resistance is primarily attributed to soft dispersoids, enhanced ductility, and increased hardness within the HEA phases. Mindi et al. [53] inspected the W addition on the wear of Ni-Mn-Cr-Mo-Wx (x = 0.2, 0.4, 0.6, 0.8, 1.0 atomic fraction) HEA. They revealed that annealed alloy powders with 0.6 atomic fractions of tungsten exhibited a maximum hardness of 644 HV5 and superior wear resistance at temperatures exceeding 1273 K. This improvement is likely due to the extensive formation of the σ-phase, MoNi3, and MnNi phases.

9.2.2.3 Fatigue Resistance

To fully unlock the potential of HEAs as engineering materials, it is crucial to investigate their deformation behavior under cyclic loading or fatigue. Fatigue is categorized into stress-controlled high-cycle and strain-controlled low-cycle fatigue, exhibiting significantly higher plastic strain per cycle. Since fatigue accounts for 90% of engineering failures, often with catastrophic consequences, understanding fatigue behavior is essential for revealing HEAs' underlying deformation and damage mechanisms. Hemphill et al. [12] investigated the high cycle fatigue behavior of as-cast and wrought $Al_{0.5}$-Co-Cr-Cu-Fe-Ni alloy through four-point bending tests at room temperature. The study revealed that the HEA exhibited fatigue characteristics comparable to steel and titanium, with a distinct endurance limit ranging from 540 to 945 MPa, depending on the specific conditions. Niendorf et al. [54] demonstrated that the increased tendency for austenite-to-martensite transformation during cyclic loading contributed to cyclic hardening, resulting in an improved fatigue life of the Fe_{50}-Mn_{30}-Co_{10}-Cr_{10} HEA. Bahadur et al. [55] investigated the low-cycle fatigue behavior of Co-Cu-Fe-Mn-Ni HEA, consisting of an FCC matrix and copper-rich nanoclusters. The study revealed that the interaction between dislocations and copper-rich nanoclusters plays a crucial role in the alloy's cyclic hardening and softening behavior, ultimately leading to failure through intergranular fracture. Tang et al. [13] examined the fatigue behavior of a cold-rolled two-phase $Al_{0.5}$-Co-Cr-Cu-Fe-Ni HEA. Their findings showed that fatigue life decreased with increasing stress. Additionally, samples from high-purity components exhibited better fatigue strength than those from commercial-grade materials, primarily due to preexisting microcracks, pores, and oxides. Xu et al. [56] investigated the $(FeCoNi)_{86}$-Al_7-Ti_7 HEA and found that aging treatment enhanced the fatigue endurance limit by 30%. This improvement in fatigue strength was attributed to the increased resistance of the $L1_2$ and $L2_1$ precipitates formed during aging. Tian et al. [57] investigated the fatigue behavior of Co-Cr-Fe-Mn-Ni HEA and observed that the fatigue strength could be significantly enhanced by fine grain size. Liu et al. [58] examined the fatigue behavior of ultrafine-grained triplex $Al_{0.3}$-Co-Cr-Fe-Ni HEA and observed excellent fatigue resistance, with a fatigue endurance limit to Ultimate Tensile Strength (UTS) ratio of 0.43. This superior performance was attributed to the formation of deformation nano twins. Liu et al. [59] reported enhanced fatigue strength in grain-refined Fe_{42}-Mn_{28}-Cr_{15}-Co_{10}-Si_5, attributed to the $\gamma \rightarrow \epsilon$ transformation promoted by the alloy's metastability. Liu et al. [60] explored the impact of nano-sized precipitates on the fatigue properties of $Al_{0.7}$-Co-Cr-Fe-Ni HEA. They discovered that coherent $L1_2$ nano-precipitates did not enhance fatigue performance, likely because these precipitates were sheared during cyclic loading. Kim et al. [61] investigated the high-cycle fatigue properties of additive-manufactured Co-Cr-Fe-Mn-Ni HEA produced by selective laser melting. The alloy's exceptional fatigue resistance was attributed to its distinct microstructural features, including heterogeneous grain structures, dislocation networks, in-situ formed oxides, and the formation of deformation twins during cyclic loading. Shams et al. [62] studied grain size's impact on the carbon-containing Co-Cr-Fe-Mn-Ni HEA's fatigue behavior. They suggested that fatigue life could be enhanced through controlled thermomechanical treatment, which produces fine-grained microstructures while preventing the precipitation of coarse carbides. Shams et al. [63] investigated the low-cycle fatigue behavior of equiatomic Co-Cr-Fe-Mn-Ni HEA. They found that the reduced fatigue performance was attributed to a strain-induced α'-martensite phase formation. Lu et al. [64] revealed the low-cycle fatigue behavior of dual-phase $Al_{0.5}$-Co-Cr-Fe-Mn-Ni HEA, and it is observed that higher yield strength and solid-solution strengthening led to higher cyclic stress resistance. He et al. [65] studied the fatigue behavior of Cr-Fe-Co-Ni-$Mo_{0.2}$ HEA; the increase in dislocation density reduces the ratcheting strain rate, helping to stabilize the structure and enhancing fatigue resistance. Lu et al. [66] concluded that Co-Cr-Fe-Mn-Ni HEA reduces low-cycle fatigue (LCF) life at 823 K. This decline is attributed to elemental segregation and grain boundary embrittlement caused by in-situ oxidation at this temperature. Jin et al. [67] investigated the fatigue behavior of Cr-Mn-Fe-Co-Ni HEA across temperatures ranging from 295 K to 873 K. They observed a shift in fracture mode from transgranular at 295 K to a more intergranular-like fracture at 873 K. This led to increased brittleness, reduced ductility, and significantly decreased fatigue life at the higher temperature.

9.2.2.4 Tensile Properties

Gali et al. [68] investigated the tensile behavior of a Cr-Mn-Fe-Co-Ni HEA across a temperature range from –469 K to 1273 K, using strain rates of 10^{-3} and 10^{-1} s^{-1} and observed that the tensile strength of the alloy is highly temperature-dependent. Similarly, Otto et al. [69] performed tensile tests on a Co-Cr-Fe-Mn-Ni HEA at an engineering strain rate of 10^{-3} s^{-1} over a temperature range of 77 to 1073 K. Their study revealed that deformation twinning, which introduces new interfaces and decreases the mean free path of dislocations during tensile testing—a "dynamic Hall–Petch" effect—resulted in significant work hardening and a notable increase in ultimate tensile strength. Chen et al. [70] suggested that the tensile properties of Co-Cr-Fe-Ni-Mn-Cx (x= 0, 0.05, 0.1, 1.5, 2.0) alloy improved with minor carbon addition (x = 0.05) due to enhanced strength and ductility from a mixed deformation mode. However, at higher carbon concentrations (x > 0.1), strength increases while ductility decreases, leading to quasi-cleavage fracture. Daoud et al. [71] studied the tensile properties of as-cast and heat-treated Al_8-Co_{17}-Cr_{17}-Cu_8-Fe_{17}-Ni_3 HEA at different temperatures and found that annealing at 1402 K for 5 hours has shown superior ductility. He et al. [72] observed that tensile properties of $(Fe\text{-}Co\text{-}Ni\text{-}Cr\text{-}Mn)_{100-x}$-$Al_x$ HEAs vary significantly with the alloy's crystalline structure, which evolves as aluminum content increases (Al = 0–20 at %). Alloys with a single FCC structure exhibit relatively low strength but high ductility, behaving like a solid solution. In the duplex FCC+BCC structure, tensile strength sharply increases, though ductility is reduced, giving the alloys composite-like behavior. As the structure transitions to a single BCC phase with higher Al concentrations, the alloys become incredibly brittle, showing a clear link between microstructural changes and mechanical performance. Huo et al. [73] proposed that the tensile properties of Co-Cr-Fe-Ni HEA improved with increasing strain rates, showing higher tensile strength and ductility. At lower strain rates, deformation was dominated by dislocation motion, while at higher strain rates, stacking faults became the primary mechanism. Jang et al. [74] suggested that the tensile properties of hot-rolled Cr-Mn-Fe-Co-Ni HEA were strongly temperature-dependent. Yield strength dropped significantly from 413 MPa at 973 K to 94 MPa at 1173 K. Ductility increased notably above 973 K but decreased beyond 1173 K. The alloy maintained a single FCC phase, and incomplete dynamic recrystallization at 973 K contributed to variations in flow behavior. Jin et al. [75] witnessed that the Fe_{20}-Co_{20}-Ni_{41}-Al_{19} eutectic HEA exhibited excellent tensile properties, with a room-temperature tensile strength of 1103 MPa and a total elongation of 18.7%. These properties surpass those of many as-cast HEAs, making it a promising candidate for industrial applications. Kireeva et al. [76] studied the tensile properties of [001]-oriented single crystals of equiatomic Co_{20}-Cr_{20}-Fe_{20}-Mn_{20}-Ni_{20} HEA and found deformation twinning at 77 K, leading to increased strain hardening compared to slip deformation at higher temperatures. The critical resolved shear stress for twinning was measured at 210 ± 10 MPa after 5% strain. Kuznetsov et al. [77] identified that hot-forged Al-Cu-Cr-Fe-Ni-Co HEA exhibited superior tensile properties compared to the as-cast condition, with a yield stress of 1040 MPa, ultimate tensile strength of 1170 MPa, and tensile ductility of 1%. The alloy displayed superplastic behavior at 1073–1273 K, achieving tensile elongation up to 860% at 1273 K. Lilensten et al. [78] saw Ti_{35}-$Zr_{27.5}$-$Hf_{27.5}$-Nb_5-Ta_5 HEA exhibited a significant transformation-induced plasticity effect, reaching a high work-hardening rate of 0.103 without compromising ductility. After tensile testing, the microstructure revealed a 65% volume fraction of stress-induced martensite α'', optimizing its mechanical properties. Shun et al. [79] suggested that the $Al_{0.3}$-Co-Cr-Fe-Ni HEA exhibited a yield drop and low work-hardening during tensile testing due to coherent nano-precipitates in the as-cast condition. However, rod-shaped precipitates developed after aging at 1173 K, resulting in increased work-hardening without a yield drop. Zhao et al. [80] observed that the Al-Co_x-Cr-Fe-Ni HEAs exhibit improved tensile properties with increased Co content, where the fracture strain rises from 5.9% to 15.4%. At the same time, the yield strength decreases from 450 MPa to 360 MPa, and the ultimate tensile strength marginally increases from 734 MPa to 739 MPa. These variations are attributed to phase structure changes and microstructure evolution resulting from the higher Co content in the deposited alloys. Yao et al. [81]

proposed that the non-equiatomic Fe-Mn-Ni-Co-Cr alloy exhibits remarkable tensile ductility, characterizing it as a genuine HEA. This alloy maintains a stable random solid solution despite having a relatively low configurational entropy. This ductility is noteworthy, given the substantial lattice distortion present in the material. Liu et al. [82] saw that the Fe-Co-Cu-Ni-Sn_x HEAs exhibit high strength and plasticity, achieving a maximum tensile strength of 633 MPa and an elongation strain of 19.8% when Sn content is between 0.05 and 0.07. Adding Sn enhances the tensile strength while maintaining a single FCC solution at lower concentrations.

9.2.2.5 Oxidation Resistance

In conventional alloys, adding suitable oxidizing elements significantly improves oxidation resistance by generating a dense and stable oxide layer on the surface at high temperatures. Liu et al. [83] examined the oxidation behavior of cast Alx-Co-Cr-Fe-Ni alloys (where x = 0.15 and 0.4) in supercritical water at temperatures of 823K and 873K, comparing their performance to that of HR3C steel. The oxide layers formed on the HEAs were identified as spinel type $(Fe, Cr)_3O_4$. Unlike the thicker oxide film observed on HR3C steel, the $Al_{0.15}$-Co-Cr-Fe-Ni and $Al_{0.4}$-Co-Cr-Fe-Ni alloys developed a thinner oxide layer with fine oxide particles, indicating superior oxidation resistance. Rao et al. [84] investigated the oxidation behavior of Alx-$Cr_{0.4}$-Cu-$Fe_{0.4}$-Mn-Ni alloys, where x varied from 0 to 0.4. In the alloy that lacked aluminum, the oxide layer was characterized by a manganese-rich outer layer and a thin, dense chromium oxide layer at the interface. However, with an increase in aluminum content, a marked reduction in the oxidation rate was observed, underscoring the vital contribution of Al_2O_3 in improving oxidation resistance. Chang and Yeh [85] reported the development of a continuous Cr_2O_3 protective oxide layer on the surface of the $Al_{0.2}$-$Co_{1.5}$-Cr-Fe-Ni-Ti HEA at a temperature of 1173 K. This oxide layer effectively reduces high-temperature oxidation, making these HEAs ideal for use in high-temperature applications. Grewal et al. [86] reported that the $Al_{0.1}$-Co-Cr-Fe-Ni HEA demonstrates a higher diffusion activation energy than traditional superalloys and steels. This is attributed to the presence of interstitial sites and the low potential energy that traps atoms within the lattice. Additionally, the mean jump rate of oxygen in the HEA is much slower than in T91, T22 steels, and pure iron, contributing to its reduced ion diffusion and enhanced oxidation resistance. Butler and Weaver [87] investigated Al_x(Ni-Co-Cr-Fe) HEAs and analyzed how varying aluminum content (up to 30 atomic percent) influences their microstructure and oxidation resistance. With increasing Al concentration, the microstructure shifted from an FCC matrix containing Ni-Al-rich regions to a BCC matrix with rich precipitates Al and Ni. The HEA containing 30 atomic percent Al demonstrated the best oxidation resistance, attributed to forming a continuous Al_2O_3 protective layer. Dabrowa et al. [88] explored the impact of varying copper content on the oxidation response of Al-Co-Cr-Cu_x-Fe-Ni HEAs at 1273 K over durations of 100 and 500 hours. They found that higher Cu content reduced the oxidation rate but compromised the adhesion of the protective α-Al_2O_3 layer. After 500 hours, the HEA with the highest copper concentration experienced complete spallation of the α-Al_2O_3 film. Using atmospheric plasma spray, Hsu et al. [89] applied Ni-$Co_{0.6}$-$Fe_{0.2}$-$Cr_{1.5}$-Si-Al-$Ti_{0.2}$ (C3) HEA coatings onto 304 stainless steel and Al_2O_3 substrates. The oxidation behavior of these coatings was examined at 1173 K, 1273 K, and 1373 K. At 1373 K, the HEA coating exhibited outstanding oxidation resistance, comparable to that of NiCrAlY. The formation of protective oxide layers, including α-Al_2O_3 and Al_2O_3/Cr_2O_3, at this temperature significantly reduced oxygen diffusion and enhanced oxidation resistance. Adomako et al. [90] developed Co-Cr-Ni, Co-Cr-Mn-Ni, and Co-Cr-Fe-Mn-Ni HEAs. They investigated their oxidation behavior through thermo-gravimetric analysis (TGA) at temperatures ranging from 1073 K to 1273 K over 24 hours. The Co-Cr-Ni alloy exhibited the highest oxidation resistance, followed by Co-Cr-Fe-Mn-Ni and Co-Cr-Mn-Ni, with Mn and Fe enhancing oxidation resistance. The formation of Cr2O3 in Co-Cr-Fe-Mn-Ni and the lower Mn content contributed to reduced oxidation rates, with activation energies measured at 108 kJ/mol for Co-Cr-Mn-Ni and 137 kJ/mol for Co-Cr-Fe-Mn-Ni. Gorr et al. [91] examined the oxidation performance of refractory high entropy alloys (RHEAs) Al-Cr-Mo-Ti-W, Al-Cr-Mo-Nb-Ti, and Al-Cr-Mo-Ta-Ti at 1273 K and 1373 K. Among these, Al-Cr-Mo-Ta-Ti showed superior oxidation resistance at both temperatures due to the development of a dense, continuous aluminum oxide scale that resulted in a minimal mass gain. In contrast, the alloys containing W and Nb exhibited moderate mass increases and complex oxidation behavior, characterized by a nonuniform oxide layer with thick, porous regions. Butler et al. [92] synthesized equimolar Nb-Ti-Zr-V and Nb-Ti-Zr-Cr HEAs using vacuum arc melting and investigated their oxidation behavior in air at 1273 K. The Nb-Ti-Zr-V alloy underwent rapid oxidation, becoming fully oxidized within 8 hours. Both alloys experienced internal oxidation, with the oxidized layer consisting of compounds such as TiO_2, V_2O_5, $TiNb_2O_7$, and $Nb_2Zr_6O_{17}$, which did not improve their oxidation resistance. Chang et al. [93] synthesized a Hf-Ta-Ti-Nb-Zr RHEA and studied how varying aluminum content influences its oxidation behavior. The RHEA without Al exhibited increased mass gain with rising temperatures and provided insufficient oxidation protection. However, adding 1% Al improved oxidation resistance between 973 K and 1173 K. At higher temperatures of 1373 K and 1573 K, the protection became less effective due to forming less dense oxide scales, facilitating oxygen diffusion into the alloy. Lu et al. [94] examined the impact of doping small amounts of Y (0.02 at.%) and Hf (0.02 at.%) into the Al-Co-Cr-Fe-Ni HEA. They observed that after 1000 hours of oxidation at 1373 K, a uniform, continuous oxide scale formed, consisting primarily of α-Al_2O_3 with traces of HfO_2 and $Y_3Al_5O_{12}$ at the grain boundaries. The oxide scale displayed a double-layered structure with smooth interfaces, suggesting strong adhesion to the substrate. Oxygen inward diffusion was found to control the scale's growth. The slow oxidation rate in the Y/Hf-doped HEA was linked to the rapid development of a protective α-Al_2O_3 layer during the early oxidation stages, enhanced by the rare earth effect and the large columnar grain structure of the oxide layer. Ye et al. [95, 96] discovered that the oxidation of $(Hf_{0.2}$-$Zr_{0.2}$-$Ta_{0.2}$-$Nb_{0.2}$-$Ti_{0.2})C$ at temperatures ranging from 1073 to 1673 K followed a parabolic rate law, indicating that the oxidation process is primarily diffusion-controlled. They noted that the inward diffusion of oxidants played a significant role in this oxidation process, as evidenced by the calculated oxidation activation energy. When the oxidation temperature was raised to 1773 K, the oxidation kinetics shifted to a parabolic-linear relationship, leading to the complete oxidation of the carbide within 60 minutes of isothermal exposure. Wang et al. [96–98] observed a thin Ti-depleted layer at the interface between the matrix and the oxide during high-temperature oxidation. They attributed the formation of this layer to the active oxidation of TiC and the subsequent diffusion of Ti-containing species. Their findings are corroborated by thermodynamic calculations, indicating that the formation of TiO is energetically favorable. The absence of a Ti-depleted layer when titanium was omitted from the high entropy carbides indicates a shift in the oxidation process towards inward oxygen diffusion through the oxide layer. This was further confirmed by detecting a dense, amorphous oxy-carbide interlayer, which acted as an effective diffusion barrier for oxygen. Although the exact oxidation mechanism is still debated, the multilayer structure of the oxide scale is a distinct feature of high entropy carbides, with Hf/Zr-rich oxides forming the outer layer and Ta/Nb-rich oxides constituting the inner layer. This suggests that preferential oxidation is influenced by the relative thermodynamic stabilities of the various oxides, with Hf demonstrating the strongest binding affinity for oxygen [99,100].

9.2.2.6 Magnetic Properties

Most HEAs contain high concentrations of metallic elements like ferromagnetic Ni, Co, and Fe, making their magnetic properties a compelling scientific topic. A simple cubic structure and a random atomic arrangement lead to low magnetic anisotropy, resulting in many HEAs exhibiting soft magnetic properties. Compared to traditional soft magnetic alloys, HEAs with nearly equal atomic ratios generally exhibit lower saturation magnetization (Ms) because of the reduced presence of elements with magnetic solid moments. Additionally, they demonstrate higher intrinsic coercivity (Hc) due to structural distortions that impede magnetic reordering [101]. Zhang et al. [102] enhanced the mechanical, electrical, and magnetic properties of Fe-Co-Ni-(AlSi)x alloys by adding Al and Si. While the saturation magnetization (Ms) decreased with increasing Al and Si content due to their non-magnetic nature, the coercivity (Hc) and electrical resistance showed a more complex relationship with composition. The alloy with x = 0.2 exhibited the best properties, including Ms of 1.5T, Hc of 1400A/m, electrical resistivity of 69.5 μΩ·cm, and plastic strain of over 50%, making it ideal for soft magnetic material applications [102–104]. Ji et al. [105] prepared Co-Cr-Fe-Mn-Ni alloy powders through mechanical alloying and consolidated them using spark plasma sintering (SPS). The as-milled powders exhibited super-paramagnetic behavior due to their ultra-fine grain size. However, after consolidation, the bulk Co-Cr-Fe-Mn-Ni HEA demonstrated a paramagnetic character featuring a saturation magnetization (Ms) of 32.49 emu/g and zero coercivity (Hc) in a magnetic field of 20,000 Oe. Schneeweiss et al. [106] investigated the temperature-dependent magnetic properties of the Co-Cr-Fe-Mn-Ni alloy. They discovered that the alloy remained paramagnetic down to 93 K, at which point it transitioned into a (super)spin glass state. At 38 K, a second magnetic transformation to a ferromagnetic-like state was observed, indicating intricate magnetic behavior with multiple phase transitions at lower temperatures. Meanwhile, Feng et al. [107] developed an Nd-based hard magnetic HEA, Nd_{20}-Pr_{20}-La_{20}-Fe_{20}-Co_{10}-Al_{10}, which consisted of an amorphous phase, HCP Nd (Pr) phase, HCP La phase, and δ-phase. This alloy exhibited an intrinsic coercivity of 280 k•A/m and a saturation magnetization of 0.117 T, slightly lower than that of Nd-based bulk amorphous alloys. Its elevated coercivity is likely attributed to the combined effects of exchange coupling and domain wall pinning. Na et al. [108] developed a hard magnetic HEA without rare earth elements. They found that incorporating Cu and Ti into Fe-Co-Ni-Al–based HEAs significantly enhanced coercivity while preserving complex magnetic characteristics. The alloy with $Cu_{0.4}$-$Ti_{0.4}$ achieved a coercivity of 1078 Oe and a maximum energy product of 2.06 MGOe, slightly outperforming Alnico 2 magnets. These results underscore the potential of HEAs as viable hard magnetic materials. Mao et al. [109] reported that $(Co\text{-}Cr\text{-}Fe\text{-}Mn\text{-}Ni)_3O_4$ powders exhibited ferromagnetic behavior at room temperature, as evidenced by a hysteresis loop. The saturation magnetization varied between 4.85 and 15.98 emu/g, while coercivity ranged from 176.55 to 21.39 Oe, depending on the synthesis temperature. Additionally, the magnetic properties of rock-salt $(Co_{0.2}\text{-}Cu_{0.2}\text{-}Mg_{0.2}\text{-}Ni_{0.2}\text{-}Zn_{0.2})O$ HEA nanocrystalline powders demonstrated long-range antiferromagnetic behavior below the Néel temperature (TN = 106 K), driven by super-exchange interactions. Jimenez-Segura et al. [110] confirmed this long-range magnetic ordering, observing antiferromagnetic behavior below 120 K, despite the structural disorder introduced by the random distribution of magnetic ions. Meisenheimer et al. [111] investigated how disorder and composition affect the exchange anisotropy in permalloy/$(Mg_{0.25}(1{-}x)Co_xNi_{0.25}(1{-}x)Cu_{0.25}(1{-}x)Zn_{0.25}(1{-}x))O$ heterostructures. Using pulsed laser deposition, they successfully fabricated an exchange bias heterostructure comprising a 3 nm ferromagnetic layer (Py), an antiferromagnetic layer, and an 80 nm entropy-stabilized oxide layer. By adjusting the oxide composition, they could tune the disorder, exchange field, and magnetic anisotropy, achieving an exchange bias at low temperatures that was 10 times greater than that of a CoO/Py control sample. Witte et al. [112] investigated the magnetic properties of HEAs composed of rare earth (RE) and transition metal (TM) elements. Their research showed that the specific TM present at the B site highly influences the magnetic behavior of the $(5A_{0.2})BO_3$ series. Notably, for B = Fe, Mossbauer spectroscopy demonstrated a significant interaction between the magnetic states of the TM and RE subsystems, leading to a substantial coercive field of 2 T at room temperature. Ma et al. [113] synthesized three high entropy (HE) spinel-type ferrite ceramics—$(Mg_{0.2}\text{-}Mn_{0.2}\text{-}Fe_{0.2}\text{-}Co_{0.2}\text{-}Ni_{0.2})Fe_2O_4$, $(Mg_{0.2}\text{-}Fe_{0.2}\text{-}Co_{0.2}\text{-}Ni_{0.2}\text{-}Cu_{0.2})Fe_2O_4$, and $(Mg_{0.2}\text{-}Fe_{0.2}\text{-}Co_{0.2}\text{-}Ni_{0.2}\text{-}Zn_{0.2})Fe_2O$, revealing synergetic dielectric and magnetic losses. Among these, $(Mg_{0.2}\text{-}Mn_{0.2}\text{-}Fe_{0.2}\text{-}Co_{0.2}\text{-}Ni_{0.2})Fe_2O_4$ exhibited superior EM wave absorption, achieving a minimum reflection loss (RLmin) of –35.10 dB at 6.78 GHz and an adequate absorption bandwidth of 7.48 GHz, highlighting the promising potential of HE MFe_2O_4 as advanced EM absorption materials.

9.2.2.7 Electrical Properties

As-cast HEAs generally exhibit electrical resistivities ranging from 100 to 220 μΩ-cm, one to two orders of magnitude greater than many conventional metals. This increased electrical resistivity in HEAs is attributed to their highly distorted lattice structure, which causes significant scattering of electron waves [114,115]. Kao et al. [116] investigated the electrical properties of Al_x-Co-Cr-Fe-Ni alloys, examining as-cast, homogenized, and deformed variants across temperatures ranging from 4.2 to 300 K. The carrier density in these alloys is found to be between 10^{22} and 10^{23} cm^{-3}, while the carrier mobility of the homogenized alloys (H-x) ranges from 0.40 to

2.61 $cm^2V^{-1}s^{-1}$. The residual electrical resistivity varies from 100 to 220 μΩ cm, and the temperature coefficient of resistivity (TCR) for the H-2.00 alloy is relatively low at 82.5 ppm/K, indicating that defects in the lattice significantly influence electrical transport. Additionally, specific compositions exhibit Kondo-like behavior, and the alloys demonstrate hole-like carrier characteristics, with ferromagnetic variants displaying an anomalous Hall Effect (AHE). Jin et al. [117] synthesized a range of Ni-based FCC equiatomic alloys, including Ni-Co, Ni-Fe, Ni-Co-Fe, Ni-Co-Cr, Ni-Co-Fe-Cr, Ni-Co-Fe-Cr-Mn, and Ni-Co-Fe-Cr-Pd, using arc melting followed by the optical floating zone method. Their investigation into the electrical properties revealed that the Ni-Co-Fe-Cr-Pd HEA exhibited the highest electrical resistivity among the alloys, suggesting that the type of elements significantly impacts electrical resistivity more than the total number of elements present. Additionally, the alloys with elevated electrical resistivity demonstrated lower temperature coefficients of resistivity (TCR). Guo et al. [118] developed a superconducting BCC $(TaNb)_{0.67}(HfZrTi)_{0.33}$ HEA through arc melting and investigated how pressure influences its superconducting properties. They found that the transition temperature to superconductivity (Tc) rises from 7.7 K to 10 K as pressure increases from 1 atm to 60 GPa. However, when the pressure is further elevated to 190.6 GPa, Tc decreases to 9 K.

9.2.2.8 Thermal Properties

The thermal properties of a material mainly include the thermal expansion coefficient and thermal conductivity, both of which are essential for material design in processing and maintaining performance at elevated temperatures, over a broad temperature range, or in thermal management applications. Antao et al. [119] investigate the production and thermal properties of the Cu-Cr-Fe-Ti-V HEA, synthesized through mechanical alloying and spark plasma sintering. Analysis of the sintered samples revealed that the alloy's elemental composition increased iron content while decreasing the proportions of other elements. Thermal measurements indicated that the Cu-Cr-Fe-Ti-V alloy possesses suitable thermal properties, making it a promising candidate for use as a thermal barrier. Yan et al. [120] worked on high entropy carbide ceramic $(Hf_{0.2}\text{-}Zr_{0.2}\text{-}Ta_{0.2}\text{-}Nb_{0.2}\text{-}Ti_{0.2})C$, synthesized by spark plasma sintering, exhibits significantly lower thermal diffusivity and conductivity compared to its binary carbide counterparts such as HfC, ZrC, TaC, and TiC. This reduction in thermal conductivity is attributed to enhanced phonon scattering caused by the distorted anion sublattice. Although the alloy exhibits lower thermal conductivity, it retains the high elastic modulus and hardness typical of binary carbide ceramics, resulting in distinct thermal properties. Zhao et al. [121] observed RE aluminate solid solution $(Nd_{0.2}\text{-}Sm_{0.2}\text{-}Eu_{0.2}\text{-}Y_{0.2}\text{-}Yb_{0.2})_4Al_2O_9$, designed for environmental barrier coatings (EBCs), exhibits promising thermal properties. It has a thermal expansion coefficient of 6.96×10^{-6} K^{-1} from 300 to 1473 K, matching well with mullite, and demonstrates good phase stability in this temperature range. The material also shows low thermal conductivity (1.50 $W{\cdot}m^{-1}{\cdot}K^{-1}$ at room temperature), attributed to lattice distortion and mass differences among its atoms. Additionally, anisotropic solid thermal expansion, driven by anisotropic chemical bonding, is observed compared to $Y_4Al_2O_9$ and $Yb_4Al_2O_9$. Chen et al. [122] suggested that the novel high entropy rare earth monosilicate $(Yb_{0.25}\text{-}Y_{0.25}\text{-}Lu_{0.25}\text{-}Er_{0.25})_2SiO_5$ demonstrates excellent phase stability and significant anisotropy in thermal expansion. Its thermal expansion coefficients (CTEs) are highly directional: $\alpha a = (2.57 \pm 0.07) \times 10^{-6}$ K^{-1}, $\alpha b = (8.07 \pm 0.13) \times 10^{-6}$ K^{-1}, and $\alpha c = (9.98 \pm 0.10) \times 10^{-6}$ K^{-1}. This anisotropy in thermal expansion is beneficial for reducing the mismatch between coating and substrate, primarily if the preferred orientation can be controlled on metal or ceramic substrates. Zhao et al. [121] observed novel cuspidine-structured RE aluminate solid solution ($Nd_{0.2}$-$Sm_{0.2}$-$Eu_{0.2}$-$Y_{0.2}$-$Yb_{0.2}$)$4Al_2O_9$ exhibits promising thermal properties for environmental barrier coatings. It has a thermal expansion coefficient of 6.96×10^{-6} K^{-1} (300–1473 K), closely matching mullite's, and demonstrates good phase stability over a wide temperature range. Additionally, it features low thermal conductivity (1.50 $W{\cdot}m^{-1}{\cdot}K^{-1}$ at room temperature). The reduced thermal conductivity can be caused by lattice distortion and variations in mass among the constituent atoms. In contrast, the observed anisotropic thermal expansion results from anisotropic chemical bonding, primarily influenced by large RE cations. Zhao et al. [123] found that the novel high entropy (HE) RE phosphate monazite ceramic ($La_{0.2}$-$Ce_{0.2}$-$Nd_{0.2}$-$Sm_{0.2}$-$Eu_{0.2}$)PO_4 exhibits excellent thermal properties, making it a likely choice for environmental and thermal barrier coatings (EBCs/TBCs). It has a thermal expansion coefficient (TEC) of 8.9×10^{-6}/°C (573–1273 K), closely matching that of Al_2O_3. Additionally, its thermal conductivity is notably low at 2.08 $W{\cdot}m^{-1}{\cdot}K^{-1}$ at room temperature, which is 42% lower than that of $LaPO_4$. These thermal characteristics and good chemical compatibility with Al_2O_3 up to 1873 K make it an ideal material for EBC/TBC applications. Chen et al. [124] proposed porous HE ($Zr_{0.2}$-$Hf_{0.2}$-$Ti_{0.2}$-$Nb_{0.2}$-$Ta_{0.2}$)C exhibits excellent thermal properties for ultra-high temperature insulation applications. It has low thermal conductivity (0.39 $W{\cdot}m^{-1}{\cdot}K^{-1}$) and low thermal diffusivity (0.74 $mm^2{\cdot}s^{-1}$) at room temperature, along with high porosity (80.99%) and good thermal stability at high temperatures. These characteristics make it a strong candidate for lightweight thermal insulation materials in extreme environments. Gild et al. [125] suggested that the synthesized HE metal diborides, including ($Hf_{0.2}$-$Zr_{0.2}$-$Ti_{0.2}$-$Ta_{0.2}$-$Mo_{0.2}$)B_2 and ($Hf_{0.2}$-$Zr_{0.2}$-$Ti_{0.2}$-$Ta_{0.2}$-$Cr_{0.2}$)B_2, exhibit significantly lower thermal conductivities, ranging from 12 to 25 W/mK, which are approximately 1/10 to 1/5 of the thermal conductivities reported for HfB_2 and ZrB_2. This reduction in thermal conductivity is attributed to the complex lattice structure of high entropy materials, which enhances phonon scattering and thereby reduces heat transfer. These low thermal conductivities make the HE metal diborides suitable for applications requiring efficient thermal insulation.

9.2.3 Thermal Stability of HEA

The equiatomic Al-Co-Cr-Fe-Ni HEA synthesized by mechanical alloying shows limited thermal stability. Differential scanning calorimetry (DSC) reveals a small thermal event near 873 K, indicating a diffusive phase transformation. Upon heat treatment, the initially disordered BCC phase partially transforms into the Ni_3Al (L12) phase at lower temperatures of 623 K. With further annealing at 848–1073 K, the alloy exhibits the evolution of a partially ordered BCC phase (B2), along with the L12 phase and a tetragonal σ-phase. These phase transformations suggest that, upon heating, the HEA transitions from a HE phase into a mix of high and medium entropy phases, showing thermal instability with multiple intermetallic phase formations [126]. Refractory HEAs represent a promising class of multi-component alloys known for their exceptional mechanical properties at elevated temperatures, making them suitable for various applications [127]. Notably, the Nb-Mo-Ta-W alloy demonstrates superior performance retention compared to tungsten after exposure to heat at 1,373 K for 3 days [128]—these remarkable properties position HEAs as innovative materials for nanoscale devices, particularly in high-stress and elevated-temperature environments. Sathiyamoorthi et al. [129] discovered that ultrafine-grained Co-Cr-Fe-Ni HEAs exhibit outstanding thermal stability when sintered compacts are subjected to high temperatures ranging from 973 K to 1173 K for extended periods of 600 hours. Specifically, samples maintained at 973 K for 600 hours showed negligible changes in hardness, while hardness decreased by 9.5% and 25% after 600 hours at 1073 K and 1173 K, respectively. $LaMgAl_{11}O_{19}$ (LMA) is a promising material for hypersonic vehicle thermal protection due to its high melting point and low density. Still, its phase instability at high temperatures reduces its emissivity below 6 μm. Three HE ceramics ($La_{0.2}$-$Nd_{0.2}$-$Gd_{0.2}$-$Sm_{0.2}$-$Pr_{0.2}$)$MgAl_{11}O_{19}$ (HE LMA-1), ($La_{0.2}$-$Nd_{0.2}$-$Gd_{0.2}$-$Sm_{0.2}$-$Dy_{0.2}$)$MgAl_{11}O_{19}$ (HE LMA-2), and La($Mg_{0.2}$-$Fe_{0.2}$-$Co_{0.2}$-$Ni_{0.2}$-$Zn_{0.2}$)$Al_{11}O_{19}$ (HE LMA-3) were developed to improve this, heat treatment experiments revealed that HE LMA-1 maintains its single-phase structure after high-temperature exposure, while secondary phases emerge in HE LMA-2 and HE LMA-3. The enhanced thermal stability of HE LMA-1 is attributed to the slight average size difference (~4%) among its constituent elements. This stability and improved thermal emissivity position HE $REMgAl_{11}O_{19}$—particularly HE LMA-1—is a promising candidate for thermal protection coatings in hypersonic vehicles [130]. Despite considerable grain refinement, the equiatomic Co-Cr-Fe-Mn-Ni HEA maintains a stable single-phase solid solution down to the atomic level following severe plastic deformation (SPD). Isochronal and isothermal annealing treatments reveal an initial increase in hardness up to 630 HV at 723 K, followed by softening at higher temperatures, indicating complex thermal stability behavior. New nanoscale phases form during annealing, facilitated by the abundant grain boundaries, which provide rapid diffusion pathways, ultimately contributing to increased hardness over extended annealing times [131]. Guan et al. [132] observed that HE metal boron carbonitride (BCN) ceramic powders, including ($Ta_{0.2}$-$Nb_{0.2}$-$Zr_{0.2}$-$Hf_{0.2}$-$W_{0.2}$)BCN, ($Ta_{0.2}$-$Nb_{0.2}$-$Zr_{0.2}$-$Hf_{0.2}$-$Ti_{0.2}$)BCN, and ($Ta_{0.2}$-$Nb_{0.2}$-$Zr_{0.2}$-$Ti_{0.2}$-$W_{0.2}$)BCN, exhibit significant thermal stability when heated to 1773 K for 30 minutes in a flowing argon atmosphere. Despite detecting metal oxides like HfO_2 and ZrO_2 due to preexisting surface oxides, the HE ceramics maintain a single FCC solid solution structure with high compositional uniformity. Additionally, some amorphous phases remain post-heat treatment, indicating the materials' resilience to thermal degradation.

9.3 CORROSION RESISTANCE OF HEAs

9.3.1 Mechanism of Corrosion Resistance

The corrosion resistance mechanism in HEAs primarily revolves around forming protective passive oxide layers on the alloy's surface. This passivation occurs due to elements like chromium, aluminum, and titanium, which have a strong affinity for oxygen and readily form stable oxide films. These oxide layers act as a barrier, preventing further interaction between the underlying metal and corrosive environments, such as acidic or saline solutions. Additionally, the unique multi-element nature of HEAs leads to a uniform microstructure, which minimizes localized corrosion sites like grain boundaries or phase separations. The synergy between elements also contributes to high electrochemical stability, enhancing corrosion resistance.

Understanding the corrosion mechanism of HEAs at the micro-scale requires the application of advanced surface analysis techniques such as transmission electron microscopy (TEM), atomic force microscopy (AFM), time-of-flight secondary ion mass spectrometry (ToF-SIMS), X-ray photoelectron spectroscopy (XPS), and Auger electron spectroscopy (AES). Additionally, integrating computational materials engineering for in-situ or simulation-based analysis is essential. To comprehensively investigate the intricate corrosion processes and mechanisms in HEAs at this scale, it is crucial to focus on the structural changes at the HEA/solution interface, as well as the composition, nanostructure, and electronic properties of the passive film formed on the HEA surface, along with the factors contributing to the breakdown of this passive film.

9.3.2 Factors Affecting Corrosion Resistance

Xue et al. [133] investigated the impact of grain size on the localized corrosion behavior of the Al_2-Cr_5-Cu_5-Fe_{53}-Ni_{35} HEA. Their findings revealed that reducing the grain size improved the alloy's resistance to localized corrosion. The impact of grain size, dislocation density, and crystal structure on the corrosion resistance of HEAs was studied by Abhinav et al. [134] in sequentially synthesized alloys such as Al-Ni-Co-Cr-Fe, Co-Ni-Fe-Cr-Al, Fe-Cr-Ni-Al-Co, base alloy, and cast alloy. The study found that increasing grain size and decreasing dislocation density improves corrosion resistance. Larger grain sizes, as seen in the base and cast alloys, resulted in better corrosion resistance, with the base alloy showing lower corrosion current (Icorr) values than the Co-Ni-Fe-Cr-Al alloy. Cast alloys exhibited superior corrosion resistance due to their large grain sizes. The phase fraction also affected the corrosion resistance, with an optimal BCC: FCC ratio (~3:2) being identified for Fe-Cr-Ni-Al-Co. This alloy performed better than others with varying FCC content, such as Co-Ni-Fe-Cr-Al (lower FCC content) and Al-Ni-Co-Cr-Fe (higher FCC content). The balance between grain size and phase fraction plays a critical role, where larger grain sizes combined with an optimal BCC: FCC ratio yield superior corrosion resistance. While grain size and dislocation density influence corrosion behavior, phase fraction has the most significant impact, with dislocation density having the least.

Anodizing treatment is commonly employed to enhance the corrosion resistance of alloys by optimizing their surface structure and composition. Lee et al. [135] demonstrated that a well-executed anodizing treatment significantly improved the pitting corrosion resistance of the Al_x-Cr-$Fe_{1.5}$-Mn-$Ni_{0.5}$ ($x = 0, 0.3, 0.5$) alloy. After undergoing potentiostatic polarization at 0.7 VSHE for 1800 seconds in a 15% H_2SO_4 solution, a stable and protective film, mainly consisting of chromium oxide, developed on the alloy's surface. Measurements from electrochemical impedance spectroscopy showed that the charge transfer resistance of Cr-$Fe_{1.5}$-Mn-$Ni_{0.5}$ and

$Al_{0.3}$-Cr-$Fe_{1.5}$-Mn-$Ni_{0.5}$ in a 0.1 M HCl solution increased by two orders of magnitude following anodization, indicating a significant enhancement in corrosion resistance.

In the Co-Cr-$Cu_{0.5}$-Fe-Ni HEA, heat treatment at 1373–1623 K for 24 hours successfully dissolved the Cu-rich phase into the FCC matrix, improving corrosion resistance [136]. Nevertheless, heat treatment of $Al_{0.5}$-Co-Cr-Fe-Ni [137] and Co-Cr-$Cu_{0.5}$-Fe-Ni [136] at lower temperatures (653–1173 K) facilitated the development of Al-rich, Cu-rich, and Cr-rich phases, which adversely affected the corrosion resistance of the alloys in a 3.5 wt. % NaCl solution. Wang et al. [138] investigated the non-equiatomic $Al_{0.6}$-Co-Fe-Ni-$Cr_{0.4}$ HEA, characterized by a dual FCC + BCC structure, which was produced through arc melting in a high-purity argon atmosphere, to evaluate its corrosion resistance over a temperature range of 823 K to 1123 K for 24 hours. The observed corrosion resistance followed the order: aged-823 K > aged-923 K > as-cast > aged-1023 K > aged-1123 K > HEA. When immersed in 3.5 wt. % NaCl solution, the B2 (α-Fe) phase (AlNi) in the alloy degraded the corrosion resistance more significantly than the BCC and FCC (Fe–Cr) phases. This suggests that phases with a higher aluminum concentration are more susceptible to corrosion, as they are more easily eroded and dissolved in NaCl solution. Chen et al. [139] produced Fe-Co-Ni-$Al_{0.2}$-$Si_{0.2}$ HEA through melt-spinning for rapid solidification and induction melting for slow cooling. The slowly cooled alloy exhibited a dendritic structure with compositional segregation of Al, Si, and Ni and the precipitation of an $L1_2$-type ordered phase within the solid-solution matrix. In contrast, rapid solidification promoted structural and compositional homogeneity. This homogeneity significantly improved the stability of the passive film, leading to a notable increase in corrosion resistance compared to the slowly cooled alloy. Additionally, the distribution of constituent elements was crucial in forming the passive film on the surface of the HEA. Mishra et al. [140] examined the corrosion behavior of annealed equiatomic Al-Cr-Fe-Mn-Ni-Ti HEA in an NaCl solution. The alloy exhibited a higher passivation region and better corrosion resistance in 0.5 M NaCl than in 1 M NaCl. This improved resistance is likely due to forming a protective layer on the alloy's surface, attributed to the presence of Cr, Ni, and Al. However, this protective layer becomes less effective at higher chloride ion concentrations, leading to reduced corrosion resistance. Guo et al. [141] fabricated an Al-Co-Cr-Fe-Ni-$Ti_{0.5}$ HEA and investigated the effect of temperature on its corrosion resistance. They found that as the heat treatment temperature increased, the corrosion resistance decreased, following the trend: 1073 K-2h > 1173 K-2h > as-cast > 1273 K-2h > 1373 K-2h > 1473 K-2h. This decrease in corrosion resistance was inversely related to the increase in BCC phase volume fraction. The reduction in corrosion resistance is attributed to the increased number of crystal and phase interfaces and the potential difference between the FCC and BCC phases. Wang et al. [142] studied the corrosion behavior of 3D-printed and aged $(Fe\text{-}Co\text{-}Ni)_{86}$-$Al_7$-$Ti_7$ samples in a 3.5 wt. % NaCl solution. Electrochemical tests revealed that the aged HEA exhibited superior corrosion resistance than the as-printed specimen. This improvement is attributed to forming a more stable and robust passive film on the aged HEA's surface, enhancing its corrosion resistance. Zhang et al. [143] showed that remelting and annealing greatly enhanced the pitting corrosion resistance of the Al-Fe-Ni-Co-Cu-Cr alloy in a chloride-containing environment. After annealing at 1273 K for 3 hours, the alloy's dendrites regrew, and the compositional disparity between the dendrites and interdendrites diminished compared to the as-cast alloy. Furthermore, the corrosion type transitioned from pitting and intergranular corrosion to primarily pitting corrosion in a 3.5 wt. % NaCl solution, with a substantial decrease in the number and size of pits. Wang et al. [144] discovered that the corrosion resistance of $(Co\text{-}Cr\text{-}Fe\text{-}Ni)_{97}Mo_3$ alloy treated at 973 K for 16 hours was superior to that of the alloy treated at 1173 K for the same duration. The higher aging temperature of 1173 K exacerbated elemental segregation, leading to an increased concentration of Mo in the interdendritic regions and the precipitation of new phases, which negatively affected the alloy's corrosion resistance. Lin et al. [145] observed that aging the $Cu_{0.5}$-Co-Cr-Fe-Ni HEA at 1523 K for 24 hours resulted in the dissolution of the Cu-rich phase into the matrix. This process reduced Cu segregation within the microstructure, leading to a lower corrosion rate of the $Cu_{0.5}$-Co-Cr-Fe-Ni alloy when exposed to a 3.5 wt. % NaCl solution. Guo et al. [146] studied the impact of heat treatment on the corrosion resistance of the Al-Co-Cr-Fe-Ni-Si HEA. The alloy treated at 1373 K for 2 hours demonstrated optimal corrosion resistance, with E_{corr} and I_{corr} values of −330.776 mV and 637.587 nA, respectively, attributed to the emergence of a weak passivation phenomenon. Fujieda et al. [147] investigated the impact of solution treatment on the pitting corrosion resistance of Co-Cr-Fe-Ni-Ti-based HEA synthesized via selective laser melting (SLM). They found that water quenching significantly enhanced the alloy's pitting corrosion resistance. Additionally, the pitting potentials of the solution-treated alloy were influenced by the size and volume fraction of very fine-ordered particles composed of Ni and Ti, with diameters in the tens of nanometers. Tan et al. [148] fabricated a novel lightweight Al_2-Nb-Ti_3-V_2-Zr HEA using vacuum hot pressing and investigated the effect of sintering temperature (1473–1823 K) on its corrosion resistance. They observed that as the sintering temperature increased from 1723 K to 1823 K, the grain size of the matrix phase grew from 30.5 μm to 47.9 μm. This coarsening of the grain boundary made the alloy more susceptible to attack by corrosive media (10 wt. % HNO_3 solution), as the wider grain boundaries reduced the density of the protective film, lowering the alloy's corrosion resistance. Wang et al. [149] prepared a non-equiatomic Co-Cr-Fe-Ni-$W_{0.5}$ HEA with dual FCC and μ phases using an electromagnetic levitation melting method and studied its corrosion resistance after annealing at temperatures ranging from 873 to 1473 K for 3 hours. The alloy demonstrated a strong ability to form a passive film; however, its corrosion resistance decreased as the volume fraction of μ phase precipitates increased. The higher amount of μ phase precipitates led to more phase interfaces and a greater potential difference, making the material more prone to galvanic corrosion between the FCC matrix and phase boundaries. Chang et al. [150] explored how different heat treatment processes and aging times influenced the corrosion behavior and properties of the passive film in the $(Fe\text{-}Co\text{-}Ni)_{86}$-$Al_7$-$Ti_7$ HEA. Data from electrochemical corrosion tests and pitting morphology analysis indicated that the sample aged for 1 hour exhibited the highest corrosion resistance. In contrast, the sample aged for 0.5 hours demonstrated the least effective performance. The sample aged for 1 hour showed the lowest degree of element segregation between the FCC phase and the $L2_1$ phase, contributing to its improved corrosion resistance. This sample also displayed the highest passivation index (n = 0.91), suggesting the quickest formation of the passive film and optimal protective effectiveness in a 3.5 wt. % NaCl solution. The main components of the passive film on the HEA in this environment included Ti, Al, and Fe oxides, with Ti oxide and Fe oxide playing the most crucial roles in enhancing corrosion resistance. The 1-hour aging period improved the stability and protective qualities of the passive film on the HEA surface, significantly increasing the cation fractions of Ti oxide, Fe oxide, and O^{2-}/OH^{-} values within the passive film. Therefore, managing the aging time helps reduce compositional segregation between the two phases, particularly for the Fe and Ti elements, strengthening the passive film's protective capacity and helping prevent severe localized corrosion.

Shang et al. [151] proposed that similar to stainless steel, the accumulation of Cr_2O_3 in the passive film of HEAs plays a vital role in their outstanding corrosion resistance. The authors investigated the corrosion behavior of Ni-Co, Ni-Co-Fe, Ni-Co-Cr, and Ni-Co-Cr-Fe alloys in 0.1 M NaCl and 0.1 M HCl solutions. Their results demonstrated that the presence of Cr in the solid solution significantly affects corrosion behavior, with the buildup of Cr oxides and hydroxides in the passive film being the main factor contributing to the observed corrosion resistance. Additionally, the moderate inclusion of elements like Mo, W, and Ti can promote the conversion of hydroxides to oxides within the passive film, thereby increasing the amount of Cr_2O_3 and improving the stability and protective properties of the passive film in HEAs. Qiu et al. [152–154] suggested that the presence of unoxidized noble metal atoms such as Al, Ti, V, and Cr within the passive film of the Al-Ti-V-Cr HEA is a crucial factor contributing to the alloy's exceptional corrosion resistance. Typically, the proportion of these unoxidized noble metals in the passive film of HEAs is higher than in stainless steel. According to the point defect model, these defects assist in the transport of current through the metallic passive film, playing a role in the reactions occurring at both the metal/film and film/solution interfaces.

Therefore, the movement of point defects within the passive film, combined with interfacial reactions, controls the passivation process and the breakdown of the film. Based on this model, a new corrosion mechanism for HEAs is proposed: unoxidized noble metal atoms obstruct the movement of point defects through the film, reducing strain and forming a barrier at the metal/film and film/solution interfaces. This mechanism effectively prevents metal ions from entering or leaving the film, thereby significantly improving the corrosion resistance of HEAs.

The vacuum melting method is commonly employed for producing bulk HEAs, typically involving four to five melting cycles to achieve uniform elemental distribution. However, this process can be quite complex, often leading to issues such as elemental segregation and shrinkage cavities that negatively impact the microstructure and properties of the HEAs. In contrast, HEAs produced via mechanical alloying exhibit stable microstructures and uniform elemental distribution but require suitable techniques to ensure effective powder consolidation. The laser cladding method is frequently utilized for fabricating HEA coatings due to its advantages, including high energy density, rapid heating and cooling rates, minimal impact on the substrate, and the ability to accommodate a wide range of HEA compositions, all of which contribute to reduced corrosion sensitivity [155]. Magnetron sputtering is a widely recognized method for producing HEA coatings. For example, this technique was used to create a gradient film on a Ti-Al-Cr-Fe-Ni alloy, illustrating its effectiveness as a viable method for advancing the development of corrosion-resistant HEAs [156]. Feng et al. [157] examined the anti-corrosion properties of a Co-Cr-Fe-Ni-$Mo_{0.2}$ HEA coating fabricated using high-speed laser cladding (HLC). Their findings showed that the coating demonstrated exceptional corrosion resistance primarily due to its uniform, fine-grain structure. Rui et al. [158] examined the corrosion resistance of a Fe-Cr-Ni-Mn-Al HEA coating fabricated using laser cladding technology. They found that the laser process parameters significantly influenced the electrochemical corrosion resistance of the coating. Specifically, as the laser energy density (Led) increased, the corrosion resistance of the HEA coatings initially improved and then declined, with the optimal corrosion resistance observed at a Led of 65 J/mm^2. In a 3.5 wt. % NaCl solution, the corrosion resistance was closely linked to the characteristics of the passive film on the surface; higher impedance (Rp) and density (n^2) of the passive film corresponded to improved corrosion resistance. The surface morphology of the coating after corrosion was consistent with the electrochemical detection results. At a low Led of 50 J/mm^2, microcrack in the coating allowed chloride ions (Cl^-) to penetrate, while a high Led of 70 J/mm^2 introduced defects like cracks and holes, facilitating Cl^- accumulation and accelerating corrosion. Dehestani et al. [159] investigated the corrosion properties of novel Fe-Co-Ni-Mo-W HEA coatings produced via the pulse current electro-deposition method, focusing on the impact of different pulse current densities (2, 6, and 10 A/dm^2). Their findings revealed that the coating deposited at a current density of 6 A/dm^2 demonstrated the best corrosion resistance, characterized by the most noble corrosion potential (Ecorr = −0.55 V) and the lowest corrosion current density (Icorr = 3.2 $\mu A/cm^2$).

9.3.3 Effect of Alloying Elements on Corrosion Behavior of HEA in Different Environments

The corrosion behavior of HEAs is significantly influenced by their alloying elements, which can enhance or diminish their resistance in various environments. Chromium, nickel, and molybdenum improve corrosion resistance by forming stable passive oxide layers that protect the alloy from further degradation, particularly in oxidizing environments. For instance, Cr promotes the formation of a chromium oxide layer, which is highly resistant to corrosion in acidic and chloride environments. Conversely, aluminum (Al) and copper (Cu) can make the alloy more susceptible to localized corrosion due to their tendency to form less protective oxides or create galvanic couples. The synergistic effects of multiple alloying elements in HEAs lead to complex interactions in different environments, such as acidic, alkaline, or chloride-rich conditions, where the composition, microstructure, and stability of passive films dictate the overall corrosion performance of the alloy. Thus, the choice and balance of alloying elements play a crucial role in tailoring the corrosion resistance of HEAs for specific applications.

9.3.3.1 Resistance of HEAs to Chloride-Induced Corrosion

HEAs, consisting of multiple elements in near-equiatomic proportions, are emerging materials known for their remarkable mechanical and corrosion-resistant properties. In chloride-rich environments, which are particularly aggressive for many metals and alloys, HEAs exhibit unique corrosion resistance characteristics. The complex chemistry and microstructure of HEAs may promote the formation of protective passive layers, enhancing their ability to resist pitting and localized corrosion. Understanding the corrosion characteristics of HEAs in chloride environments is essential for their potential applications in harsh, corrosive conditions such as marine and industrial settings.

9.3.3.1.1 Aluminum

Li et al. [160] found that Alx-Co-Cr-Fe-Ni HEAs (with x values of 1.0 and 1.3) showed more excellent thermal corrosion resistance than traditional Ni-Co-Cr-Al alloys in a Na_2SO_4 + 25% NaCl environment at 1173 K. Aluminum played a vital role in the alloy by reacting with NaCl to form $AlCl_3$, creating a protective Al_2O_3 layer. The HEA's excellent hot corrosion resistance was due to its loop-locked coherent phase interfaces within the modulated structure and the uniform distribution of aluminum and chromium, which resulted in minimal and shallow corrosion damage. Lee et al. [161] studied the corrosion mechanism of Alx-Cr-$Fe_{1.5}$-Mn-$Ni_{0.5}$ (x = 0, 0.3, and 0.5) alloys using electrochemical methods and morphological analysis. The results indicated that increasing aluminum content enhanced overall corrosion resistance in NaCl solutions, though it also increased susceptibility to pitting corrosion. Yen et al. [162] explored the impact of aluminum content on the corrosion resistance of Al_x-Co-Cu-Fe-Mn (x = 0, 0.25, 0.5, 0.75, 1) HEAs in 3.5 wt. % NaCl solution. Their findings demonstrated that increasing the Al content improved the overall corrosion resistance of the HEAs. Al-Co-Cu-Fe-Mn exhibited exceptional corrosion resistance among the five compositions, with E_{corr} and I_{corr} values measured at −0.342 V and 5.53×10^{-6} A cm^{-2}, respectively. Izadi et al. [163] investigated the influence of aluminum content on the corrosion resistance of Al_x-Co-Cr-Fe-Ni HEAs (x = 0, 0.25, 0.75, 1.25, 2). Potentiodynamic polarization tests revealed that the corrosion current densities increased with higher aluminum molar ratios. The FCC phase, known for its corrosion-resistant properties, contributed to improved polarization resistance. However, electrochemical impedance spectroscopy (EIS) results indicated that corrosion resistance declined as the aluminum content and BCC phase increased due to passive layer properties and elemental segregation. Hu et al. [164] studied the corrosion behavior of Al_x-Cu-Fe-Ni-Co-Cr (x = 0.5, 1.0, 1.5, 2.0) HEAs in 3.5 wt. % NaCl solution. Their study showed that the electrochemical corrosion of the HEAs varied with different aluminum content. For alloys with an Al molar ratio below 1.0, pitting corrosion was the predominant form, while alloys with an Al ratio above 1.0 experienced more uniform surface corrosion. Corrosion is typically initiated near surface defects, and the alloy exhibited optimal E_{corr} and I_{corr} values when the Al molar ratio was 1.0. Wang et al. [165] observed that the corrosion resistance of $Al_{0.1}Cr_{0.9}$ to $Al_{0.3}Cr_{0.7}$ HEAs with a single FCC phase gradually decreased in 3.5 wt. % NaCl solution as the Al/Cr molar ratio increased. The addition of aluminum, with its larger atomic radius, likely caused lattice distortion in the FCC phase, contributing to a reduction in corrosion resistance. Comparative analysis of the self-corrosion current densities revealed the corrosion potential of dual-phase $Al_{0.4}Cr_{0.6}$ to $Al_{0.9}Cr_{0.1}$ HEAs. The corrosion rates of these alloys were linked to the varying volume fractions of segregation phases (FCC or BCC), which influenced the protective capabilities of the passive film. When the volume fractions of segregation and matrix phases were nearly equal, the $Al_{0.7}Cr_{0.3}$ HEA exhibited the poorest corrosion resistance. Additionally, the BCC single-phase $Al_{1.0}Cr_0$ alloy had the lowest self-corrosion potential, primarily due to the absence of Cr, leading to a significant drop in corrosion resistance. Based on corrosion current densities, the corrosion resistance order was found to be $Al_{0.4}$-Co-Fe-Ni-$Cr_{0.6}$ > $Al_{0.9}$-Co-Fe-Ni-$Cr_{0.1}$ > $Al_{0.1}$-Co-Fe-Ni-$Cr_{0.9}$ > $Al_{0.2}$-Co-Fe-Ni-$Cr_{0.8}$ > $Al_{0.8}$-Co-Fe-Ni-$Cr_{0.2}$ > $Al_{0.5}$-Co-Fe-Ni-$Cr_{0.5}$ > $Al_{0.3}$-Co-Fe-Ni-$Cr_{0.7}$ > $Al_{0.7}$-Co-Fe-Ni-$Cr_{0.3}$ > $Al_{1.0}$-Co-Fe-Ni-Cr_0.

9.3.3.1.2 Boron

Zhang et al. [166] explored the corrosion behavior of boron-containing HEAs, specifically Fe-Cr-Ni-Co-B_x ($0.5 \leq x \leq 1.0$), in a 0.6 M NaCl solution. Their research revealed that while the initial addition of boron improved the corrosion resistance, further increases in boron content led to a decline in corrosion resistance. This decline in resistance may be attributed to the presence of the $(Cr, Fe)_2B$ orthorhombic phase, which is more corrosion-resistant than the FCC matrix in the alloy's microstructure. Chong et al. [167] developed B-reinforced Al-Co-Cr-Fe-Ni HEA coatings by introducing trace amounts of boron into argon gas fluxes during laser remelting (LR) and adding boron refined the grain size from over 20 μm to less than 5 μm. Boron interacted with chromium, forming Cr_2B compound particles within the BCC matrix, while Cr-depleted regions exhibited the formation of an FCC phase. In the LR coating, boron led to uncorroded Cr_2B particles on the surface, and the reduced pitting surface area of these particles enhanced the corrosion resistance of the LR-B coating.

9.3.3.1.3 Carbon

Wei et al. [168] synthesized Co-Cr-Fe-Ni-Cx (x = 0.1, 0.2, 0.3, 0.5, and 0.9 wt. %) HEAs via vacuum arc melting to explore the impact of carbon content on corrosion resistance in 3.5 wt. % NaCl solutions. With increasing carbon content, the phase structure shifted from a single FCC phase to a mix of FCC and M7C3 phases. The alloy containing 0.3 wt. % C showed the best corrosion resistance, demonstrating that an optimal carbon level can enhance the passivation film's thickness and stability, thereby preventing the dissolution of corrosion-resistant elements. Carbon also influenced the corrosion mode: low-carbon alloys (x = 0.1) showed uniform corrosion, medium-carbon alloys (x = 0.2 and 0.3) experienced intergranular corrosion, while high-carbon alloys (x = 0.5 and 0.9) developed significant carbide precipitates, leading to galvanic coupling corrosion between the FCC phases and M_7C_3 carbides. Chmielak et al. [169] investigated the impact of carbon and nitrogen on the corrosion behavior of Cr-Mn-Fe-Co-Ni HEAs in a 3% NaCl solution. Their findings revealed that although the surface corrosion resistance of these HEAs is comparable to that of austenitic steels, HEAs are more susceptible to pitting corrosion in chloride-rich environments, mainly due to the presence of precipitates.

9.3.3.1.4 Cerium

Zhang et al. [170] investigated how incorporating cerium (Ce) influenced the passivation film and corrosion resistance of non-equiatomic Fe_{40}-Mn_{20}-Cr_{20}-Ni_{20} HEAs in a 3.5 wt. % NaCl solution. The study found that the HEAs retained a single FCC phase despite adding Ce. Notably, the pitting potential (Epit) rose significantly from −58 mVSCE without Ce to +265 mVSCE with Ce, indicating improved pitting resistance due to mitigating adverse effects from elements such as oxygen and sulfur. Furthermore, the presence of Ce contributed to an increase in the thickness of the passivation film and enhanced the charge transfer resistance (Rct). Mott-Schottky analysis confirmed that Ce contributed to a denser passivation film, lowering donor density, reducing the corrosion rate, and improving overall corrosion resistance. Wu et al. [171] studied the influence of cerium (Ce) on the corrosion resistance of Fe-Co-Ni-Cu-Al HEAs in a 3.5% NaCl solution over 0.5 hours. The HEAs demonstrated excellent corrosion resistance, comparable to 304 stainless steel. Adding Ce significantly enhanced corrosion resistance, with a 31% improvement in I_{corr} observed when x = 0.09. Furthermore, the corrosion mechanism shifted from general corrosion to localized corrosion with the inclusion of Ce.

9.3.3.1.5 *Cobalt*

Zhao et al. [172] studied the corrosion behavior of Co_x-Cr-Cu-Fe-Mn-Ni HEAs in 3.5 wt. % NaCl solution with varying Co concentrations (x = 0.5, 1.0, 1.5, and 2.0 mol). The study indicated that an increase in cobalt (Co) content facilitated a phase transition from a (Fe, Cr)-rich FCC1 phase to a (Co, Cu, Ni, Mn)-rich FCC2 phase, which in turn improved corrosion resistance. Electrochemical assessments demonstrated increased corrosion potential (Ecorr) and pitting potential (E_{pit}) as the Co content increased. In contrast, the corrosion current density (i_{corr}) and average corrosion rates (V_{acr}) decreased. This trend suggests that the passivation film became denser and more stable with higher levels of cobalt. Additionally, the increased Co contributed to a thicker passivation layer, providing better protection. Research by Jiang et al. [173] indicates that cobalt (Co) is crucial in minimizing compositional and potential variations within HEA coatings. This difference reduction contributes to decreased galvanic corrosion, enhancing overall corrosion resistance.

9.3.3.1.6 *Chromium*

Yan et al. [174] examined the corrosion characteristics of $Al_{0.3}$-Cr_x-Fe-Co-Ni HEAs in a 3.5 wt. % NaCl solution, varying the chromium (Cr) content (x = 0, 0.5, 1.0, 1.5, 1.7, 2.0). Their findings revealed that as the concentration of Cr increased, the alloy underwent a phase transition from an FCC structure for x values of 0 to 1.0 to a combination of FCC and BCC/B2 structures at x values of 1.5 to 2.0, highlighting Cr's role as a stabilizing agent for the BCC phase. The results showed a decrease in corrosion current density (i_corr) with higher Cr levels, while corrosion potential (E_corr), pitting resistance, and charge transfer resistance (R_ct) improved. These enhancements were linked to the increased formation of Cr_2O_3 in the passivation film, resulting in a denser and more stable alloy surface. Furthermore, the alloy's Cr concentration of $\geq$ 30% helped prevent pitting, particularly by addressing Cr deficiencies in the BCC phase. Heidari et al. [175] observed similar findings, where increasing Cr content in Ti0.5-Mo0.5-Co-Ni-Mn-Cr1.5 (x = 0.5, 1, 1.5) HEAs reduced surface corrosion. No pits were detected on the alloy surface, indicating that the HEAs exhibited optimal corrosion resistance at higher Cr concentrations. Gao et al. [176] demonstrated that the oxide film formed by Cr significantly improved the corrosion resistance of the Cr-Nb-Ta-Ti-V HEA when exposed to a Na_2SO_4 + 25 wt. % NaCl molten salt environment at 1173 K. Zhang et al. [177] developed a Cr-Fe-Ni_2–based multi-component alloy system to study its corrosion behavior in a 3.5 wt. % NaCl solution. The alloy exhibited localized corrosion, but electrochemical tests revealed that the $CrFeNi_2$-based alloys had higher E_{pit} values and lower i_{corr}, indicating improved corrosion resistance. This enhancement was attributed to the formation of compact and stable protective oxide/hydroxide films.

9.3.3.1.7 Copper

Wang et al. [178] investigated the corrosion behavior of Al-Cr-Fe-Ni_3-Cu_x (x = 0, 0.2, 0.4, 0.6, 0.8, 1.0) HEAs in a 3.5 wt. % NaCl solution. They employed potentiodynamic polarization and electrochemical impedance spectroscopy (EIS) for the analysis. The alloy's microstructure evolved from a biphasic FCC + B2 eutectic structure to a dendritic form. The electrochemical findings indicated a broad passivation region in the anodic polarization curves without showing a clear activation-passivation transition. The pitting potential (Epit) declined as the copper content increased, with the optimal corrosion resistance at x = 0.4. At this composition, the corrosion rate was measured at 0.0201 mm/y, and the corrosion current density (icorr) was recorded at 135 nA/cm^2. Kato et al. [179] studied the corrosion behavior of Co-Cr-Cu_x-Fe-Ni HEAs in a 1 M NaCl solution, with copper contents of 0, 1, 3, and 5 at.%. They observed that at 1 at.% Cu, the pitting potential (Epit) increased due to copper deposition on the anode, which suppressed active dissolution. However, when the Cu content reached 3 at.% or more, the dissolution of CuO in the passive film at higher potentials caused a reduction in Epit. In a separate study, Chen et al. [180] studied the anti-corrosion behavior of $Al_{0.3}$-Cu_x-Co-Cr-Fe-Ni HEAs in a 3.5 wt. % NaCl solution at 298 K. Adding Cu increased the volume fraction of Cu- and Al-rich precipitates, disrupting the passivation film's uniformity and reducing the alloy's local corrosion resistance. Meng et al. [181] examined the corrosion behavior of Co-Cr-Fe-Ni-Cu_x HEA coatings in a 3.5 wt. % NaCl solution. They found that increasing the Cu content led to copper polarization at grain boundaries, which heightened local galvanic coupling corrosion and diminished the overall corrosion resistance of the alloy. Similarly, Mukanov et al. [182] explored the impact of varying Cu content (0–20% in 5% increments) on the corrosion properties of Co-Cr-Cu-Fe-Ni HEAs. Corrosion tests in 3.5% NaCl showed a reduction in corrosion resistance at lower Cu concentrations (5–10%), likely due to the formation of Cu_2O oxide, which was less protective compared to Cr_2O_3. However, at higher Cu concentrations (15–20%), galvanic corrosion was mitigated as the coarsening of FCC2 grains reduced the contact area between the cathode (FCC2) and the anode (FCC1). Peng et al. [183] investigated the electrochemical corrosion behavior of lightweight HEAs Al-Cr-Ti-$V_{0.5}$-Cu_x (x = 0 to 1.0 with 0.2 difference) in a 0.6 M sodium chloride solution. The results showed that increasing copper content led to a decrease in corrosion resistance. This decline was attributed to copper polarization between the dendrites, forming a Cu-rich phase. As the Cu content rose, this polarization effect intensified, creating significant potential differences between Cu-rich and Cu-poor regions, increasing susceptibility to galvanic coupling corrosion in the dendrites.

Hsu et al. [184] studied the corrosion properties of Fe-Co-Ni-Cr-Cu_x HEAs in a 3.5% sodium chloride solution. Their findings revealed that the alloys became more prone to localized corrosion as copper content increased. This was due to the formation of Cu-rich interdendritic areas and Cu-depleted (Cr-rich) dendrites, which resulted in a notable potential difference. The galvanic effect from this difference led to preferential corrosion along the interdendritic regions in the copper-containing HEAs. Liu et al. [185] studied Co_{1-x}-Cr-Fe-Ni_{1+x}-Cu_Y HEAs and observed the formation of distinct Cu-rich and Cu-poor phases, with a high concentration of copper at the boundary zone. Increasing the copper content in a 3.5 wt. % NaCl solution led to diminished corrosion resistance. This decline resulted from forming a weakly protective Cu_2O film on the Cu-rich phase, compromising the overall passive layer. The exposure of the Cu-rich phase and boundary zone to the electrolyte, coupled with the potential difference between these regions, resulted in preferential corrosion of the Cu-rich phase. Huang et al. [186] examined the effect of copper content on the corrosion resistance of Al-Cu_x-Ni-Ti-$Zr_{0.75}$ HEA films. They found that adding copper slightly weakened the passivation properties of the films, leading to a gradual reduction in corrosion resistance with increasing Cu content. Despite this, the corrosion current density remained low, between 10^{-6} and 10^{-7} A/cm^2.

9.3.3.1.8 Gallium

Sun et al. [187] examined the corrosion behavior of Co-Fe-$Mn_{1\cdot2}$-Ni HEA in a 3.5 wt. % sodium chloride solution. The results showed that this alloy exhibited a higher corrosion potential (Ecorr), lower corrosion current density (icorr), and increased pitting potential (Epit) when compared to 304 stainless steel. The passivated Co-Fe-$Mn_{1\cdot2}$-Ni-$Ga_{0\cdot8}$ demonstrated improved surface corrosion resistance due to forming a shielding oxide layer with uniform elements such as nickel and cobalt across the alloy's surface.

9.3.3.1.9 Germanium

Yao et al. [188] investigated the impact of Ge content on the corrosion resistance of Zr–1Nb–xGe (x = 0–0.20, wt. %) in deionized water at 633 K/18.6 MPa and in superheated steam at 673 K/10.3 MPa and 773 K/10.3 MPa. They found that the corrosion resistance of the alloy improved with increasing Ge content under all conditions. Additionally, the dissolution of Ge in the α-Zr matrix significantly contributed to enhancing the corrosion resistance of the Zr–1Nb alloy. Liu et al. [189] synthesized Co-Cr-Fe-Ni-Ge_x (x = 0 to 0.3 with 0.1 variation) HEAs using vacuum arc melting to investigate the influence of germanium content on the anti-corrosion properties of Co-Cr-Fe-Ni HEAs in a 3.5 wt. % sodium chloride solution. Following homogenization, the HEAs exhibited a single-phase FCC solid solution structure and improved corrosion resistance, particularly against pitting. Notably, the incorporation of germanium led to the formation of GeO_2, which contributed to enhanced corrosion resistance and strengthened the protective properties of the passivation film.

9.3.3.1.10 Lanthanum

Sun et al. [190] inspected the corrosion behavior of Cr-Mn-Fe-Ni-La_x (x = 0 and 0.1) HEAs in Sodium chloride solution of various concentrations. Both alloys displayed a single FCC phase, and the addition of trace amounts of lanthanum did not affect this phase. In the NaCl solution, passivation did not occur; instead, increasing the concentration led to localized electrochemical inhomogeneities between the dendrites and inter-dendrites, resulting in selective corrosion. The introduction of lanthanum increased the manganese content while reducing nickel in the passivation film. Although higher manganese oxide content is generally associated with enhanced corrosion resistance, the increase in manganese due to lanthanum inclusion in the Cr-Mn-Fe-Ni-La_x HEAs negatively impacted their overall corrosion resistance.

9.3.3.1.11 Manganese

Torbati-Sarraf et al. [191] demonstrated that adding Mn decreased the concentration of Cr, a key passivation element, in the protective film. Additionally, the pitting resistance of the Cr-Fe-Co-Ni HEA in chloride solutions was notably reduced due to the elevated dissolution rates of manganese and chromium oxides. Wong et al. [192] investigated the impact of Mn addition on the electrochemical corrosion behavior of $Al_{0.3}$-Co-Cr-Fe-Ni-Mn_x ($x = 0, 0.1, 0.3$) HEAs. Their findings revealed that including Mn resulted in a slight reduction in corrosion resistance.

9.3.3.1.12 Molybdenum

Dai et al. [193] examined the impact of molybdenum addition on the corrosion performance of as-cast Fe-Co-Cr-Ni-Mo_x ($x = 0, 0.1, 0.3, 0.6$) HEAs in a chloride-rich environment. The Mo_0 and $Mo_{0.1}$ alloys displayed a single FCC solid solution structure, while the $Mo_{0.3}$ alloy formed a Cr-$Fe_{2.32}$-Mo-Ni phase rich in chromium and molybdenum. In the case of the $Mo_{0.6}$ alloy, a two-phase structure emerged, characterized by a larger volume of the Cr-$Fe_{2.32}$-Mo-Ni phase and a matrix deficient in chromium and molybdenum. Potentiodynamic polarization tests conducted in 0.25 M and 1 M NaCl solutions exhibited typical passivation behavior, broad passivation zones, and low passive current densities (ipass), indicating the excellent corrosion resistance of these alloys. Boakye et al. [194] investigated the corrosion resistance of Co-Cr-Fe-Ni-$Mo_{0.85}$ HEA coatings in CO_2 and H_2S environments at 523 K. Their findings demonstrated that this HEA outperformed several commercially available coatings, including WC_{10}-Co_4-Cr, Cr_{14}-Fe_5-Si_5-B_3-Ni_{Bal}, and Cr_3-C_2-Mo_{20}-W_{10}-Mn_5-B_5-Si_2-Fe_{Bal}, in terms of corrosion resistance. The enhanced performance of the Co-Cr-Fe-Ni-$Mo_{0.85}$ coating was attributed to the alloy's ability to passivate and the slow diffusion of elements from the HEA matrix. Godlewska et al. [195] examined the corrosion behavior of Al-Cr-Fe_2-Ni_2-Mo_x ($x = 0.00, 0.05, 0.10, 0.15$) HEAs in a 3.5 wt. % NaCl solution. Their study found that minor additions of molybdenum enhanced the corrosion resistance of the Al-Cr-Fe_2-Ni_2 alloy by reducing the formation of pits. The equimolar alloy, Al-Co-Cr-Fe-Ni, which exhibited an FCC phase, showed a more noble pitting potential, lower corrosion current density, and reduced corrosion rate than the BCC phase. Wang et al. [144] investigated how molybdenum content and aging temperatures affect the corrosion behavior of alloys in a 3.5% NaCl solution. Their findings demonstrated that increasing the Mo content from 1% to 2% led to a gradual enhancement in corrosion resistance, as indicated by polarization curves and electrochemical impedance spectroscopy, particularly at elevated aging temperatures. However, the corrosion resistance of the $(Co\text{-}Cr\text{-}Fe\text{-}Ni)_{97}Mo_3$ alloy significantly declined when aged at 1173 K. X-ray photoelectron spectroscopy

(XPS) analysis showed that the passive film primarily comprised Cr_2O_3 and MoO_3, with the intensity of these components increasing as the Mo content rose, thereby improving the stability of the passive film. Wang et al. [196] investigated the corrosion resistance of $Co_{1.5}$-Cr-Fe-$Ni_{1.5}$-$Ta_{0.1}$-Mo_x (x = 0, 0.1, 0.3, 0.5) HEAs. Their potentiodynamic testing indicated that the $Mo_{0.3}$ alloy exhibited the most favorable corrosion performance, showcasing exceptional stability of the passive film in a 3.5 wt. % NaCl solution.

9.3.3.1.13 Nickel

Wu et al. [197] explored the corrosion resistance of Fe-Co-Cr-Al-Cu-Ni_x (x = 0.5 to 1.5 with 0.5 difference) HEA coatings in a 3.5 wt. % NaCl solution. Their electrochemical analysis revealed that as Ni content increased, the corrosion resistance initially improved but then declined. Despite the absence of passivation zones in the potentiodynamic polarization curves, the lower i_{corr} values and corrosion rates (r_{corr}) indicated superior corrosion resistance of the HEA coatings compared to the Cu substrate. Similarly, Zemanate et al. [198] investigated the impact of varying Ni concentrations on Al-Cr-Fe-Co-Ni_x (x = 1.0, 1.5, 2.0) HEAs using potentiodynamic polarization and electrochemical impedance spectroscopy. Their findings demonstrated excellent resistance to localized corrosion, with low corrosion current densities and high pitting potentials. The formation of protective films was dependent on Ni content, influenced by the FCC/BCC phase composition ratio. The Al_2-Cr-Fe-Co-Cu-Ti-Ni_x HEA coating demonstrates excellent corrosion resistance in 1 mol/L NaOH and 3.5% NaCl solutions. As the nickel content increases, there is an initial enhancement in corrosion resistance; however, this improvement is followed by a decline at higher nickel levels. This behavior can be attributed to the high corrosion resistance of Ni, though its small atomic radius leads to significant lattice distortion at higher concentrations. This distortion affects the alloy's microstructure, thereby influencing its corrosion resistance. Among the compositions, the Ni1.0 alloy demonstrates the best overall corrosion resistance [199].

9.3.3.1.14 Niobium

Wen et al. [200] examined how a dual-phase microstructure, consisting of FCC and Laves phases, influences the corrosion and tribo-corrosion properties of $Ni_{1.5}$-Cr-Co-$Fe_{0.5}$-$Mo_{0.1}$-Nb_x (x = 0.55 hypoeutectic, 0.68 eutectic, and 0.8 hyper eutectic) HEA coatings in a 3.5 wt. % NaCl solution. In this setup, the FCC phase served as the anodic site. The passive regions of the coatings increase with Nb content. Mott–Schottky analysis indicated that the improved performance of the passive film was primarily attributed to a reduction in point-defect density. Wang et al. [201] investigated the corrosion behavior and underlying mechanisms of Cr-Fe-Co-Ni-Nb_x HEAs featuring hypo- and hypereutectic (x = 0.38, x = 0.54) structures, respectively. Both alloys exhibited a dual-phase structure consisting of FCC and Laves phases. When subjected to electrochemical polarization in a 3.5 wt. % NaCl solution at room temperature, the corrosion behavior of the two HEAs was similar, although the hypereutectic HEA demonstrated slightly improved corrosion resistance. The passive films formed on both alloys contained substantial amounts of chromium and niobium oxides, with no significant concentration differences, contributing to their high and comparable corrosion resistance. Liu et al. [202] investigated the corrosion resistance of Fe-Co-Cr-Ni-Nb_x (x = 0, 0.15, 0.33, 0.5) HEAs in a 3.5 wt. % NaCl solution, comparing their performance to that of 316 L stainless steel. The HEAs exhibited a two-phase structure consisting of FCC and Laves phases, with an increasing content of Laves phase corresponding to higher niobium percentages. The HEAs and 316 L stainless steel revealed passivation performance in the NaCl solution, as evidenced by the potentiodynamic polarization curves. The HEAs showed lower passive current density (ipass) values and broader passivation zones than the stainless steel, representing that adding niobium facilitated the formation of protective passivation films and enhanced pitting resistance. While an optimal amount of niobium enhanced corrosion resistance, excessive niobium led to compositional segregation, adversely affecting corrosion performance.

9.3.3.1.15 Silicon

Yang et al. [203] studied the influence of silicon content on the microstructure and corrosion behavior of $Al_{0.2}$-Co-Cr-$Fe_{1.5}$-Ni-Si_x (x = 0, 0.1, 0.2, 0.3) HEAs produced through vacuum arc melting. Their findings showed that increasing silicon content led to a transition from a single FCC phase to a dual-phase structure with both FCC and BCC phases. However, adding silicon caused compositional segregation within the BCC phase, compromising homogeneity. Potentiodynamic polarization tests revealed that silicon enhanced pitting resistance and decreased the alloy's re-passivation ability. Among the alloys, the $Al_{0.2}$-Co-Cr-$Fe_{1.5}$-Ni-$Si_{0.1}$ composition exhibited the most stable passivation, evidenced by its highest corrosion potential (Ecorr: –256 mVAg/AgCl) and lowest corrosion current density (icorr: 217 nA/cm^2). This suggests that a small addition of Si improves corrosion resistance in single-phase FCC HEAs. Still, when Si content reaches 0.2% or more, the BCC phase promotes galvanic coupling corrosion and selective dissolution due to inhomogeneous composition, reducing corrosion resistance. Gu et al. [204] examined the anti-corrosion performance of as-cast $Al_{0.3}$-Co-Cr-Fe-Ni HEAs and observed that adding silicon promoted the development of a BCC phase. This BCC phase acted as a micro-anode, providing protective effects for the FCC phase, which functioned as a micro-cathode. As the Si content increased, the volume fraction of the BCC phase also grew, contributing to enhanced corrosion resistance in the silicon-containing alloys. Furthermore, the reduction in grain size associated with higher Si content further boosted the corrosion resistance of the alloys. Bi et al. [205] studied the corrosion and passivation behavior of Al-Nb-Ti-Zr-Si_x HEAs with varying Si content in a simulated seawater environment at room temperature. The findings revealed that increasing Si content reduced the average grain size, likely due to the inhibitory effect on grain growth caused by the precipitation of the Zr_5Al_3 phase. The passive films on the HEAs were primarily composed of TiO_2, Nb_2O_5, and ZrO_2. Preferential corrosion occurred in the Zr_5Al_3 phase regions, and the addition of Si decreased the stability of the passive film, leading to a decline in corrosion resistance. Lonski et al. [206] explored the corrosion resistance of a dual-phase Al-Co-Ni-Fe-Ti alloy, finding that adding silicon significantly impacted the alloy's phase composition. XRD outcomes indicated that, without Si, the alloy comprised coexisting L21 and BCC phases. However, introducing Si led to Ti-rich HCP and Al-rich B2 phases. The Si-free Al-Co-Fe-Ni-Ti alloy exhibited strong corrosion resistance, with a polarization resistance of about 38.6 kΩcm^2 and a corrosion current density of 0.77 μA/cm^2. Additionally, the alloy showed a high saturation magnetization of 29.83 emu/g. Ji et al. [207] found that the corrosion resistance of as-cast Al_{35}-Mg_{30-x}-Zn_{30}-Cu_5-Si_x HEAs in a 3.5 wt. % NaCl solution improved as the Si/Mg ratio increased. Corrosion was primarily initiated at the grain boundaries and progressively extended into the eutectic and intermetallic phases rich in Mg.

9.3.3.1.16 Tin

Muangtong et al. [208] examined the effect of Sn addition on the microstructure and corrosion properties of Co-Cr-Fe-Ni-Sn HEA. The alloy demonstrated excellent corrosion resistance in a 0.6 M sodium chloride solution. The introduction of Sn enhanced the consistency of the passivation film by forming two protective layers, SnO_2 and Cr_2O_3, on the alloy surface, which inhibited corrosion. During damage, the FCC phase underwent galvanic coupling corrosion, causing ions such as Co^{3+}, Cr^{3+}, Fe^{2+}, and Ni^{2+} to dissolve. This dissolution allowed the passivation films to regenerate, further improving the alloy's corrosion resistance. Chang et al. [209] demonstrated that Co-Cr-Fe-Ni-Sn_x HEAs transitioned from a single FCC phase to a dendritic structure consisting of both FCC and Ni_3Sn_2 phases as the Sn content increased. Electrochemical impedance spectroscopy (EIS), polarization tests, and scanning Kelvin probe force microscopy (SKPFM) results indicated that with higher Sn content, the FCC phase, acting as the anode, reduced in the area while the cathodic Ni_3Sn_2 phase expanded. Consequently, the Co-Cr-Fe-Ni-$Sn_{0.5}$ and Co-Cr-Fe-Ni-$Sn_{1.0}$ alloys demonstrated improved pitting resistance in a 3.5 wt. % NaCl solution compared to the alloy without Sn.

9.3.3.1.17 Titanium

Titanium is recognized for its capacity to form a durable and compact oxide layer, which promotes passivation. This oxide layer reinforces the surface film, greatly enhancing the material's corrosion resistance. Wu et al. [210] inspected the corrosion resistance of the Al-Cr-Fe-Co-Ni-Ti0.5 HEA in a 3.5 wt. % sodium chloride solution. The addition of Ti significantly enhanced the alloy's corrosion resistance, particularly against pitting corrosion, as indicated by lower I_{corr} and I_{pit} values, alongside increased Rp and Rct. This improvement was attributed to Ti's role in forming a more robust passive film, which inhibited point defect transport and prevented the breakdown of the passive layer. Zhou et al. [211] investigated the corrosion behavior of $(Fe\text{-}Cr\text{-}Co\text{-}Ni)_{100-x}$-$Ti_x$ (x = 0, 1, 3, 5) HEAs in a 1 M sodium chloride solution. The alloys displayed a single FCC structure after solid solution treatment at 1373 K. With Ti addition, the i_{pass} decreased, and the passivation range expanded, representing that Ti improved the corrosion resistance of the alloy in environments rich in chloride and hydrogen ions. Wang et al. [212] noted that pitting corrosion was the primary form of degradation observed in Co-Cr-Fe-Ni-Ti_x HEAs when subjected to 3.5 wt. % NaCl solution. The HEA exhibited a higher corrosion potential and lower corrosion current density than other HEAs. They increased Ti content and further enhanced corrosion resistance, mainly due to the high energy barriers at the Cr_2O_3/

TiO_2 interfaces, which contributed to the improved protective properties of the alloy. Pratskova et al. [213] identified that in NaCl solution, $Al_{0.5}$-Co-Cr-Fe-$Ni_{1.6}$-$Ti_{0.7}$ HEA exhibited a wide range of passivation potentials ($\Delta E = 0.6–0.7$ V). However, despite this broad passivation range, the alloy experienced active dissolution and significant corrosion due to its low corrosion potential and high corrosion current. Qiu et al. [154] studied the corrosion behavior of the Al_x-Co-Cr-Fe-Ni-Ti_y HEA system with varying compositions, showing that these alloys exhibited superior resistance to general corrosion compared to stainless-steel 304 L in a 0.6 M NaCl solution. However, cyclic potentiodynamic polarization tests revealed a susceptibility to pitting corrosion once passivation breakdown occurred. Prolonged immersion tests showed the formation of very fine pitting as the primary form of corrosion, though it was not extensive. Although there was a heterogeneous microstructure and significant elemental segregation, no significant evidence of micro-galvanic corrosion or selective phase dissolution was observed. Han et al. [214] investigated the corrosion resistance of Al-Cr-Fe-Ni-$Mo_{0.5}$-Ti_x ($x = 0, 0.25, 0.4, 0.5$) HEAs and found that increasing Ti content improved the protective nature of the passive film. The charge transfer resistance (Rct) increased significantly from 229 $k\Omega \cdot cm^2$ for the Ti-free alloy to 635 $k\Omega \cdot cm^2$ for the alloy with $Ti_{0.5}$. Additionally, the pitting potential (Epit) rose from 90 mV to 410 mV as Ti content increased, indicating that Ti addition greatly enhanced the corrosion resistance of the Al-Cr-Fe-Ni-$Mo_{0.5}$-Ti_x HEA. Ozturk et al. [215] explored the impact of Ti addition on the corrosion behavior of the Co-Cu-Fe-Ni-Mn-Tix alloy in a 3.5% NaCl solution. They observed a decline in corrosion resistance with increasing Ti content. For the Ti-free Co-Cu-Fe-Ni-Mn alloy, the Ecorr and Icorr values were -0.322 V and 6.30×10^{-7} A/cm², respectively, whereas for the Co-Cu-Fe-Ni-Mn-$Ti_{0.5}$ alloy, these values shifted to -0.982 V and 1.16×10^{-3} A/cm². While the Ti-free alloy experienced pitting corrosion, adding Ti formed Ti-rich interdendritic zones, resulting in galvanic corrosion.

9.3.3.1.18 Vanadium

Xiang et al. [216] reported that Mo-Nb-Fe-Cr-Ti and Mo-Nb-Fe-V-Ti alloys demonstrated exceptional corrosion resistance when tested in a deaerated 1 mol/L NaCl solution. The influence of vanadium on the corrosion resistance of as-cast Al-Fe-Ni-Cr-Co-$Ti_{0.5}V_x$ ($x = 0, 0.5, 1, 1.5, 2$) alloys was studied in a 3.5 M NaCl solution by Han et al. [217]. The presence of Cr, Ni, Ti, and Al elements in the alloys promotes passivation, which enhances the corrosion resistance. Increasing vanadium content alters the distribution of Cr and Ti, as vanadium predominantly accumulates in the interdendritic regions. This modification leads to an increase in the corrosion potential of the alloys and a reduction in passive current. As a result, the alloys exhibit a protective effect in the saline solution.

9.3.3.2 Acidic Corrosion Behavior of HEAs

The primary constituent elements within the alloy strongly influence the anti-corrosion properties of HEAs. Alloying components such as Al, Cu, Mo, B, and Cr can lead to microstructural changes and elemental distribution variations, significantly impacting the corrosion behavior of HEAs.

9.3.3.2.1 Aluminum

Gu et al. [218] examined the corrosion resistance of Al_x-(Co-Cr-Fe-Ni)$_{100-x}$ (x = 0 to 20 with a variation of 5) HEAs in a 0.5 M H_2SO_4 solution. Their results indicated that increasing the aluminum content shifted the alloy's microstructure from an FCC to a BCC structure. This shift caused chromium and iron to become concentrated in the FCC phase, while aluminum and nickel accumulated in the BCC phase, decreasing the uniform distribution of elements. Potentiodynamic polarization curves showed a typical activation-passivation transition (from −0.3 to −0.2 V), followed by passivation (0–0.9 V) and over-passivation (above 0.9 V). When the aluminum content exceeded 5%, the E_{corr} decreased, and the i_{corr} increased with increased Al content. The formation of an aluminum-rich, chromium-poor, nickel-depleted BCC phase, which has less work function than the FCC phase, enhanced corrosion resistance. Tang et al. [219] found that aluminum alloying significantly influences the passive film properties of HEAs. In Al-containing HEAs, higher aluminum concentrations typically form more porous passive films, reducing corrosion resistance. Additionally, increasing the aluminum content promotes the development of the BCC phase in Al_x-Co-Cr-Fe-Ni and Al_x-Cr-$Fe_{1.5}$-Mn-$Ni_{0.5}$ HEA systems. The passive films generated on these BCC-structured alloys are generally less stable and more liable to corrosion than those found on FCC-structured HEAs with lower aluminum ratios (e.g., x ≤ 0.3). Liu et al. [220] investigated the corrosion resistance of Al_x-Co-Cr-Cu-Fe-Ni HEAs (x = 0, 0.5, 1, 1.5, 2) and found that the Al_2-Co-Cr-Cu-Fe-Ni alloy exhibited the highest corrosion rate in acidic solutions. Additionally, this alloy showed the most susceptibility to corrosion in NaOH solution, indicating a peak in corrosion tendency under alkaline conditions. Kao et al. [221] investigated the corrosion behavior of Al_x-Co-Cr-Fe-Ni alloys in H_2SO_4 using potentiodynamic polarization, electrochemical impedance spectroscopy (EIS), and weight loss tests at temperatures ranging from 293 to 338 K. Their findings indicated that aluminum adversely affects corrosion resistance in H_2SO_4 at temperatures exceeding 300 K by forming a porous and less protective oxide film on the surface of the alloy. Fu et al. [222] studied the corrosion behavior and film properties of Fe-Co-Cr-Ni-Al_x (x = 0.1, 0.3) HEAs in 0.5 M H_2SO_4 and 0.5 M HCl solutions through electrochemical measurements. They found that increasing the aluminum content enhanced the corrosion resistance of Fe-Co-Cr-Ni-Al_x in H_2SO_4 by increasing both the chromium oxide content and the thickness of the passive film. However, in the HCl solution, the effect was detrimental, as chloride ions tended to interact with aluminum-containing metastable ion complexes, accelerating the dissolution of the HEA.

9.3.3.2.2 Boron

Lee et al. [223] studied the corrosion behavior of boron-containing HEAs Al0.5-Co-Cr-Cu-Fe-Ni-Bx (with x varying from 0 to 15.4 at. %) in a 1 N H_2SO_4 solution. Their research showed that these alloys demonstrated negative hysteresis, which suggested resistance to localized corrosion. However, increased boron content increased the i_{corr} and decreased E_{rp}.

9.3.3.2.3 Cerium

Zhang et al. [224] investigated how cerium (Ce) affects the localized corrosion behavior of non-equiatomic Fe_{40}-Mn_{20}-Cr_{20}-Ni_{20} HEAs in a 0.5 M H_2SO_4 solution. Their findings indicated that adding an optimal amount of Ce improved the alloy's corrosion resistance by preventing the formation of MnS compounds susceptible to corrosion and encouraging the development of inclusions with spherical morphology. Additionally, Ce enhanced the density of the passivation film and improved the $Cr_2O_3/Cr(OH)_3$ ratio, further reinforcing the protective properties of the alloy.

9.3.3.2.4 Cobalt

The Al_2-Cr-Fe-Co_x-Cu-Ni-Ti HEA coating exhibits excellent corrosion resistance in a 0.5 M H_2SO_4 solution, primarily due to the formability of the cobalt (Co) element and the protective passive film that develops on the alloy's surface [225]. Qiu et al. [226] found that the Al_2-Co_x-Cr-Cu-Fe-Ni-Ti coatings exhibit excellent corrosion resistance in both H_2SO_4 and HCl solutions due to three key factors: the formability of the cobalt element, the passivation film on the alloy surface, and the alloy's microstructure. Increasing the cobalt content further improved the corrosion resistance of HEA coatings in a 0.5 M HCl solution. Cyclic polarization tests revealed no pitting corrosion in 0.5 M H_2SO_4 for the $Co_{2.0}$ coating, while the $Co_{0.5}$ coating showed slight pitting in 0.5 M HCl, likely due to chloride ions (Cl^-) in the solution.

9.3.3.2.5 Chromium

Xiao et al. [227] examined the corrosion properties of Al-Co-Cu-Fe-Ni-Cr HEAs in a 0.5 M H_2SO_4 solution. They found that including chromium (Cr) promoted a phase change from FCC to BCC, enhancing the Al-Co-Cu-Fe-Ni HEAs' corrosion resistance.

9.3.3.2.6 Copper

Ren et al. [228] investigated the corrosion behavior of Cu-containing HEAs (Cu-Cr-Fe-Ni-Mn) using immersion and potentiodynamic polarization tests in 1 M H_2SO_4. The study revealed that the Cu-Cr_2-Fe_2-Ni_2-Mn_2 alloy, with a lower copper content, exhibited minimal segregation and significantly higher corrosion resistance. In contrast, the Cu_2-Cr-Fe_2-Ni_2-Mn_2 alloy, with a higher copper content, showed notable Cu segregation, leading to the lowest corrosion resistance. Chen et al. [229] investigated how Cu concentration and passivation behavior affect the corrosion resistance of $Al_{0.3}$-Cu_x-Co-Cr-Fe-Ni HEAs with varying Cu concentrations (x = 0.3, 0.4, 0.5, 0.6). As the Cu content increased, Al- and Cu-rich precipitates gradually formed, reducing corrosion resistance and weakening the passive film. Passivation treatment in 30 wt. % HNO_3 and 1 wt. % NaOH solutions enhanced the Cr^{3+} content in the passive film of $Al_{0.3}$-$Cu_{0.3}$-Co-Cr-Fe-Ni, significantly improving its corrosion resistance by lowering the likelihood of pitting. However, the treatment proved less effective at higher copper levels, as the increased presence of Al- and Cu-rich precipitates interfered with forming a uniform passive film and exacerbated galvanic corrosion.

9.3.3.2.7 Manganese

Hsu et al. [230] examined how varying manganese (Mn) content affects the corrosion behavior of Fe-Cr-Ni-Co-Mn_x alloy in a 0.5 M H_2SO_4 solution. Following a

24-hour heat treatment at 1473 K, the alloys developed a single FCC phase, minimizing the potential for galvanic corrosion between different phases. Their findings indicated that reducing the Mn content slowed the corrosion rate of the Fe-Cr-Ni-Co-Mn_x alloys. Compositional analysis showed that the passivation films were rich in chromium but lacked nickel, with the Cr_2O_3 content varying due to competitive oxidation between manganese and chromium. As the Mn content decreased, corrosion resistance improved due to a higher concentration of Cr_2O_3 and a reduction in Mn-related oxide/sulfide inclusions within the passivating layer.

9.3.3.2.8 Molybdenum

Chou et al. [231] examined the electrochemical properties of $Co_{1.5}$-Cr-Fe-$Ni_{1.5}$-$Ti_{0.5}$-Mo_x HEAs in three aqueous environments simulating acidic, marine, and basic conditions at 298 K. Potentiodynamic polarization tests in H_2SO_4 and NaOH solutions revealed that the Mo-free alloy exhibited superior corrosion resistance to the Mo-containing alloys. Shang et al. [232] investigated the corrosion behavior of Co-Cr-Fe-Ni-Mox alloys in H_2SO_4 solution, finding that Co-Cr-Fe-Ni-$Mo_{0.2}$ and Co-Cr-Fe-Ni-$Mo_{0.4}$ exhibited the best corrosion resistance, suggesting an optimal balance between Mo content and r-phase formation. XPS analysis of the alloys revealed that chromium and molybdenum are vital contributors to corrosion resistance, with molybdenum mainly present as MoO_3.

9.3.3.2.9 Nickel

Huang et al. [233] found that adding nickel negatively impacted the corrosion resistance of Co-Cr-Ni-Al-Ti HEAs in a 0.5 M H_2SO_4 solution. This was due to the promotion of Cr^{3+} and NiS formation on the corroded surface, which diminished the alloy's overall resistance to corrosion. It was observed that in Al_2-Cr-Fe-Co-Cu-Ti-Nix HEAs [234], the alloy with Ni content ($x = 1$) exhibited a lower corrosion current density (icorr) in neutral and alkaline solutions. Still, the icorr increased as the Ni content rose. It was reported that Ni and Al form an Al-Ni enriched B_2 phase, which can negatively affect corrosion resistance in Ni-containing HEAs. Additionally, Ni was found to form FCC and BCC phases with other alloying elements, contributing to poor corrosion resistance. This is likely due to Ni's smaller atomic size than other elements, causing lattice distortion in high Ni-content alloys, making them more susceptible to corrosion.

9.3.3.2.10 Niobium

Wu et al. [145] found that the corrosion behavior of Co-Cr-Fe-Ni-Nb_x high entropy alloy coatings (HEACs) in 0.5 M H_2SO_4 solution was comparable to that of 316 stainless steel (SS), but with stronger corrosion resistance. The $Nb_{0.3}$ HEAC exhibited the best acid corrosion resistance with a corrosion potential (E_{corr}) of −164 mV and a corrosion current density (i_{corr}) of 6.40×10^{-7} A/cm^2. This superior performance was attributed to forming a stable, dense passivation film on its surface. Unlike the intergranular corrosion observed in other coatings, corrosion in $Nb_{0.3}$ HEAC mainly occurred in the dendritic region, where the FCC phase, acting as the anode, coexisted with the Laves phase. Tsau et al. [235] found that adding niobium (Nb) to Cr-Fe-Co-Ni-Nb_x alloys slightly reduced their corrosion resistance due to the development of dual-phased dendritic microstructures. However, the Nb addition also helped stabilize and expand the passivation regions of these alloys in acidic environments. After polarization tests in 1 M deaerated H_2SO_4 solution, the FCC phase of the alloys was more prone to corrosion than the Laves phase.

9.3.3.2.11 Nitrogen

Chen et al. [236] studied the effect of nitrogen (N) on the corrosion behavior of Fe-Co-Cr-Ni HEAs in a 0.5 M H_2SO_4 solution. Introducing nitrogen reduced the average grain size from 58 μm to 44 μm, resulting in a refined, inhomogeneous microstructure. Electrochemical tests revealed that Fe-Co-Cr-Ni-N HEA had the lowest corrosion current density (icorr of 58.9 $\mu A/cm^2$) and passivation current density (ipass of 6.1 $\mu A/cm^2$), indicating the formation of a stable and dense passivation film. The enhanced corrosion resistance was linked to the passivation film's composition, with XPS analysis showing a significant increase in Cr_2O_3 content, significantly improving the alloy's corrosion resistance.

9.3.3.2.12 Titanium

In the Alx-Co-Cr-Fe-Ni-Ti_{1-x} series tested in H_2SO_4 solution, the Al-0.5 alloy demonstrated the highest corrosion resistance, followed by Al-1.0 and Al-0.8. Adding a small amount of titanium was insufficient to mitigate the adverse effects of elemental segregation and the biphasic structure on corrosion resistance. A notable enhancement in corrosion resistance was achieved only when titanium's atomic content exceeded aluminum's [237]. Qiu et al. [238] found that increasing the Ti content in Al_2-Cr-Fe-Ni-Co-Cu-Ti_x HEAs promotes the formation of a BCC phase to some extent. Compared to Q235 steel, the free-corrosion current density of these HEA coatings was reduced by one to two orders of magnitude, and the free-corrosion potential shifted to more positive values. As the Ti content increased, the corrosion resistance of the Al_2-Cr-Fe-Co-Cu-Ni-Ti_x HEA coatings improved in a 0.5 mol/L HNO_3 solution. Lin et al. [239] investigated the corrosion behavior of Crx-Fe-Ni-Cu0.5-Ti0.5 HEAs (x = 0, 0.3, 0.5, 0.7, 0.9, 1.2). The potentiodynamic polarization curves revealed that the Fe-$Cr_{0.7}$-Ni-$Cu_{0.5}$-$Ti_{0.5}$ alloy exhibited excellent corrosion resistance, with a corrosion potential of −0.181 V and a self-corrosion current density of 5.455×10^{-7} A/cm^2 in 0.5 M H_2SO_4 solution. XPS analysis showed that forming Cr_2O_3, TiO_2, and Fe oxides on the alloy's surface created strong passive films, reducing ionic diffusion and enhancing corrosion resistance, which was the key mechanism for its anti-corrosive properties. In their study, Xing et al. [240], a (Co-Cr-Fe-Ni)$_{94}$-$Ti_{1.5}$-$Al_{4.5}$ HEA coating was prepared using plasma cladding, featuring a single FCC phase structure and uniform microstructure. Adding Ti and Al enhanced the formation of a more compact passive film by reducing the proportion of porous hydroxides. This improvement increased the film's impedance, resulting in better corrosion resistance. After 4 hours of potentiostatic polarization in a 0.5 M H_2SO_4 solution, a stable passive film formed, with a corrosion current density of only 175 nA/cm^2, much lower than that of 304 stainless steel and pure titanium. This demonstrates that minor additions of Ti and Al can significantly enhance corrosion resistance, particularly in sulfuric acid, without forming harmful secondary phases.

9.3.3.2.13 Yttrium

Wang et al. [241] found that Y enhances corrosion resistance by promoting the formation of a thick passivation film that firmly adheres to the alloy surface. Meng et al. [242] studied Fe-$Cr_{0.7}$-Ni-$Cu_{0.5}$-$Ti_{0.5}$-Y_x (x = 0, 0.02, 0.04, 0.06, 0.08, 0.1) HEAs, exploring how Y affects their microstructure and corrosion properties in 0.5 M H_2SO_4 solution. They observed that increasing Y content reduced the FCC phase while promoting the formation of Cu_5Y, causing significant segregation. Electrochemical tests revealed that a small Y addition (x = 0.06) enhanced the formation of a complete passivation film and improved corrosion resistance. However, excessive Y content led to Cu_5Y formation, disrupting microstructural uniformity and increasing localized corrosion.

9.3.3.2.14 Zirconium

Yao et al. [243] studied the corrosion behavior of Al-Co-Cr-Fe-Ni-Zr_x alloys in 0.5 M H_2SO_4 solution and observed galvanic corrosion in both the dendritic and interdendritic zones. In these regions, the Al- and Ni-rich, Cr-depleted B2 phase served as the anode, while the Fe-Cr-rich BCC phase and Zr-rich Laves phase functioned as cathodes. The corrosion resistance of the alloys initially improved and then decreased with increasing Zr content. While Zr addition at 0.1, 0.3, and 0.5 mol enhanced corrosion resistance, Zr0.3 exhibited the poorest performance due to severe galvanic corrosion of the B2 phases. In contrast, Zr0.1 exhibited the highest corrosion resistance, attributed to the segregation of the Zr-rich Laves phase along the grain boundaries.

9.3.4 Corrosion Testing Methods of HEAs

9.3.4.1 Electrochemical Techniques

Electrochemical techniques are widely used to study the corrosion behavior of HEAs because they provide detailed insights into corrosion kinetics, mechanisms, and the formation of passive films. These methods allow for real-time measurement of corrosion rates and the evaluation of corrosion resistance. Here's an in-depth explanation of the commonly used electrochemical techniques for testing the corrosion resistance of HEAs.

9.3.4.1.1 Potentiodynamic Polarization (PDP)

Potentiodynamic polarization is one of the most frequently used electrochemical techniques for studying the corrosion behavior of HEAs. In this test, the working electrode's potential (the HEA) is scanned linearly or logarithmically while measuring the current response. The PDP curve provides valuable information about the alloy's corrosion resistance, including:

Corrosion Potential (Ecorr): The potential at which the current is minimal, representing the tendency of the alloy to corrode. A more positive Ecorr indicates better corrosion resistance.

Corrosion Current Density (icorr): Obtained from the Tafel extrapolation method, icorr is proportional to the corrosion rate. A lower icorr suggests a slower corrosion rate.

Passivation Range: A stable low-current range signifies the formation of a passive film that safeguards the surface from additional corrosion.

Pitting and Overpassivation: Observing the breakdown of the passive film (pitting potential) and overpassivation currents helps assess the alloy's susceptibility to localized corrosion.

For HEAs, potentiodynamic polarization is essential for identifying passive films and evaluating how alloying elements like Cr, Al, and Ti influence the passivation process in various acidic environments, such as H_2SO_4.

9.3.4.1.2 Electrochemical Impedance Spectroscopy (EIS)

EIS is a powerful technique for assessing the electrochemical properties of passive films and corrosion rates in HEAs. In an EIS test, a small alternating current (AC) perturbation is applied over a range of frequencies, and the impedance response of the system is measured. The resulting data is typically presented as Nyquist or Bode plots, which provide information on:

Charge Transfer Resistance (Rct) represents the resistance to electron transfer reactions at the alloy-electrolyte interface. A higher Rct indicates better corrosion resistance, suggesting a lower corrosion rate.

The Capacitance of the Passive Film (Cdl): Related to the double-layer capacitance, this provides insight into the quality and thickness of the passive film. A lower Cdl is often associated with a thicker and more protective passive layer.

Film Impedance: EIS can also measure the passive film's impedance, reflecting its resistance to ionic diffusion and stability over time.

EIS is beneficial for long-term corrosion studies of HEAs, as it can detect changes in the passive film over time and in different environments, providing detailed insights into how the alloying elements affect film stability.

9.3.4.1.3 Cyclic Polarization

Cyclic polarization is a variation of potentiodynamic polarization used to evaluate the susceptibility of HEAs to pitting corrosion and repassivation susceptibility of HEAs to pitting corrosion and repassivation. The method involves sweeping the potential forward past the corrosion potential (E_{corr}) into the passive region and then reversing the scan back toward the original potential.

This technique provides the following critical information:

Pitting Corrosion Potential (Epit): The potential at which localized corrosion (pitting) initiates. A more positive Epit suggests higher resistance to pitting corrosion.

Repassivation Potential (Erp): The potential at which the current starts to decrease on the reverse scan, indicating the repassivation of the material.

Hysteresis Loop: A large loop indicates a greater tendency for pitting and poor repassivation, while a narrow or nonexistent loop suggests better corrosion resistance and faster repassivation.

For HEAs, cyclic polarization helps assess the localized corrosion behavior, particularly in environments like chloride-containing or acidic solutions, where pitting is more likely to occur.

9.3.4.1.4 Open Circuit Potential (OCP)

Open circuit potential (OCP) is the working electrode's potential when no external current is applied. It is often measured before other electrochemical tests to establish the corrosion potential of the HEA in a given environment. OCP measurements help in:

1. Monitoring the stabilization of the passive film over time.
2. Comparing the inherent corrosion resistance of different HEAs or alloying compositions.
3. Assessing the impact of passivating elements (such as Cr and Ti) on corrosion behavior.

OCP is typically recorded to ensure the HEA's surface reaches a steady-state potential, indicating a stable passive film formation.

9.3.4.1.5 Potentiostatic Polarization

In potentiostatic polarization, a constant potential is applied to the HEA, and the resulting current is monitored over time. This method helps evaluate the growth and stability of the passive film and understand long-term corrosion behavior.

Current Decay: A decrease in current over time indicates the formation of a stable and protective passive film, while an increasing or fluctuating current suggests film breakdown and corrosion.

Passive Film Growth: Potentiostatic tests provide insights into how alloying elements contribute to the development and thickening of the passive film in HEAs.

For HEAs, potentiostatic polarization is instrumental when investigating passive films formed in acidic solutions like H_2SO_4, where passive film stability is critical for corrosion resistance.

9.3.4.1.6 Weight Loss Method (Complementary to Electrochemical Testing)

Although primarily a gravimetric method, weight loss testing often complements electrochemical techniques to provide a broader understanding of corrosion rates. In this method, the HEA is immersed in an acidic solution like H_2SO_4 for a specified period, and the mass loss due to corrosion is measured. The corrosion rate can then be calculated based on the mass loss, surface area, and exposure time. This method helps validate electrochemical test results and provides a more practical estimate of corrosion behavior in real-world conditions.

9.3.4.2 Immersion Test

The immersion test is a widely used method for assessing the corrosion resistance of materials, including HEAs, in various environments. It involves exposing the material to a corrosive solution for a specified period and measuring the resulting corrosion effects, such as weight loss, surface degradation, and localized attack. Here is a detailed explanation of the procedure for conducting an immersion corrosion test:

1. **Preparation of the Sample**
   Material Selection: Choose the material to be tested, such as a HEA or another metallic material. Ensure that the sample is representative of the bulk alloy or coating that will be used in actual applications.
   Sample Dimensions: Cut the sample into standardized dimensions. Typically, flat or cylindrical specimens are used for immersion testing. The dimensions should be chosen to allow uniform exposure to the corrosive medium.
   Surface Preparation: The sample's surface must be cleaned and polished before the test to remove any oxides, scale, or contaminants that could influence the results. The following steps are commonly performed:
   Mechanical Polishing: Use abrasive papers (typically coarse to fine grits) to smooth the surface.
   Degreasing: Clean the sample with acetone, ethanol, or methanol to remove oils, grease, or organic contaminants.
   Rinsing: Rinse the sample with distilled or deionized water to eliminate residues.
   Drying: Air-dry the sample or use a clean cloth or air blower to remove water droplets.
2. **Weighing the Sample**
   Initial Weight Measurement: Accurately weigh the prepared sample using a high-precision analytical balance. This weight will be the baseline for calculating the material loss due to corrosion. Record the initial weight (W_initial) in grams.
   Surface Area Measurement: Measure the total exposed surface area of the sample. This information is required to calculate the rate of corrosion. The surface area can be calculated based on the specimen's geometry (e.g., length, width, thickness for rectangular specimens).
3. **Preparation of the Corrosive Solution**
   Selection of Corrosive Medium: Choose the corrosive environment for the test. For example, a 0.5 M H_2SO_4 (sulfuric acid) solution is often used to test HEAs' acid corrosion resistance. The solution concentration should be relevant to the intended application of the alloy.
   Solution Preparation: Prepare the corrosive medium by diluting the acid to the desired concentration using distilled or deionized water. Ensure that the pH and concentration are precisely controlled for repeatability.
   Temperature Control: If the test is to be conducted at a specific temperature (e.g., elevated or reduced temperatures), use a temperature-controlled bath or environmental chamber to maintain a consistent temperature throughout the test.

4. **Immersion of the Sample**

   Test Setup: Place the prepared sample in the corrosive solution in a container, ensuring it is fully immersed. Use materials like glass or plastic containers chemically resistant to the test solution to avoid contamination.

   Test Duration: The immersion time can vary based on the desired test parameters, ranging from a few hours to several days, weeks, or months, depending on the study's objectives. Typical immersion times for laboratory tests range from 24 hours to several weeks.

   Solution Agitation: Sometimes the solution is stirred or agitated to simulate real-world conditions where flow or turbulence may influence corrosion rates. Alternatively, static immersion tests are conducted without agitation.

5. **Monitoring the Test**

   Regular Observations: During the test, it is crucial to monitor the sample for signs of visible corrosion, such as surface discoloration, pitting, or uniform degradation. Photos may be taken at regular intervals to document the progress of corrosion.

   Solution pH Monitoring: Regularly check and record the pH of the corrosive medium, as it can change over time due to reactions with the sample. Replenish or adjust the solution if necessary to maintain consistent test conditions.

6. **Post-Test Procedures**

   Sample Removal: At the end of the test period, remove the sample from the solution using non-metallic tweezers to avoid contamination or additional wear on the surface.

   Cleaning the Sample: Rinse the sample thoroughly with distilled or deionized water to remove any remaining corrosive solution.

   Corrosion Product Removal: If required, remove the corrosion products (rust, scale, etc.) from the sample's surface. Standard cleaning procedures include:

   Mechanical Cleaning: Use soft brushes or cloths to scrub away loose corrosion products gently.

   Chemical Cleaning: Some standards (e.g., ASTM G1) recommend using chemical methods (such as immersing the sample in a mild acid solution) to remove adherent corrosion products. This step must be done carefully to avoid removing the base material.

7. **Final Weighing**

   Final Weight Measurement: After cleaning and drying the sample, weigh it again using the analytical balance. Record the final weight (W_final). The difference between the initial and final weights represents the mass loss due to corrosion.

8. **Data Analysis**

   Weight Loss Calculation: The weight loss is calculated as the difference between the initial and final weight of the sample:

$$\text{Weight Loss} = W_{\text{initial}} - W_{\text{final}}$$

Corrosion Rate Calculation: The rate of corrosion can be evaluated by the following formula, expressed in terms of millimeters per year (mm/y):

$$\text{Corrosion Rate} = \frac{K \times \Delta w}{A \times T \times D}$$

Where:

K is a constant (typically 8.76×10^4 for corrosion rate in mm/y)
ΔW is the weight loss (g)
A is the surface area of the specimen (cm^2)
T is the exposure time (hours)
D is the density of the alloy (g/cm^3)

9. **Visual and Microscopic Examination**

Surface Inspection: Examine the sample's surface for evidence of localized corrosion, including pitting, crevice corrosion, or uniform surface degradation. Take high-resolution photos to document the type and extent of corrosion.

Microscopy Analysis: Use optical or scanning electron microscopy (SEM) to analyze the corrosion morphology at a finer scale. This helps in understanding whether the corrosion is uniform, localized, or related to microstructural features of the HEA (e.g., phase boundaries, grain boundaries, or inclusions).

10. **Reporting the Results**

Corrosion Rate: Report the corrosion rate in terms of mm/y or mpy (mils per year) based on the weight loss and calculated data.

Corrosion Morphology: Include details of the type of corrosion observed, such as pitting, intergranular corrosion, or uniform corrosion.

Comparison with Standards: Compare the test results with established standards (e.g., ASTM standards) or the corrosion performance of reference materials such as stainless steel or other alloys.

9.3.4.3 Salt Spray Test

The salt spray test (also known as the salt fog test or neutral salt spray test) is a widely used accelerated corrosion test method that assesses the corrosion resistance of materials, coatings, or alloys such as HEAs. The test simulates harsh environmental conditions by exposing samples to a saline mist or fog in a controlled environment to observe their corrosion behavior over a short period. This test helps HEAs evaluate their ability to resist corrosion in marine-like environments.

Here is an in-depth explanation of the salt spray test procedure for HEAs:

1. **Purpose of the Salt Spray Test**

The main goal of the salt spray test is to accelerate corrosion processes that naturally occur over extended periods in real-world environments, especially those rich in chloride ions (like marine or coastal atmospheres). The test provides data about:

Coating performance (e.g., passivation layers, protective films)

Surface treatments (e.g., anodizing, galvanizing)

Material corrosion resistance (how quickly a material corrodes or resists attack)

HEA protective mechanisms (effectiveness of alloying elements in corrosion resistance)

2. **Test Standards**

Several standards govern the salt spray test, with the most commonly used being:

ASTM B117: The standard method for operating salt spray (fog) apparatus.
ISO 9227: Corrosion tests in artificial atmospheres — Salt spray tests.
JIS Z 2371: Japanese standard for salt spray testing.

HEAs often follow these standards to ensure reproducibility and consistency across tests.

3. **Preparation of the Sample**

Material Selection: Choose the HEA sample that you want to test. This can be a bulk material, a coated specimen, or a treated surface.

Sample Dimensions: Samples can be flat or curved and should fit within the test chamber without obstruction.

Surface Preparation: Clean the sample surface to remove contaminants that might affect the test outcome:

Mechanical Cleaning: If required, mechanically polish the surface to remove oxides or rough patches.

Chemical Cleaning: Use solvents like acetone or alcohol to remove oils, grease, and surface residues.

Coating/Pre-Treatment: If the HEA is coated (e.g., anodized, electroplated), the coating should be applied before testing.

4. **Preparation of the Salt Solution**

Salt Composition: Prepare a 5% sodium chloride (NaCl) solution by weight, dissolved in deionized or distilled water. The pH of the solution is adjusted to be neutral (around 6.5 to 7.2), as per ASTM B117 or ISO 9227 standards.

Solution Filtration: The salt solution is filtered to ensure that no impurities or solid particles interfere with the atomization process in the test chamber.

pH Control: Ensure the pH stays neutral throughout the test by regularly checking and adjusting it if necessary.

5. **Test Chamber Setup**

Salt Spray Cabinet: The samples are placed in a specialized salt spray chamber, which is designed to maintain consistent temperature, humidity, and spray conditions. The chamber typically consists of:

Atomizing Nozzle: This nozzle sprays the salt solution as a fine mist or fog inside the chamber.

Temperature Control: The chamber maintains a constant temperature of 35°C ± 2°C (95°F ± 3.6°F), a standard temperature for salt spray testing.

Salt Spray Concentration: The salt solution is atomized to create a continuous fog, and the spray rate is controlled to produce 1–2 mL of solution per hour per 80 cm^2 of surface area.

Sample Positioning: Samples are mounted at an angle (typically 15°–30° from the vertical) to allow uniform exposure to the spray without the pooling of solution on the sample surface.

6. **Conducting the Test**

Test Duration: The exposure duration depends on the material type and the desired corrosion level. For HEAs, which are often tested against conventional alloys (like stainless steel or aluminum alloys), the exposure time can range from 24 hours to several weeks, depending on the test objectives.

Short Tests: 24–96 hours for initial assessments.

Longer Tests: Up to 1000 hours or more for thorough corrosion analysis.

Constant Spraying: The chamber continuously sprays the salt solution to maintain a humid, saline atmosphere. The condensation and drainage of the saline fog simulate the corrosive effects of marine and coastal environments.

7. **Monitoring the Test**

Regular Observations: Inspect the samples periodically (if possible) for early signs of corrosion. Observations include surface rust, pitting, blistering, delamination, and other forms of degradation.

Salt Solution Monitoring: The concentration and pH of the salt solution are periodically measured to ensure they remain consistent with the testing parameters.

Temperature and Humidity Control: The temperature inside the chamber must be maintained at 35°C (±2°C), and the relative humidity should remain near 100%.

8. **Post-Test Procedure**

Sample Removal: At the end of the test, carefully remove the samples from the chamber without disturbing the corrosion products or surface features.

Rinsing: Some standards recommend rinsing the samples with deionized water to remove loose salt deposits without scrubbing or cleaning the corrosion products.

Drying: Allow the samples to air-dry or use a mild heat source to ensure the surface is dry for examination.

9. **Corrosion Evaluation**

Visual Inspection: Examine the samples under good lighting conditions to assess the degree of corrosion. Common signs include:

Rust Formation: For ferrous HEAs.

Pitting: Localized corrosion in small pits or cavities.

Blistering or Peeling: For coated surfaces.

Discoloration or Oxide Films may indicate surface passivation or protective oxide formation.

Photography: Document the results with high-resolution photographs, especially where pitting, cracking, or rust is present.

10. **Quantitative Measurements**

Mass Loss Measurement: Weigh the sample before and after the test to quantify the mass loss due to corrosion:

Initial Weight ($W_{initial}$): The weight of the cleaned sample before the test.

Final Weight (W_{final}): The weight of the sample after corrosion testing and removal of loose corrosion products (if applicable).

Mass Loss: The difference between the initial and final weights indicates material loss due to corrosion.

Pitting Depth: If pitting occurs, measure the depth of the pits using microscopy or profilometry.

Coating Integrity: For coated HEAs, measure the degree of coating breakdown, including blistering or peeling.

11. **Microscopic Examination**

Optical Microscopy: Examine the corrosion products, grain boundaries, and surface degradation under an optical microscope to gain more detailed insights into the corrosion mechanisms.

SEM/EDS Analysis: SEM and energy-dispersive X-ray spectroscopy (EDS) are used to analyze corrosion product composition and phase distributions. This can help determine which phases in the HEA contribute to corrosion resistance (or susceptibility).

Cross-Sectional Analysis: For an in-depth evaluation, sample cross-sections can be prepared and analyzed to understand the corrosion penetration and the integrity of protective layers.

12. **Corrosion Rate Calculation**

The corrosion rate can be calculated based on mass loss using the formulae mentioned in section 3.5.2.

13. **Reporting the Results**

Corrosion Rate: Report the corrosion rate in mm/year or mils per year (mpy).

Corrosion Morphology: Include detailed descriptions of the morphology (e.g., uniform corrosion, pitting, intergranular attack).

Coating Degradation: For coated HEAs, describe any blistering, peeling, or cracks in the coating.

Photographs: Attach photographs taken during and after the test to support the findings.

Comparison with Control Materials: If tested alongside standard materials like stainless steel, compare the corrosion behavior of HEAs with the control materials.

9.3.4.4 High-Temperature Corrosion Testing

High-temperature corrosion testing is a crucial method for evaluating the performance and durability of materials, including HEAs, in extreme environments where oxidation, sulfidation, and other corrosive processes occur at elevated temperatures. HEAs, characterized by their distinctive multi-principal-element compositions, frequently demonstrate enhanced high-temperature corrosion resistance relative to conventional alloys. This property has positioned them as a key area of research for high-temperature applications, including gas turbines, aerospace components, and power plants.

A comprehensive explanation of high-temperature corrosion testing for HEAs, including the test methods, procedures, and the evaluation of corrosion mechanisms, is mentioned below.

1. **Purpose of High-Temperature Corrosion Testing**

   The main objective of high-temperature corrosion testing is to assess the resistance of HEAs when exposed to oxidizing, sulfidizing, carburizing, or other corrosive environments at elevated temperatures. This testing simulates the real-world conditions that materials may face in high-temperature industrial applications, providing insight into:

   Oxidation resistance (formation of protective oxide layers)
   Sulfidation resistance (protection against sulfur-containing atmospheres)
   Hot corrosion resistance (resistance to combined oxidation and salt attack, such as Na_2SO_4)
   Material degradation mechanisms (phase transformation, interdiffusion, etc.)
   Protective coating performance (if the HEA is coated)

2. **Test Standards**

   High-temperature corrosion testing typically follows recognized standards to ensure repeatability and comparability of results. Common standards include:

   ASTM G111: Standard guide for corrosion testing in high-temperature environments.
   ASTM G54: Standard practice for static oxidation tests in dry air at high temperatures.
   ASTM G116: Guide for conducting wire-on-ring wear tests for high-temperature environments.
   ISO 17245: Test methods for high-temperature corrosion testing in aggressive environments.

   These standards guide sample preparation, test duration, temperature control, and environmental conditions.

3. **Types of High-Temperature Corrosion Testing**

   High-temperature corrosion testing can be classified into critical methods based on the corrosive environment. The specific method chosen depends on the operational conditions in which the HEA will be used.

   a. **Oxidation Testing**

      Purpose: Assess the development of oxide scales on the alloy's surface when subjected to high temperatures in an oxygen-rich environment (air, O_2).
      Temperature Range: Typically between 873 K and 1473 K.
      Test Gases: Dry air, pure oxygen, or controlled oxygen partial pressure.
      Mechanism: As temperature increases, HEAs undergo oxidation where elements like Cr, Al, or Ti form protective oxide layers (e.g., Cr_2O_3, Al_2O_3) that prevent further oxidation.
      Procedure: The sample is heated to a predetermined temperature, held for a fixed time, and then cooled. Weight change is measured to assess the extent of oxidation (gain in mass due to oxide formation).

b. **Sulfidation Testing**

Purpose: Determines the resistance of HEAs to sulfur-containing atmospheres, such as in coal-fired power plants or petrochemical industries.

Test Environment: The sample is exposed to a sulfur-rich gas (e.g., SO_2, H_2S) or liquid environment.

Mechanism: Sulfidation occurs when sulfur reacts with alloying elements to form sulfides, typically less protective than oxides.

Temperature Range: Between 773 K and 1273 K.

Evaluation: Post-test analysis focuses on forming surface sulfides, pitting, and cracking due to sulfidation.

c. **Hot Corrosion Testing**

Purpose: Simulates aggressive environments such as those found in gas turbines, where molten salts like sodium sulfate (Na_2SO_4) react with high-temperature components.

Test Environment: A molten salt layer (e.g., Na_2SO_4, V_2O_5) is deposited on the sample's surface. The sample is then exposed to an oxidizing atmosphere (often containing O_2, SO_2, or steam) at high temperatures.

Mechanism: The salts melt and penetrate the oxide scale, causing accelerated oxidation, sulfidation, and potential spallation of protective layers.

Temperature Range: Between 973 K and 1273 K.

Procedure: The sample is coated with salt and heated to a high temperature, and the degradation is monitored over time.

4. **Preparation of the Sample**

Material Selection: Choose the HEA sample, which can be in bulk form or coated, depending on the study. The choice of HEA is based on its application, such as Fe-Co-Cr-Ni-based HEAs for oxidation resistance.

Sample Dimensions: Common shapes include small plates, rods, or disks. Sample size and geometry must be compatible with the test equipment.

Surface Preparation: Surface roughness can affect corrosion results. Samples are polished to a smooth finish to ensure consistency. Cleaning is performed using solvents like acetone or alcohol to remove surface contaminants.

5. **Test Procedure**

a. **Isothermal Oxidation Test Procedure**

Pre-Test Weighing: Weigh the prepared sample to record the initial mass ($W_{initial}$).

Heating: Place the sample in a furnace or high-temperature reactor. Heat the sample to the target temperature (e.g., 1173 K) in a controlled environment (air, O_2, or mixed gases).

Exposure Time: Hold the sample at the set temperature for a specific duration, typically 1–100 hours or more, depending on the test.

Cooling: After exposure, cool the sample either in-situ or remove it for rapid cooling.

Post-Test Weighing: After cooling, weigh the sample again to determine the final mass (W_{final}).

Mass Gain Measurement: The increase in weight due to oxide formation is calculated using:

$$\text{Weight Loss} = W_{initial} - W_{final}$$

Corrosion Rate Calculation: The corrosion rate can be expressed in weight change per unit area (mg/cm^2).

b. **Cyclic Oxidation Testing**

Procedure: In cyclic oxidation, samples undergo repeated heating and cooling cycles (e.g., 1-hour heating at 1273 K followed by 15-minute cooling). This simulates real-world applications where materials experience temperature fluctuations.

Objective: This test evaluates the ability of the oxide layer to adhere to the substrate under thermal stresses.

6. **Post-Test Analysis**

Weight Change Measurements: The change in weight due to oxidation or other reactions gives a direct measure of material degradation. A weight gain typically indicates the formation of protective oxides, while a loss may suggest spallation or volatile oxide formation.

X-Ray Diffraction (XRD): Identifies the crystalline phases of oxides or other corrosion products that form on the surface.

Scanning Electron Microscopy (SEM): Used to examine the microstructure of the corroded surface, showing oxide layer thickness, cracks, spallation, and grain boundary oxidation.

Energy-Dispersive X-Ray Spectroscopy (EDS): Determines the elemental composition of corrosion products, helping to understand the formation of protective scales (e.g., Cr_2O_3, Al_2O_3).

Optical Microscopy: Provides an overview of the surface morphology and scale adherence.

Cross-Sectional Analysis: Provides insight into the depth of oxide penetration, revealing protective layer integrity or internal corrosion.

7. **High-Temperature Corrosion Mechanisms in HEAs**

Protective Oxide Formation: HEAs often rely on forming stable, protective oxides such as Cr_2O_3 or Al_2O_3. For example, Cr-rich HEAs form dense, continuous oxide scales that limit further oxygen diffusion and corrosion.

Lattice Distortion Effect: The multi-principal element nature of HEAs results in lattice distortion, which may hinder the diffusion of corrosive species (oxygen, sulfur) and enhance the formation of stable oxide layers.

Elemental Segregation: At high temperatures, some elements in HEAs may segregate to grain boundaries, forming more corrosion-resistant phases (e.g., Laves phases or intermetallics).

Phase Stability: Depending on their composition, HEAs can exhibit different phases (FCC, BCC, etc.). The stability of these phases at high temperatures impacts corrosion resistance.

8. **Data Interpretation and Reporting**
   Oxidation Kinetics: Plot the weight change vs. time to observe oxidation kinetics. The rate of oxidation typically follows parabolic or logarithmic behavior in protective oxides.
   Oxide Scale Analysis: Report oxide scales' thickness, composition, and adherence.
   Corrosion Mechanism: Based on the microstructural and chemical analysis, determine whether corrosion is dominated by oxidation, sulfidation, or other mechanisms.
   Comparative Performance: For HEAs, compare the corrosion performance against conventional high-temperature alloys like Inconel, stainless steel, or superalloys.

9.3.5 Applications of Corrosion-Resistant HEA

Corrosion-resistant HEAs have attracted significant interest due to their exceptional resistance to various corrosive environments. Combining multiple principal elements in HEAs creates unique properties like high strength, toughness, and superior corrosion resistance, making them suitable for demanding applications in various industries. Below are the key areas where corrosion-resistant HEAs find wide applications:

1. **Aerospace Industry**
   High-Temperature Corrosion Resistance: HEAs are used in aerospace components like turbine blades, exhaust nozzles, and combustion chamber liners where high-temperature oxidation and hot corrosion are significant concerns. The ability of HEAs to withstand high-temperature environments and resist degradation from oxygen, sulfur, and molten salts makes them ideal for use in jet engines and rocket propulsion systems.
   Lightweight Components: HEAs are also being investigated as lightweight materials with improved corrosion resistance for aircraft structures exposed to extreme environments at high altitudes.
2. **Power Generation and Energy Sector**
   Gas Turbines: In power plants, HEAs can be used in gas turbine components such as blades, vanes, and heat shields, which operate in high-temperature environments and are exposed to corrosive gases like sulfur oxides and vanadium pentoxide.
   Nuclear Reactors: HEAs with radiation and high corrosion resistance are being explored in nuclear reactor components, especially those that come into contact with coolants like molten salts, water, or liquid metals. HEAs can replace conventional materials from irradiation-induced damage and corrosion in nuclear reactors.
   Fossil Fuel Plants: HEAs offer better resistance to high-temperature oxidation, sulfidation, and carburization, making them suitable for superheater tubes, steam generators, and other components in coal-fired power plants.

Fuel Cells: Corrosion-resistant HEAs can be used in fuel cells and electrolyzers, where exposure to harsh chemical environments and high temperatures can degrade conventional materials.

3. **Marine and Offshore Industries**

   Corrosion from Saltwater: HEAs exhibit excellent resistance to pitting, crevice, and general corrosion in marine environments, making them ideal for naval vessels, offshore oil rigs, and underwater structures. HEAs are particularly useful for components exposed to saltwater, such as ship hulls, propellers, and marine fasteners.

   Anti-Biofouling: HEAs can also resist biofouling (accumulation of microorganisms on metal surfaces) in oceanic environments, reducing marine infrastructure maintenance costs.

4. **Chemical and Petrochemical Industries**

   Corrosive Chemical Environments: HEAs offer superior corrosion resistance to acids (such as sulfuric and hydrochloric acid), alkalis, and other aggressive chemicals, making them suitable for chemical reactors, storage tanks, pipelines, and heat exchangers. HEAs are also resistant to intergranular corrosion and stress corrosion cracking.

   Catalyst Materials: Some HEAs exhibit catalytic properties that make them effective in chemical reactions, such as in producing ammonia, hydrogenation reactions, and catalytic converters.

5. **Oil and Gas Industry**

   Sour Gas (H_2S) Resistance: HEAs demonstrate excellent corrosion resistance in environments containing hydrogen sulfide (H_2S), making them suitable for sour gas applications. They can be used in oil and gas exploration equipment such as pipelines, valves, and drilling tools, where traditional alloys suffer from sulfide stress cracking.

   Downhole Components: HEAs are also being developed for downhole applications, such as drill bits and tubing used in oil and gas extraction, where the material is subjected to high temperatures and corrosive fluids.

6. **Automotive Industry**

   Exhaust Systems: HEAs with good resistance to high-temperature oxidation can be used in exhaust manifolds, catalytic converters, and turbochargers, which are exposed to high temperatures and corrosive gases from combustion.

   Corrosion-Resistant Structural Components: In electric vehicles, HEAs can be applied to components such as battery enclosures, electric motor housings, and cooling systems, where resistance to corrosion and mechanical wear is critical for durability.

7. **Biomedical Applications**

   Implants and Prosthetics: HEAs that demonstrate exceptional corrosion resistance, biocompatibility, and high strength are promising options for biomedical implants, including joint replacements (hip and knee implants) and dental implants. HEAs reduce the risk of metal ion release and wear-related issues in traditional implant materials.

Surgical Tools: HEAs' corrosion resistance and high durability make them ideal for surgical tools and equipment that are frequently sterilized and exposed to bodily fluids.

8. **Additive Manufacturing**

   3D-Printed Components: Corrosion-resistant HEAs are used in additive manufacturing (3D printing) to produce complex geometries and customized parts for various applications, including aerospace, biomedical, and marine industries. Their resistance to oxidation and corrosion makes them suitable for fabricating components exposed to extreme environments.

9. **Electronics Industry**

   Corrosion-Resistant Coatings: HEAs are being explored as protective coatings for electronic components that operate in harsh environments, such as semiconductor devices and sensors in aerospace, military, and industrial applications. HEAs prevent degradation from humidity, salt, and temperature variations.

   Conductive Materials: Some HEAs exhibit good electrical conductivity and corrosion resistance, making them suitable for use in connectors, wiring, and other components where both properties are critical.

10. **Construction and Infrastructure**

    Corrosion-Resistant Steel: HEAs can be used to improve the corrosion resistance of structural steel in bridges, buildings, and other infrastructure, especially in regions with harsh environmental conditions, such as coastal areas or industrial zones.

    Coatings for Reinforced Concrete: HEAs can be applied as protective coatings for rebar in reinforced concrete structures to prevent corrosion from moisture and salt ingress, thus prolonging the life span of infrastructure.

11. **Defence and Military Applications**

    Armors and Protective Coatings: HEAs with high corrosion and wear resistance are being considered for use in military vehicles, aircraft, and naval vessels. They can also be applied as protective coatings to prevent corrosion in harsh field conditions.

    High-Temperature Components: In missile and defense systems, HEAs can be used for components that operate at extreme temperatures and require resistance to oxidation and corrosion.

12. **Renewable Energy**

    Wind Turbines: HEAs with corrosion resistance and mechanical strength can be used for wind turbine blades, offshore wind platforms, and structural components where exposure to saltwater and atmospheric corrosion is a concern.

    Geothermal Energy: HEAs can be applied in geothermal energy systems where components are exposed to corrosive fluids, such as brine, at high temperatures.

13. **Mining and Metallurgy**

    Mining Equipment: HEAs can be used for equipment and tools operating in abrasive and corrosive environments. This includes pumps, valves, and drilling tools exposed to acidic and sulfur-containing environments.

14. **Food and Beverage Industry**
Processing Equipment: HEAs can be applied in food processing equipment that requires high corrosion resistance to acids, salts, and cleaning chemicals. They ensure long-lasting performance in environments that require frequent sanitization and exposure to corrosive substances like vinegar or brine.

9.4 RADIATION RESISTANCE OF HEAs

The radiation resistance of HEAs is a fascinating area of research due to the unique composition and properties of these materials, which enhance performance in extreme environments, including those involving high radiation levels. Radiation resistance refers to a material's ability to withstand damage caused by radiation exposure, typically involving ionizing radiation such as neutrons, ions, or electrons. This resistance is crucial for materials used in nuclear reactors, space exploration, or any environment where radiation exposure can degrade material properties over time.

9.4.1 Radiation Resistance Mechanism in HEA

9.4.1.1 Atomic Arrangement

Contrary to pure metals and binary alloys, where uniform neighboring atoms generally surround atoms, HEAs are adjacent to surrounding atoms of dissimilar sizes, as illustrated in Figure 9.1. This irregular atomic environment is expected to significantly hinder the diffusion of vacancies or atoms, unlike the relatively straightforward diffusion seen in pure metals and binary alloys [244]. The variation in surrounding atoms causes each atomic site in an HEA to have different lattice potential energy. As a result, in an irradiated HEA, atoms do not undergo direct displacement but rather encounter fluctuating lattice potential energy along their diffusion paths, as illustrated in Figure 9.2, where circles indicate deep traps. More significant fluctuations in potential energy lead to higher diffusion barriers, believed to decrease irradiation-induced defects in HEAs.

Likewise, Lu et al. [245] performed molecular dynamics simulations and experimental investigations, indicating that the substantial site-to-site lattice distortions and compositional complexities in bulk HEAs adjust the diffusion pathways of irradiation-induced defects, including network dislocations, dislocation loops, and

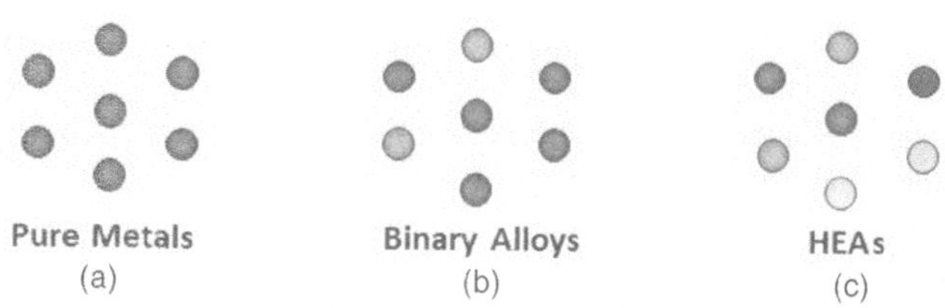

FIGURE 9.1 A comparison of the atomic neighbors surrounding an atom in different material systems: (a) pure metals, (b) binary alloys, and (c) HEAs.

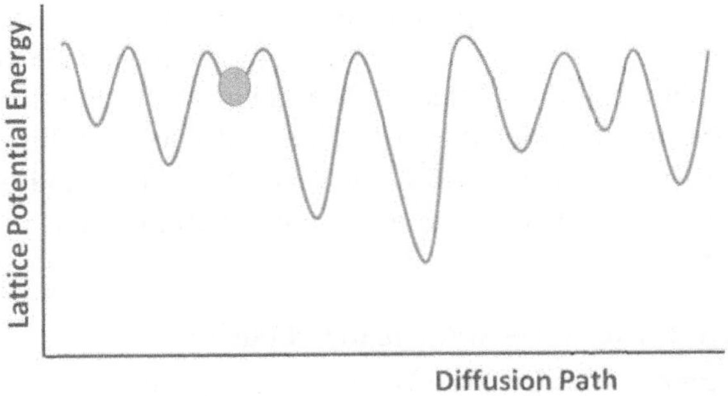

FIGURE 9.2 Diagram illustrating the variation of lattice potential energy along the diffusion pathway of an atom within the lattice.

vacancies. Their findings indicate that small clusters of self-interstitial atoms in pure metals and conventional alloys can migrate over long distances in a 1D manner without encountering many vacancies, leading to high vacancy super-saturation and substantial void swelling. In contrast, in HEAs, defect clusters migrate randomly in 3D, suppressing rapid interstitial movement and reducing void swelling. As a result, HEAs exhibit improved resistance to irradiation damage and void swelling.

9.4.1.2 Compositional Complexity

Several researchers have proposed that the complex elemental makeup of HEAs plays a crucial role in enhancing their resistance to irradiation. A prevalent strategy involves incrementally adding alloying elements and observing the consequent changes in radiation damage to investigate the relationship between compositional complexity and radiation resistance. This approach has been widely used to examine the Cantor alloy's irradiation behavior and various nickel-containing single-phase alloys. For example, research on various nickel-containing equiatomic alloys supports this theory. Using both experimental methods and simulations, Granberg et al. [246] assessed the radiation resistance of pure Ni, Ni-Fe, and Ni-Co-Cr bulk alloys, concluding that alloying reduces dislocation movement and mobility, thereby alleviating irradiation-induced damage. In addition, Zhang et al. [247] investigated the impact of compositional complexity on bulk alloys, specifically Ni, Ni-Fe, Ni-Co, and Ni-Co-Fe-Cr. They noted a noteworthy reduction in the step height of the alloys as compositional complexity increased. Moreover, the volume swelling in bulk Ni-Co-Fe-Cr-Mn was lower than in pure Ni [248].

9.4.1.3 Atomic-Scale Pressure

Another theory proposed by Egami et al. [249] focuses on the atomic-level pressure exerted on the atoms within bulk HEAs. Through first-principles calculations using density functional theory, they demonstrate that substantial atomic-level pressures exist in HEAs, resulting not only from variations in intrinsic atomic sizes but also from charge transfer among the elements, which reflects differences in electronegativity. For instance, the atomic-level pressures in the Co-Fe-Ni alloy are measured at −2.11 GPa for Co, −15.07 GPa for Fe, and 18.75 GPa for Ni. Similarly, in the

Co-Cr-Fe-Ni alloys, the atomic-level pressures are 9.01 GPa for Co, –37.55 GPa for Cr, –5.74 GPa for Fe, and 35.16 GPa for Ni. Positive values indicate compression, while negative values represent tension. The atomic-level pressures in HEAs are greater than those in conventional alloys. The local volume strain associated with these pressures is approximately ±0.1 for Co-Fe-Ni and around ±0.2 for Co-Cr-Fe-Ni. This atomic-level pressure promotes the amorphization of the alloy when exposed to particle irradiation.

9.4.1.4 Defective Vacancies and Localized Fusion

When high-energy particles irradiate HEAs, the atoms of the constituent elements are displaced from their equilibrium positions, forming different forms of vacancy-interstitial defects or even amorphization [250]. Concurrently, these high-energy particles impart substantial heat energy, causing localized thermal spikes. This thermal influx triggers localized melting and rapid crystallization processes. The interaction of these phenomena generates considerable atomic stress from forming multiple Frenkel pairs (interstitials and vacancy defects) within the HEA, which obstructs dislocation movement and consequently improves the material's irradiation resistance [251]. In a related study, Wang et al. [252] performed a thermodynamic analysis, which demonstrated that the equilibrium vacancy concentrations and their clusters in HEAs are significantly higher than metals and simple binary alloys. This increased concentration of equilibrium vacancies can efficiently accommodate the interstitial atoms generated by radiation damage. Their results show that the equilibrium vacancy concentration in HEAs is significantly greater than in pure elements, highlighting a strong tendency for the recombination of interstitials and vacancies.

Experimental findings on sputtered Zr-Hf-Nb HEA coatings demonstrate that BCC phase stability is maintained after irradiation with 10 displacements per atom (dpa) at 298 K. Furthermore, the structural changes induced by irradiation are influenced by both the irradiation temperature and dosage [253]. Additional studies on bulk Co-Cu-Cr-Fe-Ni indicate that the FCC structure exhibits remarkable phase stability when subjected to fast electron irradiation over a broad temperature range from 298 to 773 K, up to 40 dpa. Similarly, Al_x-Co-Cr-Fe-Ni bulk HEAs have been observed to experience less volume swelling compared to conventional nuclear materials after being irradiated with 3 MeV Au ions [254].

9.4.2 Radiation Damage in HEAs

9.4.2.1 Radiation-Induced Defects in HEAs

Radiation leads to defect clusters, such as dislocation loops, in HEAs. The exceptional properties of HEAs contribute to the restriction of defect mobility during irradiation, effectively preventing the migration and diffusion of these defects. Research has demonstrated that the Fe-Ni-Mn-Cr bulk HEA displayed smaller defect clusters with a higher density than the Fe-Cr-Ni ternary alloy after being irradiated at temperatures between 673 and 973 K [255]. The phase structure significantly influences the irradiation resistance of high entropy materials (HEMs). Density functional theory (DFT) calculations on V-Ta-Cr-W indicate that the more significant lattice distortion found in BCC HEAs may lead to enhanced irradiation resistance compared to FCC HEAs [256]. In a study involving Al_x-Co-Cr-Fe-Ni (x = 0.1, 0.75, 1.5) bulk HEAs, the samples with disordered BCC and FCC phases showed a reduced number of defect clusters following irradiation with 3 MeV Au^+ ions compared to those with an ordered B2-phase structure [257]. The stacking fault energy (SFE) influences the type and morphology of radiation-induced defects, as lower SFE suppresses the formation of larger dislocation loops in HEAs. Lu et al. [258] irradiated bulk Ni-Fe, Ni-Co-Fe, Ni-Co-Fe-Cr, and Ni-Co-Fe-Cr-Mn single-phase alloys with 3 MeV Ni^+ ions at 773 K. In their study, the Ni-Co-Fe-Cr-Mn alloy exhibited smaller and denser radiation-induced dislocation loops than the binary Ni-Fe alloy. Additionally, increasing the compositional complexity postponed the growth of dislocation loops. Yang et al. [259] observed that the increased lattice distortion in bulk Ni-Co-Fe-Cr-Pd HEA contributed to delayed dislocation growth during irradiation. Additionally, the low stacking fault energy in Fe-Co-Ni-Cr-$Ti_{0.2}$ bulk HEAs inhibited the transformation of faulted dislocation loops into perfect dislocation loops, a feature that was rarely observed in non-irradiated HEAs [260]. Temperature is a critical factor affecting the formation of radiation-induced dislocation loops, as different mechanisms of radiation-induced segregation occur across various temperature ranges. Chen et al. [261] proposed that at low irradiation temperatures, the high entropy effect and sluggish diffusion in $Al_{0.3}$-Co-Cr-Fe-Ni and Co-Cr-Fe-Mn-Ni bulk HEAs had a limited impact on the development of dislocation loops.

Each element in HEAs plays a distinct role in radiation-induced damage. He et al. [262] examined element segregation in Cr-Fe-Co-Ni, Cr-Fe-Co-Ni-Mn, and Cr-Fe-Co-Ni-Pd bulk HEAs under 1250 kV electron irradiation at 673 K, reaching doses of up to 1 dpa. Their findings indicated that Cr, Fe, Mn, and Pd were prone to depletion, while Co and Ni tended to accumulate at the radiation-induced dislocation loops. The first study on bulk high entropy ceramics was conducted by Wang et al. [263], who explored the microstructural response of ($Zr_{0.25}$-$Ta_{0.25}$-$Nb_{0.25}$-$Ti_{0.25}$)C HEC under irradiation with 3 MeV Zr^{2+} ions (20 dpa) at temperatures of 298 K, 573 K, and 773 K. The lattice parameter increased by 0.22% after irradiation at 298 K and by 0.19% at 773 K, with dislocation loops being the primary irradiation-induced microstructures. The high entropy effect, along with the compositional complexity and strong local lattice distortion from Zr, Ta, Nb, and Ti atoms occupying the anion positions, effectively suppressed the growth of these dislocation loops.

9.4.2.2 Radiation-Induced Cavities in HEAs

The creation of cavities within the microstructure of materials is an inevitable consequence of atomic displacement during irradiation. As atoms are moved from their original sites, these cavities accumulate due to the diffusion of displaced atoms, leading to macroscopic issues like helium bubble formation and volume swelling. The sluggish diffusion effect is often cited as a critical factor in reducing radiation damage in high entropy materials (HEMs). For example, bulk Alx-Co-Cr-Fe-Ni and Hf-Nb-Ta-Ti-Zr alloys [264] exhibit significantly less volume swelling than conventional alloys when subjected to similar irradiation conditions. Furthermore, the Ni-Co-Fe-Cr-Mn bulk HEA shows a swelling resistance 40 times greater than pure nickel, with iron demonstrating a more pronounced suppressive effect on swelling than chromium and cobalt in nickel-based alloys. Yang et al. [259] developed Ni-Co-Fe-Cr-Pd bulk HEAs that exhibit improved swelling resistance compared to their Ni-Co-Fe-Cr-Mn counterparts, attributing this enhancement to lattice distortion and an increased barrier for defect migration.

When metallic materials are exposed to neutron irradiation, transmutation produces He+ ions, which can lead to the accumulation of vacancies and the formation of helium bubbles. For instance, after irradiation with 275 keV He+ ions at 673 K, the Fe-Co-Ni-Cr-$Ti_{0.2}$ bulk HEAs develop helium bubbles in the peak damage zone [260]. The slow diffusion rate of vacancies in HEAs is crucial in mitigating radiation damage from He+ ions. For example, the density of helium bubbles in the Ti_2-Zr-Hf-$V_{0.5}$-$Mo_{0.2}$ bulk HEA is lower than that in traditional alloys under similar irradiation conditions [265]. After He ion irradiation at 723 K, the average size of helium bubbles in the Cr-Mn-Fe-Co-Ni bulk HEA is smaller than that in pure nickel and 304 stainless steel. Under exposure to 6 keV He+ ions, the growth of helium bubbles in Fe-Cr-Mn-Ni HEAs is slower than in AISI 348 steel, which can be attributed to the sluggish diffusion effect in bulk HEAs [266]. Chen et al. [267] noted that the volume fraction of bubbles in the Fe-Co-Ni-Cr bulk HEA was lower than that of pure nickel under identical irradiation conditions. They recommended that the sluggish diffusion effect of He+ ions is only practical at elevated temperatures for Fe-Co-Ni-Cr bulk HEAs, with helium diffusion affected by the specific energetics of point defects [268]. A similar suppression of He+ ion diffusion has been observed in HEA and high entropy ceramic (HEC) films.

Pu et al. [269] conducted experiments on nano-crystalline $Al_{1.5}$-Co-Cr-Fe-Ni HEA films irradiated at 300 K with 60 keV He+ ions and found no evidence of helium bubble formation, even when the He concentration reached 8.50 at. %. Similarly, no helium bubbles were detected in Al-Cr-Mo-Nb-Zr/(Al-Cr-Mo-Nb-Zr)N multilayer coatings after exposure to 400 keV He+ ions [270]. HEC coatings have demonstrated structural and chemical stability even under high irradiation doses. For instance, Komarov et al. [271] reported that (Ti, Hf, Zr, V, Nb)N coatings maintained structural and phase stability when irradiated with 500 keV helium ions. Since irradiation resistance is closely linked to thermal effects influenced by thermal conductivity and material thickness, differences in behavior between coatings/films and bulk HEAs are expected [272].

9.4.3 Radiation-Induced Changes in the Mechanical Characteristics of HEA

While irradiation can lead to thermal annealing, it produces radiation-induced dislocation loops, voids, defect clusters, and precipitates, contributing to material hardening and embrittlement. However, HEAs typically show less fluctuation in mechanical properties than traditional materials due to their enhanced ability to tolerate defect formation. Moreover, mechanisms like self-healing and defect recombination in HEAs can mitigate increases in hardness following irradiation. For instance, the nano-indentation hardness of bulk HEAs such as Ti_2-Zr-Hf-$V_{0.5}$-$Mo_{0.2}$ and Hf-Nb-Ta-Ti-Zr exhibited minimal alterations after exposure to irradiation. Additionally, the hardness of as-deposited W_{38}-Ta_{36}-Cr_{15}-V_{11} HEA films experienced a slight increase following thermal annealing and irradiation with 1 MeV Kr^{+2} ions at doses up to 1.6 dpa at 1073 K [273]. The bulk HEA $V_{2.5}$-$Cr_{1.2}$-W-Mo-$Co_{0.04}$ demonstrated remarkable resistance to hardening and embrittlement when subjected to 5 MeV Au^{+} ion irradiation at doses up to 42 dpa [274]. Li et al. [275] investigated the effects of neutron irradiation on Ni-Fe-Mn-Cr bulk HEA at 333 K up to 0.1 and 1 dpa, noting mechanical property changes comparable to those observed in austenitic stainless steels. The hardness increase in $Al_{0.3}$-Co-Cr-Fe-Ni and Co-Cr-Mn-Fe-Ni bulk HEAs was somewhat more significant than that of 316H stainless steel after irradiation with 1 MeV krypton ions at 573 K [276]. These results indicate that the mechanical behavior of high entropy materials under irradiation is influenced not only by sluggish diffusion and high configurational entropy but also by additional factors, such as the formation of dislocation loops over short distances, akin to behaviors seen in stainless steel. Moschetti et al. [277,278] examined the room-temperature mechanical properties of Ti-Zr-Nb-Hf-Ta RHEAs after ion irradiation, demonstrating their resistance to helium irradiation damage. The study found that the strength and ductility of Ti-Zr-Nb-Hf-Ta RHEAs remained well-preserved post-irradiation. Specifically, the hardness increased by only 13.9%, while elongation improved by 4.5%, indicating strong resistance to helium-induced hardening and embrittlement. Nanoscale helium bubbles were observed within the grain boundaries, which activated a grain boundary gliding mechanism, thereby preserving ductility. These helium bubbles are thought to weaken the grain boundaries, allowing for movement or sliding under stress.

9.4.4 Experimental Techniques for Evaluating Radiation Resistance of HEA

Evaluating the radiation resistance of HEAs requires specialized experimental techniques that help understand how these materials perform under irradiation. The essential experimental methods for assessing the radiation resistance of HEAs are mentioned below.

9.4.4.1 Ion Irradiation Experiments

Ion irradiation is widely used to simulate radiation damage because it allows precise control over irradiation conditions such as dose, dose rate, and temperature. Heavy ions (e.g., Fe, Au, or Xe) are often used to simulate the effects of neutron or alpha-particle radiation faster and more controlled than traditional reactor-based neutron irradiation. Key features include:

Controlled Dose: The ion beam can be tuned to deliver specific doses to replicate different levels of radiation exposure.

Surface vs. Bulk Effects: Ion irradiation typically affects only the near-surface region of materials, so results may not fully represent bulk behavior.

Advantages: The primary advantage of ion irradiation is that it allows rapid simulation of radiation damage without requiring access to nuclear reactors.

Techniques used for analyzing irradiated samples:

Transmission Electron Microscopy (TEM): Used to study microstructural changes like dislocation loops, voids, and precipitates. TEM helps in quantifying defect densities.

Scanning Electron Microscopy (SEM): Provides detailed imaging of the irradiated surfaces, showing morphological changes like blistering, void formation, and swelling.

Focused Ion Beam (FIB): Often coupled with SEM or TEM to create site-specific thin samples for further analysis, allowing for precise sectioning of irradiated areas.

9.4.4.2 Neutron Irradiation

Neutron irradiation provides a more realistic environment for evaluating radiation resistance, especially for nuclear applications where materials are exposed to fast neutron flux. The experimental setup typically involves:

In-Reactor Testing: HEA samples are exposed to neutron flux in test reactors over a specific period. These experiments are expensive and time-consuming but crucial for realistic damage assessments.

Displacement per Atom (DPA): A standard metric used to quantify damage in neutron irradiation experiments. DPA estimates the number of atomic displacements caused by radiation.

Post-Irradiation Examination (PIE): After neutron irradiation, materials are analyzed for radiation-induced changes, such as swelling, phase stability, and mechanical property degradation.

9.4.4.3 Mechanical Testing of Irradiated Samples

Post-irradiation mechanical testing provides critical data about how irradiation affects HEAs' strength, ductility, and toughness.

Nano-Indentation: This technique measures hardness and elastic modulus changes in irradiated materials at a micro- or nanoscale. It helps assess localized radiation damage and mechanical property changes in small-scale regions of the material.

Tensile Testing: Mechanical tensile tests are performed to measure the stress-strain behavior of irradiated HEAs. These tests show how radiation affects yield strength, ultimate tensile strength, and elongation.

Fracture Toughness Testing: Fracture toughness experiments provide insights into how irradiation impacts the material's resistance to crack propagation. Radiation-induced embrittlement is a common concern in materials subjected to high-dose irradiation.

9.4.4.4 Positron Annihilation Spectroscopy (PAS)

PAS is used to detect vacancies and small clusters of defects, which can be difficult to observe using other methods. In irradiated materials, vacancy-type defects form large numbers, and PAS provides a sensitive method for identifying and quantifying these defects.

Slow Positron Beam: A slow positron beam can probe different depths in the material, giving information about defect distributions near the surface or deeper into the bulk.

Positron Lifetime Measurement: This technique helps estimate the size and concentration of defect clusters by measuring the time between positron injection and annihilation.

9.4.4.5 X-Ray Diffraction (XRD)

XRD is a non-destructive technique that evaluates changes in the crystal structure of HEAs due to radiation exposure. Critical phenomena that can be detected include:

Lattice Parameter Changes: Radiation can cause lattice expansion or contraction due to defects like interstitials and vacancies.

Phase Transformation: Some HEAs might undergo phase changes under irradiation, and XRD is sensitive to such structural modifications.

In-situ XRD during irradiation is another variant, where materials are analyzed in real time, giving insights into the radiation-induced phase transformations.

9.4.4.6 Atom Probe Tomography (APT)

APT provides atomic-scale compositional analysis, allowing researchers to observe radiation-induced segregation of elements or clustering. This is particularly useful in HEAs because they are multi-element systems, and radiation can cause preferential migration of specific components.

Advantages of APT:

i. Provides 3D atom-by-atom reconstruction of the irradiated material.
ii. It allows the detection of nanoclusters and solute segregation, which is common in irradiated alloys.

9.4.4.7 Transmission Electron Microscopy (TEM)

TEM extensively studies the radiation-induced defects at the nanoscale in HEAs. Key observations include:

Dislocation Loops: TEM allows researchers to directly visualize dislocation loops created when atoms are displaced from their original lattice sites.

Voids and Helium Bubbles: Radiation often leads to the forming of voids or bubbles (particularly in helium-irradiated samples). TEM provides direct imaging of these features and their impact on material properties.

9.4.4.8 Thermal Desorption Spectroscopy (TDS)

TDS is beneficial for studying how gases like helium or hydrogen, introduced during radiation, are trapped in and released from the alloy. This is critical for HEAs exposed to fusion or fission environments because gas retention can lead to swelling or embrittlement.

Working principle: The irradiated material is heated in a vacuum chamber, and the desorbed gases are analyzed to understand how radiation affects gas absorption and retention in the alloy.

9.4.4.9 Small-Angle Neutron Scattering (SANS)

SANS is used to analyze the formation of radiation-induced defects such as voids, precipitates, or clusters, especially at the nanometer scale. It is a complementary technique to TEM and provides information about defect sizes, shapes, and volume fractions.

Applications in HEAs:

It helps understand how HEAs resist radiation-induced void swelling, a primary concern in nuclear materials.

9.4.4.10 Differential Scanning Calorimetry (DSC)

DSC is used to analyze the thermal stability of HEAs after irradiation. It can detect phase transitions and changes in heat capacity due to radiation-induced structural modifications.

9.4.4.11 Nano-Indentation in Radiation Resistance Studies

For HEAs, nano-indentation is a powerful tool to assess how radiation affects the material's mechanical properties. In radiation resistance studies, nano-indentation helps in:

1. *Evaluating Radiation-Induced Hardening*: After radiation exposure, materials tend to harden due to the formation of defect clusters like dislocation loops or vacancy clusters. Nano-indentation is sensitive to these changes, showing increased hardness in the irradiated zones.
2. *Studying Radiation-Induced Softening or Annealing Effects*: Some materials may soften upon irradiation due to microstructural changes (e.g., grain boundary mobility or defect recombination). Nano-indentation can detect such changes by measuring the reduced hardness in irradiated regions.
3. *Depth Profiling*: Since ion irradiation only affects a surface layer (typically in the micrometer range), nano-indentation can perform a depth-resolved study to map how mechanical properties vary with depth. This is crucial for understanding the distribution of radiation-induced defects.
4. *Comparing Irradiated and Non-Irradiated Zones*: By measuring irradiated and non-irradiated regions, nano-indentation directly compares how radiation alters the mechanical behavior of HEAs.

Experimental Setup and Procedure

1. *Sample Preparation*: Samples are usually polished to ensure a smooth surface, as roughness can affect the precision of nano-indentation measurements.
2. *Indentation Process*: The indenter is pressed into the sample with an increasing load, and the indentation depth is recorded. Once the desired maximum load is reached, the indenter is retracted, and the unloading curve is recorded.
3. *Analysis*: The load-displacement data is analyzed using models such as the Oliver-Pharr method to extract hardness and elastic modulus. Depth profiling or grid mapping can be performed to study the variation of mechanical properties over a specific region or depth.

Challenges and Limitations

Small-Scale Nature: Nano-indentation results reflect the behavior of a tiny material region, which may not fully represent bulk properties.

Surface Roughness Sensitivity: Surface roughness and sample preparation can influence the accuracy of measurements.

Complexity in Radiation-Damaged Materials: Radiation-induced defects, phase transformations, and residual stresses can complicate the interpretation of nano-indentation data. Multiple complementary techniques, such as TEM or PAS, are often used alongside nano-indentation.

9.4.4.12 Molecular Dynamic Simulations

Molecular dynamics (MD) simulations are widely used to evaluate the radiation resistance of HEAs, a class of materials known for their unique combination of properties, including high strength, corrosion resistance, and the ability to withstand extreme environments, such as radiation. Radiation resistance refers to the material's ability to maintain its structural integrity and properties when exposed to energetic particles such as neutrons, ions, or electrons, which can displace atoms from their positions and create defects within the material.

Why Use Molecular Dynamics Simulations for Radiation Resistance Studies?

1. *Atomic-Scale Insights*: MD simulations allow researchers to study the behavior of individual atoms during and after radiation events. They provide detailed information on how radiation interacts with the material at the atomic level, such as the formation and evolution of defects like vacancies, interstitials, and dislocations.
2. *Defect Dynamics*: Radiation causes cascades of atomic displacements, leading to defect formation. MD simulations can track the formation of point defects (vacancies and interstitials), dislocation loops, voids, and clusters, providing insights into how these defects evolve, migrate, and interact.
3. *Short Time and Length Scales*: Radiation damage occurs on very short scales (picoseconds to nanoseconds) and length scales (nano-meters to micro-meters). MD simulations are ideal for capturing these processes due

to their ability to model the motion of atoms over short time scales and small spatial regions.

4. *Complex Compositions of HEAs*: HEAs include five or more principal elements in near-equal atomic percentages, resulting in complex chemical environments. MD simulations can help evaluate how such complex compositions influence radiation damage resistance, defect formation, and healing mechanisms.

Critical Steps in MD Simulations for Radiation Resistance of HEAs

1. **Initial Setup of the Material System**

   Atomic Structure: The initial configuration of the HEA is typically generated by creating a crystal structure with a random distribution of atoms from different elements (to mimic the random nature of HEAs). Common crystal structures include FCC or BCC.

   Interatomic Potentials: Accurate interatomic potentials are crucial for reliable MD simulations. For HEAs, embedded atom method (EAM) or modified embedded atom method (MEAM) potentials are commonly used to describe the interactions between atoms. These potentials must account for the diverse chemical environment and interactions between constituent elements.

   Simulation Box: A periodic simulation box is created to avoid edge effects, ensuring that the behavior of atoms at the boundaries of the simulation is continuous with their periodic images.

2. **Introduction of Radiation Damage**

   Primary Knock-On Atom (PKA): Radiation damage is simulated by introducing a high-energy atom (the PKA) into the system. This atom is given a large initial velocity, simulating the impact of an energetic particle (e.g., a neutron or ion) displacing an atom in the lattice.

   Collision Cascades: The PKA transfers energy to neighboring atoms, setting off a chain of atomic displacements known as a collision cascade. This process can create a variety of radiation-induced defects, such as vacancies (where atoms are missing) and interstitials (where displaced atoms are inserted into non-equilibrium positions).

   Cascade Simulations: MD simulations track the cascade of displacements over time (typically up to a few picoseconds), capturing the number of displaced atoms, defect formation, and the spatial distribution of these defects.

3. **Defect Analysis**

   Defect Formation: The simulation analyzes the defect structure after the collision cascade subsides. The number of vacancies, interstitials, and other defects are quantified.

   Defect Clustering: Radiation damage can lead to defect clusters, such as voids (vacancy clusters) or dislocation loops (grouped interstitials). MD simulations allow researchers to study these clusters' size, distribution, and stability.

Radiation-Induced Hardening: Defects generated by radiation can obstruct dislocation motion, leading to radiation-induced material hardening. MD simulations can indirectly evaluate this by examining how defects impact the material's mechanical response.

4. **Thermal Annealing and Defect Recovery**

   Defect Migration: Defects can migrate through the lattice after the initial radiation event. MD simulations help track defects' movement and whether they recombine (e.g., an interstitial returning to a vacancy, annihilating both defects).

   Self-Healing Mechanisms: HEAs are known for their self-healing properties, where certain defects spontaneously recombine and eliminate damage. MD simulations provide insights into the likelihood and mechanisms of self-healing in HEAs under radiation.

   Thermal Annealing Simulations: To simulate recovery processes, the system can be heated, and the mobility of defects and their potential to annihilate are studied. These simulations help understand how radiation damage evolves at elevated temperatures.

5. **Mechanical Response Post-Irradiation**

   Stress-Strain Simulations: MD simulations can be used to assess the mechanical behavior of HEAs after radiation damage. By applying external stress, the stress-strain response of the irradiated material is calculated. Radiation-induced hardening or embrittlement can be inferred from changes in yield strength, ductility, or fracture behavior.

   Dislocation-Defect Interaction: MD can simulate how dislocations interact with radiation-induced defects, such as voids or dislocation loops. These interactions can strongly influence the mechanical behavior of irradiated HEAs.

6. **Comparison between Irradiated and Non-Irradiated HEAs**

   Irradiated vs. Pristine: One of the strengths of MD simulations is that they can directly compare irradiated and non-irradiated materials. This provides a clear understanding of how radiation affects the mechanical properties, defect density, and structural integrity of the HEA.

   Dose Dependence: MD simulations can be repeated for different levels of radiation doses (i.e., different PKA energies or multiple cascades) to study how radiation damage accumulates and how material properties evolve with increasing radiation exposure.

Key Outcomes of MD Simulations for Radiation Resistance in HEAs

1. *Defect Formation Resistance*: MD simulations help determine how resistant HEAs are to the formation of radiation-induced defects. Due to the high configurational entropy and the complex atomic arrangement in HEAs, these materials often show lower defect densities than conventional alloys.
2. *Defect Mobility and Recovery*: HEAs exhibit a high degree of defect mobility and self-recovery, meaning that radiation-induced defects are more

likely to migrate and recombine, reducing long-term damage. MD simulations help quantify the migration energies of defects and their recombination probabilities.

3. *Radiation-Induced Hardening*: MD simulations show how radiation-induced defects can pin dislocations, causing radiation-induced hardening. This effect is significant for understanding how HEAs maintain strength in radiation-rich environments like nuclear reactors.
4. *Superior Radiation Resistance of HEAs*: MD simulations have shown that HEAs often exhibit superior radiation resistance compared to traditional materials. This is attributed to the complex atomic arrangement, which leads to fewer dislocation loops and voids, and to the ability of the material to heal itself through defect recombination.
5. *Synergistic Effects of Composition*: MD simulations reveal how the unique multi-element composition of HEAs affects radiation damage. For example, HEAs' different atomic sizes and bonding energies create barriers to defect movement, enhancing radiation resistance.

Case Studies and Examples

1. *MD Simulations of FCC HEAs*: Studies on FCC HEAs, such as Ni-Co-Fe-Cr-Mn, have shown that these materials exhibit a high degree of defect recombination and lower void formation than conventional alloys like stainless steel. MD simulations have been used to quantify the number of surviving defects after radiation and assess the material's hardness and mechanical response.
2. *Radiation Resistance of BCC HEAs*: MD simulations on BCC HEAs show they exhibit fewer dislocation loops and are less prone to embrittlement after irradiation. These materials also show a reduced tendency for phase separation or amorphization, common issues in conventional materials under radiation.
3. *Effects of Temperature*: MD simulations have also studied how radiation resistance in HEAs changes with temperature. Elevated temperatures can enhance defect mobility and promote recombination, thereby improving the self-healing capabilities of the material.

Challenges and Limitations

1. *Time and Length Scales*: MD simulations typically operate on short scales (nanoseconds) and small spatial scales (nanometers). Long-term radiation damage effects, such as void swelling and long-range diffusion of defects, may not be fully captured in MD simulations.
2. *Interatomic Potentials*: MD simulations' authenticity depends on the interatomic potentials' quality. For HEAs, where multiple elements are involved, developing accurate potentials that capture all possible interactions is challenging.

3. *Computational Resources*: Simulating large systems with millions of atoms or studying high-dose radiation events requires significant computational power, which can limit the scope and duration of MD simulations.

9.4.5 Applications of Radiation-Resistant HEAs

9.4.5.1 Nuclear Energy Applications

One of the most critical applications of radiation-resistant HEAs is in the nuclear energy sector, where materials are exposed to high neutron fluxes and must endure prolonged radiation exposure. Nuclear reactors, especially next-generation reactors, operate under extreme conditions of high temperature, corrosive environments, and intense radiation, which conventional materials often cannot withstand over long periods.

Critical Applications in Nuclear Energy:

1. Nuclear Reactor Core Components:
   Fuel Cladding: The material surrounding nuclear fuel rods, known as fuel cladding, must have high radiation resistance to maintain its integrity during reactor operation. HEAs are being investigated as replacements for traditional materials like zirconium alloys, which suffer from radiation-induced embrittlement and corrosion.
   Structural Materials: The components that support the core, such as pressure vessels and control rod mechanisms, need to be both mechanically strong and radiation-resistant. HEAs offer excellent strength retention, phase stability, and low defect accumulation, making them suitable candidates.
   Radiation-Induced Swelling Resistance: Traditional materials used in reactor cores are prone to radiation-induced swelling, where vacancies form, leading to a volume increase and weakening of the material. Due to their complex atomic structures, HEAs exhibit a lower tendency for such swelling, which helps maintain dimensional stability in critical components.
2. Next-Generation Reactors:
   Fast Breeder Reactors (FBRs): These reactors generate higher neutron fluxes than conventional reactors, leading to more intense radiation exposure. Due to their defect recombination capabilities and resistance to radiation-induced phase transformations, HEAs are ideal for use in high-radiation environments.
   Molten Salt Reactors (MSRs): In MSRs, where the reactor coolant is a molten salt, materials must resist radiation and withstand the molten salt's corrosive nature. With their inherent corrosion resistance, HEAs are being explored for both the structural components and the containers for molten salts.
3. Fusion Reactors:

   In fusion, materials are exposed to intense neutron bombardment at very high temperatures, which can cause severe radiation damage. With their superior radiation resistance and ability to retain mechanical strength at high temperatures, HEAs are being explored in plasma-facing components, first walls, and blankets.

Advantages of HEAs in Nuclear Applications:

Reduced Radiation-Induced Hardening: HEAs show less radiation-induced hardening than traditional materials, meaning they maintain their ductility and do not become excessively brittle under irradiation.

Thermal Stability: HEAs remain stable at high temperatures, which is critical for nuclear reactor operation, where radiation and heat can degrade materials over time.

Resistance to Radiation-Induced Amorphization: Traditional alloys, especially those containing metals like zirconium, can become amorphous under radiation. Due to their high entropy and stable crystalline structure, HEAs resist such phase changes, maintaining their mechanical properties.

9.4.5.2 Space Exploration and Aerospace Applications

Materials used in space exploration are subjected to extreme conditions, including cosmic radiation, composed of high-energy particles such as protons, alpha particles, and heavy ions. These materials must be radiation-resistant, thermally stable, and capable of withstanding high and low temperatures.

Critical Applications in Space Exploration:

1. Spacecraft Components:

   Structural Components: Components of spacecraft, satellites, and space probes are continuously bombarded by cosmic radiation. With their superior radiation resistance, HEAs offer an advantage over conventional materials by reducing radiation damage accumulation over time. This leads to longer mission life spans and improved safety.

   Shielding Materials: Protecting sensitive electronics and human passengers from radiation is critical in space missions. HEAs could be used as radiation shielding materials, as their ability to absorb and disperse high-energy particles without forming significant defects makes them ideal candidates.

2. Deep Space Missions:

   Deep space exploration missions, such as those to Mars or beyond, require materials that can endure the harsh radiation environment outside Earth's protective magnetosphere. HEAs offer enhanced resistance to cosmic radiation, ensuring the structural integrity of spacecraft over extended periods.

   Solar Wind and Cosmic Radiation Resistance: Unlike in low Earth orbit, where radiation is partially mitigated by Earth's atmosphere, deep space missions encounter higher levels of solar wind and cosmic radiation. HEAs can minimize radiation-induced degradation with their unique defect resistance and self-healing properties.

3. Nuclear-Powered Spacecraft:
   Using nuclear reactors or radioisotope thermoelectric generators (RTGs) for power in space exploration requires radiation- and corrosion-resistant materials. HEAs can be utilized in the structural components of these power systems, ensuring long-term stability under radiation and heat.

9.4.5.3 Medical Applications

In nuclear medicine and medical imaging, materials are often exposed to radiation for diagnostic or therapeutic purposes. These materials must resist radiation-induced degradation while maintaining biocompatibility and mechanical strength.

Critical Applications in Medicine:

1. Radiation Therapy Equipment:
   Materials used in radiation therapy machines, such as linear accelerators (linacs), must withstand continuous exposure to high-energy X-rays and electrons. HEAs offer an advantage in these machines due to their resistance to radiation-induced embrittlement and ability to maintain structural integrity over time.
2. Implants in Radiation Environments:
   Implants, such as pacemakers, joint replacements, or stents, may be exposed to radiation during diagnostic imaging (e.g., X-rays, CT scans) or therapeutic procedures (e.g., cancer radiation therapy). HEAs, which resist radiation damage, can be used in these implants to ensure long-term stability and minimize the risk of radiation-induced material degradation or alteration of mechanical properties.
3. Radiation-Resistant Coatings:
   HEAs can be applied as radiation-resistant coatings on medical instruments or devices in radiation therapy or diagnostic imaging. These coatings would protect sensitive components from radiation-induced damage, extending their operational lifetimes.

9.4.5.4 Defense and Military Applications

In defense and military applications, materials are often subjected to radiation from nuclear weapons or nuclear-powered systems. Materials in such environments must resist radiation while maintaining structural and mechanical integrity.

Critical Applications in Defense:

1. Nuclear Submarines:
   The reactor cores of nuclear submarines are exposed to both radiation and corrosive seawater. HEAs, with their dual radiation and corrosion resistance, could be used in reactor pressure vessels, heat exchangers, and steam generators, improving the longevity and reliability of the systems.
2. Armor and Shielding:
   In nuclear warfare scenarios, armor or protective shielding materials must endure radiation exposure. With their ability to absorb and disperse

radiation damage without suffering catastrophic failure, HEAs are being investigated for use in radiation-resistant armor.

3. Nuclear-Powered Systems:

   Military applications, such as nuclear-powered aircraft carriers, drones, and other vehicles, require materials that can withstand long-term exposure to radiation while maintaining mechanical performance. HEAs are excellent candidates for structural components in these vehicles, ensuring durability under extreme conditions.

9.4.5.5 High-Radiation Environments in Industry

Specific industrial processes generate high levels of radiation, requiring materials that can resist radiation-induced damage over long periods.

Critical Applications in Industrial Environments:

1. Nuclear Waste Storage:

   The storage of nuclear waste involves exposure to radiation for extended periods. Containers or casks for high-level waste must be radiation-resistant to prevent degradation over time. With their stable crystal structures and low defect accumulation, HEAs are being considered for nuclear waste containment materials.
2. Radiation-Tolerant Robotics:

   Robots used in high-radiation environments, such as nuclear reactors for maintenance and inspection, require materials that can endure radiation exposure without becoming brittle or failing mechanically. HEAs can be used to construct these robots to extend their operational lifetimes in radiation-heavy environments.
3. Radiation-Resistant Sensors and Electronics:

   Sensors and electronic components in high-radiation environments, such as nuclear reactors, medical equipment, or space, must be protected from radiation damage. HEAs can be applied as radiation-resistant materials for housings or coatings to protect these sensitive components.

9.5 SYNERGISTIC EFFECTS OF CORROSION AND RADIATION RESISTANCE OF HEAs

Irradiation affects the corrosion of core structural materials in at least three distinct ways.

1. The dose of irradiation and the associated temperature play significant roles in the microstructural changes observed in irradiated alloys. These changes include the development of dislocation loops, cavities, helium bubbles, and precipitates. Such microstructural modifications impact the kinetics of oxidation by influencing the diffusion rates of oxygen vacancies and metal cations. Consequently, these defects may enhance oxidation processes at both the interfaces of oxide/liquid and metal/oxide.

2. Irradiation-induced elemental segregation at grain boundaries, characterized by the reduction of Cr, Fe, and Mo and the accumulation of Si, Ni, and P, is believed to influence the intergranular oxidation behavior of alloys significantly.
3. Irradiating water causes its breakdown into free radicals and oxidizing agents, increasing the aqueous environment's corrosion potential. When exposed to high-temperature water, this heightened corrosion potential leads to the dissolution of the chromium-rich inner oxide layer in alloys, resulting in an increased corrosion rate.

The most effective approach for acquiring irradiation corrosion data is simulating the irradiated water environment. Autoclave systems have successfully replicated primary circuit conditions, operating at temperatures between 553 and 598 K and pressures ranging from 7.1 to 16.2 MPa. In high-pressure sterilizers with recirculation loops, ion-exchange resins are used to monitor water chemistry, while an in situ H2 sensor measures dissolved hydrogen by detecting the thermal conductivity of the water. Neutron irradiation experiments are challenging due to extended cycle times, high costs, and radioactivity, which limit research. In contrast, ion irradiation offers several benefits, including high damage rates, precise control over parameters, the absence of radioactivity, and cost-effectiveness. As a result, ion implantation is commonly used in laboratories to simulate neutron irradiation, aiding fundamental research, radiation damage modeling, and understanding key mechanisms [279].

9.5.1 Impact of Irradiation Defects on Corrosion

HEAs are materials of multiple principal elements, typically five or more, in nearly equiatomic proportions. This unique composition gives HEAs a range of excellent properties, such as high strength, exceptional thermal stability, and impressive corrosion resistance. However, irradiation-induced defects can significantly influence their corrosion behavior under extreme environments, such as high radiation fields (nuclear reactors, space applications).

Understanding how irradiation affects the corrosion resistance of HEAs is crucial for their use in nuclear and space environments, where both irradiation damage and corrosion are major concerns.

9.5.1.1 Irradiation-Induced Defects

When HEAs are exposed to irradiation (such as neutron, proton, or heavy ion bombardment), several types of defects can form in the material. These defects include:

Vacancies: Atoms are knocked out of their positions, leaving empty spaces.

Interstitials: Atoms displaced from their lattice positions occupy interstitial sites.

Dislocation Loops: Complex structures formed by clusters of vacancies or interstitials.

Void Formation: Irradiation can lead to the coalescence of vacancies into larger voids, which can change the material's density.

Amorphization: In some cases, severe irradiation can lead to the loss of crystallinity in localized areas of the material.

These defects affect not only the mechanical integrity of the alloy but also its interaction with corrosive environments.

9.5.1.2 Effect of Irradiation on Corrosion in HEAs

Irradiation-induced defects can enhance corrosion in HEAs due to several factors:

9.5.1.2.1 Accelerated Diffusion and Grain Boundary Corrosion

Defect Clustering and Segregation: The vacancies and interstitials generated by irradiation act as diffusion paths for atoms, increasing atomic mobility. This can lead to the segregation of elements at grain boundaries or defect clusters, weakening the local chemistry balance. Grain boundaries are already more susceptible to corrosion due to their higher energy state, and irradiation-induced defects further exacerbate this susceptibility.

Radiation-Induced Segregation (RIS): Specific elements can segregate to grain boundaries or dislocation loops due to irradiation. RIS can disrupt the protective passivation layer in HEAs, where elements are present in nearly equiatomic proportions. For instance, the selective enrichment or depletion of key corrosion-resistant elements (such as Cr or Al) at the surface can lead to localized corrosion (pitting or intergranular attack).

9.5.1.2.2 Formation of Defect Sites as Corrosion Initiation Points

Vacancies and Voids as Corrosion Nuclei: The vacancies and voids created by irradiation can act as nucleation sites for corrosion. These defects provide easy access points for corrosive agents, such as oxygen, water, or chloride ions, to penetrate the material.

Dislocation Loop-Induced Micro-Cracks: Dislocation loops may evolve into micro-cracks under the stress of irradiation and subsequent thermal cycling. These micro-cracks can facilitate the ingress of corrosive species deep into the material, leading to stress-corrosion cracking (SCC).

9.5.1.2.3 Amorphization and Loss of Passivity

Amorphization Impact: In extreme cases, irradiation can cause localized amorphization, destroying the crystalline structure. An amorphous region can have drastically different chemical properties and is often less corrosion-resistant. The loss of crystallinity impairs the formation of stable passive oxide layers, which are critical for the corrosion resistance of many alloys.

Degradation of Passivation Layers: In many HEAs, corrosion resistance is enhanced by forming protective oxide layers (e.g., Cr_2O_3 in Cr-containing alloys). Irradiation can damage these layers, leading to their breakdown. The passivation layer's breakdown makes the alloy surface more reactive and susceptible to uniform and localized corrosion.

9.5.1.3 Factors Influencing the Impact of Irradiation on Corrosion

Several factors influence how irradiation defects affect the corrosion of HEAs:

9.5.1.3.1 Alloy Composition

The unique multi-principal element composition of HEAs can mitigate or exacerbate the effects of irradiation on corrosion. HEAs with elements known for their corrosion resistance, like chromium, can maintain protective oxide layers despite irradiation. However, irradiation-induced segregation can lead to the depletion of these elements at the surface or grain boundaries, lowering corrosion resistance.

9.5.1.3.2 Irradiation Conditions

Dose and Dose Rate: The total radiation dose and dose rate affect the density and type of defects formed. Higher doses lead to more severe defect formation, increasing the potential for irradiation-assisted corrosion.

Irradiation Temperature: At higher temperatures, irradiation effects can be different. For example, higher temperatures can promote the annealing of defects and enhance the diffusion of segregating species, resulting in radiation-induced segregation at grain boundaries.

Type of Radiation: Different types of radiation (neutrons, heavy ions, or protons) produce different defect structures. Heavy ions, for instance, create denser cascades of damage than neutrons, leading to more severe impacts on corrosion.

9.5.1.3.3 Environmental Conditions

Corrosive Media: Irradiation effects are significantly influenced by the surrounding environment. Irradiation defects can accelerate pitting corrosion in a highly aggressive medium like chloride solutions. Similarly, in water-based environments (nuclear reactors), irradiation can promote stress corrosion cracking and general dissolution of the alloy.

9.5.1.4 Mitigation Strategies

To improve the resistance of HEAs to irradiation-induced corrosion, several strategies can be employed:

Optimizing Alloy Composition: HEAs with higher concentrations of corrosion-resistant elements like Cr or Al may show enhanced stability against irradiation effects. Alloying with elements that reduce radiation-induced segregation can also help maintain surface chemistry.

Surface Treatments and Coatings: Applying protective coatings or modifying surface treatments before exposure can help protect the alloy from irradiation damage and subsequent corrosion.

Annealing Treatments: Post-irradiation annealing can help recover defects and reduce their impact on corrosion. Thermal treatments can heal vacancies and dislocation loops, restoring some of the alloy's properties.

Grain Size Engineering: Reducing grain size can limit the extent of irradiation-induced damage, as smaller grains provide more grain boundaries, which act as sinks for defects, reducing the concentration of defects in the bulk material.

9.6 CHALLENGES AND FUTURE DIRECTIONS

9.6.1 Challenges in the Development of Corrosion and Radiation-Resistant HEAs

HEAs are a unique class of materials composed of five or more principal elements in near-equiatomic proportions. The novel concept behind HEAs allows them to possess enhanced properties such as excellent mechanical strength, thermal stability, and corrosion resistance. These features make HEAs potential candidates for applications in extreme environments, such as nuclear reactors or space exploration, where materials face radiation damage and corrosive environments simultaneously. However, developing HEAs that are both corrosion and radiation-resistant presents a variety of scientific and technical challenges.

9.6.1.1 Understanding the Complex Composition-Property Relationships

9.6.1.1.1 Multi-Principal Element Effects

One key factor distinguishing HEAs from traditional alloys is their multi-principal element nature. This composition complexity leads to several effects that can either enhance or hinder corrosion and radiation resistance:

Entropic Stabilization: The high configurational entropy of HEAs stabilizes a simple phase structure, such as FCC or BCC, rather than forming brittle intermetallic compounds. This straightforward phase structure can improve radiation resistance by reducing phase instability under irradiation. However, not all compositions are equally stable, and fine-tuning the alloy's composition is critical.

Sluggish Diffusion: In HEAs, atomic diffusion is slower due to the nuclear size mismatch and varying bond strengths between different elements. While sluggish diffusion can help reduce void swelling under irradiation and hinder corrosion by slowing down the degradation process, it also makes phase transformations and recovery after radiation exposure more complex, complicating the design of HEAs.

9.6.1.1.2 Elemental Segregation and Radiation-Induced Segregation (RIS)

In HEAs, each element plays a different role in contributing to properties like corrosion resistance (e.g., Cr or Al form protective oxides) or radiation resistance (e.g., Ni or Mo). Achieving a balance between these properties is not straightforward because:

Elemental Segregation: During processing (e.g., solidification, thermomechanical treatment), certain elements may segregate at grain boundaries or other microstructural features. This segregation can locally decrease corrosion resistance by depleting elements that form protective oxides or increasing the susceptibility to grain boundary corrosion.

Radiation-Induced Segregation (RIS): Irradiation can cause elements to segregate differently within the alloy. For example, elements that provide corrosion resistance (like Cr) may migrate away from grain boundaries or surfaces under irradiation, making localized areas more vulnerable to corrosion. Understanding and mitigating RIS remains a critical challenge in HEA development.

9.6.1.1.3 Synergistic Effects

The interaction between the different elements in HEAs can lead to unexpected behaviors. For instance:

Corrosion vs. Radiation Resistance Trade-Offs: Elements that improve radiation resistance (such as Mo or W) may promote passivation layer breakdown in corrosive environments, while those that enhance corrosion resistance (such as Cr) may decrease radiation stability. Balancing these effects to achieve corrosion and radiation resistance remains a significant design challenge.

Phase Stability: The phase stability of HEAs under irradiation and in corrosive environments is highly composition-dependent. Complex interactions between alloying elements can lead to phase transformations under extreme conditions, affecting radiation resistance and corrosion behavior.

9.6.1.2 Developing Microstructural Stability Under Dual-Stress Conditions

9.6.1.2.1 Irradiation Effects on Microstructure

Radiation exposure can lead to various microstructural changes in HEAs that influence their corrosion resistance:

Void Swelling: Radiation induces the formation of vacancies and interstitials, leading to void swelling, which degrades the material's mechanical properties and creates regions where corrosion can initiate.

Dislocation Loops and Defect Clustering: Irradiation generates dislocation loops and defect clusters, which alter the microstructure and provide diffusion paths for corrosive species. These defects can be nucleation points for corrosion, especially in high-temperature environments.

Amorphization: In some cases, intense radiation can lead to localized amorphization, where crystalline structures are lost, making the alloy more susceptible to corrosion.

9.6.1.2.2 Grain Boundary and Interface Engineering

Grain boundaries in HEAs can act as sinks for irradiation-induced defects, helping to heal radiation damage. However, they are also highly susceptible to corrosion, especially in aggressive environments like high-temperature water (nuclear reactors) or chloride-rich environments. Challenges related to grain boundary engineering include:

Grain Boundary Corrosion: Radiation can accelerate grain boundary corrosion by promoting elemental segregation or creating defect clusters at these boundaries.

Nanocrystalline Structures: Reducing the grain size to the nanoscale may improve radiation resistance by providing more defect sinks. However, the increased grain boundary area can increase susceptibility to corrosion. This trade-off must be managed carefully when designing HEAs.

9.6.1.3 Environmental Factors and Their Impact on HEAs

9.6.1.3.1 Aggressive Corrosive Environments

HEAs must withstand various corrosive environments, such as:

Aqueous Corrosion (e.g., Nuclear Reactor Coolants): Water-based environments, particularly those involving high temperatures, can exacerbate the formation of localized corrosion sites, especially when combined with radiation. In nuclear reactors, the breakdown of passivation layers and the formation of radiation-induced defects can lead to stress corrosion cracking (SCC).

High-Temperature Corrosion (e.g., Space Applications): Oxidation can occur rapidly in high-temperature environments. Protective oxide layers can break down due to radiation-induced changes in surface chemistry. HEAs must be designed to maintain their protective oxide layers even under thermal cycling and irradiation.

Chloride-Induced Pitting: Many applications expose materials to chloride ions (e.g., marine environments or specific industrial processes). Chloride ions can penetrate the passivation layer, causing pitting corrosion exacerbated by radiation-induced microstructural defects.

9.6.1.3.2 Radiation Environment Challenges

Neutron Radiation (Nuclear Applications): Neutrons are highly penetrating and cause significant lattice damage in materials. HEAs designed for nuclear reactors must endure neutron bombardment without losing structural integrity or corrosion resistance.

Heavy Ion and Proton Irradiation (Space Applications): In space, materials are exposed to cosmic rays and heavy ions, which cause concentrated damage in small regions, leading to severe local corrosion. HEAs must be engineered to distribute radiation damage across their structure without compromising corrosion resistance.

9.6.1.4 Material Design and Optimization Challenges

9.6.1.4.1 Alloy Design via Combinatorial Methods

One of the advantages of HEAs is the vast compositional space available for exploration. However, this also presents a significant challenge:

High Dimensionality of Composition Space: With five or more principal elements, the number of possible combinations is enormous. Traditional trial-and-error methods for optimizing composition are impractical. Researchers must rely on advanced computational techniques, such as machine learning or high-throughput experimentation, to efficiently explore this vast compositional space.

Limited Data on Corrosion and Radiation Resistance: While there has been significant research on the mechanical properties of HEAs, more work needs to be done to understand their corrosion and radiation resistance. Developing databases that correlate HEA compositions with their performance under extreme environments is essential for guiding alloy design.

9.6.1.4.2 Processing and Fabrication Challenges

Uniformity in Microstructure: A significant challenge is to achieve a uniform microstructure, especially in multi-phase HEAs. Heterogeneities, such as phase separations or compositional variations, can make localized regions more vulnerable to radiation damage and corrosion.

Surface Treatments and Coatings: Surface treatments like passivation, nitriding, or coating application may be necessary to improve the anti-corrosion behavior of HEAs. However, these treatments must be compatible with the alloy's response to irradiation.

9.6.1.5 Testing and Evaluation Under Realistic Conditions

9.6.1.5.1 Challenges in Simulating Combined Corrosion and Radiation Environments

Testing HEAs under realistic combined stress conditions (simultaneous corrosion and radiation) is extremely difficult and expensive:

Radiation Facilities: Simulating the neutron flux experienced in a nuclear reactor or the heavy ion bombardment in space requires specialized facilities, which are often limited in number and availability.

Accelerated Corrosion Testing: Testing HEAs' long-term corrosion behavior under irradiated conditions requires accelerated testing methods that can simulate years of exposure in a short time. Developing reliable and predictive accelerated testing protocols is challenging.

9.6.1.5.2 In-Situ Monitoring of Damage Evolution

Advanced Characterization Techniques: The ability to monitor the evolution of irradiation-induced defects and corrosion in real-time is crucial for understanding how these processes interact. Techniques such as TEM, APT, and in-situ electrochemical analysis under irradiation are necessary but require advanced equipment and expertise.

9.6.2 Emerging Trends and Innovations

The development of HEAs has entered a new phase, where emerging trends and innovations in their design, characterization, and application are opening up exciting possibilities. The key emerging trends and innovations driving the future of HEAs are mentioned below.

9.6.2.1 Advanced Alloy Design Approaches

9.6.2.1.1 Combinatorial Alloy Design and High-Throughput Experimentation

One of the significant innovations in HEA development is the use of combinatorial alloy design powered by high-throughput experimentation and computational methods:

Exploring Complex Compositional Spaces: Traditional alloy design methods cannot keep pace with the vast compositional space that HEAs present, as five or more elements offer an exponential number of possible combinations. High-throughput experimentation allows researchers to quickly test different compositions for desirable properties, such as strength, corrosion resistance, or thermal stability.

Automation and Robotics in Alloy Synthesis: Innovations in robotics and automation enable the simultaneous rapid synthesis of thousands of different alloy compositions. This combinatorial approach accelerates the discovery of novel HEAs by quickly screening potential candidates, followed by detailed testing of promising compositions.

Data-Driven Alloy Design: Machine learning (ML) and artificial intelligence (AI) are increasingly integrated with materials science. By training algorithms on existing data from high-throughput experiments, ML models can predict the properties of new HEA compositions, guiding alloy design more efficiently. These data-driven approaches allow researchers to identify patterns in how different elemental combinations affect phase stability, mechanical properties, and other vital factors.

9.6.2.1.2 Computational Materials Science and Predictive Modeling

Another key trend is the use of computational methods, particularly for predicting and optimizing the behavior of HEAs:

Density Functional Theory (DFT) and Molecular Dynamics (MD) Simulations: DFT and MD simulations have become critical tools for predicting HEAs' phase stability, mechanical properties, and thermodynamic behavior. These methods allow researchers to model atomic-scale interactions in HEAs and predict how specific compositions will perform under different conditions, such as high temperatures or radiation exposure.

Phase Diagram Calculations and CALPHAD (CALculation of PHAse Diagrams): CALPHAD modeling is applied to HEAs to predict phase diagrams and stability. This is particularly important for multi-element systems where phase boundaries are poorly understood. With CALPHAD, researchers can estimate the thermodynamic properties of new HEAs, enabling more targeted alloy design.

Integrated Computational Materials Engineering (ICME): ICME is an emerging approach that combines multiple computational tools to design materials. ICME integrates computational modeling, experimental data, and processing simulations for HEAs to accelerate the alloy development process. This approach allows the simultaneous optimization of radiation, ductility, strength and corrosion resistance properties.

9.6.2.2 Innovative Manufacturing Techniques for HEAs

9.6.2.2.1 Additive Manufacturing (3D Printing)

Additive manufacturing (AM) is revolutionizing the production of HEAs, enabling the fabrication of complex components with tailored microstructures:

Customization of Alloy Composition: With AM, it is possible to fine-tune the composition of an alloy layer-by-layer during printing, allowing for gradient alloys or multi-material structures. This opens up new possibilities for functionally graded HEAs that exhibit varying properties in different regions of the same component.

Control Over Microstructure: AM allows precise control over cooling rates, influencing HEAs' microstructure and phase distribution. This can lead to enhanced properties, such as increased strength or improved corrosion resistance, due to refined microstructures and reduced segregation of elements.

In-Situ Alloying: AM processes such as laser metal deposition (LMD) allow in-situ alloying, where elemental powders are mixed and melted during printing. This makes it easier to experiment with novel HEA compositions without the need for traditional casting methods.

9.6.2.2.2 Severe Plastic Deformation (SPD) Techniques

Severe plastic deformation techniques, such as high-pressure torsion (HPT) and equal channel angular pressing (ECAP), are being applied to HEAs to refine their microstructures and enhance mechanical properties:

Nanostructuring HEAs: SPD can produce HEAs with ultrafine or nanocrystalline grain sizes, significantly improving strength, hardness, and wear resistance. Nanostructured HEAs are particularly interesting for applications in extreme environments requiring high strength and thermal stability.

Improving Radiation Resistance: Reducing grain size through SPD provides more grain boundaries and sinks for radiation-induced defects like vacancies and interstitials. This enhances the radiation resistance of HEAs, making them suitable for nuclear and space applications.

9.6.2.3 Tailoring Properties for Specific Applications

9.6.2.3.1 High-Temperature HEAs for Aerospace and Energy Sectors

HEAs are increasingly being designed for high-temperature applications, where traditional alloys tend to degrade or lose strength:

Refractory HEAs: These alloys are composed of refractory elements such as tungsten, tantalum, and molybdenum, known for their ability to withstand high temperatures without losing structural integrity. Refractory HEAs exhibit excellent mechanical properties at temperatures exceeding 1273 K, making them potential candidates for turbine blades, space exploration, and energy generation.

Oxidation and Creep Resistance: Innovations in alloy composition, mainly adding elements like aluminum and chromium, have improved the oxidation resistance of high-temperature HEAs. Additionally, optimizing microstructure through alloy design helps improve creep resistance, which is critical for long-term performance in high-stress environments.

9.6.2.3.2 Corrosion-Resistant HEAs for Harsh Environments

Protective Oxide Formation: By incorporating elements such as chromium, aluminum, or titanium, HEAs can form protective oxide layers that enhance their corrosion resistance. These oxides prevent aggressive elements, such as chloride ions in marine environments or high-temperature steam in nuclear reactors, from penetrating the surface and causing corrosion.

Passivation Layer Stability: Ongoing research focuses on the stability of passivation layers in HEAs under harsh conditions. Emerging trends include designing alloys that maintain or regenerate their protective oxide layers under cyclic environmental conditions, ensuring long-term corrosion resistance in acidic and alkaline environments.

9.6.2.4 Designing HEAs for Extreme Radiation Environments

9.6.2.4.1 Radiation-Resistant HEAs for Nuclear Applications

One of the most promising areas of innovation is the design of HEAs that can withstand extreme radiation environments, such as those found in nuclear reactors or space:

Reduced Void Swelling: Radiation exposure leads to defects like vacancies, which cluster to form voids. These voids cause swelling and mechanical degradation. HEAs have shown reduced void swelling due to their sluggish diffusion and high-density grain boundaries, which absorb radiation-induced defects more effectively than traditional alloys.

Radiation-Induced Phase Stability: Emerging research shows that specific HEA compositions remain stable under high doses of radiation. This stability is due to the complex lattice structure of HEAs, which can tolerate the formation of irradiation defects without undergoing significant phase transformation. Ongoing research is focused on identifying the optimal compositions for such applications.

Self-Healing Mechanisms: Some HEAs are designed with "self-healing" properties, where the alloy's structure can naturally recover from irradiation-induced damage. This involves selecting elements that encourage defect recombination or introducing features like grain boundaries and phase interfaces that act as defect sinks.

9.6.2.5 Biocompatible and Bio-Inspired HEAs

9.6.2.5.1 HEAs for Biomedical Applications

Biocompatible HEAs are a growing area of interest, particularly for medical implants and devices. Their combination of high strength, wear resistance, and corrosion resistance makes them ideal candidates for long-term use in the human body:

Nickel-Free HEAs: Since nickel is a common allergen, researchers are developing nickel-free HEAs with elements such as titanium, zirconium, and tantalum, which exhibit excellent biocompatibility and corrosion resistance. These alloys are being explored for applications like bone implants, joint replacements, and stents.

Bio-Inspired HEAs: Researchers are taking inspiration from biological materials to design HEAs with enhanced toughness and strength. These bio-inspired HEAs aim to combine lightweight design with extraordinary mechanical properties by mimicking the hierarchical structure of bone or other biological tissues.

9.6.2.6 Environmental and Sustainable Innovations

9.6.2.6.1 Green HEA Development

As environmental concerns continue to grow, innovations are focusing on the development of "green" HEAs:

Sustainable Element Selection: Researchers are exploring the use of abundant and non-toxic elements in HEA compositions to reduce the environmental impact of their production. Replacing critical or rare elements, such as cobalt or tungsten, with more abundant alternatives is a crucial focus of this research.

Energy-Efficient Manufacturing: Innovations in manufacturing processes, such as cold spraying or energy-efficient additive manufacturing, are being developed to reduce the energy consumption and carbon footprint associated with HEA production.

9.6.2.6.2 Recyclability and Circular Economy

HEAs are being designed with recyclability in mind to promote a circular economy. By selecting elements that can be easily separated and recycled, researchers aim to minimize waste and ensure that HEAs can be reused sustainably.

9.6.3 Future Research Directions

As industries increasingly demand materials that can withstand harsh operating conditions—particularly in nuclear, aerospace, marine, and chemical processing applications—research on HEAs is rapidly advancing. The future of HEAs, especially those focused on corrosion and radiation resistance, holds significant promise. Below are several key research directions that will shape the future landscape of HEAs:

9.6.3.1 Design and Computational Modeling

9.6.3.1.1 Machine Learning and Data-Driven Approaches

Accelerated Alloy Discovery: ML techniques can be employed to predict the properties of new HEAs based on their composition. By leveraging existing databases of material properties, researchers can rapidly identify promising candidates for specific applications, such as corrosion and radiation resistance.

Optimization of Composition and Microstructure: Computational models can help optimize the composition and microstructure of HEAs to achieve desired properties, such as enhanced corrosion resistance or improved radiation resistance. By simulating the influence of different alloying constituents and heat treatments, researchers can tailor HEAs for specific applications.

9.6.3.1.2 Thermodynamic and Kinetic Modeling

Phase Stability Predictions: Advanced thermodynamic modeling tools, like CALPHAD, can predict phase stability and transformations in HEAs under various temperature and pressure conditions. Understanding these behaviors is crucial for developing materials reliably performing under extreme conditions.

Diffusion Studies: Investigating diffusion mechanisms in HEAs through kinetic modeling will provide insights into how alloying elements interact, particularly regarding radiation-induced defect formation and migration, essential for understanding long-term performance in nuclear applications.

9.6.3.2 Synthesis and Processing Techniques

9.6.3.2.1 Innovative Manufacturing Methods

Additive Manufacturing (AM): Research into additive manufacturing techniques, such as 3D printing, could revolutionize HEA production by enabling complex geometries and compositions that are difficult to achieve with traditional methods. Optimizing AM parameters for HEAs to enhance their mechanical properties and corrosion resistance will be crucial for industrial applications.

Powder Metallurgy: This technique can be further explored for synthesizing HEAs with controlled microstructures and improved properties. Optimizing sintering processes and parameters will help enhance the density and performance of HEAs.

9.6.3.2.2 Heat Treatment and Surface Engineering

Microstructural Optimization: Further investigation into the effects of heat treatments (e.g., solution treatment, aging) on the microstructure and properties of HEAs can lead to improved performance, particularly regarding corrosion resistance and radiation stability.

Coatings and Surface Treatments: Developing advanced coatings or surface treatments for HEAs could enhance their resistance to corrosion and wear. Research into methods like thermal spray, chemical vapor deposition (CVD), and laser surface melting can significantly improve surface properties.

9.6.3.3 Corrosion Mechanisms and Characterization

9.6.3.3.1 Understanding Corrosion Mechanisms

In-Depth Corrosion Studies: Future research should focus on understanding the fundamental corrosion mechanisms of HEAs in various environments (e.g., aqueous, high-temperature, and gaseous environments). This includes investigating localized corrosion phenomena such as pitting, crevice corrosion, and stress corrosion cracking.

Electrochemical Characterization: Employing advanced electrochemical techniques (e.g., potentiodynamic polarization, electrochemical impedance spectroscopy) will aid in characterizing the corrosion resistance of HEAs and understanding the influence of microstructure and composition on corrosion behavior.

9.6.3.3.2 Long-Term Performance Evaluation

Accelerated Testing Protocols: Developing standardized testing protocols for evaluating HEAs' long-term corrosion and radiation resistance will be essential for industry adoption. These protocols can help predict the performance of HEAs in real-world applications and guide their selection for specific environments.

9.6.3.4 Radiation Damage Studies

9.6.3.4.1 Mechanisms of Radiation Damage

Defect Formation and Evolution: Future research should focus on the fundamental mechanisms of radiation damage in HEAs, including the formation of defects such as vacancies, interstitials, and dislocation loops. Understanding these mechanisms will inform the design of more radiation-resistant alloys.

Microstructural Stability Under Radiation: Investigating how the unique microstructures of HEAs respond to radiation exposure will provide insights into their performance in nuclear environments. Studies can explore how different alloying elements and phase combinations influence radiation resistance.

9.6.3.4.2 Ion Irradiation Experiments

Simulating Radiation Environments: Ion irradiation experiments can mimic the effects of neutron radiation and provide valuable data on how HEAs perform under radiation exposure. This research will be crucial for developing materials suitable for long-term use in nuclear reactors and space applications.

9.6.3.5 Multifunctional Applications

9.6.3.5.1 Combining Properties for Advanced Applications

Integration of Multiple Functions: Future research could explore the potential of developing multifunctional HEAs that combine corrosion resistance, radiation resistance, and other desirable properties (e.g., shape memory and superplasticity). Such materials could find applications in advanced systems, including intelligent structures and adaptive materials.

Biomedical Applications: Investigating HEAs for biomedical applications, such as implants and prosthetics, could leverage their resistance to corrosion and mechanical properties while ensuring biocompatibility. Research should focus on tailoring the composition of HEAs for specific biomedical requirements.

9.6.3.6 Real-World Testing and Validation

9.6.3.6.1 Field Testing and Performance Evaluation

Pilot Projects and Case Studies: Conducting pilot projects and field tests to assess the performance of HEAs in practical applications will be crucial for demonstrating their capabilities and gaining industry acceptance. This research can provide valuable feedback on material behavior under actual operating conditions.

Collaboration with Industry Partners: Engaging in partnerships with industry stakeholders to test HEAs in practical applications can accelerate the adoption of these materials. Such collaborations can help identify specific challenges and areas for improvement in HEA development.

9.7 SUMMARY

1. HEAs represent a transformative leap in materials science, offering unparalleled opportunities to develop advanced materials capable of withstanding extreme conditions, such as corrosion and radiation. Through this comprehensive chapter, we have traced the evolution of HEAs, their classification, processing techniques, and phase formation, underscoring the flexibility of these alloys in various industrial applications.
2. At the core of HEA research is the unique composition and microstructural complexity that distinguishes these alloys from conventional materials. The theory of the high entropy effect, which influences a wide array of material properties such as hardness, resistance to wear, fatigue strength, and thermal stability, forms the foundation for understanding the superior mechanical properties of HEAs. Furthermore, HEAs' ability to maintain thermal stability even at elevated temperatures makes them highly apt for high-performance applications in aerospace, nuclear energy, and defense sectors.
3. One of the key highlights of this chapter is the discussion of corrosion resistance in HEAs. The complex interplay of alloying elements, such as Al, Cr, and Mo, is vital in improving HEA's resistance to chloride-induced and acidic corrosion. We also examined various electrochemical techniques used to evaluate corrosion resistance, along with environmental testing methods like salt spray and high-temperature corrosion tests, which are critical for assessing the durability of HEAs in practical applications.
4. Radiation resistance, another critical attribute of HEAs, was explored through atomic arrangement, compositional complexity, and defective vacancies. The chapter presented various experimental techniques such as ion and neutron irradiation experiments, mechanical testing of irradiated

samples, and advanced characterization methods like APT and PAS. These techniques are essential in understanding how HEAs respond to radiation-induced damage and their applicability in high-radiation environments such as nuclear reactors, space exploration, and defense technologies.

5. The synergistic effects of corrosion and radiation resistance in HEAs were also discussed, emphasizing how irradiation-induced defects can influence corrosion behavior and exploring strategies to mitigate these effects. This dual resistance makes HEAs highly versatile for multifunctional applications.
6. Despite their promising potential, challenges remain in developing HEAs for corrosion and radiation resistance. Understanding the complex composition-property relationships, maintaining microstructural stability under dual-stress conditions, and optimizing materials for specific environmental factors are significant hurdles. Additionally, innovative manufacturing techniques, advanced alloy design approaches, and real-world testing methods must be further refined to unlock the full potential of HEAs.
7. Looking to the future, the chapter identified emerging trends such as biocompatible and bio-inspired HEAs, sustainable and environmentally friendly innovations, and the application of computational modeling to design next-generation alloys. These research avenues point to the broader multifunctional possibilities of HEAs, ranging from nuclear energy and aerospace to medical and defense applications. Ongoing research and development in this field will lead to the creation of advanced, high-performance materials that can withstand extreme conditions, ensuring their long-term relevance in essential industries for years to come.

REFERENCES

1. Cantor, B. "Stable and metastable multicomponent alloys." Annales de chimie-Science des materiaux 32, no. 3, (2007): pp. 245–256.
2. Garip, Y., Ergin, N., Ozdemir, O. "Resistance sintering of CoCrFeNiAlx (x= 0.7, 0.85, 1) high entropy alloys: Microstructural characterization, oxidation and corrosion properties." Journal of Alloys and Compounds 877, (2021): pp. 160–180.
3. Huang, M., Jiang, J., Wang, Y., Liu, Y., Zhang, Y. "High temperature and short-term oxidation behavior of CoCrCu1. 2FeNi high entropy alloy in solid and semi-solid state." Corrosion Science 193, (2021): p. 109897.
4. Cantor, B., Chang, I.T.H., Knight, P., Vincent, A.J.B. "Microstructural development in equiatomic multicomponent alloys." Materials Science and Engineering: A 375, (2004): pp. 213–218.
5. Yeh, J.W., Chen, S.K., Lin, S.J., Gan, J.Y., Chin, T.S., Shun, T.T., Tsau, C.H., Chang, S.Y. Nanostructured high-entropy alloys with multiple principal elements: Novel alloy design concepts and outcomes. Advanced Engineering Materials 6, no. 5, (2004), pp. 299–303.
6. Zhang, Y. et al. "Microstructures and properties of high-entropy alloys." Progress in Materials Science 61, (2014): pp. 1–93.
7. Zhou, Z., Wang, L., Zhao, X., Wu, J., Zhang, F., Pi, J. "Effects of W addition on the corrosion behaviors of FeCoNiCrMn high entropy alloy composites in the 3.5 wt. % NaCl solution." Surfaces and Interfaces 23, (2021): p. 100956.

8. Diao, H., Santodonato, L.J., Tang, Z., Egami, T., Liaw, P.K. "Local structures of high-entropy alloys (HEAs) on atomic scales: An overview." JOM 67, (2015): pp. 2321–2325.
9. Shi, Y., Yang, B., Liaw, P.K. "Corrosion-resistant high-entropy alloys: A review." Metals 7, no. 2 (2017): 43.
10. Senkov, O.N., Senkova, S.V., Woodward, C. "Effect of aluminum on the microstructure and properties of two refractory high-entropy alloys." Acta Materialia68, (2014): pp. 214–228.
11. Li, Z., Pradeep, K.G., Deng, Y., Raabe, D., Tasan, C.C. "Metastable high-entropy dual-phase alloys overcome the strength-ductility trade-off." Nature 534, (2016): pp. 227–230.
12. Hemphill, M.A., Yuan, T., Wang, G.Y., Yeh, J.W., Tsai, C.W., Chuang, A., Liaw, P.K. "Fatigue behavior of Al0.5CoCrCuFeNi high entropy alloys." Acta Materialia 60, (2012): pp. 5723–5734.
13. Tang, Z., Yuan, T., Tsai, C.-W., Yeh, J.-W., Lundin, C.D., Liaw, P.K. "Fatigue behavior of a wrought Al0.5CoCrCuFeNi two-phase high-entropy alloy." Acta Materialia 99, (2015): pp. 247–258.
14. Seifi, M., Li, D., Yong, Z., Liaw, P.K., Lewandowski, J.J. "Fracture toughness and fatigue crack growth behavior of as-cast high-entropy alloys." JOM 67, (2015): pp. 2288–2295.
15. Gludovatz, B., Hohenwarter, A., Catoor, D., Chang, E.H., George, E.P., Ritchie, R.O. "A fracture-resistant high-entropy alloy for cryogenic applications." Science 345, (2014): pp. 1153–1158.
16. Zhang, Z., Mao, M.M., Wang, J., Gludovatz, B., Zhang, Z., Mao, S.X., George, E.P., Yu, Q., Ritchie, R.O. "Nanoscale origins of the damage tolerance of the high-entropy alloy CrMnFeCoNi." Nature Communications 6, (2015): p. 10143.
17. Santodonato, L.J., Zhang, Y., Feygenson, M., Parish, C.M., Gao, M.C., Weber, R.J., Neuefeind, J.C., Tang, Z., Liaw, P.K. "Deviation from high-entropy configurations in the atomic distributions of a multi-principal-element alloy." Nature Communications 6, (2015): p. 5964.
18. Das, P., Nandan, R., Pandey, P.M. "A review on corrosion properties of high entropy alloys fabricated by additive manufacturing." Transactions of the Indian Institute of Metals 75, no. 10 (2022): pp. 2465–2476.
19. MacDonald, B.E., Fu, Z., Zheng, B., Chen, W., Lin, Y., Chen, F., Zhang, L. et al. "Recent progress in high entropy alloy research." JOM 69, (2017): pp. 2024–2031.
20. Murty, B.S., Yeh, J.W., Ranganathan, S., "Synthesis and processing," In: High entropy alloys, Elsevier, 2014, pp. 77–89.
21. Gao, M. C., Yeh, J.W., Liaw, P.K. Zhang, Z.Y., High entropy alloys. Cham: Springer International Publishing, 2016.
22. Miracle, D., Senkov, O. "A critical review of high entropy alloys and related concepts." Acta Materialia 122, (2017): pp. 448–511.
23. He, J.Y., Wang, H., Huang, H.L., Xu, X.D., Chen, M.W., Wu, Y., Liu, X.J., Nieh, T.G., An, K., Lu, Z.P. "A precipitation-hardened high-entropy alloy with outstanding tensile properties." Acta Materialia 102, (2016): pp. 187–196.
24. Guo, S., Hu, Q., Ng, C., Liu, C.T. "More than entropy in high-entropy alloys: Forming solid solutions or amorphous phase." Intermetallics 41, (2013): pp. 96–103.
25. Zhang, Y., Lu, Z.P., Ma, S.G., Liaw, P.K., Tang, Z., Cheng, Y.Q., Gao, M.C. "Guidelines in predicting phase formation of high-entropy alloys." MRS Communications 4, (2014): pp. 57–62.
26. Borkar, T., Gwalani, B., Choudhuri, D., Mikler, C.V., Yannetta, C.J., Chen, X., Ramanujan, R.V., Styles, M.J., Gibson, M.A., Banerjee, R. "A combinatorial assessment of AlxCrCuFeNi2 (0< x < 1.5) complex concentrated alloys: Microstructure, microhardness, and magnetic properties." Acta Materialia 116, (2016): pp. 63–76.

27. Gao, M.C., Yeh, J.-W., Liaw, P.K., Zhang, Y., eds. High-entropy alloys: Fundamentals and applications. Springer, 2016.
28. Fu, Y., Li, J., Luo, H., Du, C., Li, X. "Recent advances on environmental corrosion behavior and mechanism of high-entropy alloys." Journal of Materials Science & Technology 80, (2021): pp. 217–233.
29. Wu, J.-M., Lin, S.-J., Yeh, J.-W., Chen, S.-K., Huang, Y.-S., Chen, H.-C. "Adhesive wear behavior of AlxCoCrCuFeNi high-entropy alloys as a function of aluminum content." Wear 261, (2006): pp. 513–519.
30. Li, Y., Yang, Z., Ma, Z., Bai, Y., Wu, C., Li, J. "Effect of element V on the as-cast microstructure and mechanical properties of Al0. 4Co0. 5VxFeNi high entropy alloys." Journal of Alloys and Compounds 911, (2022): p. 165043.
31. Cichocki, K., Bała, P., Kozieł, T., Cios, G., Schell, N., Muszka, K. "Effect of Mo on phase stability and properties in FeMnNiCo high-entropy alloys." Metallurgical and Materials Transactions A 53, no. 5 (2022): pp. 1749–1760.
32. Zhang, F., Xiang, C., Han, E.-H., Zhang, Z. "Effect of Nb content on microstructure and mechanical properties of Mo0. 25V0. 25Ti1. 5Zr0. 5Nb x high-entropy alloys." Acta Metallurgica Sinica (English Letters) 35, no. 10 (2022): pp. 1641–1652.
33. Huang, T., Jiang, H., Lu, Y., Wang, T., Li, T. "Effect of Sc and Y addition on the microstructure and properties of HCP-structured high-entropy alloys." Applied Physics A 125, (2019): pp. 1–5.
34. Guo, Z., Zhang, A., Han, J., Meng, J. "Effect of Si additions on microstructure and mechanical properties of refractory NbTaWMo high-entropy alloys." Journal of Materials Science 54, (2019): pp. 5844–5851.
35. Huang, L., Wang, X., Jia, F., Zhao, X., Huang, B., Ma, J., Wang, C. "Effect of Si element on phase transformation and mechanical properties for FeCoCrNiSix high entropy alloys." Materials Letters 282, (2021): p. 128809.
36. Ye, X., Wang, T., Xu, Z., Liu, C., Wu, H., Zhao, G., Fang, D. "Effect of Ti content on microstructure and mechanical properties of CuCoFeNi high-entropy alloys." International Journal of Minerals, Metallurgy and Materials 27, (2020): pp. 1326–1331.
37. Chen, Q., Zhou, K., Jiang, L., Lu, Y., Li, T. "Effects of Fe content on microstructures and properties of AlCoCrFe x Ni high-entropy alloys." Arabian Journal for Science and Engineering 40, (2015): pp. 3657–3663.
38. Churyumov, A.Y., Pozdniakov, A.V., Bazlov, A.I., Mao, H., Polkin, V.I., Louzguine-Luzgin, D.V. "Effect of Nb addition on microstructure and thermal and mechanical properties of Fe-Co-Ni-Cu-Cr multiprincipal-element (high-entropy) alloys in as-cast and heat-treated state." JOM 71, (2019): pp. 3481–3489.
39. Zhu, M., Zhang, M., Yao, L., Jie, Z.Q., Liu, Y., Li, K., Jian, Z. "Effect of V content on phase formation and mechanical properties of the CoFeNiMnVx high-entropy alloys." Journal of Materials Engineering and Performance 31, no. 4 (2022): pp. 1–8.
40. Pauzi, S.S.M., Darham, W., Ramli, R., Harun, M.K., Talari, M.K. "Effect of Zr addition on microstructure and properties of FeCrNiMnCoZr x and Al 0.5 FeCrNiMnCoZr x high entropy alloys." Transactions of the Indian Institute of Metals 66, (2013): pp. 305–308.
41. Chandrakar, R., Kumar, A., Chandraker, S., Rao, K.R., Chopkar, M. "Microstructural and mechanical properties of AlCoCrCuFeNiSix (x= 0 and 0.9) high entropy alloys." Vacuum 184, (2021): p. 109943.
42. Liu, M., Xu, W., Zhang, S., Wang, Z., Wang, Z., Wang, B., Wang, D., Li, F. "Microstructures and hardnesses of AlCoCr0. 5FexNi2. 5 high entropy alloys with equal valence electron concentration." Journal of Alloys and Compounds 824, (2020): p. 153881.
43. Tong, C.J., Chen, M.R., Chen, S.K., Yeh, J.W., Shun, T.T., Lin, S.J., Chang, S.Y. "Mechanical performance of the AlxCoCrCuFeNi high-entropy alloy system with

multiprincipal elements." Metallurgical and Materials Transactions A 36, (2005): pp. 1263–1271.

44. Hsu, C.Y., Yeh, J.W., Chen, S.K., Shun, T.T. "Wear resistance and high-temperature compression strength of Fcc CuCoNiCrAl0.5Fe alloy with boron addition." Metallurgical and Materials Transactions A 35, (2004): pp. 1465–1469.
45. Chuang, M.H., Tsai, M.H., Wang, W.R., Lin, S.J., Yeh, J.W. "Microstructure and wear behavior of AlxCo1.5CrFeNi1.5 tiy high-entropy alloys." Acta Materialia 59, (2011): pp. 6308–6317.
46. Chen, M.R., Lin, S.J., Yeh, J.W., Chen, S.K., Huang, Y.S., Tu, C.P. "Microstructure and properties of Al0.5CoCrCuFeNiTix (x = 0–2.0) high-entropy alloys." Materials Transactions 47, (2006): pp. 1395–1401.
47. Hsu, C.Y., Sheu, T.S., Yeh, J.W., Chen, S.K. "Effect of iron content on wear behavior of AlCoCrFexMo0.5Ni high entropy alloys." Wear 268, (2010): pp. 653–659.
48. Chen, M.R., Lin, S.J., Yeh, J.W., Chen, S.K., Huang, Y.S., Chuang, M.H. "Effect of vanadium addition on the microstructure, hardness, and wear resistance of Al0.5CoCrCuFeNi high-entropy alloy." Metallurgical and Materials Transactions A 37, (2006): pp. 1363–1369.
49. Moazzen, P., Toroghinejad, M.R., Zargar, T., Cavaliere, P. "Investigation of hardness, wear and magnetic properties of NiCoCrFeZrx HEA prepared through mechanical alloying and spark plasma sintering." Journal of Alloys and Compounds 892, (2022): p. 161924.
50. Xiaotao, L., Wenbin, L., Lijuan, M., Jinling, L., Jing, L., Jianzhong, C. "Effect of boron on the microstructure, phase assemblage and wear properties of Al0.5CoCrCuFeNi high entropy alloy." Rare Metal Materials and Engineering 45, (2006): pp. 2201–2207.
51. Gorban, V.F., Krapivka, N.A., Karpets, M.V., Kostenko, A.D., Samelyuk, A.N., Kantsyr, E.V. "Effect of temperature on wear behavior of high-entropy alloys." Journal of Friction and Wear 38, (2017): pp. 292–296.
52. Yadav, S., Kumar, A., Biswas, K. "Wear behavior of high entropy alloys containing soft dispersoids (Pb, Bi)." Materials Chemistry and Physics 210, (2018): pp. 222–232.
53. Mindi, N.K., Alam, S.N., Dutta, K. "Effect of tungsten on microstructure and mechanical properties of novel NiMnCrMoW x high-entropy alloys developed by mechanical alloying." Journal of Materials Engineering and Performance (2024): pp. 1–18.
54. Niendorf, T., Wegener, T., Li, Z., Raabe, D. "Unexpected cyclic stress-strain response of dual-phase high-entropy alloys induced by partial reversibility of deformation." Scripta Materialia 143, (2018): pp. 63–67.
55. Bahadur, F., Biswas, K., Gurao, N.P. "Micro-mechanisms of microstructural damage due to low cycle fatigue in CoCuFeMnNi high entropy alloy." International Journal of Fatigue 130, (2020): p. 105258.
56. Xu, L., Jia, Y., Wang, Z., Wu, S., Jia, Y., Geng, C., Peng, J., Tan, X., Wang, G. "Dual precipitate simultaneous enhancement of tensile and fatigue strength in (FeCoNi) 86Al7Ti7 high-entropy alloy fabricated using selective laser melting." Journal of Materials Science & Technology 148, (2023): pp. 90–104.
57. Tian, Y.Z., Sun, S.J., Lin, H.R., Zhang, Z.F. "Fatigue behavior of CoCrFeMnNi high-entropy alloy under fully reversed cyclic deformation." Journal of Materials Science & Technology 35, no. 3 (2019): pp. 334–340.
58. Liu, K., Komarasamy, M., Gwalani, B., Shukla, S., Mishra, R.S. "Fatigue behavior of ultrafine grained triplex Al0. 3CoCrFeNi high entropy alloy." Scripta Materialia 158, (2019): pp. 116–120.
59. Liu, K., Nene, S.S., Frank, M., Sinha, S., Mishra, R.S. "Metastability-assisted fatigue behavior in a friction stir processed dual-phase high entropy alloy." Materials Research Letters 6, no. 11 (2018): pp. 613–619.

60. Liu, K., Gwalani, B., Komarasamy, M., Shukla, S., Wang, T., Mishra, R.S. "Effect of nano-sized precipitates on the fatigue property of a lamellar structured high entropy alloy." Materials Science and Engineering: A 760, (2019): pp. 225–230.
61. Kim, Y.-K., Baek, M.-S., Yang, S., Lee, K.-A. "In-situ formed oxide enables extraordinary high-cycle fatigue resistance in additively manufactured CoCrFeMnNi high-entropy alloy." Additive Manufacturing 38, (2021): p. 101832.
62. Shams, S.A.A., Kim, G., Won, J.W., Kim, J.N., Kim, H.S., Lee, C.S. "Effect of grain size on the low-cycle fatigue behavior of carbon-containing high-entropy alloys." Materials Science and Engineering: A 810, (2021): p. 140985.
63. Shams, S.A.A., Jang, G., Won, J.W., Bae, J.W., Jin, H., Kim, H.S., Lee, C.S. "Low-cycle fatigue properties of CoCrFeMnNi high-entropy alloy compared with its conventional counterparts." Materials Science and Engineering: A 792, (2020): p. 139661.
64. Lu, K., Knöpfle, F., Chauhan, A., Jeong, H.T., Litvinov, D., Walter, M., Kim, W.J., Aktaa, J. "Low-cycle fatigue behavior and deformation mechanisms of a dual-phase Al0. 5CoCrFeMnNi high-entropy alloy." International Journal of Fatigue 163, (2022): p. 107075.
65. He, H., Wang, B., Ma, D., Stoica, A.D., Wu, Z., Lan, S., Naeem, M., Wang, X.-L. "In situ neutron diffraction study of fatigue behavior of CrFeCoNiMo0. 2 high entropy alloy." Intermetallics 139, (2021): p. 107371.
66. Lu, K., Chauhan, A., Litvinov, D., Aktaa, J. "Temperature-dependent cyclic deformation behavior of CoCrFeMnNi high-entropy alloy." International Journal of Fatigue 160, (2022): p. 106863.
67. Jin, M., Hosseini, E., Holdsworth, S.R., Pham, M.-S. "Thermally activated dependence of fatigue behaviour of CrMnFeCoNi high entropy alloy fabricated by laser powder-bed fusion." Additive Manufacturing 51, (2022): p. 102600.
68. Gali, A., George, E.P. "Tensile properties of high-and medium-entropy alloys." Intermetallics 39, (2013): pp. 74–78.
69. Otto, F., Dlouhý, A., Somsen, C., Bei, H., Eggeler, G., George, E.P. "The influences of temperature and microstructure on the tensile properties of a CoCrFeMnNi high-entropy alloy." Acta Materialia 61, no. 15 (2013): pp. 5743–5755.
70. Chen, J., Yao, Z., Wang, X., Lu, Y., Wang, X., Liu, Y., Fan, X. "Effect of C content on microstructure and tensile properties of as-cast CoCrFeMnNi high entropy alloy." Materials Chemistry and Physics 210, (2018): pp. 136–145.
71. Daoud, H.M., Manzoni, A., Völkl, R., Wanderka, N., Glatzel, U. "Microstructure and tensile behavior of Al 8 Co 17 Cr 17 Cu 8 Fe 17 Ni 33 (at.%) high-entropy alloy." JOM 65, (2013): pp. 1805–1814.
72. He, J.Y., Liu, W.H., Wang, H., Wu, Y., Liu, X.J., Nieh, T.G., Lu, Z.P. "Effects of Al addition on structural evolution and tensile properties of the FeCoNiCrMn high-entropy alloy system." Acta Materialia 62, (2014): pp. 105–113.
73. Huo, W., Zhou, H., Fang, F., Hu, X., Xie, Z., Jiang, J. "Strain-rate effect upon the tensile behavior of CoCrFeNi high-entropy alloys." Materials Science and Engineering: A 689, (2017): pp. 366–369.
74. Jang, M.J., Praveen, S., Sung, H.J., Bae, J.W., Moon, J., Kim, H.S. "High-temperature tensile deformation behavior of hot rolled CrMnFeCoNi high-entropy alloy." Journal of Alloys and Compounds 730, (2018): pp. 242–248.
75. Jin, X., Zhou, Y., Zhang, L., Du, X., Li, B. "A novel Fe20Co20Ni41Al19 eutectic high entropy alloy with excellent tensile properties." Materials Letters 216, (2018): pp. 144–146.
76. Kireeva, I.V., Chumlyakov, Y.I., Pobedennaya, Z.V., Vyrodova, A.V., Karaman, I. "Twinning in [001]-oriented single crystals of CoCrFeMnNi high-entropy alloy at tensile deformation." Materials Science and Engineering: A 713, (2018): pp. 253–259.

77. Kuznetsov, A.V., Shaysultanov, D.G., Stepanov, N.D., Salishchev, G.A., Senkov, O.N. "Tensile properties of an AlCrCuNiFeCo high-entropy alloy in as-cast and wrought conditions." Materials Science and Engineering: A 533, (2012): pp. 107–118.
78. Lilensten, L., Couzinié, J.-P., Bourgon, J., Perrière, L., Dirras, G., Prima, F., Guillot, I. "Design and tensile properties of a bcc Ti-rich high-entropy alloy with transformation-induced plasticity." Materials Research Letters 5, no. 2 (2017): pp. 110–116.
79. Shun, T.-T., Du, Y.-C. "Microstructure and tensile behaviors of FCC Al0. 3CoCrFeNi high entropy alloy." Journal of Alloys and Compounds 479, no. 1-2 (2009): pp. 157–160.
80. Zhao, S., Xin, D., Chen, X., Stasic, J., Trtica, M., Siddiquee, A.N., Mohan, S. "Microstructure and enhanced tensile properties of AlCoxCrFeNi high entropy alloys with high Co content fabricated by laser melting deposition." Journal of Alloys and Compounds 917, (2022): p. 165403.
81. Yao, M.J., Pradeep, K.G., Tasan, C.C., Raabe, D. "A novel, single phase, non-equiatomic FeMnNiCoCr high-entropy alloy with exceptional phase stability and tensile ductility." Scripta Materialia 72–73, (2014): pp. 5–8.
82. Liu, L., Zhu, J.B., Zhang, C., Li, J.C., Jiang, Q. "Microstructure and the properties of FeCoCuNiSnx high entropy alloys." Materials Science and Engineering: A 548, (2012): pp. 64–68.
83. Liu, Y.X., Cheng, C.Q., Shang, J.L., Wang, R., Li, P., Zhao, J. "Oxidation behavior of high-entropy alloys AlxCoCrFeNi (x¼0.15, 0.4) in supercritical water and comparison with HR3C steel." Transactions of Nonferrous Metals Society of China 25, (2015c): pp. 1341–1351.
84. Rao, Z., Wang, X., Wang, Q., Liu, T., Chen, X., Wang, L., Hui, X. "Microstructure, mechanical properties, and oxidation behavior of AlxCr0.4CuFe0.4MnNi high entropy alloys." Advanced Engineering Materials 19, (2017): p. 1600726.
85. Chang, Y.J., Yeh, A.C. "The evolution of microstructures and high temperature properties of AlxCo1.5CrFeNi1.5Tiy high entropy alloys." Journal of Alloys and Compounds 653, (2015): pp. 379–385.
86. Grewal, H.S., Sanjiv, R.M., Arora, H.S., Kumar, R., Ayyagari, A., Mukherjee, S., Singh, H. "Activation energy and high temperature oxidation behavior of multi-principal element alloy." Advanced Engineering Materials 19, (2017): pp. 1–5.
87. Butler, T.M., Alfano, J.P., Martens, R.L., Weaver, M.L. "High temperature oxidation behavior of Al-Co-Cr-Ni-(Fe or Si) multicomponent high entropy alloys." JOM 67, (2015): pp. 246–259.
88. Dąbrowa, J., Cieslak, G., Stygar, M., Mroczka, K., Berent, K., Kulik, T., Danielewski, M. "Influence of Cu content on high temperature oxidation behavior of AlCoCrCuxFeNi high entropy alloys (x¼0; 0.5; 1)." Intermetallics 84, (2017): pp. 52–61.
89. Hsu, W.L., Yang, Y.C., Chen, C.Y., Yeh, J.W. "Thermal sprayed high-entropy NiCo0.6Fe0.2Cr1.5SiAlTi0.2 coating with improved mechanical properties and oxidation resistance." Intermetallics 89, (2017): pp. 105–110.
90. Adomako, N.K., Kim, J.H., Hyun, Y.T. "High-temperature oxidation behaviour of low entropy alloy to medium- and high-entropy alloys." Journal of Thermal Analysis and Calorimetry 133, (2018): pp. 13–26.
91. Gorr, B., Muller, F., Azim, M., Christ, H.J., Muller, T., Chen, H., Heilmaier, M. "High-temperature oxidation behavior of refractory high-entropy alloys: Effect of alloy composition." Oxidation of Metals 88, (2017): pp. 339–349.
92. Butler, T.M., Chaput, K.J., Dietrich, J.R., Senkov, O.N. "High temperature oxidation behaviors of equimolar NbTiZrV and NbTiZrCr refractory complex concentrated alloys (RCCAs)." Journal of Alloys and Compounds 729, (2017): pp. 1004–1019.
93. Chang, C.H., Titus, M.S., Yeh, J.W. "Oxidation behavior between 700 and 1300°C of refractory TiZrNbHfTa high-entropy alloys containing Al." Advanced Engineering Materials (2018): p. 1700948.

94. Lu, J., Chen, Y., Zhang, H., Ni, N., Li, L., He, L., Mu, R., Zhao, X., Guo, F. "Y/Hf-doped AlCoCrFeNi high-entropy alloy with ultra oxidation and spallation resistance." Corrosion Science 166, (2020): p. 108426.
95. Ye, B., Wen, T., Liu, D., Chu, Y. "Oxidation behavior of (Hf0. 2Zr0. 2Ta0. 2Nb0. 2Ti0. 2) C high-entropy ceramics at 1073-1473 K in air." Corrosion Science 153, (2019): pp. 327–332.
96. Ye, B., Wen, T., Chu, Y. "High-temperature oxidation behavior of (Hf0. 2Zr0. 2Ta0. 2Nb0. 2Ti0. 2) C high-entropy ceramics in air." Journal of the American Ceramic Society 103, no. 1 (2020): pp. 500–507.
97. Wang, H., Han, X., Liu, W., Wang, Y. "Oxidation behavior of high-entropy carbide (Hf0. 2Ta0. 2Zr0. 2Ti0. 2Nb0. 2) C at 1400–1600 C." Ceramics International 47, no. 8 (2021): pp. 10848–10854.
98. Wang, Y., Zhang, R., Zhang, B., Skurikhina, O., Balaz, P., Araullo-Peters, V., Reece, M.J. "The role of multi-elements and interlayer on the oxidation behaviour of (Hf-Ta-Zr-Nb) C high entropy ceramics." Corrosion Science 176, (2020): p. 109019.
99. Backman, L., Gild, J., Luo, J., Opila, E.J. "Part I: Theoretical predictions of preferential oxidation in refractory high entropy materials." Acta Materialia 197, (2020): pp. 20–27.
100. Backman, L., Gild, J., Luo, J., Opila, E.J. "Part II: Experimental verification of computationally predicted preferential oxidation of refractory high entropy ultra-high temperature ceramics." Acta Materialia 197, (2020): pp. 81–90.
101. Gao, M.C., Miracle, D.B., Maurice, D., Yan, X., Zhang, Y., Hawk, J.A. "High-entropy functional materials." Journal of Materials Research 33, no. 19 (2018): pp. 3138–3155.
102. Zhang, Y., Zuo, T.T., Cheng, Y.Q., Liaw, P.K. "High-entropy alloys with high saturation magnetization, electrical resistivity and malleability." Scientific Reports 3, no. 1 (2013): 1455.
103. Zuo, T.T., Li, R.B., Ren, X.J., Zhang, Y. "Effects of Al and Si addition on the structure and properties of CoFeNi equal atomic ratio alloy." Journal of Magnetism and Magnetic Materials 371, (2014): pp. 60–68.
104. Li, P., Wang, A., Liu, C.T. "Composition dependence of structure, physical and mechanical properties of FeCoNi (MnAl) x high entropy alloys." Intermetallics 87, (2017): pp. 21–26.
105. Ji, W., Wang, W., Wang, H., Zhang, J., Wang, Y., Zhang, F., Fu, Z. "Alloying behavior and novel properties of CoCrFeNiMn high-entropy alloy fabricated by mechanical alloying and spark plasma sintering." Intermetallics 56, (2015): pp. 24–27.
106. Schneeweiss, O., Friák, M., Dudová, M., Holec, D., Šob, M., Kriegner, D., Holý, V. et al. "Magnetic properties of the CrMnFeCoNi high-entropy alloy." Physical Review B 96, no. 1 (2017): 014437.
107. Feng, X., Zheng, R., Wu, Z., Zhang, Y., Li, Z., Tan, X., Xu, H. "Study on a new high-entropy alloy Nd20Pr20La20Fe20Co10Al10 with hard magnetic properties." Journal of Alloys and Compounds 882, (2021): p. 160640.
108. Na, S.-M., Lambert, P.K., Jones, N.J. "Hard Magnetic properties of FeCoNiAlCuXTiX based high entropy alloys." AIP Advances 11, no. 1 (2021).
109. Mao, A., Quan, F., Xiang, H.-Z., Zhang, Z.-G., Kuramoto, K., Xia, A.-L. "Facile synthesis and ferrimagnetic property of spinel (CoCrFeMnNi) 3O4 high-entropy oxide nanocrystalline powder." Journal of Molecular Structure 1194, (2019): pp. 11–18.
110. Jimenez-Segura, M.P., Takayama, T., Bérardan, D., Hoser, A., Reehuis, M., Takagi, H., Dragoe, N. "Long-range magnetic ordering in rocksalt-type high-entropy oxides." Applied Physics Letters 114, no. 12 (2019).
111. Meisenheimer, P.B., Kratofil, T.J., Heron, J.T. "Giant enhancement of exchange coupling in entropy-stabilized oxide heterostructures." Scientific Reports 7, no. 1 (2017): 13344.

112. Witte, R., Sarkar, A., Velasco, L., Kruk, R., Brand, R.A., Eggert, B., Ollefs, K., Weschke, E., Wende, H., Hahn, H. "Magnetic properties of rare-earth and transition metal based perovskite type high entropy oxides." Journal of Applied Physics 127, no. 18 (2020).
113. Ma, J., Zhao, B., Xiang, H., Dai, F.-Z., Liu, Y., Zhang, R., Zhou, Y. "High-entropy spinel ferrites MFe2O4 (M= Mg, Mn, Fe, Co, Ni, Cu, Zn) with tunable electromagnetic properties and strong microwave absorption." Journal of Advanced Ceramics 11, no. 5 (2022): pp. 754–768.
114. Chen, S.-K., Kao, Y.-F. "Near-constant resistivity in 4.2-360 K in a B2 Al2. 08CoCrFeNi." AIP Advances 2, no. 1 (2012).
115. Chou, H.P., Chang, Y.S., Chen, S.K., Yeh, J.W. "Microstructure, thermophysical and electrical properties in AlxCoCrFeNi (0 <= x <= 2) high-entropy alloys." Materials Science and Engineering: B 163, (2009): pp. 184–189.
116. Kao, Y.F., Chen, S.K., Chen, T.J., Chu, P.C., Yeh, J.W., Lin, S.J. "Electrical, magnetic, and Hall properties of AlxCoCrFeNi high-entropy alloys." Journal of Alloys and Compounds 509, (20110): pp. 1607–1614.
117. Jin, K., Sales, B.C., Stocks, G.M., Samolyuk, G.D., Daene, M., Weber, W.J., Zhang, Y., Bei, H. "Tailoring the physical properties of Ni-based single-phase equiatomic alloys by modifying the chemical complexity." Scientific Reports 6, (2016): p. 20159.
118. Guo, J., Wang, H., von Rohr, F., Wang, Z., Cai, S., Zhou, Y., Yang, K., Li, A., Jiang, S., Wu, Q., Cava, R.J., Sun, L. "Robust zero resistance in a superconducting high-entropy alloy at pressures up to 190 GPa." Proceedings of the National Academy of Sciences of the United States of America 114, (2017): p. 13144.
119. Antão, F., Dias, M., Correia, J.B., Galatanu, A., Galatanu, M., Mardolcar, U.V., Myakush, A. et al. "The effects of mechanical alloying on the physical and thermal properties of CuCrFeTiV alloy." Materials Science and Engineering: B 263, (2021): p. 114805.
120. Yan, X., Constantin, L., Lu, Y., Silvain, J.F., Nastasi, M., Cui, B. "(Hf0. 2Zr0. 2Ta0. 2Nb0. 2Ti0. 2) C high-entropy ceramics with low thermal conductivity." Journal of the American Ceramic Society 101, no. 10 (2018): pp. 4486–4491.
121. Zhao, Z., Xiang, H., Chen, H., Dai, F.-Z., Wang, X., Peng, Z., Zhou, Y. "High-entropy (Nd 0.2 Sm 0.2 Eu 0.2 Y 0.2 Yb 0.2) 4 Al 2 O 9 with good high temperature stability, low thermal conductivity, and anisotropic thermal expansivity." Journal of Advanced Ceramics 9, (2020): pp. 595–605.
122. Chen, H., Xiang, H., Dai, F.-Z., Liu, J., Zhou, Y. "High entropy (Yb0. 25Y0. 25Lu0. 25Er0. 25) 2SiO5 with strong anisotropy in thermal expansion." Journal of Materials Science & Technology 36, (2020): pp. 134–139.
123. Zhao, Z., Chen, H., Xiang, H., Dai, F.-Z., Wang, X., Peng, Z., Zhou, Y. "(La0. 2Ce0. 2Nd0. 2Sm0. 2Eu0. 2) PO4: A high-entropy rare-earth phosphate monazite ceramic with low thermal conductivity and good compatibility with Al2O3." Journal of Materials Science & Technology 35, no. 12 (2019): pp. 2892–2896.
124. Chen, H., Xiang, H., Dai, F.-Z., Liu, J., Lei, Y., Zhang, J., Zhou, Y. "High porosity and low thermal conductivity high entropy (Zr0. 2Hf0. 2Ti0. 2Nb0. 2Ta0. 2) C." Journal of Materials Science & Technology 35, no. 8 (2019): pp. 1700–1705.
125. Gild, J., Wright, A., Quiambao-Tomko, K., Qin, M. Tomko, J. A., bin Hoque, Md S., Braun, J. L. et al. "Thermal conductivity and hardness of three single-phase high-entropy metal diborides fabricated by borocarbothermal reduction and spark plasma sintering." Ceramics International 46, no. 5 (2020): pp. 6906–6913.
126. Shivam, V., Basu, J., Pandey, V.K., Shadangi, Y., Mukhopadhyay, N.K. "Alloying behaviour, thermal stability and phase evolution in quinary AlCoCrFeNi high entropy alloy." Advanced Powder Technology 29, no. 9 (2018): pp. 2221–2230.

127. Senkov, O.N., Wilks, G.B., Miracle, D.B., Chuang, C.P., Liaw, P.K. "Refractory high-entropy alloys." Intermetallics 18, no. 9 (2010): pp. 1758–1765.
128. Zou, Y., Ma, H., Spolenak, R. "Ultrastrong ductile and stable highentropy alloys at small scales." Nature Communications 6, (2015): p. 7748.
129. Sathiyamoorthi, P., Basu, J., Kashyap, S., Pradeep, K.G., Kottada, R.S. "Thermal stability and grain boundary strengthening in ultrafine-grained CoCrFeNi high entropy alloy composite." Materials & Design 134, (2017): pp. 426–433.
130. Zhu, H., Liu, L., Xiang, H., Dai, F.-Z., Wang, X., Ma, Z., Liu, Y., Zhou, Y. "Improved thermal stability and infrared emissivity of high-entropy REMgAl11O19 and LaMAl11O19 (RE= La, Nd, Gd, Sm, Pr, Dy; M= Mg, Fe, Co, Ni, Zn)." Journal of Materials Science & Technology 104, (2022): pp. 131–144.
131. Schuh, B., Mendez-Martin, F., Völker, B., George, E.P., Clemens, H., Pippan, R., Hohenwarter, A. "Mechanical properties, microstructure and thermal stability of a nanocrystalline CoCrFeMnNi high-entropy alloy after severe plastic deformation." Acta Materialia 96, (2015): pp. 258–268.
132. Guan, J., Li, D., Yang, Z., Wang, B., Cai, D., Duan, X., He, P., Jia, D., Zhou, Y. "Synthesis and thermal stability of novel high-entropy metal boron carbonitride ceramic powders." Ceramics International 46, no. 17 (2020): pp. 26581–26589.
133. Xue, L., Ding, Y., Pradeep, K.G., Case, R., Castaneda, H., Paredes, M. "The grain size effect on corrosion property of Al2Cr5Cu5Fe53Ni35 high-entropy alloy in marine environment." Corrosion Science 208, (2022): p. 110625.
134. Parakh, A., Vaidya, M., Kumar, N., Chetty, R., Murty, B.S. "Effect of crystal structure and grain size on corrosion properties of AlCoCrFeNi high entropy alloy." Journal of Alloys and Compounds 863, (2021): p. 158056.
135. Lee, C.P., Chen, Y.Y., Hsu, C.Y., Yeh, J.W., Shih, H.C. "Enhancing pitting corrosion resistance of AlxCrFel. 5MnNi0. 5 high-entropy alloys by anodic treatment in sulfuric acid." Thin Solid Films 517, no. 3 (2008): pp. 1301–1305.
136. Raza, A., Abdulahad, S., Kang, B., Ryu, H.J., Hong, S.H. "Corrosion resistance of weight reduced AlxCrFeMoV high entropy alloys." Applied Surface Science 485, (2019): pp. 368–374.
137. Zhang, J.J., Yin, X.L., Dong, Y., Lu, Y.P., Jiang, L., Wang, T.M., Li, T.J. "Corrosion properties of AlxCoCrFeNiTi0· 5 high entropy alloys in 0· 5M H2SO4 aqueous solution." Materials Research Innovations 18, no. sup4 (2014): S4-756.
138. Wang, S., Zhao, Y., Xu, X., Cheng, P., Hou, H. "Evolution of mechanical properties and corrosion resistance of Al0. 6CoFeNiCr0. 4 high-entropy alloys at different heat treatment temperature." Materials Chemistry and Physics 244, (2020): p. 122700.
139. Chen, C., Zhang, H., Fan, Y., Wei, R., Zhang, W., Wang, T., Zhang, T. et al. "Improvement of corrosion resistance and magnetic properties of FeCoNiAl0. 2Si0. 2 high entropy alloy via rapid-solidification." Intermetallics 122, (2020): p. 106778.
140. Mishra, R.K., Sahay, P.P., Shahi, R.R. "Alloying, magnetic and corrosion behavior of AlCrFeMnNiTi high entropy alloy." Journal of Materials Science 54, (2019): pp. 4433–4443.
141. Guo, W.H., Li, J., Qi, M., Xu, Y., Ezatpour, H.R. "Effects of heat treatment on the microstructure, mechanical properties and corrosion resistance of AlCoCrFeNiTi0. 5 high-entropy alloy." Journal of Alloys and Compounds 884, (2021): p. 161026.
142. Wang, Y., Li, G., Li, H., Mu, Y., Sun, K., Zhao, W., Li, D., Jia, Y., Wang, G. "Effect of aging on corrosion resistance of (FeCoNi) 86Al7Ti7 high entropy alloys." Corrosion Science 227, (2024): p. 111717.
143. Zhang, X., Guo, J., Zhang, X., Song, Y., Li, Z., Xing, X., Kong, D. "Influence of remelting and annealing treatment on corrosion resistance of AlFeNiCoCuCr high entropy alloy in 3.5% NaCl solution." Journal of Alloys and Compounds 775, (2019): pp. 565–570.

144. Wang, W., Wang, J., Sun, Z., Li, J., Li, L., Song, X., Wen, X., Xie, L., Yang, X. "Effect of Mo and aging temperature on corrosion behavior of (CoCrFeNi) 100-xMox high-entropy alloys." Journal of Alloys and Compounds 812, (2020): p. 152139.
145. Lin, C.-M., Tsai, H.-L., Bor, H.-Y. "Effect of aging treatment on microstructure and properties of high-entropy Cu0. 5CoCrFeNi alloy." Intermetallics 18, no. 6 (2010): pp. 1244–1250.
146. Guo, W., Li, J., Qi, M., Xu, Y., Ezatpour, H.R. "Effects of heat treatment on the microstructure, wear behavior and corrosion resistance of AlCoCrFeNiSi high-entropy alloy." Intermetallics 138, (2021): p. 107324.
147. Fujieda, T., Chen, M., Shiratori, H., Kuwabara, K., Yamanaka, K., Koizumi, Y., Chiba, A., Watanabe, S. "Mechanical and corrosion properties of CoCrFeNiTi-based high-entropy alloy additive manufactured using selective laser melting." Additive Manufacturing 25, (2019): pp. 412–420.
148. Tan, X.-R., Zhao, R.-F., Ren, B., Zhi, Q., Zhang, G.-P., Liu, Z.-X. "Effects of hot pressing temperature on microstructure, hardness and corrosion resistance of Al2NbTi3V2Zr high-entropy alloy." Materials Science and Technology 32, no. 15 (2016): pp. 1582–1591.
149. Wang, P., Wang, Y., Cui, F., Yang, X., Pan, A., Wu, W. "Microstructural evolution, mechanical properties and corrosion resistance of CoCrFeNiW0. 5 high entropy alloys with various annealing heat treatment." Journal of Alloys and Compounds 918, (2022): p. 165602.
150. Cheng, H., Luo, H., Wang, X., Li, X. "Microstructure and corrosion resistance properties of (FeCoNi) 86Al7Ti7 high-entropy alloy via different aging time." Corrosion Science 226, (2024): p. 111670.
151. Shang, X.L., Wang, Z.J., He, F., Wang, J.C., Li, J.J., Yu, J.K. "The intrinsic mechanism of corrosion resistance for FCC high entropy alloys." Science China Technological Sciences 61, (2018): pp. 189–196.
152. Zhao, Y., Wang, M., Cui, H., Zhao, Y., Song, X., Zeng, Y., Gao, X., Lu, F., Wang, C., Song, Q. "Effects of Ti-to-Al ratios on the phases, microstructures, mechanical properties, and corrosion resistance of Al2-xCoCrFeNiTix high-entropy alloys." Journal of Alloys and Compounds 805, (2019): pp. 585–596.
153. Qiu, Y., Thomas, S., Gibson, M.A., Fraser, H.L., Pohl, K., Birbilis, N. "Microstructure and corrosion properties of the low-density single-phase compositionally complex alloy AlTiVCr." Corrosion Science 133, (2018): pp. 386–396.
154. Qiu, Y., Thomas, S., Fabijanic, D., Barlow, A.J., Fraser, H.L., Birbilis, N. "Microstructural evolution, electrochemical and corrosion properties of AlxCoCrFeNiTiy high entropy alloys." Materials & Design 170, (2019): p. 107698.
155. Shon, Y., Joshi, S.S., Katakam, S., Rajamure, R.S., Dahotre, N.B. "Laser additive synthesis of high entropy alloy coating on aluminum: Corrosion behavior." Materials Letters 142, (2015): pp. 122–125.
156. Zhang, Y., Yan, X.H., Ma, J., Lu, Z.P., Zhao, Y.H. "Compositional gradient films constructed by sputtering in a multicomponent Ti–Al–(Cr, Fe, Ni) system." Journal of Materials Research 33, no. 19 (2018): pp. 3330–3338.
157. Ma, X., Sun, Y., Cheng, W., Chong, Z., Huang, L., Meng, A., Jiang, L. "Effect of high-speed laser cladding on microstructure and corrosion resistance of CoCrFeNiMo0. 2 high-entropy alloy." Journal of Central South University 29, no. 10 (2022): pp. 3436–3446.
158. Rui, H., Meiping, W., Chen, C., Xiaojin, M., YuLing, G. "Effects of laser energy density on microstructure and corrosion resistance of FeCrNiMnAl high entropy alloy coating." Optics & Laser Technology 152, (2022): p. 108188.
159. Dehestani, M., Sharafi, S., Khayati, G.R. "The effect of pulse current density on the microstructure, magnetic, mechanical, and corrosion properties of high-entropy alloy coating Fe–Co–Ni–Mo–W, achieved through electro co-deposition." Intermetallics 147, (2022): p. 107610.

160. Li, L., Lu, J., Liu, X., Dong, T., Zhao, X., Yang, F., Guo, F. "AlxCoCrFeNi high entropy alloys with superior hot corrosion resistance to Na2SO4+ 25% NaCl at 900 °C." Corrosion Science 187, (2021): p. 109479.
161. Lee, C., Chang, C., Chen, Y., Yeh, J., Shih, H. "Effect of the aluminium content of AlxCrFe1.5MnNi0.5 high-entropy alloys on the corrosion behaviour in aqueous environments." Corrosion Science 50, (2008): pp. 2053–2060.
162. Yan, Y., Fang, L., Tan, Y., Tao, X., Ouyang, Y., Du, Y. "Mechanical properties and corrosion resistance of AlxCoCuFeMn high-entropy alloys." Journal of Materials Research and Technology 24, (2023): pp. 5250–5259.
163. Izadi, M., Soltanieh, M., Alamolhoda, S., Aghamiri, S.M.S., Mehdizade, M. "Microstructural characterization and corrosion behavior of AlxCoCrFeNi high entropy alloys." Materials Chemistry and Physics 273, (2021): p. 124937.
164. Hu, R., Du, J., Zhang, Y., Ji, Q., Zhang, R., Chen, J. "Microstructure and corrosion properties of AlxCuFeNiCoCr (x= 0.5, 1.0, 1.5, 2.0) high entropy alloys with Al content." Journal of Alloys and Compounds 921, (2022): p. 165455.
165. Wang, S., Zhao, Y., Cheng, P., Guo, Q., Xu, X., Hou, H. "Study on the mechanical properties and corrosion resistance of AlxCoFeNiCr1− x high-entropy alloys." Materials Research Express 6, no. 12 (2020): 1265e2.
166. Zhang, C., Chen, G.J., Dai, P.Q. "Evolution of the microstructure and properties of laser-clad FeCrNiCoBx high-entropy alloy coatings." Materials Science and Technology 32, (2016): pp. 1–7.
167. Chong, Z., Sun, Y., Cheng, W., Han, C., Ma, X., Meng, A., Jiang, L. "Trace boronizing strengthened AlCoCrFeNi high-entropy alloy coating manufactured by laser remelting: Enhanced wear and corrosion resistances." Materials Letters 330, (2023): p. 133314.
168. Wei, X., Zhang, L., Zhang, F., Zhang, C., Jia, Q., Sun, K., Duan, D., Jiang, H., Li, G. "Effect of carbon addition on the microstructure and corrosion resistance of the CoCrFeNi high-entropy alloy." Corrosion Science (2024): p. 111965.
169. Chmielak, L., Mujica Roncery, L., Niederhofer, P., Weber, S., Theisen, W. "CrMnFeCoNi high entropy alloys with carbon and nitrogen: Mechanical properties, wear and corrosion resistance." SN Applied Sciences 3, (2021): pp. 1–12.
170. Zhang, Z.-C., Lan, A.-D., Zhang, M., Qiao, J.-W. "Effect of Ce on the pitting corrosion resistance of non-equiatomic high-entropy alloy Fe40Mn20Cr20Ni20 in 3.5 wt. % NaCl solution." Journal of Alloys and Compounds 909, (2022): p. 164641.
171. Wu, Z., Li, B., Chen, M., Yang, Y., Zheng, R., Yuan, L., Li, Z., Tan, X., Xu, H. "Tailoring magnetic property and corrosion resistance of FeCoNiCuAl high-entropy alloy with Ce additive." Journal of Alloys and Compounds 901, (2022): p. 163665.
172. Zhao, R.-F., Ren, B., Cai, B., Liu, Z.-X., Zhang, G.-P., Zhang, J. "Corrosion behavior of CoxCrCuFeMnNi high-entropy alloys prepared by hot pressing sintered in 3.5% NaCl solution." Results in Physics 15, (2019): p. 102667.
173. Jiang, D., Cui, H., Zhao, X., Chen, H., Ma, G., Song, X., Han, Y. "Effect of potential difference in CoxCrNiMoCB high-entropy alloy coatings on corrosion resistance improvement." Materials Letters 330, (2023): p. 133299.
174. Yan, X., Guo, H., Yang, W., Pang, S., Wang, Q., Liu, Y., Liaw, P.K., Zhang, T. "Al0.3CrxFeCoNi high-entropy alloys with high corrosion resistance and good mechanical properties." Journal of Alloys and Compounds 860, (2021): p. 158436.
175. Heidari, E., Atapour, M., Obeydavi, A. "The effect of Cr-content on the corrosion behavior of Ti0.5Mo0.5CoNiMnCrx high-entropy alloy thin films deposited by direct current magnetron sputtering." Journal of Alloys and Compounds 976, (2024): p. 173265.
176. Gao, Y., Chong, K., Liu, C., Cao, Y., Wu, D., Zou, Y. "Electrochemical corrosion and high temperature hot corrosion behavior of NbTaTiV and CrNbTaTiV high entropy alloy." Journal of Materials Research and Technology 28, (2024): pp. 216–234.

177. Zhang, M., Shi, X., Li, Z., Xu, H., Li, G. "Corrosion behaviors and mechanism of CrFeNi2 based high-entropy alloys." Corrosion Science 207, (2022): p. 110562.
178. Wang, J., Jiang, H., Chang, X., Zhang, L., Wang, H., Zhu, L., Qin, S. "Effect of Cu content on the microstructure and corrosion resistance of AlCrFeNi3Cux high entropy alloys." Corrosion Science 221, (2023): p. 111313.
179. Kato, M., Nishimoto, M., Muto, I., Sugawara, Y. "Role of Cu in corrosion resistance of CoCrCuFeNi medium-entropy alloys: Importance of compositional change and thickening of oxide films." Corrosion Science 213, (2023): p. 110982.
180. Chen, X., Qian, H., Luo, Y., Yang, B., Cui, T., Zhang, D. "Effects of Cu-content and passivation treatment on the corrosion resistance of Al0.3CuxCoCrFeNi highentropy alloys." Journal of Alloys and Compounds 920, (2022): p. 165956.
181. Meng, C., Song, Z., Qiu, X., Wang, G., Wu, C., Ren, X., Zhuang, W., Wang, X. "Effect of Cu content on microstructure and properties of CoCrFeNiCux high-entropy alloy coatings prepared by induction cladding." Journal of Alloys and Compounds 934, (2023): p. 167896.
182. Mukanov, S., Loginov, P., Fedotov, A., Bychkova, M., Antonyuk, M., Levashov, E. "The effect of copper on the microstructure, wear and corrosion resistance of CoCrCuFeNi high-entropy alloys manufactured by powder metallurgy." Materials 16, no. 3 (2023): 1178.
183. Peng, Z., Li, B., Luo, Z., Chen, X., Tang, Y., Yang, G., Gong, P. "A lightweight AlCrTiV0. 5Cu x high-entropy alloy with excellent corrosion resistance." Materials 16, no. 7 (2023): 2922.
184. Hsu, Y.-J., Chiang, W.-C., Wu, J.-K. "Corrosion behavior of FeCoNiCrCux high-entropy alloys in 3.5% sodium chloride solution." Materials Chemistry and Physics 92, no. 1 (2005): pp. 112–117.
185. Liu, F., Song, Q., Chen, R., Wang, C., Sun, J. "Effect of Co, Ni, Cu content on phase composition, microstructure and corrosion resistance of Co1-xCrFeNi1+ xCuy series high-entropy alloys." Vacuum 210, (2023): p. 111830.
186. Huang, K., Wang, G., Qing, H., Chen, Y., Guo, H. "Effect of Cu content on electrical resistivity, mechanical properties and corrosion resistance of AlCuxNiTiZr0. 75 high entropy alloy films." Vacuum 195, (2022): p. 110695.
187. Sun, X., Du, L., Lan, H., Cui, J., Wang, L., Li, R., Liu, Z., Liu, J., Zhang, W. "Mechanical, corrosion and magnetic behavior of a CoFeMn1.2NiGa0.8 high entropy alloy." Journal of Material Science & Technology 73, (2021): pp. 139–144.
188. Yao, M., Wu, X., Huang, J., Zhang, J., Zhang, X., Zhou, B. "Study on the corrosion resistance of Zr–1Nb–x M (M= S, Cu, Ge, Bi) alloys." Oxidation of Metals 84, (2015): pp. 647–659.
189. Liu, H.-C., Tsai, C.-W. "Effect of Ge addition on the microstructure, mechanical properties, and corrosion behavior of CoCrFeNi high-entropy alloys." Intermetallics 132, (2021): p. 107167.
190. Sun, Y., Wang, Z., Yang, H., Lan, A., Qiao, J. "Effects of the element La on the corrosion properties of CrMnFeNi high entropy alloys." Journal of Alloys and Compounds 842, (2020): p. 155825.
191. Torbati-Sarraf, H., Shabani, M., Jablonski, P.D., Pataky, G.J., Poursaee, A. "The influence of incorporation of Mn on the pitting corrosion performance of CrFeCoNi High Entropy Alloy at different temperatures." Materials & Design 184, (2019): p. 108170.
192. Wong, S.-K., Shun, T.-T., Chang, C.-H., Lee, C.-F. "Microstructures and properties of Al0. 3CoCrFeNiMnx high-entropy alloys." Materials Chemistry and Physics 210, (2018): pp. 146–151.
193. Dai, C., Zhao, T., Du, C., Liu, Z., Zhang, D. "Effect of molybdenum content on the microstructure and corrosion behavior of FeCoCrNiMox high-entropy alloys." Journal of Material Science & Technology 46, (2020): pp. 64–73.

194. Boakye, G.O., Straume, E.O., Gunnarsson, B.G., Kovalov, D., Karlsdottir, S.N. "Corrosion behaviour of HVOF developed Mo-based high entropy alloy coating and selected hard coatings for high temperature geothermal applications." Materials & Design 235, (2023): p. 112431.
195. Godlewska, E.M., Mitoraj-Królikowska, M., Czerski, J., Jawańska, M., Gein, S., Hecht, U. "Corrosion of Al (Co) CrFeNi high-entropy alloys." Frontiers in Materials 7, (2020): p. 566336.
196. Wang, K., Zhu, Y., Wang, P., Li, X., Malomo, B., Yang, L. "Enhancing corrosion resistance in CoCrFeNiTa high entropy alloys via Mo addition." Electrochimica Acta 480, (2024): p. 143951.
197. Wu, C., Zhang, S., Zhang, C., Zhang, H., Dong, S. "Phase evolution and properties in laser surface alloying of FeCoCrAlCuNix high-entropy alloy on copper substrate." Surface and Coatings Technology 315, (2017): pp. 368–376.
198. Zemanate, A.M., Jorge, A.M. Júnior, Andreani, G.F.L., Roche, V., Cardoso, K.R. "Corrosion behavior of AlCoCrFeNix high entropy alloys." Electrochimica Acta 441, (2023): p. 141844.
199. Qiu, X.-W., Liu, C.-G. "Microstructure and properties of Al2CrFeCoCuTiNix high-entropy alloys prepared by laser cladding." Journal of Alloys and Compounds 553, (2013): pp. 216–220.
200. Wen, X., Cui, X., Jin, G., Liu, Y., Zhang, Y., Zhang, X., Liu, E., Tian, H., Fang, Y. "Corrosion and tribo-corrosion behaviors of nano-lamellar Ni1.5CrCoFe0.5Mo0.1Nbx eutectic high-entropy alloy coatings: The role of dual-phase microstructure." Corrosion Science 201, (2022): p. 110305.
201. Wang, W., Yuan, S., Chen, J., Li, X., Niu, Z., Wei, R., Wang, T., Zhang, T., Guan, S., Li, F. "Corrosion in Cr-Fe-Co-Ni-Nb hypoeutectic and hypereutectic high entropy alloys." Materials Today, Communications 31, (2022): p. 103612.
202. Liu, C., Gao, Y., Chong, K., Guo, F., Wu, D., Zou, Y. "Effect of Nb content on the microstructure and corrosion resistance of FeCoCrNiNbx high-entropy alloys in chloride ion environment." Journal of Alloys and Compounds 935, (2023): p. 168013.
203. Yang, H., Liu, X., Li, A., Li, R., Xu, S., Zhang, M., Yu, P., Yu, S., Jiang, M., Huo, C. "Effect of silicon addition on the corrosion resistance of Al0.2CoCrFe1.5Ni high-entropy alloy in saline solution." Journal of Alloys and Compounds 964, (2023): p. 171226.
204. Gu, X., Zhuang, Y., Huang, D. "Corrosion behaviors related to the microstructural evolutions of as-cast Al0.3CoCrFeNi high entropy alloy with addition of Si and Ti elements." Intermetallics 147, (2022): p. 107600.
205. Bi, D., Chang, Y., Luo, H., Pan, Z., Zhao, Q., Cheng, H., Wang, X. et al. "Corrosion behavior and passive film characteristics of AlNbTiZrSix high-entropy alloys in simulated seawater environment." Corrosion Science 224, (2023): p. 111530.
206. Łoński, W., Spilka, M., Kądziołka-Gaweł, M., Gębara, P., Radoń, A., Warski, T., Łoński, S., Barbusiński, K., Młynarek-Żak, K., Babilas, R. "Microstructure, magnetic properties, corrosion resistance and catalytic activity of dual-phase AlCoNiFeTi and AlCoNiFeTiSi high entropy alloys." Journal of Alloys and Compounds 934, (2023): p. 167827.
207. Ji, C., Ma, A., Jiang, J. "Mechanical properties and corrosion behavior of novel Al-Mg-Zn-Cu-Si lightweight high entropy alloys." Journal of Alloys and Compounds 900, (2022): p. 163508.
208. Muangtong, P., Rodchanarowan, A., Chaysuwan, D., Chanlek, N., Goodall, R. "The corrosion behaviour of CoCrFeNi-x (x= Cu, Al, Sn) high entropy alloy systems in chloride solution." Corrosion Science 172, (2020): p. 108740.
209. Chang, W., Wang, X., Qian, H., Chen, X., Lou, Y., Zhou, M., Guo, D., Kwok, C.T., Tam, L.M., Zhang, D. "Effect of Sn addition on microstructure, hardness and corrosion behavior of CoCrFeNiSnx high entropy alloys in chloride environment." Corrosion Science 227, (2024): p. 111808.

210. Wu, M., Setiawan, R.C., Li, D. "Benefits of passive element ti to the resistance of AlCrFeCoNi high-entropy alloy to corrosion and corrosive wear." Wear 492, (2022): p. 204231.
211. Zhou, Z., Li, H., Liu, Y., Niu, T., Li, H. "Effect of Ti on the corrosion behavior of (FeCrCoNi) 100-xTix alloy." Corrosion Science 209, (2022): p. 110807.
212. Wang, X., Liu, Q., Huang, Y., Xie, L., Xu, Q., Zhao, T. "Effect of ti content on the microstructure and corrosion resistance of CoCrFeNiTix high entropy alloys prepared by laser cladding." Materials 13, no. 10 (2020): 2209.
213. Pratskova, S., Samoilova, O., Ageenko, E., Shaburova, N., Moghaddam, A.O., Trofimov, E. "Corrosion resistance of Al x CoCrFeNiM (M= Ti, V, Si, Mn, Cu) high entropy alloys in NaCl and H2SO4 solutions." Metals 12, no. 2 (2022): 352.
214. Han, Z., Chen, N., Lu, S., Luan, H., Peng, R., Xu, H., Shao, Y., Peng, Z., Yao, K. "Structures and corrosion properties of the AlCrFeNiMo0. 5Tix high entropy alloys." Materials and Corrosion 69, no. 5 (2018): pp. 641–647.
215. Öztürk, S., Alptekin, F., Önal, S., Sünbül, S.E., Şahin, Ömer, İçin, K. "Effect of titanium addition on the corrosion behavior of CoCuFeNiMn high entropy alloy." Journal of Alloys and Compounds 903, (2022): p. 163867.
216. Xiang, C., Zhang, Z.-M., Fu, H.-M., Han, E.-H., Wang, J.-Q., Zhang, H.-F., Hu, G.-D. "Microstructure, mechanical properties, and corrosion behavior of MoNbFeCrV, MoNbFeCrTi, and MoNbFeVTi high-entropy alloys." Acta Metallurgica Sinica (English Letters) 32, (2019): pp. 1053–1064.
217. Han, L.-X., Wang, C.-M., Sun, H.-F. "Microstructure and anticorrosion property of high-entropy alloy AlFeNiCrCoTi0. 5Vx." Materials Transactions 57, no. 7 (2016): pp. 1134–1137.
218. Gu, X.-H., Yan, H.-J., Zhang, Q.-H., Meng, X.-Z., Wu, L.-K., Cao, F.-H. "Microstructure characterization and corrosion behavior of alx (CoCrFeNi) 100-x (x= 0, 5, 10, 15, 20) high entropy alloys in 0.5 M H2SO4 solution." Journal of Alloys and Compounds 944, (2023): p. 169247.
219. Tang, Z., Huang, L., He, W., Liaw, P.K. "Alloying and processing effects on the aqueous corrosion behavior of high-entropy alloys." Entropy 16, no. 2 (2014): pp. 895–911.
220. Liu, Y.Y., Chen, Z., Shi, J.C., Wang, Z.Y., Zhang, J.Y. "The effect of Al content on microstructures and comprehensive properties in AlxCoCrCuFeNi high entropy alloys." Vacuum 161, (2019): pp. 143–149.
221. Kao, Y.-F., Lee, T.-D., Chen, S.-K., Chang, Y.-S. "Electrochemical passive properties of AlxCoCrFeNi (x= 0, 0.25, 0.50, 1.00) alloys in sulfuric acids." Corrosion Science 52, no. 3 (2010): pp. 1026–1034.
222. Fu, Y., Dai, C., Luo, H., Li, D., Du, C., Li, X. "The corrosion behavior and film properties of Al-containing high-entropy alloys in acidic solutions." Applied Surface Science 560, (2021): p. 149854.
223. Lee, C.P., Chen, Y.Y., Hsu, C.Y., Yeh, J.W., Shih, H.C. "The effect of boron on the corrosion resistance of the high entropy alloys Al0. 5CoCrCuFeNiB x." Journal of the Electrochemical Society 154, no. 8 (2007): C424.
224. Zhicheng, Z., Aidong, L., Min, Z., Junwei, Q. "Effect of Ce on the localized corrosion behavior of non-equiatomic high-entropy alloy Fe40Mn20Cr20Ni20 in 0.5 M H2SO4 solution." Corrosion Science 206, (2022): p. 110489.
225. Qiu, X.-w, Wu, M.-j, Liu, C.-g, Zhang, Y.-p, Huang, C.-x "Corrosion performance of Al2CrFeCoxCuNiTi high-entropy alloy coatings in acid liquids." Journal of Alloys Compounds 708, (2017): pp. 353–357.
226. Qiu, X., Wu, M., Liu, C., Zhang, Y., Huang, C. "Corrosion performance of Al2CrFeCoxCuNiTi high-entropy alloy coatings in acid liquids." Journal of Alloys and Compounds 708, (2017): pp. 353–357.

227. Xiao, D., Zhou, P., Wu, W., Diao, H., Gao, M., Song, M., Liaw, P. "Microstructure, mechanical and corrosion behaviors of AlCoCuFeNi-(Cr, Ti) high entropy alloys." Materials & Design 116, (2017): pp. 438–447.
228. Ren, B. et al. "Corrosion behavior of CuCrFeNiMn high entropy alloy system in 1 M sulfuric acid solution." Materials and Corrosion 63, (2011): pp. 828–834.
229. Chen, X., Qian, H., Lou, Y., Yang, B., Cui, T., Zhang, D. "Effects of Cu-content and passivation treatment on the corrosion resistance of Al0. 3CuxCoCrFeNi high-entropy alloys." Journal of Alloys and Compounds 920, (2022): p. 165956.
230. Hsu, K.-M., Chen, S.-H., Lin, C.-S.J.C.S. "Microstructure and corrosion behavior of FeCrNiCoMnx (x= 1.0, 0.6, 0.3, 0) high entropy alloys in 0.5 M H2SO4." Corrosion Science 190, (2021): p. 109694.
231. Chou, Y.L., Yeh, J.W., Shih, H.C. "The effect of molybdenum on the corrosion behaviour of the high-entropy alloys Co1. 5CrFeNi1. 5Ti0. 5Mox in aqueous environments." Corrosion Science 52, no. 8 (2010): pp. 2571–2581.
232. Shang, X.-L., Wang, Z.-J., Wu, Q.-F., Wang, J.-C., Li, J.-J., Yu, J.-K. "Effect of Mo addition on corrosion behavior of high-entropy alloys CoCrFeNiMo x in aqueous environments." Acta Metallurgica Sinica (English Letters) 32, (2019): pp. 41–51.
233. Huang, L., Wang, X., Zhao, X., Wang, C., Yang, Y.J.M.C. "Analysis on the key role in corrosion behavior of CoCrNiAlTi-based high entropy alloy." Materials Chemistry, Physics 259, (2021): p. 124007.
234. Kao, Y.F., Lee, T.D., Chen, S.K., Chang, Y.S. "Electrochemical passive properties of alxcocrfeni (x = 0, 0.25, 0.50, 1.00) alloys in sulfuric acids." Corrosion Science 52, (2010): pp. 1026–1034.
235. Tsau, C.-H., Yeh, C.-Y., Tsai, M.-C. "The effect of Nb-content on the microstructures and corrosion properties of CrFeCoNiNbx high-entropy alloys." Materials 12, no. 22 (2019): 3716.
236. Chen, W., Zhou, R., Li, W., Chen, Y.-H., Chou, T.-H., Wang, X., Liu, Y., Zhu, Y., Huang, J. "Effect of interstitial carbon and nitrogen on corrosion of FeCoCrNi multi-principal element alloys made by selective laser melting." Journal of Materials Science & Technology 148, (2023): pp. 52–63.
237. Jiang, S., Lin, Z., Xu, H., Sun, Y. "Studies on the microstructure and properties of AlxCoCrFeNiTi1-x high entropy alloys." Journal of Alloys and Compounds 741, (2018): pp. 826–833.
238. Qiu, X.W., Zhang, Y.P., Liu, C.G. "Effect of Ti content on structure and properties of Al2CrFeNiCoCuTix high-entropy alloy coatings." Journal of Alloys and Compounds 585, (2014): pp. 282–286.
239. Lin, C., Meng, L., Tan, M., Ke, L., Qi, T. "Microstructures and properties of Crx-FeNiCu0. 5Ti0. 5 high-entropy alloys for corrosion resistance." Intermetallics 153, (2023): p. 107781.
240. Xing, B., Ding, Q., Jin, B., Zuo, X., Zhang, N., Yin, S. "Corrosion resistance and passivation behavior of CoCrFeNi-TiAl high-entropy alloy coatings in acidic solutions." Journal of Thermal Spray Technology 31, no. 5 (2022): pp. 1673–1682.
241. Wang, X., Gao, Y., Li, K., Yan, J., Li, Y., Feng, J. "Effect of yttrium on the corrosion behaviour of 09CrCuSb alloy in concentrated sulphuric acid." Corrosion Science 69, (2013): pp. 369–375.
242. Meng, L., Lin, C., Fang, S., Ke, L., Tan, M., Qi, T. "Designing FeCr0.7NiCu0.5Ti0.5 high entropy alloy doped with rare earth yttrium: Effects on microstructure and passive film." Journal of Alloys and Compounds 967, (2023): p. 171772.
243. Yao, Y., Jin, Y., Gao, W., Liang, X., Chen, J., Zhu, S. "Corrosion behavior of AlFeCrCoNiZr x high-entropy alloys in 0.5 M sulfuric acid solution." Metals 11, no. 9 (2021): 1471.

244. Yeh, J.W. "Physical metallurgy of high-entropy alloys." JOM 67, (2015): pp. 2254–2261.
245. Lu, C., Niu, L., Chen, N., Jin, K., Yang, T., Xiu, P., Zhang, Y., Gao, F., Bei, H., Shi, S. et al. "Enhancing radiation tolerance by controlling defect mobility and migration pathways in multicomponent single-phase alloys." Nature Communications 7, (2016): p. 13564.
246. Tunes, M.A., Vishnyakov, V.M., Donnelly, S.E. "Synthesis and characterisation of high-entropy alloy thin films as candidates for coating nuclear fuel cladding alloys." Thin Solid Films 649, (2018): pp. 115–120.
247. Zhang, Y., Stocks, G.M., Jin, K., Lu, C., Bei, H., Sales, B.C., Wang, L., Béland, L.K., Stoller, R.E., Samolyuk, G.D. et al. "Influence of chemical disorder on energy dissipation and defect evolution in concentrated solid solution alloys." Nature Communications 6, (2015): p. 8736.
248. Jin, K., Lu, C., Wang, L.M., Qu, J., Weber, W.J., Zhang, Y., Bei, H. "Effects of compositional complexity on the ion-irradiation induced swelling and hardening in Ni-containing equiatomic alloys." Scrpta Materialia 119, (2016): pp. 65–70.
249. Egami, T., Ojha, M., Khorgolkhuu, O., Nicholson, D.M., Stocks, G.M. "Local electronic effects and irradiation resistance in high-entropy alloys." JOM 67, (2015): pp. 2345–2349.
250. Nagase, T., Anada, S., Rack, P.D., Noh, J.H., Yasuda, H., Mori, H., Egami, T. "Electron-irradiation-induced structural change in Zr-Hf-Nb alloy." Intermetallics 26, (2012): pp. 122–130.
251. Nagase, T., Rack, P.D., Noh, J.H., Egami, T. "In-situ TEM observation of structural changes in nano-crystalline CoCrCuFeNi multicomponent high-entropy alloy (HEA) under fast electron irradiation by high voltage electron microscopy (HVEM)." Intermetallics 59, (2015): pp. 32–42.
252. Wang, Z., Liu, C.T., Dou, P. "Thermodynamics of vacancies and clusters in high-entropy alloys." Physical Review Materials 1, (2017): p. 043601.
253. Nagase, T., Anada, S., Rack, P.D., Noh, J.H., Yasuda, H., Mori, H., Egami, T. "MeV electron-irradiation-induced structural change in the bcc phase of Zr–Hf–Nb alloy with an approximately equiatomic ratio." Intermetallics 38, (2013): pp. 70–79.
254. Xia, S.Q., Yang, X., Yang, T.F., Liu, S., Zhang, Y. "Irradiation resistance in AlxCoCrFeNi high entropy alloys." JOM 67, (2015): pp. 2340–2344.
255. Kiran Kumar, N.A.P., Li, C., Leonard, K.J., Bei, H., Zinkle, S.J. "Microstructural stability and mechanical behavior of FeNiMnCr high entropy alloy under ion irradiation." Acta Materialia 113, (2016): pp. 230–244.
256. Zhao, S. "Defect properties in a VTaCrW equiatomic high entropy alloy (HEA) with the body centered cubic (Bcc) structure." Journal of Material Science & Technology 44, (2020): pp. 133–139.
257. Xia, S., Gao, M.C., Yang, T., Liaw, P.K., Zhang, Y. "Phase stability and microstructures of high entropy alloys ion irradiated to high doses." Journal of Nuclear Materials 480, (2016): pp. 100–108.
258. Lu, C., Yang, T., Jin, K., Gao, N., Xiu, P., Zhang, Y., Gao, F., Bei, H., Weber, W.J., Sun, K. et al. "Radiation-induced segregation on defect clusters in single-phase concentrated solid-solution alloys." Acta Materialia 127, (2017): pp. 98–107.
259. Yang, T.-n, Lu, C., Velisa, G., Jin, K., Xiu, P., Zhang, Y., Bei, H., Wang, L. "Influence of irradiation temperature on void swelling in NiCoFeCrMn and NiCoFeCrPd." Scripta Materialia 158, (2019): pp. 57–61.
260. Chen, D., Tong, Y., Wang, J., Han, B., Zhao, Y.L., He, F., Kai, J.J. "Microstructural response of He+ irradiated FeCoNiCrTi0.2 high-entropy alloy." Journal of Nuclear Materials 510, (2018): pp. 187–192.

261. Chen, W.Y., Kirk, M.A., Hashimoto, N., Yeh, J.W., Liu, X., Chen, Y. “Irradiation effects on Al0.3CoCrFeNi and CoCrMnFeNi high-entropy alloys, and 316H stainless steel at 500°C.” Journal of Nuclear Materials 539, (2020): p. 152324.
262. He, M.R., Wang, S., Shi, S., Jin, K., Bei, H., Yasuda, K., Matsumura, S., Higashida, K., Robertson, I.M. “Mechanisms of radiation-induced segregation in CrFeCoNi-based single-phase concentrated solid solution alloys.” Acta Materialia 126, (2017): pp. 182–193.
263. Wang, F., Yan, X., Wang, T., Wu, Y., Shao, L., Nastasi, M., Lu, Y., Cui, B. “Irradiation damage in (Zr0.25Ta0.25Nb0.25Ti0.25)C high-entropy carbide ceramics.” Acta Materialia 195, (2020): pp. 739–749.
264. Chang, S., Tseng, K.-K., Yang, T.-Y., Chao, D.-S., Yeh, J.-W., Liang, J.-H. “Irradiation-induced swelling and hardening in HfNbTaTiZr refractory high-entropy alloy.” Materials Letters 272, (2020): p. 127832.
265. Lu, Y., Huang, H., Gao, X., Ren, C., Gao, J., Zhang, H., Zheng, S., Jin, Q., Zhao, Y., Lu, C. et al. “A promising new class of irradiation tolerant materials: Ti2ZrHfV0.5Mo0.2 high-entropy alloy.” Journal of Material Science & Technology 35, (2019): pp. 369–373.
266. Tunes, M.A., Le, H., Greaves, G., Schön, C.G., Bei, H., Zhang, Y., Edmondson, P.D., Donnelly, S.E. “Investigating sluggish diffusion in a concentrated solid solution alloy using ion irradiation with in situ TEM.” Intermetallics 110, (2019): p. 106461.
267. Chen, D., Tong, Y., Li, H., Wang, J., Zhao, Y.L., Hu, A., Kai, J.J. “Helium accumulation and bubble formation in FeCoNiCr alloy under high fluence He+ implantation.” Journal of Nuclear Materials 501, (2018): pp. 208–216.
268. Chen, D., Zhao, S., Sun, J., Tai, P., Sheng, Y., Zhao, Y., Yeli, G., Lin, W., Liu, S., Kai, W. et al. “Diffusion controlled helium bubble formation resistance of FeCoNiCr high-entropy alloy in the half-melting temperature regime.” Journal of Nuclear Materials 526, (2019): p. 151747.
269. Pu, G., Lin, L., Ang, R., Zhang, K., Liu, B., Liu, B., Peng, T., Liu, S., Li, Q. “Outstanding radiation tolerance and mechanical behavior in ultra-fine nanocrystalline Al1.5CoCrFeNi high entropy alloy films under he ion irradiation.” Applied Surface Science 516, (2020): p. 146129.
270. Zhang, W., Wang, M., Wang, L., Liu, C.H., Chang, H., Yang, J.J., Liao, J.L., Yang, Y.Y., Liu, N. “Interface stability, mechanical and corrosion properties of AlCrMoNbZr/(AlCrMoNbZr)N high-entropy alloy multilayer coatings under helium ion irradiation.” Applied Surface Science 485, (2019): pp. 108–118.
271. Komarov, F.F., Pogrebnyak, A.D., Konstantinov, S.V. “Radiation resistance of high-entropy nanostructured (Ti, Hf, Zr, V, Nb)N coatings.” Technical Physics 60, (2015): pp. 1519–1524.
272. Pickering, E.J., Carruthers, A.W., Barron, P.J., Middleburgh, S.C., Armstrong, D.E.J., Gandy, A.S. “High-entropy alloys for advanced nuclear applications.” Entropy 23, (2021): p. 98.
273. El-Atwani, O., Li, N., Li, M., Devaraj, A., Baldwin, J.K.S., Schneider, M.M., Sobieraj, D., Wróbel, J.S., Nguyen-Manh, D., Maloy, S.A., Martinez, E. “Outstanding radiation resistance of tungsten-based high-entropy alloys.” Science Advances 5, (2019): p. eaav2002.
274. Patel, D., Richardson, M.D., Jim, B., Akhmadaliev, S., Goodall, R., Gandy, A.S. “Radiation damage tolerance of a novel metastable refractory high entropy alloy V2.5Cr1.2WMoCo0.04.” Journal of Nuclear Materials 531, (2020): p. 152005.
275. Li, C., Hu, X., Yang, T., Kiran Kumar, N.A.P., Wirth, B.D., Zinkle, S.J. “Neutron irradiation response of a Co-free high entropy alloy.” Journal of Nuclear Materials 527, (2019): p. 151838.

276. Chen, W.Y., Liu, X., Chen, Y., Yeh, J.W., Tseng, K.K., Natesan, K. "Irradiation effects in high entropy alloys and 316H stainless steel at 300°C." Journal of Nuclear Materials 510, (2018): pp. 421–430.
277. Moschetti, M., Xu, A., Schuh, B., Hohenwarter, A., Couzinié, J.-P., Kruzic, J.J., Bhattacharyya, D., Gludovatz, B. "On the room temperature mechanical properties of an ion-irradiated TiZrNbHfTa refractory high entropy alloy." JOM 72, (2020): pp. 130–138.
278. Moschetti, M., Xu, A., Hohenwarter, A., Wei, T., Davis, J., Short, K., Thorogood, G.J., Kong, C., Couzinié, J.-P., Bhattacharyya, D. et al. "The influence of phase formation on irradiation tolerance in a nanocrystalline TiZrNbHfTa refractory high-entropy alloy." Advanced Engineering Materials (2023): p. 2300863.
279. Was, G.S., Busby, J.T., Allen, T. et al. "Emulation of neutron irradiation effects with protons: Validation of principle." Journal of Nuclear Materials 300, (2002): pp. 198–216.

Section IV

Applications of High Entropy Alloys

10 High Entropy Alloys as Alternatives to Ni-Based Superalloys

Ajit Panigrahi, Pradyut Sengupta and Tapas Kumar Bhanj

This chapter brings out a comparison of various processing techniques, the evolution of phase and microstructure, phase stability with temperature, mechanical properties at room temperature and high temperature (≥ 800°C), creep resistance of alloys, oxidation resistance, and wear resistance between superalloys and high entropy alloys, in particular refractory-based HEAs (RHEAs) elucidating the future potential of RHEAs for high-temperature applications. It also describes about the consideration of the gray areas in the advancement, processing, and characterization of RHEA alloy systems for their applicability in replacing Ni-based superalloys.

10.1 INTRODUCTION

The growing need of high-temperature structural materials in the domains of aerospace, metallurgical engineering, etc. warrants faster development of new materials with superior performance. At present, Ni-based superalloys are widely used for high-temperature applications owing to their outstanding comprehensive properties at temperatures as high as 1000°C. However, the use of nickel-based superalloys is not recommended when the working temperature rises above 1300°C due to the lower melting point of the precipitation phase and its primary constituents. Therefore, they are unable to satisfy the demands of higher-temperature applications. In 2004, a new class of alloy based on multiple principal elements has been developed, which is termed "high entropy alloys (HEAs)." These alloys show superior properties in many respects compared to Ni-based superalloys and therefore, researchers see a future potential in HEAs for high-temperature applications. With this brief background, a concise overview of superalloys and HEAs with an emphasized focus on RHEAs – a subset of HEA, is presented.

Conventionally, superalloys are defined as metallic alloys that can be used for high-temperature applications, often in excess of 0.7 of the absolute melting temperature. Based on composition, they are classified into iron, cobalt, and nickel-based superalloys. The evolution and applicability of superalloys were driven by the two most important factors, such as high-temperature mechanical performance, and inherent resistance to deteriorating factors like oxidation, wear, creep, etc. Iron, cobalt, and iron-cobalt-based superalloys are considered as earlier development of

DOI: 10.1201/9781003664093-14

superalloys while nickel-based superalloys belong to the latest development, which are strengthened by two major mechanisms: solid solution strengthening and precipitation hardening [1, 2]. The composition of Ni-based superalloys usually contains Al and Ti that aid the formation of Ni_3Al and Ni_3Ti intermetallic phases, popularly denoted by γ'-precipitates growing in austenitic Ni-matrix (γ). The crystal lattice structure of γ and γ' phases is face-centered cubic and cubic, respectively. Being in the same crystal structure family with similar lattice parameters, a lower value of misfit between these phases results in coherency. Hence, the microstructure remains stable even at elevated temperatures. Again, the formed γ' phase offers resistance to dislocation motion causing an increase in strength. Thus, the aim is to develop the finer γ'-precipitates by subjecting the superalloy to solutionizing temperature followed by ageing at lower temperatures. The γ'-precipitate enhances the high-temperature strength in superalloys. Based on the type of Ni-based superalloys, the alloying elements are 10–20% Cr, up to 8% Al/Ti, 5–10% C, 2–9% Mo, 1–5% Nb, and minor content of Ta, C, W, Zr, Ru, Re, Hf, B, etc. [3]. Elements Cr and Al improve the oxidation resistance by the formation of stable oxide phases like chromia and alumina. The elements C, Mo, Nb, W, Cr, and Ta are aiding in carbide formation causing the grain boundary strengthening in polycrystalline superalloys. Rhenium, an expensive element is generally incorporated to improve the creep strength by promotion of cuboidal form of γ'-precipitate in the matrix and is helpful in the reduction of diffusion rate. B and Zr enhance the cohesive strength of grains which improves the creep strength along with ductility [3]. Owing to superb high temperature properties, Ni-based superalloys usually find applications in critical areas of aerospace and industrial gas turbine blades, where the component is exposed to both high temperature as well as severe loading conditions. In 1939, the service life of turbine blade of jet engine was 10–25 hours at operating temperature of 650°C. With the advent of technology and continual research efforts by scientists and engineers in developing the materials, modern jet engines' turbine can sustain up to 50,000 hours or more at a service temperature of 1150°C. In addition, the weight of modern aircrafts has been tremendously increased. The typical weight of aircraft engine lies within a range of 2000–8500 kg in which metal/alloys constitute 85–95% of its total weight [4]. The major challenge of commercial aircrafts is to increase the thermal efficiency of engine in order to reduce carbon footprint. Therefore, there is an insatiable demand to enhance the turbine inlet temperature for improved operating efficiency, which urges researchers to continue their search of high-temperature structural materials.

Recently, HEAs were developed, which are based on the multiple principal elements (five or more) in either equimolar ratio or the variation of elements in the range of 5 to 35 at. % [5, 6]. After the discovery of HEAs in 2004, simultaneously by Yeh et al. [7] and by Cantor et. al.[8], these alloys received significant attention because of their unique properties, such as high strength, superior hardness, excellent phase stability and strength at elevated temperatures, oxidation resistance, etc. [9–12]. The remarkable combination of properties inspired the scientific community to explore and to develop several HEAs by varying the compositions through various techniques. These intriguing properties can be ascribed to four core effects, such as high configurational entropy, lattice distortion, sluggish diffusion, and cocktail effect [10]. For an alloy system, a change in Gibbs free energy (ΔG_{mix}) of mixing is defined as:

$$\Delta G_{mix} = \Delta H_{mix} - T\Delta S_{mix} \tag{10.1}$$

where ΔH_{mix} and ΔS_{mix} refer to enthalpy of mixing and entropy of mixing, respectively, and T is the absolute temperature. For a fixed enthalpy of mixing (ΔH_{mix}), a higher entropy of mixing will lead to a decrease in Gibbs free energy favoring the phase stability. For a random solid solution, the configurational entropy (ΔS_{Conf}) per mole is given as:

$$\Delta S_{Conf} = -R\sum_{i}^{n} x_i ln x_i \tag{10.2}$$

where R is the universal gas constant (= 8.314 $JK^{-1}mol^{-1}$) and x_i is the molar fraction of ith component. For the equiatomic alloys having n elements, the configurational entropy reaches the maximum value, which can be written as:

$$\Delta S_{Conf} = -R\left(\frac{1}{n}ln\frac{1}{n} + \frac{1}{n}ln\frac{1}{n} + \frac{1}{n}ln\frac{1}{n}\cdots\cdots n\,times\right) = Rlnn \tag{10.3}$$

For the equiatomic alloys having ternary, quaternary, and quinary elements, the ΔS_{Conf} becomes 1.1R, 1.39R, and 1.61R, respectively. Based on the entropy values, alloys systems are classified into three categories [13]: low ($\Delta S_{Conf} < 1$), medium ($1R \leq \Delta S_{Conf} < 1.5R$) and high entropy alloys ($\Delta S_{Conf} \geq 1.5R$). The higher configurational entropy (> 1.5R) results in the decrease of ΔG_{mix} and therefore, favors the formation of solid-solution phases.

Since 2004, several works have been carried out on HEAs consisting of transition elements such as Cu, Cr, Co, Ni, Fe, Mn, etc. In the year 2010, Senkov et al. [14] were the first to prepare HEA based on refractory elements, and coined the term "refractory high entropy alloys." Thus, HEAs can be classified into two major categories based on constituting elements: i) transition metal-based high entropy alloys (TMHEAs) and ii) refractory high entropy alloys (RHEAs). The terms HEA and TMHEA are invariably used. Unless specified, HEA always indicates the elements are transition metals. TMHEAs show very high room-temperature compressive strength > 1500 MPa [7, 15]. However, similar to conventional alloys, a rapid decrease in strength occurs at temperatures above 0.6 T_m due to softening [16]. On the other hand, RHEAs constitute the elements with higher melting points greater than 1600°C, such as Mo, Nb, Hf, Ta, Ti, Zr, Cr, W, etc. Senkov et al. introduced MoNbTaW and MoNbTaWV alloys and demonstrated superior high-temperature mechanical properties compared to Ni-based superalloys [17]. Since then, the research of HEAs has shifted the polarity towards RHEAs and it is proposed that RHEAs have the potential to replace nickel-based superalloys in the future.

To evaluate the potential applicability of RHEAs, this chapter provides a comprehensive comparison of properties between Ni-based superalloys and RHEAs. In the subsequent sections, fabrication technologies, evolution of microstructure and phase, the thermal stability of phase, high-temperature strength, creep resistance, oxidation and wear resistance for both Ni-based superalloys and RHEAs are outlined. In the last section, the summary and future scope of the research on RHEA is presented.

10.2 PREPARATION TECHNIQUES

The preparation techniques and heat treatments are key to the variation of microstructure, which eventually controls the mechanical properties. The melting points of Ni-based superalloys, such as IN625 and IN718, are below 1400°C [18]. Thus, they are primarily prepared using vacuum induction melting (VIM). The prepared ingots are remelted using secondary melting techniques such as vacuum arc remelting, electro-slag remelting (ESR), etc. to decrease the macro-segregation – improving the chemical homogeneity of the alloy. Further, it is subjected to subsequent deformation, such as hot forging, hot-rolling extrusion, etc. to fabricate wrought alloys, which causes the breakdown of cast structures, termed the conversion process. In addition, the deformation processes reduce the solidification defects and porosity. IN718 is a classic example of the wrought form of Ni-based superalloy which is used for discs of aircraft engines. For manufacturing of complex-shaped components with integrated designs, including turbine blades and vanes, investment casting – a lost wax process is followed. The heat treatment of the Ni-based superalloys includes multi-stepped homogenization (at 1260–1340°C, 90–95% of melting point) and aging treatment. The main aim of these treatments is to produce finer coherent γ'-precipitate in a γ-matrix without the formation of undesired topological close-packed (TCP) phase eliminating chemical micro-segregation. To prepare the Ni-based single crystal SX super alloys, an as-cast polycrystalline ingot is placed in a ceramic crucible with SX seed and directionally solidified in a Bridgeman furnace [4]. Liu et al. fabricated Ni-based single crystal (SX) superalloy Ni-Co-Cr-Mo-Ta-Re-Al-Ti and reported comparable mechanical properties to Rene'N5 and CMSX-4 [19]. A more advanced technology – additive manufacturing (AM) route – is also getting significant attention to prepare complicated structures with precision. Up to now, AM was successfully adopted to prepare parts of Ni-based superalloys, such as IN625 [20] and IN718 [21]. However, hot-cracking of the component due to faster solidification, the entrapment of porosity, and mechanical integrity of the produced part [21] are key issues of the AM route, which require technological intervention from researchers and manufacturers.

In recent years, a large volume of work in RHEAs has led to the exploration of different methods to prepare the RHEAs. These alloys are widely prepared by the melting and casting route. The second choice of the fabrication route is through powder metallurgy processing. In less than a decade's time, AM has also been targeted to prepare components of RHEAs. However, a bulk of the research work utilizes the melting route as a mature processing technique. Since the elements have high melting points, arc melting processing is the viable option for processing. Due to the poor oxidation resistance of refractory metals, the preparation of RHEAs involves the melting of refractory elements in vacuum conditions or a high pure inert gas medium. The ingot obtained during this process is melted several times (e.g., five to seven times) in order to minimize the compositional segregation during solidification. The vacuum arc melting (VAM) was adopted to prepare most of RHEAs [22]. Many a time, RHEAs are homogenized to break the dendritic structure. Also, hot isostatic pressing of VAM-processed RHEAs is a common practice followed to reduce porosity [10, 23–25]. Another processing method is through the powder metallurgy route which involves the mixing of refractory elements, and mechanical alloying followed by spark plasma sintering (SPS) or hot forging – consolidation techniques to prepare RHEAs. The advantages of this process are elimination of compositional segregation, ultrafine to nanograin sizes with extended solid solubility, etc. The downside of this process is contamination of the powder mixture during ball milling, which may not lead to the formation of RHEAs or would cause adverse effects on mechanical properties. A few research groups have succeeded to prepare RHEAs by powder metallurgy route [26–29]. Here, at this point, it is worth to mention that RHEAs prepared by this route always ends up with the formation of precipitates upon sintering. Recently, additive manufacturing (AM) technologies such as selective laser melting, laser metal deposition, electron beam melting, etc. are adopted to fabricate RHEA parts. WTaMoNb and TiZrNbHfTa RHEAs were successfully prepared by Zhang et al. [30] and Dobbelstein et al. [31], respectively, using laser processing techniques. The use of magnetron sputtering technique to deposit a thin film of RHEA on various substrates is also reported in [32, 33].

10.3 EVOLUTION OF PHASE AND MICROSTRUCTURE

The phase and microstructure play significant roles in deciding the mechanical properties. Table 10.1 and Figure 10.1 summarize the phase constitution of superalloys and RHEAs with respect to processing conditions. Ni-based superalloys are mainly characterized by the precipitation of γ'-phase in austenitic γ-matrix. The carbide phases are also present as MC, M_6C, $M_{23}C_6$, M_7C_3 (M = Metal) at the interface of the grain boundary. Depending on the composition and heat treatment, γ" phase (Ni_3Nb) forms. Aging for a longer duration is often associated with the formation of a detrimental δ-phase. The refractory alloying elements in excess can result in the formation of an undesirable TCP phase. The crystal structures of various TCP phase are σ-tetragonal (FeCr, CrNiMo), μ-rhombohedral (Mo_6Co_7, Co_2W_6), and P-orthorhombic ($Mo_{42}Cr_{18}Ni_{40}$). The morphology of TCP phases is plate or needle-like, commonly formed at the matrix interface. In addition, the brittle Laves phases such as Fe_2Nb, Fe_2Mo, Co_2Ta, etc. might also be formed, which disappear by homogenization treatment. The existence of these Laves phases reduces the creep rupture strength and enhances the crack formation. A service exposed IN625 (60,000 hours at 600°C) shows intragranular Nb-carbides along with γ'' (Ni_3Nb) having BCT crystal structure and Ni_2(Cr, Mo) intermetallic phase [34]. Aging at 650°C for 10 hours elicits the dissolution of Ni_2(Cr, Mo) and ellipsoid form of γ"-phase distributed evenly in the matrix, whereas aging for 500 hours causes the formation of needle-shaped δ-phase (Ni_3Nb) having an orthorhombic crystal structure, which is incoherent with the matrix. Bhujangrao et al. [35] compared the microstructure and mechanical properties of IN718 prepared by wire arc additive manufacturing (WAAM) technique with the wrought form and observed

TABLE 10.1
Summarization of Occurrence of Phases in Ni-Based Super Alloys and RHEAs with Respect to Compositions and Fabricating Techniques

Alloy	Preparation Method	Phase	Reference
Inconel 625	C, Service exposed	γ (FCC) and γ'' (Ni_3Nb), Nb-carbide (BCT) at grain boundaries, Ni_2(Cr, Mo) –intermetallic phase	[34]
Inconel 625	C, Service exposed + aging at 650°C for 1–500 h	γ and γ'' (Ni_3Nb), Nb-carbide at grain boundaries, dissolution of Ni_2(Cr,Mo) – intermetallic phase up to 10 h. Longer exposure (500 h) causes formation of orthorhombic δ-phase	[34]
Inconel 718	C + HT+ hot deformation	γ and γ'-phase ($L1_2$), MC at grain boundary, minor content of δ-phase	[35]
Inconel 718	AM (WAAM)	γ and γ'-phase, MC at grain boundary, minor content of δ-phase in addition to Laves phase	[35]
Rene' N5	DS + HT	Cuboids of γ'-phase (60–70%) with thin layer of γ-matrix	[36, 39]
CMSX-4	DS+HT	γ'-phase (70%) with thin layer of γ-matrix	[40, 41]
$Al_{0.4}Hf_{0.6}NbTaTiZr$	C+HIP+A	BCC	[42]
$Al_{20}Cr_{10}Nb_{15}Ti_{20}V_{25}Zr_{10}$	C	B2+C14 Laves+Zr_5Al_3	[37]
$Al_{0.5}NbTa_{0.5}TiZr$	PM (CIP, sintering)	BCC	[28]
$Al_{0.5}NbTa_{0.5}TiZr$	PM (CIP, sintering + hot forging)	BCC (Ti, Nb, Ta)-rich matrix and +HCP (Zr, Al)-rich precipitates	[28]
AlCrMoTi	C+A	BCC	[43]
$AlMo_{0.5}NbTa_{0.5}TiZr$	C+HIP+A	BCC1+BCC 2+Al_4Zr_5	[42]
AlMoNbTi	C+A	BCC	[43]
AlNbTiV	C+A	BCC	[44]
CrHfNbTiZr	C, C+A	BCC + Cr_2Nb+Cr_2Hf Laves	[45]
$CrMo_{0.5}NbTa_{0.5}TiZr$	C+HIP+A	BCC1+BCC2+Laves	[23]
CrMoNbTaTiVWZr	C	BCC	[46]
CrMoNbTaTiVZr	C	BCC	[46]
CrMoNbTaVW	PM (SPS)	C15 Laves $(Cr,V)_2(Ta,Nb)$ and Ta_2VO_6, C14 + BCC matrix	[26]
CuMoTaVW	MS	BCC	[32]
$HfMo_{0.5}NbSi_{0.3}TiV_{0.5}$	C	BCC+ M_5Si_3	[47]
$HfMo_{0.5}NbSi_{0.5}TiV_{0.5}$	C	BCC+ M_5Si_3	[47]
$HfMo_{0.5}NbSi_{0.7}TiV_{0.5}$	C	BCC+ M_5Si_3	[47]
$HfMo_{0.5}NbTiV_{0.5}$	C	BCC	[47]
HfMoNbTaTiZr	C	BCC	[22]
HfMoNbTiZr	C	BCC	[48]
HfMoTaTiZr	C	BCC	[22]
$HfNbSi_{0.5}TiV$	C	BCC+M_5Si_3	[49]

(Continued)

TABLE 10.1 *(Continued)*
Summarization of Occurrence of Phases in Ni-Based Super Alloys and RHEAs with Respect to Compositions and Fabricating Techniques

Alloy	Preparation Method	Phase	Reference
HfNbTaTiVZr	C	BCC	[50]
HfNbTaTiZr	C+HIP+A	BCC	[25]
HfNbTaZr	C	BCC	[51]
HfNbTiVZr	C+A	BCC+ V_2Hf Laves	[45]
MoNbTaTiV	C	BCC	[52]
MoNbTaTiVZr	C	BCC	[46]
MoNbTaTiZr	C, C+A	BCC1+BCC 2	[53]
MoNbTaVW	C+HIP+A	BCC	[17]
MoNbTaW	C+HIP+A	BCC	[17]
MoNbTaW	AM (SLM)	BCC	[30]
MoTaTiVZr	C	BCC	[17]
MoTiVZr	C	BCC	[46]
NbTaTiZr	C	BCC	[53]

A – Homogenization annealing, AM – Additive Manufacturing, C-Melting and Casting, CIP – Cold Isostatic Pressing, DS – Directional Solidification, HIP – Hot Isostatic Pressing, HT – Heat treatment, MS – Magnetron sputtering, PM – Powder Metallurgy, SPS – Spark Plasma Sintering.

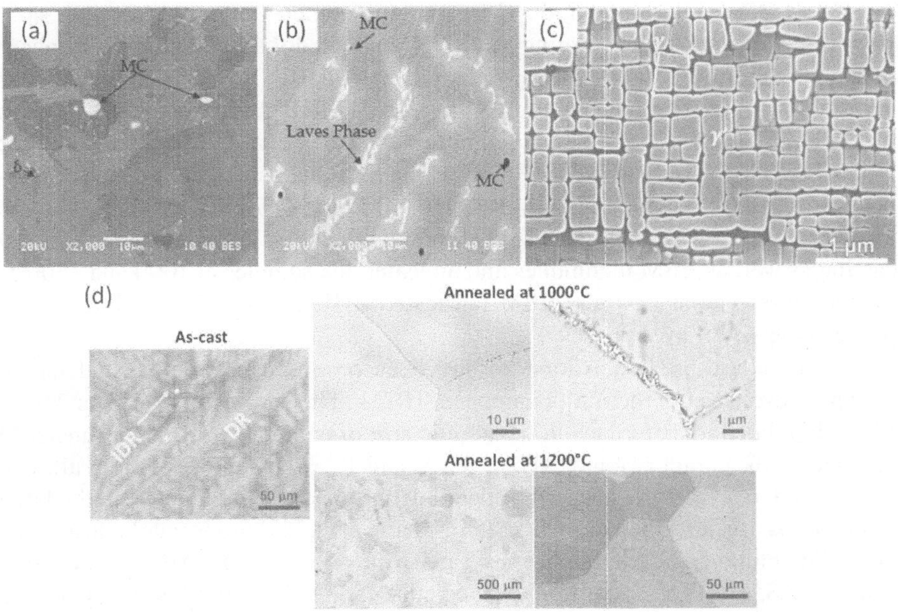

FIGURE 10.1 Microstructure of (a) IN718 in wrought form [35], (b) IN718 prepared by WAAM [35], (c) Ni-based single crystal superalloy [19], and (d) HfNbTaTiZr RHEA in as-cast and annealing condition [61].

formation of Laves phase in addition to MC and minor content of δ-phase (Figure 10.1(a) and (b)). In Ni-based SX superalloy, a higher volume fraction (60–70%) of γ' -phase in the cuboid form is beneficial for improvement in mechanical strength and creep rupture strength [36]. The microstructure of Ni-based SX superalloy is presented in Figure 10.1(c). As γ' phase is not penetrable by dislocations, mobility of dislocation is channelized through a thin γ-matrix layer. Upon exposure to stress at high temperature for a longer duration, the microstructure becomes rafted (thinning of γ'-phase and eventually fused to form an elongated structure). The rafted microstructure is not desirable as this causes reduction in creep strength significantly [36].

Based on the evolution of phase, RHEAs can be categorized into three types. The first one is RHEAs with a single BCC solid solution phase. The elements of RHEAs, such as W, Nb, Mo, Ta, Cr, and V are of BCC crystal structure. The other elements that are often added to the above are Ti, Hf, and Zr, which exhibit hexagonal crystal structures at room temperature. However, at high temperatures, the BCC phase is structurally stable. The fabrication of RHEA requires a very high temperature for its melting. Thus, at high temperature, all the elements are of BCC crystal structure and show higher mutual solubility, which upon solidification forms the BCC solid solution phase. The second category includes RHEAs having multiphase systems, such as B2, Laves phase, and intermetallic precipitates aside from the BCC-solid solution phase. $Al_{20}Cr_{10}Nb_{15}Ti_{20}V_{25}Zr_{10}$ when cast by VAM results in the hypoeutectic microstructure consisting of a softer B2 phase along with harder C14 Laves phase [37]. Upon annealing at 1200°C, Zr_5Al_3 intermetallic forms. Cao et al. [28] prepared $TiNbTa_{0.5}ZrAl_{0.5}$ RHEA by cold isostatic pressing followed by sintering at 1300°C for 16 hours under vacuum. After that, samples were subjected to hot forging at 1200°C and cited that the parent BCC phase decomposes into BCC (Ti, Nb, Ta)-rich matrix and HCP (Zr, Al)-rich precipitates upon hot forging. The hot forged $TiNbTa_{0.5}ZrAl_{0.5}$ RHEA shows a yield strength of 1740 MPa and 12% strain under compression. Long et al. [26] consolidated the NbMoTaWVCr by SPS at 1400–1700°C. Upon SPS, the parent BCC phase decomposes to form C15 (cubic) and C14 (hexagonal) Laves phase $(Cr, V)_2(Ta, Nb)$ and Ta_2VO_6 particles. Katz-Demyanetz et al. [38] produced $Al_{0.5}CrMoNbTa_{0.5}$ alloy by casting as well as EBM techniques and annealed the samples at 1000 and 1300°C for 24 hours. The as-cast and EBM fabricated RHEA shows BCC solid solution phase along with Laves phases.

Various phase prediction models have been proposed in order to understand the phase evolution in the multicomponent HEAs. The parameters, such as ΔH_{mix}, ΔS_{mix}, δ (related to difference in the atomic size of components), the parameter Ω (defined by the combination of ΔH_{mix}, ΔS_{mix}, and T_m = the melting point of alloy in K) are used to predict the phase. The detail definition of the parameters and related calculations are described elsewhere [54]. Zhang et al. [55] proposed the formation of a solid-solution phase in HEAs must satisfy the criteria $-20 \le \Delta H_{mix} \le 5$, $\delta \le 6.4$, and $12 \le \Delta S_{mix} \le 17.5$. According to Yang et al. [56], $\Omega \ge 1.1$ and $\delta \le 6.6$ favors the formation of a solid-solution phase. Guo et al. [57] suggested the criteria $-11.6 \le \Delta H_{mix} < 3.2$ and $\delta \le 6.6$ favors the solid-solution phase formation. The ΔH_{mix},

ΔS_{mix}, and δ of RHEAs are summarized in [58]. Considering RHEAs containing four to seven in equiatomic ratio, the value of ΔH_{mix} varies between –6.6 to –3.6 kJ/mol and ΔS_{mix} extend from 8.9 to 16.1 $JK^{-1}.mol^{-1}$, whereas δ values lie within the range of 1.9–4.8%. The parameter Ω for RHEAs is always larger than 1.1 [53]. All the parameters within the range indicate favorable formation of the solid solution phase. Another parameter is the valence electron concentration (VEC), which predicts the stability of the phase in terms of face centered cubic (FCC) or body centered cubic (BCC). HEAs with VEC $\geq$ 8 and VEC < 6.87 stabilize FCC and BCC as major phase, respectively [59]. The VEC of RHEAs such as HfNbTaTiZr, NbTaTiVZr, and NbTaTiZr are 4.4, 4.6, and 4.5, respectively – suggesting stabilization of the BCC solid solution phase [53].

Owing to the high melting temperature of constituting elements of RHEAs, these are prepared largely by VAM. Upon solidification, constitutional segregation usually occurs, which in turn results in the formation of dendritic and interdendritic regions. Senkov et al. were the first to investigate RHEAs, fabricated two alloys MoNbTaW, and MoNbTaVW by VAM, and reported that incorporation of V leads to finer grain size [14]. The grain size of MoNbTaW and MoNbTaVW are 200 and 80 μm, respectively. The alloy MoNbTaW showed W-enriched dendrite and Nb and Mo-enriched interdendritic regions while the concentration of Ta is nearly the same in both regions. On the other hand, for the V-incorporated MoNbTaW RHEA, dendritic regions are enriched with W and interdendritic regions are enriched with Mo, Nb, and V. Similarly, the RHEA MoNbTaTiZr shows two BCC solid-solution phases, attributing to dendritic (Ta, Nb, and Mo-rich) and interdendritic (Ti and Zr-rich) regions [60]. Many a time, the homogenization annealing of RHEAs is employed at higher temperatures for a relatively longer time to break the as-cast structure. For example, the microstructure after annealing of AlMoNbTi alloy at 1300°C for 20 hours shows polycrystalline grains [43]. Also, annealing at smaller duration in some RHEA systems leads to typical grain structure. For instance, Vlasák et al. [61] fabricated four RHEAs, NbTaTiZr, HfNbTaTiZr, HfNbTiVZr, and HfNbTaTiVZr, by VAM and compared the microstructure in as-cast condition and upon annealing at 1200°C for 2 hours. They reported the presence of dendritic and interdendritc structures in NbTaTiZr RHEA for both conditions. However, incorporation of Hf into NbTaTiZr resulted in complete conversion of dendritic structure to grains (see Figure 10.1(d)) after annealing at 1200°C for 2 hours holding time. Sometimes, after casting, the sample is subjected to hot isostatic pressing followed by annealing to get uniform grains. For example, Senkov et al. [24] processed HfNbTaTiZr alloy by HIP and homogenization at 1200°C for 24 hours to obtain the grains of 100–200 μm. However, it is important to note that, homogenization does not always lead to the formation of a typical grain structure – it is strongly dependent on the composition and homogenization temperature; however, compositional segregation might be present across the sample.

10.4 THERMAL STABILITY OF PHASE AND COEFFICIENT OF THERMAL EXPANSION

For the material to be used in high-temperature applications, the thermal stability of phase and thermal expansion play significant roles, which decide the applicability of the materials. The phase transformation with temperature leads to the change of material property and can induce the non-combability between parent and evolved phases. Therefore, it is desirable that the phase must be stable from RT to operating temperature. Another important aspect is related to the dimensional aspect of the material. With temperature, a material expands due to intrinsic lattice expansion and may cause the generation of vacancies. The dilation strain aggravates thermally induced stress into the material and therefore, may cause catastrophic failure. Ni-based superalloys are most often used as substrate material for turbine blades. In order to protect from harsh environments of high temperature, it is often coated with a ceramic layer of yttria-stabilized zirconia, termed the top-coat. As the difference in coefficient of thermal expansion (CTE, α_l) between Ni-based super alloys and top-coat is large, bond coat layer NiCrYAl is traditionally employed, providing excellent adhesion between substrate and top-coat. In addition, it offers compatibility with both substrate and top-coat. On the other hand, a material exhibiting a low CTE value is preferred for its use in high-temperature conditions. Ni-based single crystal superalloys show excellent phase stability with temperature compared to Ni-based polycrystalline counterparts owing to the higher volume fraction of γ'-phase and prevention of Laves phase formation. Liu et al. [19] studied thermal stability of 3Ta6.4Al1Ti, 0Ta6.4Al2Ti, 3Ta6Al1Ti, and 0Ta6Al2Ti alloys, subjected to a temperature of 1050°C for 500 hours and reported no formation of TCP, which indicates that these alloys are thermally stable.

HEAs, on the other hand, owing to four core effects, show excellent phase stability with respect to temperature. RHEAs show higher phase stability at significantly high temperatures. Thermal investigations of MoNbHfZrTi RHEA using differential scanning calorimetry (DSC) in the range of RT to 1470°C show no endothermic or exothermic peaks pertaining to phase change [48]. HfNbTaTiZr, even after homogenization annealing >1000°C, shows no variation of microstructure and phase compared to the as-cast condition [25]. Behera et al. investigated the phase stability of TiTaNbMoZr RHEA with respect to temperature using synchrotron X-ray radiation and reported the phase stability of the BCC phase is up to 900°C [60]. Beyond ~725°C, vacancies are generated. They showed a lower CTE value (CTE = 7.6–10.9 × 10^{-6}°C^{-1}) compared to Ni-based superalloys (CTE = 11.6–18.2 × 10^{-6}°C^{-1} [62–64]) in the temperature range of 75–1000°C, indicating better thermal stability of TiTaNbMoZr RHEAs. The CTE at room temperature of TiTaNbMoZr was observed to be lower than HfNbZr and HfNbTiTaZr RHEAs [65]. An in-situ X-ray diffraction analysis of the RHEA $Cr_{10}Mo_{25}Ta_{25}Ti_{15}V_{25}$ up to 800°C depicted a stability of the BCC phase until 600°C with no other phases [66]. After that, the formation of another BCC#1 phase can be seen along with the parent phase. However, XRD of heat-treated samples (at 1200°C) show that only the BCC phase is present. These two findings suggest that the formation of the BCC#1 phase is diffusion-controlled, which disappears during furnace cooling.

10.5 MECHANICAL PROPERTIES AT ROOM TEMPERATURE AND HIGH TEMPERATURE

10.5.1 At Room Temperature

The mechanical properties of a material are dependent on its composition, microstructure, and phases. As processing methods and the interplay of various heat treatment methods impact the evolution of microstructure, the change in mechanical properties is an outcome. Figure 10.2 displays the hardness, yield stress, and plastic strain values of Ni-based superalloys along with that of RHEAs (the mechanical properties with respect to alloy and references are summarized in Table 10.2). It is observed that Ni-based superalloys have lower values of hardness compared to most RHEAs. Due to the complexity of alloy design and evolved microstructure, the RHEAs exhibit a wide range of mechanical strength. The hardness of superalloys is restricted to 440 HV whereas, RHEAs like AlMoNbTi, $CrMo_{0.5}NbTa_{0.5}TiZr$, CrMoNbTaVW, $HfMo_{0.5}NbSi_{0.3}TiV_{0.5}$, $HfMo_{0.5}NbSi_{0.5}TiV_{0.5}$, $HfMo_{0.5}NbSi_{0.7}TiV_{0.5}$, HfMoNbTaTiZr, HfMoTaTiZr, HfNbTaTiVZr, and MoNbTaVW show the hardness

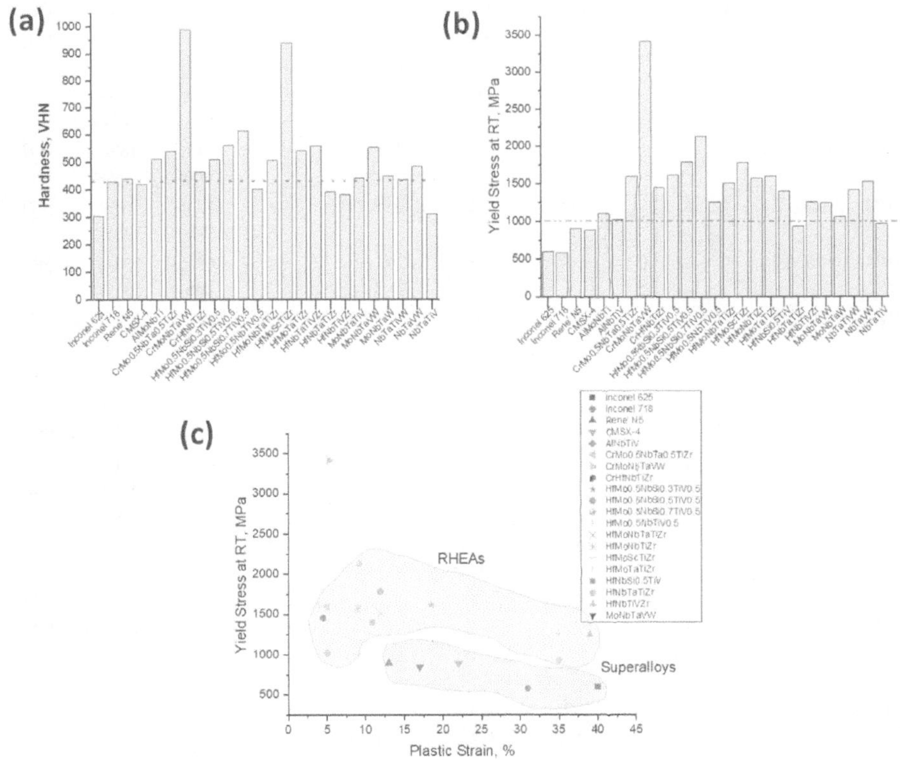

FIGURE 10.2 Comparison of mechanical properties between superalloys and RHEAs: (a) Hardness in VHN, (b) yield stress at RT, and (c) yield stress-plastic strain at RT. Data obtained from [17, 19, 22, 23, 25, 26, 35, 40, 41, 43–45, 47–50, 52, 69–75]

TABLE 10.2
Summary of the Mechanical Properties (Hardness, Yield Stress at RT, 800°C, 1000°C, and 1200°C) of Ni-Based Superalloy and RHEAs with Respect to Preparation Techniques

				σ_{YS} in MPa (ε_p in %)				
Alloy	Preparation	Density, g cm^{-3}	Hardness HV	RT	800°C	1000°C	1200°C	Ref.
Inconel 625	Hot forged + rolled	8.4	304	598 (40%)	392 (61%)	98 (78%)	-	[70]
Inconel 718	C+Hot forged + rolled	8.2	430	580 (31%)	650 (18%)	-	150 (82%)	[35, 71, 72]
Rene' N5	DS+HT	8.4	440	900 (13%)	850 (12%)	600 (17.5%)	-	[19, 73]
CMSX-4	DS+HT	8.7	420	888 (22%)	916 (22.1)	598	200	[40, 41, 74]
AlMoNbTi	C+A	6.46	510	1100	500	540	200	[43]
AlNbTiV	C+A	5.59		1020 (5%)	-	685 (> 40%)	158 (> 40%)	[44]
$CrMo_{0.5}NbTa_{0.5}TiZr$	C+HIP+A	7.98	539	1595 (5%)	983 (5.5%)	546 (> 50%)	170 (> 50%)	[23]
CrMoNbTaVW	PM (SPS)	11.5	990	3416 (5.3%)	-	-	-	[26]
CrHfNbTiZr	C, C+A	7.94	450-480	1322-1457 (4.5%)	-	-	-	[45]
$HfMo_{0.5}NbSi_{0.3}TiV_{0.5}$	C	8.75	508	1617 (18.5%)	-	398 (> 35%)	166 (> 35%)	[47]
$HfMo_{0.5}NbSi_{0.5}TiV_{0.5}$	C	8.60	560	1787 (11.9%)	-	614 (> 35%)	188 (> 35%)	[47]
$HfMo_{0.5}NbSi_{0.7}TiV_{0.5}$	C	8.49	612	2134 (9.2%)	-	673 (> 35%)	235 (> 35%)	[47]
$HfMo_{0.5}NbTiV_{0.5}$	C	9.02	403	1260 (> 35%)	-	368 (> 35%)	60 (> 35%)	[47]
HfMoNbTaTiZr	C	9.97	505	1512 (12%)	1007 (23%)	814 (> 30)	556 (> 30)	[22]
HfMoScTiZr	C	9.28	940	1778 (27%)	1118 (35%)	963 (> 42%)	498 (> 42%)	[69]
HfMoNbTiZr	C	8.68	-	1575 (9%)	825 (> 60%)	635 (> 60%)	187 (> 60%)	[48]
HfMoTaTiZr	C	10.24	542	1600 (4%)	1045 (19%)	855 (> 30)	404 (> 30)	[22]
$HfNbSi_{0.5}TiV$	C	8.6	-	1399 (10.9%)	875 (> 50)	240 (> 50)	-	[49]
HfNbTaTiVZr	C	9.36	559	-	-	-	-	[50]
HfNbTaTiZr	C+HIP+A	9.89	391	929 (35%)	535 (40%)	295 (48%)	92 (50%)	[25]
HfNbTiVZr	C, C+A	7.73	380-385	1120–1253 (39%)	-	-	-	[45]
MoNbTaTiV	C	9.29	443	-	-	-	-	[52]
MoNbTaVW	C+HIP+A	12.29	553	1246 (1.7%)	846 (17%)	842 (19%)	735 (7.5)	[17]
MoNbTaW	C+HIP+A	13.6	449	1058 (2.6%)	552 (> 22%)	548 (> 22%)	506 (> 22%)	[17]
NbTaTiVW	C	11.02	438	1420	-	-	-	[75]

(Continued)

TABLE 10.2 ***(Continued)***
Summary of the Mechanical Properties (Hardness, Yield Stress at RT, 800°C, 1000°C, and 1200°C) of Ni-Based Superalloy and RHEAs with Respect to Preparation Techniques

				σ_{YS} in MPa (ε_p in %)				
Alloy	**Preparation**	**Density, g cm^{-3}**	**Hardness HV**	**RT**	**800°C**	**1000°C**	**1200°C**	**Ref.**
NbTaVW	C	12.65	483	1530	-	-	-	[75]
NbTaTiV	C	12.25	311	965	-	-	-	[75]

A – Homogenization annealing, C – Melting and Casting, DS – Directional Solidification, HIP – Hot Isostatic Pressing, HT – Heat treatment, PM – Powder Metallurgy, SPS – Spark Plasma Sintering.

values greater than 500 HV. Similarly, all RHEAs elicit higher values of yield stress (greater than 1000 MPa) at RT– considerably higher than superalloys. Here, it is worth mentioning that, only a few tensile studies in RHEAs have been reported at RT [67, 68]. Therefore, a complete assessment of mechanical strength in terms of tensile data for high temperature is not possible and hence, for comparison purposes, compressive mechanical characteristics of RHEAs are presented in Table 10.2. From Figure 10.2, it is also noted that the highest yield stress value (3416 MPa) and highest hardness (990 VHN) are reported in CrMoNbTaVW (prepared by SPS) with very poor ductility (5.3%) [26]. The strength values at RT are reported as the highest among RHEA systems. This high value of strength CrMoNbTaVW is attributed to fine grain size (1.24 μm) and homogenous dispersion of Ta_2VO_6 particles. The second-highest hardness of ~940 VHN was reported for Sc incorporated HfMoTaZr [69]. The substitution of elements in RHEA affects significantly the balance between strength and ductility. The incorporation of V and Cr to HfNbTiZr RHEA increases compressive strength [45]. The Cr addition promotes the Laves phase formation (C15:Cr_2Nb and Cr_2Hf) while V addition causes the formation of the C14 Laves phase (V_2Hf). In the V-containing RHEA, the σ_y varies between 1120–1253 MPa while ε_p is ~39% owing to different annealing conditions. The addition of Cr into HfNbTiZr leads to an increase of σ_y to 1322–1457 MPa with considerably lower ε_p ~4.5%. From Figure 10.2(c), it can be concluded ductility is relatively less in comparison to Ni-based superalloys except for two RHEA systems HfNbTaTiZr (ε_p > 35%) and $HfMo_{0.5}NbTiV_{0.5}$ (ε_p > 35%). The addition of elements like Hf, Ti, and Zr is known to improve ductility while Al addition causes an increase in strength with poor ductility [47]. As Hf addition to NbTaTiZr alloy promotes the formation of a single-phase solid solution, enhanced ductility is expected [61]. Huang et al. [67] varied Ta concentration in as-cast $HfTa_xTiZr$ (x= 0.4, 0.5, 0.6, 1) and investigated tensile behavior. They reported a decrease in Ta concentration promotes ductilization of the alloy along with an increase in true stress from 1100 to 1500 MPa. For $HfTa_{0.4}TiZr$ the ε_p is around 27%, whereas HfTaTiZr exhibits ε_p of 4%. In the case of AlNbTiV, σ_y is 1020 MPa, and ε_p = 5%. Liu et al. [47] investigated the effect of Si addition (x = 0,0.3,0.5, 0.7) on the microstructure and mechanical properties of $HfMo_{0.5}NbSi_xTiV_{0.5}$ RHEA. They reported that with an increase of Si content from

Si = 0 to Si = 0.7%., the microhardness and compressive σ_y at RT increase considerably from 403 to 612 VHN and from 1260 to 2134 MPa, respectively concomitant with the decrease in ductility (ε_p = 35% to 9.2%). With the addition of Si, M_5Si_3 phase forms, which refines the microstructure by restricting grain growth resulting in high compressive strength. A lower value of mechanical strength is exhibited by NbTaTiV RHEA, represented by hardness of 311 VHN and compressive yield stress of 965 MPa. Incorporation of W in place of Ti is found to be enhancing hardness and compressive yield stress to 483 VHN and 1530 MPa, receptively.

10.5.2 At High Temperature

Inconel variants usually show peak yield stress around 720 MPa at a temperature range of 650–700°C and Ni-based single crystal superalloys show the highest stress of ~900 MPa close to 800°C. Due to high strength values at high temperatures, Ni-based superalloys are most preferred for high-temperature applications. However, they cannot be used for temperatures exceeding 1300°C. The high-temperature strength of superalloys deteriorates generally after 800°C. Beyond 1200°C, the strength is insufficient to withstand high temperature and load. For example, CMSX-4 (3 wt. % Re, 6 wt. % W) and Rene' N5 alloys (3 wt. % Re, 5 wt. % W) exhibit superior high-temperature performance up to 750–800°C [19, 40]. Beyond that, strength drops owing to a decrease in strength of γ and γ'-phases which are not able to endure the movement of dislocations. On the contrary, RHEAs show superior high temperature strength compared to Ni-based superalloys, and therefore, these alloys attract the researchers' attention having potential candidates for high-temperature applications. Table 10.2 and Figure 10.3 elicit a comparison of yield stress versus plastic strain of superalloys vis-à-vis RHEAs at temperatures of 800°C. At this juncture, it is important to mention that almost all RHEAs have superior yield stress than superalloys at RT. However, at 800°C, only a few RHEAs, such as $CrMo_{0.5}NbTa_{0.5}TiZr$ (σ_{YS} = 983 MPa), HfMoNbTaTiZr (σ_{YS} = 1007 MPa), HfMoScTiZr (σ_{YS} = 1118 MPa), and HfMoTaTiZr (σ_{YS} = 1045 MPa), show higher yield stress over CMSX-4 (916 MPa). The plastic strain of some RHEAs is comparable to the superalloys, except for $CrMo_{0.5}NbTa_{0.5}TiZr$. The yield stress of CrMoNbTaVW RHEA at RT (σ_{YS} = 3416 MPa) is reported to be the highest; however, mechanical properties at high temperatures are not reported. RHEAs exhibit good ductility in the range of 800-1200°C. From Table 10.2 and Figure 10.3, it is observed that Hf containing RHEAs HfMoScTiZr, HfMoNbTaTiZr, and HfMoTaTiZr show a better combination of yield stress and ductility at 800°C than superalloys. Figure 10.4 shows yield stress comparison between superalloys and RHEAs at 1000 and 1200°C. At the temperature of 1000°C, Inconel 625 exhibits the lowest yield stress value (98 MPa) while Ni-based SX superalloys show a yield stress of ~600 MPa. The RHEAs like AlNbTiV (σ_{YS} = 685 MPa), $HfMo_{0.5}NbSi_{0.5}TiV_{0.5}$ (σ_{YS} = 614 MPa), $HfMo_{0.5}NbSi_{0.7}TiV_{0.5}$, ($\sigma_{YS}$ 673 MPa) HfMoNbTaTiZr (σ_{YS} = 814 MPa), HfMoScTiZr (σ_{YS} = 963 MPa), HfMoNbTiZr (σ_{YS} = 635 MPa), HfMoTaTiZr (σ_{YS} = 845 MPa), and MoNbTaVW (σ_Y = 842 MPa) possess superior mechanical characteristics than superalloys. With an increase of temperature to 1200°C, yield stress of Ni-based SX alloys decrease to 200 MPa, whereas RHEAs $HfMo_{0.5}NbSi_{0.7}TiV_{0.5}$ (σ_{YS} = 235 MPa),

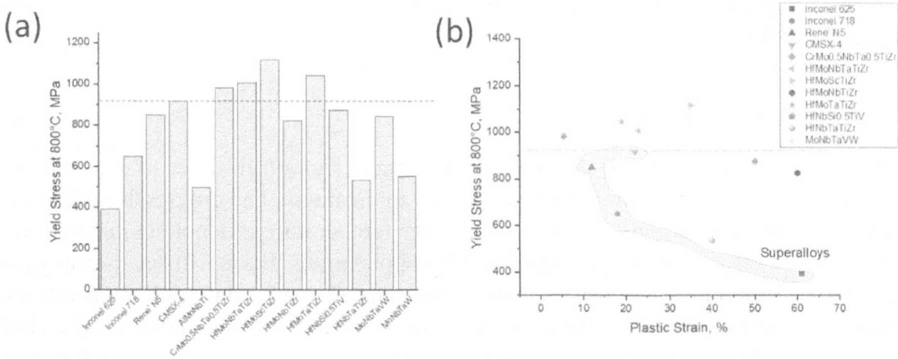

FIGURE 10.3 Comparison of (a) yield stress between superalloys and RHEAs at 800°C and (b) yield stress-plastic stain plot at 800°C. Data obtained from [17, 19, 22, 23, 25, 35, 40, 41, 43, 48, 49, 69–74].

HfMoNbTaTiZr (σ_{YS} = 556 MPa), HfMoScTiZr (σ_{YS} = 498 MPa), HfMoTaTiZr (σ_{YS} = 404 MPa), MoNbTaVW (σ_{YS} = 735 MPa), and MoNbTaW (σ_{YS} = 506 MPa) exhibit exceptionally high-temperature strength. The compressive yield strength and softening resistance of HfMoScTiZr RHEA were reported to be higher than Ni-based superalloys for all temperature conditions (RT, 800, 1000, and 1200°C). The higher strength is attributed to the presence of Mo-rich and Si-rich precipitates in the BCC matrix [64]. Senkov et al. reported the strength-retention in MoNbTaVW and MoNbTaW RHEAs at high temperatures, which can be attributed to the BCC-phase stability of the alloys over 1400°C [17]. Juan et al. demonstrated a good balance of strength and plasticity in HfMoNbTaTiZr alloy – better than that of HfMoTaTiZr for RT – 1200°C. The enhanced compressive properties with temperature were due to the formation of Mo-rich (100–500 nm in size) and (Zr, Hf)-rich precipitates (size ~100 nm) in the interdendritic regions in addition to increased defect density with deformation [22]. A superior high-temperature strength of RHEAs over superalloys

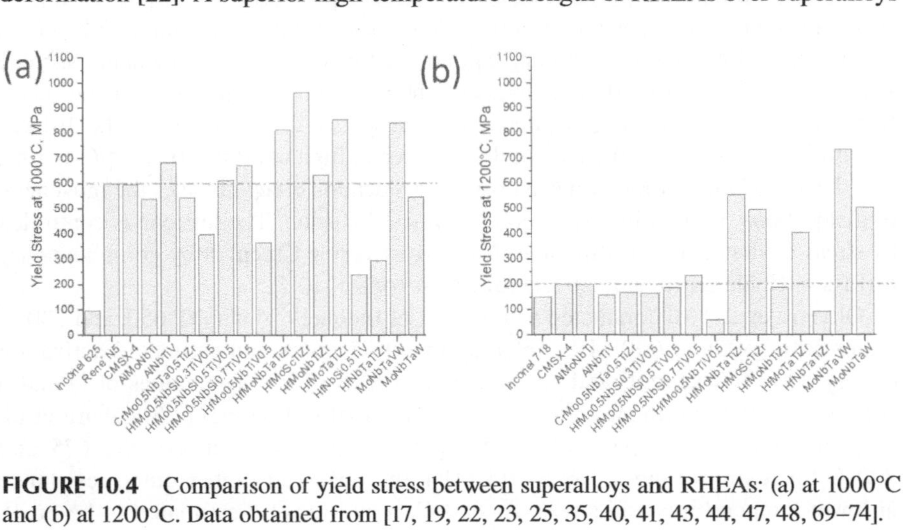

FIGURE 10.4 Comparison of yield stress between superalloys and RHEAs: (a) at 1000°C and (b) at 1200°C. Data obtained from [17, 19, 22, 23, 25, 35, 40, 41, 43, 44, 47, 48, 69–74].

suggests that RHEAs have a great potential to serve as high-temperature materials beyond 1200°C.

10.6 CREEP BEHAVIOR

The creep resistance of the alloys is an important criterion in order to evaluate the applicability of the alloy for high-temperature applications. The progressive deformation of a material at a constant stress for a given temperature is denoted as creep [76]. This phenomenon is of utmost importance in applications like aerospace, power generation, and automotive industries, where materials are exposed to high temperatures along with sustained loads. The experiment uses a tensile specimen that is subjected to a constant load at a given temperature, the strain values are recorded as a function of time. A conventional creep curve has three stages. The first stage (primary stage) is assigned to the rapid elongation of the material followed by a steady rise of creep rate ($d\varepsilon/dt$ or $\dot{\varepsilon}$) in the second stage and finally, the third stage refers to the enhancement of creep rate with time until fracture. The creep strength of the materials of superalloys is largely dependent on the microstructure, volume fraction of disordered γ and ordered γ'-phase. External conditions of temperature and stress affect the microstructures, which subsequently decide the dislocation motion, sub-grain formation, and grain boundary sliding leading to different creep mechanisms. Creep of the metal/alloys involves three different deformation processes; slip, sub-grain formation, and grain boundary sliding. With temperature, new slip systems might become operative. In addition, the source of dislocation, the interaction of dislocation-dislocation, dislocation-precipitate, and dislocation-intermetallic compound at elevated temperatures plays a significant role in controlling the movement of dislocations. Creep also leads to the arrangement of dislocations to form low-angle grain boundaries, called sub-grains. These formed sub-grains provide a less effective barrier to the mobility of dislocations. Grain boundary sliding is another characteristic, that occurs in polycrystalline metals/alloys during creep resulting in poor creep performance. A significantly lower creep strength in polycrystalline Ni-based superalloys with equiaxed microstructure is due to the generation of the microcracks at grain boundaries. Therefore, the complete elimination of high-angle grain boundary aids improvement in the creep strength [36]. Owing to different processes, the mechanism of the creep can be classified as follows: dislocation glide, dislocation creep, and diffusion creep. The dislocation glide is the main mechanism of deformation at high constant stress ($\sigma/G > 10^{-2}$, G – shear modulus). Another mechanism of dislocation creep involves the movement of dislocations overcoming barriers by vacancy diffusion. The last one is controlled by stress-mediated atomic diffusion (Nabbarro-Herring Creep) or by grain boundary diffusion (Coble creep) at relatively low stress ($\sigma/G > 10^{-4}$).

Oliveira et al. [70] conducted a creep test of Inconel 625 at 600, 650, and 700°C under 500, 550, and 600 MPa, interrupting at 75 hours and estimated activation energy (Q = 407 kJ/mol) and stress exponent (n = 11.76). The values of Q and n suggest that the dominant creep mechanism is dislocation creep. Sugahara et al. [71] performed the creep test of IN718 superalloy at temperatures of 650, 675, and 700°C for a stress range of 625–814 MPa and observed higher creep resistance at lower temperatures and stress. At 650°C, n and Q were estimated to be 36.48

and 512.97 kJ/mol, suggesting creep mechanism is mediated by dislocation climb process. Post–creep test, the fractography of samples showed ductile fracture for the temperatures of 650 and 700°C and intergranular fracture at 675°C. They demonstrated creep failure of IN718 after 225 hours when subjected to 650°C at a stress of 700 MPa. With increasing the stress value to 814 MPa, the creep rupture occurs at 4 hours of exposure. Again, raising the creep test temperature to 700°C at stress values of 625, 700, and 750 MPa, the sample experiences rupture after 16.28, 5.10, and 1.70 hours, respectively. On the other hand, creep properties of RHEAs have not been evaluated yet because of relative immaturity of RHEA systems. To date, only one published article is reported towards the creep study of HfNbTaTiZr RHEA. Gadelmeier et al. [77] investigated the creep behavior of HfNbTaTiZr RHEA at 980 and ~1100°C and compared it with that of CMSX-4 [78] (see Figure 10.5 (a–c)). They demonstrated poor creep resistance of the RHEA system. The minimum stress required to achieve a creep rate of $10^{-7}s^{-1}$ at 980°C in CMSX-4 ($\sigma = 225$ MPa) was found to be 25 times higher than HfNbTaTiZr RHEA ($\sigma = 9$ MPa). At higher temperature of ~1100°C, the stress values significantly drop to achieve creep and the difference in stress values between CMSX-4 and RHEA becomes wider. CMSX-4 was observed to be 70 times stronger than RHEAs. The microstructure of HfNbTaTiZr RHEA after the creep test revealed decomposition of the phase at 980°C under stress and the precipitates rich in Nb, Ti, and Ta were noticed. The possible explanation for this poor creep performance of RHEA was attributed to single-phase BCC in polycrystalline RHEAs in comparison to precipitate-strengthened CMSX-4 single crystal having 65% $L1_2$ γ'-phase and thus warranted investigation of single crystal RHEAs for a creep study.

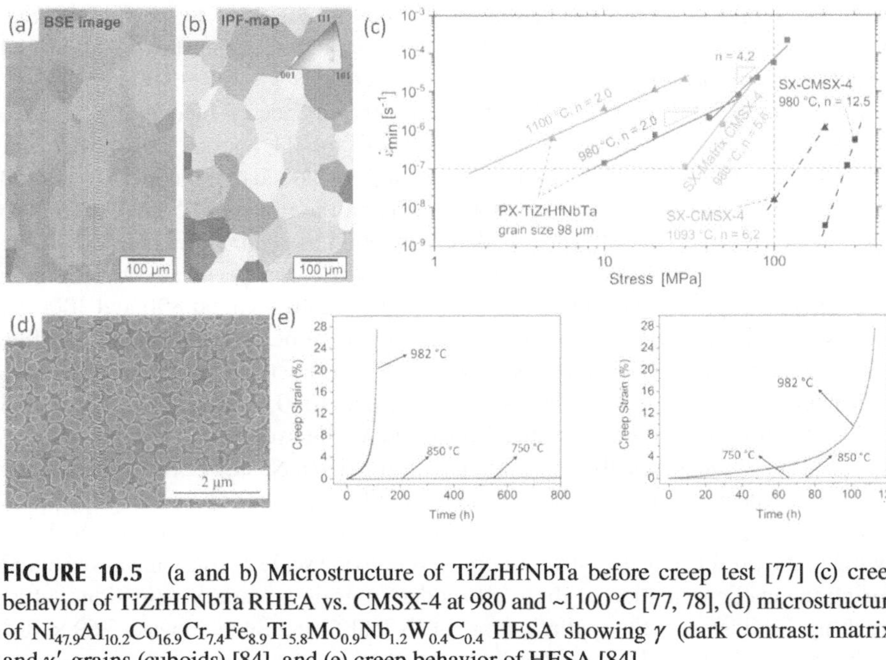

FIGURE 10.5 (a and b) Microstructure of TiZrHfNbTa before creep test [77] (c) creep behavior of TiZrHfNbTa RHEA vs. CMSX-4 at 980 and ~1100°C [77, 78], (d) microstructure of $Ni_{47.9}Al_{10.2}Co_{16.9}Cr_{7.4}Fe_{8.9}Ti_{5.8}Mo_{0.9}Nb_{1.2}W_{0.4}C_{0.4}$ HESA showing γ (dark contrast: matrix) and γ'-grains (cuboids) [84], and (e) creep behavior of HESA [84].

Attributing to the design of RHEAs, there are 31,000 possible RHEA systems [79]. However, only about a hundred HEAs have been prepared to date. One of the major issues in RHEAs is the preparation of larger bulk samples using existing techniques. Owing to the high melting points of elements and poor oxidation resistance of RHEA systems, these are often prepared with vacuum arc melting conditions, which limits the sample dimension required for tensile, creep study. Therefore, the innovation in the fabrication techniques needs to be upgraded for producing bulk RHEAs. In addition, RHEA with a single crystal should be prepared in order to assess the high-temperature strength along with creep behavior. The creep behavior in transition metal-based HEAs, such as $Al_xCoCrFeNi$ [80], $Al_xCoCrFeMnNi$ [81], and CoCrFeMnNi [82, 83], have been investigated and researchers have demonstrated poor creep behavior compared to superalloys. Recently, a new class of HEAs has been developed, termed "high entropy superalloys (HESAs)," which aims to improve high-temperature creep properties by generating $L1_2$ γ'-phase in the FCC matrix, just like superalloys (refer Figure 10.5 (d)). Tsao et al. [84] have investigated the creep behavior of $Ni_{47.9}Al_{10.2}Co_{16.9}Cr_{7.4}Fe_{8.9}Ti_{5.8}Mo_{0.9}Nb_{1.2}W_{0.4}C_{0.4}$ HSEA at 982, 850, and 750°C under 150 MPa stress. The creep failure of this HESA at 982°C was ~114 hours, which is higher than that of the first-generation superalloys. The creep life of NX-188 DS under 982°C at 138 MPa stress is 58 hours, and that of Rene′ 80 under 982°C at 145 MPa stress is 118 hours.

10.7 OXIDATION BEHAVIOR

Table 10.3 summarizes the oxidation behavior of Ni-based superalloys, TM-HEA and RHEA, over a wide range of temperature and time. The variation in mass gain, oxide layer thickness and the oxidation kinetics is governed by several factors, including the chemical composition, phase, microstructure of the starting alloy, and oxidation conditions (temperature, holding time, etc.).

Greene et al. [85] examined the oxidation performance of Inconel 718 at three different temperature regimes, viz., 900–1300°C for 24 hours, 1201–1341°C for 24 hours, and 700–950°C for 9–14 days1 in air and concluded that Inconel 718 offered diffusion-controlled parabolic oxidation kinetics in the range of 950–1300°C up to few weeks. However, catastrophic oxidation is found to occur in Inconel 718 above 1300°C. An isothermal oxidation study of Inconel 625 was performed by Malafaia et al. [86] by exposing the samples in air at 120 and 240 hours at 800 and 1000°C. The oxide layer was rich in chromia, and the formation of complex compounds like $MnCr_2O_4$ and $(Nb,Mo)Ni_3$ were also noticed. The respective oxide layer thickness after exposure at 120 and 240 hours of exposure at 1000°C was 7.9 ± 1.0 µm and 15.5 ± 0.9 µm. In addition to isothermal oxidation, the cyclic oxidation behavior of Ni-based superalloys was also reported in literature [87]. N5 alloy offered better oxidation resistance compared to Rene 80 and IN738LC as the former is Al_2O_3 former. In contrast, significant oxide scale growth was noticed in Rene 80 and IN738LC owing to the presence of Ti. Furthermore, severe oxide scale spallation was observed during the cyclic oxidation of these alloys due to grain boundary oxidation and its cracking. For instance, the spallation rate of IN738LC and Rene 80 was -3.69×10^{-3} $mg^2/(cm^4h)$ and -2.3×10^{-2} $mg^2/(cm^4h)$, respectively.

TABLE 10.3
The Oxidation Temperature, Holding Time, Mass Gain, Oxidation Rate Constant, Phase(s) Formed After Oxidation, and Oxide Layer Thickness of the Alloys Prepared by Different Methods

Alloy System	Preparation Method	Phases/Structure	Oxidation Temperature	Oxidation Time	Mass Gain	Oxidation Rate Constant (Based on Mass Gain)	Phase(s) Formed After Oxidation	Oxide Layer Thickness	Ref.
Inconel 625	Cast, forged and annealed	γ-Ni, γ′-$Ni_3(Al,Ti)$, δ-$Ni_3(Nb,Mo)$, FCC	800 and 1000°C	120 and 240 h	–	–	M_2O_3, γ-Ni, $(Nb,Mo)Ni_3$	7.9 ± 1.0 μm (1000 °C/120 h), 15.5 ± 0.9 μm (1000 °C/240 h)	[86]
Inconel 718			900–1300°C	Up to 24 h	5.124 mg/cm² (1200°C, 24 h), 13.68 mg/cm² (1300°C, 24 h)	1.94×10^{-4} mg²/(cm⁴ s) (for 1200°C) 2.42×10^{-3} mg²/(cm⁴ s) (for 1300°C)	–	–	[85]
			1201–1341°C (1474–1614 K)	Up to 24 h	4.778 mg/cm² (1201°C, 0.92 day), 194.205 mg/cm² (1341°C, 0.93 day)	2.87×10^{-4} mg²/(cm⁴ s) (1201°C, 0.92 day), 4.67×10^{-1} mg²/(cm⁴ s) (1341°C, 0.93 day)			
RR3000			1100	Up to 100 h	–	–	NiO, $CrTaO_4$, $NiCoAl_2O_4$, $NiCr_2O_4$, Al_2O_3, $NiAl_2O_4$ (after 50 h)	–	[105]
CMSX-4			1100	Up to 100 h	–	–	NiO, $TaCrO_4$, $NiCoAl_2O_4$, $NiCr_2O_4$, Al_2O_3, TiO_2 (after 100 h)	–	[105]

(Continued)

TABLE 10.3 *(Continued)*

The Oxidation Temperature, Holding Time, Mass Gain, Oxidation Rate Constant, Phase(s) Formed After Oxidation, and Oxide Layer Thickness of the Alloys Prepared by Different Methods

Alloy System	Preparation Method	Phases/Structure	Oxidation Temperature	Oxidation Time	Mass Gain	Oxidation Rate Constant (Based on Mass Gain)	Phase(s) Formed After Oxidation	Oxide Layer Thickness	Ref.
CMSX-4			788, 871, 954, and 1010°C	1000 – 10000 h	–	0.000128 mg^2/cm^4h (788°C), 0.000287 mg^2/cm^4h (871°C), 0.000414 mg^2/cm^4h (954°C), 0.003000 mg^2/cm^4h (1010°C)	–	–	[106]
825 Ni-based alloy			1000°C	100 h	~ 3 mg/cm^2 (1000 °C/100 h)	–	Cr_2O_3, $Cr_{1.3}Fe_{0.7}O_3$, $MnCr_2O_4$, $NiMnO_4$, $(Cr_{0.88}Ti_{0.12})O_3$, TiO_2, Matrix (1000°C/100 h)		[88]
2205 Duplex stainless steel			1000°C	100 h	~ 4 mg/cm^2 (1000 °C/100 h)	–	Cr_2O_3, Fe_3O_4, $MnCr_2O_4$, Matrix (1000 °C/100 h)		[88]
AlCoCrFeNi	Arc melting		800 – 1000°C	100 h	~ 2.5 mg/cm^2 (1000 °C/100 h)	–	Cr_2O_3, $Cr_{1.3}Fe_{0.7}O_3$, $NiFe_2O_4$, Matrix (1000°C/100 h)		[88]
$Al_8CoCrFeNi$	Arc melting	FCC, BCC, B2	1050°C	100 h		~ 2.5 × 10^{-11} g^2/cm^4 s (1 – 30 h), 2.5 × 10^{-12} g^2/cm^4 s (30 – 100 h)	Cr_2O_3, $NiCr_2O_4$, AlN, Al_2O_3, B2		[89]

$Al_{10}CoCrFeNi$	Arc melting	FCC, BCC, B2	1050°C	100 h		~ 2.4 × 10^{-11} g^2/cm^4 s (1 – 19 h), ~1.8 × 10^{-12} g^2/cm^4 s (19 – 100 h)	Cr_2O_3, $NiCr_2O_4$, AlN, Al_2O_3, B2		[89]
$Al_{12}CoCrFeNi$	Arc melting	FCC, BCC, B2	1050°C	100 h		~ 1.9 × 10^{-11} g^2/cm^4 s (1 – 19 h), ~ 4.6 × 10^{-13} g^2/cm^4 s (19 – 100 h)	Cr_2O_3, $NiCr_2O_4$, AlN, Al_2O_3, B2		[89]
$Al_{15}CoCrFeNi$	Arc melting	FCC, BCC, B2	1050°C	100 h		~ 4.7 × 10^{-11} g^2/cm^4 s (2 – 18 h), ~4.2 × 10^{-12} g^2/cm^4 s (18 – 100 h)	Cr_2O_3, $NiCr_2O_4$, AlN, Al_2O_3, B2		[89]
$Al_{20}CoCrFeNi$	Arc melting	BCC, B2	1050°C	100 h		~ 8.1 × 10^{-12} g^2/cm^4 s (1 – 11 h), ~1.6 × 10^{-12} g^2/cm^4 s (12 – 50 h)	Cr_2O_3, $NiCr_2O_4$, AlN, Al_2O_3, B2		[89]
$Al_{30}CoCrFeNi$	Arc melting	BCC, B2	1050°C	100 h		~ 1.9 × 10^{-12} g^2/cm^4 s (2 – 10 h),	Cr_2O_3, $NiCr_2O_4$, AlN, Al_2O_3, B2		[89]
CoCrFeMnNi	Vacuum induction casting	FCC	900°C	24 h	1.76 mg/cm^2	0.13 mg^2/cm^4h	Cr_2O_3 and Mn_2O_3		[90]
CoCrFeMnNi	Vacuum induction casting	FCC	1000°C	24 h	4.45 mg/cm^2	0.83 mg^2/cm^4h	Cr_2O_3 $(Mn, Cr)_3O_4$ and Mn_3O_4		[90]
CoCrFeMnNi	Vacuum induction casting	FCC	1100°C	24 h	9.08 mg/cm^2	3.44 mg^2/cm^4h	Cr_2O_3 $(Mn, Cr)_3O_4$ and Mn_3O_4		[90]
$Hf_{0.5}Nb_{0.5}Ta_{0.5}Ti_{1.5}Zr$	Arc melting		600, 700, 750, 800, 850, 900, 950, 1000, and 1100°C	1–5 h	Catastrophic oxidation was reported, the alloy disintegrated into powder	–	$(Hf_{0.5}Nb_{0.5}Ta_{0.5}Ti_{1.5}Zr)O_8$ (when oxidized at 800°C) $(Hf_{0.5}Nb_{0.5}Ta_{0.5}Ti_{1.5}Zr)O_8$ + $Nb_2Zr_6O_{17}$ (when oxidized at 900°C and above)	Oxide layer peeled off	[107]

(Continued)

TABLE 10.3 *(Continued)*
The Oxidation Temperature, Holding Time, Mass Gain, Oxidation Rate Constant, Phase(s) Formed After Oxidation, and Oxide Layer Thickness of the Alloys Prepared by Different Methods

Alloy System	Preparation Method	Phases/Structure	Oxidation Temperature	Oxidation Time	Mass Gain	Oxidation Rate Constant (Based on Mass Gain)	Phase(s) Formed After Oxidation	Oxide Layer Thickness	Ref.
$Al_{0.5}Cr_{0.25}Nb_{0.5}Ta_{0.5}Ti_{1.5}$	Arc melting	BCC	800°C	Up to 100 h	1.255 mg/cm^2	0.18 mg^2/(cm^4 s)	BCC, TiO_2,	–	[108]
$Al_{0.5}Cr_{0.25}Nb_{0.5}Ta_{0.5}Ti_{1.5}$	Arc melting	BCC	1100°C	Up to 100 h	10.591 mg/cm^2		Nb_2O_5, TiN, Ti_2AlNb (when	–	[108]
$Al_{0.75}Cr_{0.25}Nb_{0.5}Ta_{0.5}Ti_{1.5}$	Arc melting	BCC	800°C	Up to 100 h	–	0.17 mg^2/(cm^4 s)	oxidized at	–	[108]
$Al_{0.75}Cr_{0.25}Nb_{0.5}Ta_{0.5}Ti_{1.5}$	Arc melting	BCC	1100°C	Up to 100 h	8.243 mg/cm^2	–	800°C),	–	[108]
$Al_{0.5}Cr_{0.25}Nb_{0.5}Ta_{0.5}Ti_{1.5}Zr_{0.01}$	Arc melting	BCC	800°C	Up to 100 h	0.803 mg/cm^2	0.15 mg^2/(cm^4 s)	α-Al_2O_3, BCC, TiO_2, Nb_2O_5,	–	[108]
$Al_{0.5}Cr_{0.25}Nb_{0.5}Ta_{0.5}Ti_{1.5}Zr_{0.01}$	Arc melting	BCC	1100°C	Up to 100 h	9.653 mg/cm^2		$TiNb_2O_7$, TiN, Ti_2AlNb, Ti_2N (when oxidized at 1100°C)	–	[108]
$Al_{0.5}CoCrFeNiTi_{0.5}$	Arc melting	BCC1 (FeCr rich), BCC2 (FeCr lean), FCC, σ	700, 900, 1100°C	100 h	~ 2.8 × 10^{-3} g/cm^2 (1100°C/100 h)	9.00 × 10^{-12} g^2/(cm^4 s) (1100 °C/100 h)	–	Spinel, $(Al_{0.9}Cr_{0.1})_2O_3$	[92]
$Al_{1.0}CoCrFeNiTi_{0.5}$	Arc melting	BCC1, BCC2	700, 900, 1100°C	100 h	~ 1.1 × 10^{-3} g/cm^2 (1100°C/100 h)	1.43 ×10^{-12} g^2/(cm^4 s) (1100 °C/100 h)	–	Cr_2O_3, $(Al_{0.9}Cr_{0.1})_2O_3$	[92]
$Al_{1.5}CoCrFeNiTi_{0.5}$	Arc melting	BCC1, BCC2	700, 900, 1100°C	100 h	~ 8 × 10^{-4} g/cm^2 (1100°C/100 h)	1.03 ×10^{-12} g^2/(cm^4 s) (1100 °C/100 h)	–	TiO_2, $(Al_{0.9}Cr_{0.1})_2O_3$	[92]
AlCoCrFeNi	Arc melting	A2 and/or B2, FCC	1100°C	Up to 100 h	–	8.9 × 10^{-12} g^2/(cm^4 s)	Cr_2O_3, B2/A2, FCC, AlN, Al_2O_3, Spinel (AB_2O_4)	2.3 ± 0.8 µm (after 100 h)	[109]
$Al_2CoCrFeNi$	Arc melting	A2 and/or B2, sigma phase	1100°C	Up to 100 h	–	2.3 × 10^{-12} g^2/(cm^4 s)	B2/A2, Al_2O_3	1.9 ± 0.2 µm (after 100 h)	[109]

$Al_2CoCr_{1.5}FeNi$	Arc melting	A2 and/or B2, sigma phase	1100°C	Up to 100 h	–	–	Cr_2O_3, B2/A2, Al_2O_3, Spinel (AB_2O_4)	3.9 ± 1.8 μm (after 100 h)	[109]
AlCoCrNi	Arc melting	A2 and/or B2	1100°C	Up to 100 h	–	2×10^{-11} $g^2/(cm^4\ s)$	Cr_2O_3, B2/A2, FCC, AlN, Al_2O_3, Spinel (AB_2O_4)	11 ± 3 μm (after 100 h)	[109]
$AlCoCr_{0.5}Ni$	Arc melting	A2 and/or B2	1100°C	Up to 100 h	–	4×10^{-12} $g^2/(cm^4\ s)$	Cr_2O_3, B2/A2, FCC, Al_2O_3, Spinel (AB_2O_4)	7 ± 1 μm (after 100 h)	[109]
$Al_0CoCrCuFeNi$	Arc melting	FCC	1000°C	100 h	~ 1.3×10^{-3} g/cm^2	3.24×10^{-12} $g^2/(cm^4\ s)$, 2.29×10^{-12} $g^2/(cm^4\ s)$	Al_2O_3, Cr_2O_3, spinel oxides	–	[97]
$Al_{0.5}CoCrCuFeNi$	Arc melting	FCC	1000°C	100 h	~ 7×10^{-4} g/cm^2	9.76×10^{-13} $g^2/(cm^4\ s)$, 7.11×10^{-13} $g^2/(cm^4\ s)$		–	[97]
$Al_{1.0}CoCrCuFeNi$	Arc melting	BCC (major), FCC (minor)	1000°C	100 h	~ 3×10^{-4} g/cm^2	1.46×10^{-13} $g^2/(cm^4\ s)$		–	[97]
$Al_{1.5}CoCrCuFeNi$	Arc melting	BCC (major), FCC (minor)	1000°C	100 h	~ 2.1×10^{-4} g/cm^2	9.165×10^{-14} $g^2/(cm^4\ s)$		–	[97]
$Al_{2.0}CoCrCuFeNi$	Arc melting	BCC (major), FCC (minor)	1000°C	100 h	~ 2×10^{-4} g/cm^2	7.73×10^{-14} $g^2/(cm^4\ s)$		–	[97]
$Al_5Co_{35}Cr_{25}Fe_{15}Ni_{20}$	Arc melting	FCC	1000°C	48 h	~ 2 mg/cm^2	–	–	–	[98]
$Al_5Co_{25}Cr_{20}Fe_{15}Ni_{35}$	Arc melting	FCC	1000°C	48 h	~ 1.7 mg/cm^2	–	–	–	[98]
AlCrMoTaTi	Arc melting, annealing	A2/B2, A15, C14, Laves	1000°C	24–300 h		3.34×10^{-2} [$mg^2/cm^4\ h^2$] (0.4–4.5 h), 2.97×10^{-3} [$mg^4/cm^8\ h^4$] (4.5–100 h)	$CrTaO_4$, Al_2O_3, Cr_2O_3, TiO_2	~ 6 μm (after 100 h), ~ 15 μm (after 100 h)	[96]
AlCrMoNbTi	Arc melting, annealing	A2/B2, A15, C14, Laves	1000°C	24–100 h		1.24×10^{-1} [$mg/cm^2\ h$] (0.4–8.2 h), 3.45×10^{-1} [$mg/cm^2\ h$] (8.2–100 h)	Rutile, Al_2O_3, Cr_2O_3 and Nb_2O_5	~ 900 μm (after 100 h)	[96]

(Continued)

TABLE 10.3 *(Continued)*
The Oxidation Temperature, Holding Time, Mass Gain, Oxidation Rate Constant, Phase(s) Formed After Oxidation, and Oxide Layer Thickness of the Alloys Prepared by Different Methods

Alloy System	Preparation Method	Phases/Structure	Oxidation Temperature	Oxidation Time	Mass Gain	Oxidation Rate Constant (Based on Mass Gain)	Phase(s) Formed After Oxidation	Oxide Layer Thickness	Ref.
AlCrMoNb	Arc melting, annealing	A2/B2, C14, Laves	1000°C	24–100 h		6.95×10^{-2} [mg²/cm⁴ h²] (0.4–7.6 h), 1.12×10^{-1} [mg/cm h] (7.6–24 h)	Rutile, Al_2O_3, Cr_2O_3 and Nb_2O_5	~ 750 µm (after 24 h)	[96]
AlCrMoTa	Arc melting, annealing	A2/B2, A15, C14, Laves	1000°C	24–100 h		1.91×10^{-4} [mg⁴/cm⁸ h⁴] (0.4–3 h), 7.96×10^{-3} [mg²/cm⁴ h²] (3–48 h)	Cr_2O_3, Al_2O_3, Ta_2O_5 and $CrTaO_4$	~ 30 µm (after 48 h)	[96]
$Al_0HfMoNbTi$	Arc melting	BCC	1000°C	Up to 20 h	Mass loss (when exposed to 10 h, 20 h)	–	–	57 µm (after 1 h)	[99]
$Al_{0.25}HfMoNbTi$	Arc melting	BCC + B2	1000°C	Up to 20 h	~ 22 mg/cm² (after 10 h), ~ 18 mg/cm² (after 20 h)	–	–	89 µm (after 1 h)	[99]
$Al_{0.50}HfMoNbTi$	Arc melting	B2	1000°C	Up to 20 h	~ 32 mg/cm² (after 10 h), ~ 15 mg/cm² (after 20 h)	–	–	18 µm (after 1 h)	[99]
$Al_{0.75}HfMoNbTi$	Arc melting	B2	1000°C	Up to 20 h	~ 34 mg/cm² (after 10 h), ~ 35 mg/cm² (after 20 h)	–	–	117 µm (after 1 h)	[99]
$Al_{1.0}HfMoNbTi$	Arc melting	B2	1000°C	Up to 20 h	~ 75 mg/cm² (after 10 h), ~ 36 mg/cm² (after 20 h)	–	–	142 µm (after 1 h)	[99]

AlCrMoNbTi	Arc melting, annealing	B2, Al(Mo,Nb)$_3$, Cr_2Nb	1000°C	Up to 48 h	~ 10 mg/cm^2 (after 48 h)	1.2×10^{-1} mg/(cm^2 h) (up to 8.5 h), 3.5×10^{-1} mg/(cm^2 h) (8.5–48 h)	–	–	[100]
AlCrMoNbTi0.5Y	Arc melting, annealing	B2, Cr_2Nb, Al_2Y	–	Up to 48 h	~ 12 mg/cm^2 (after 48 h)	2.0×10^{-1} mg^2/(cm^4 h) (up to 21 h), 6.4×10^{-2} mg^n/(cm^{2n} h) (18 – 48 h)	–	–	[100]
AlCrMoNbTi1.0Y	Arc melting, annealing	B2, Cr_2Nb, Al_2Y	–	Up to 48 h	~ 26 mg/cm^2 (after 48 h)	8.9×10^{-1} mg^4/(cm^8 h) (up to 18 h), 9.0×10^{-2} mg^n/(cm^{2n} h) (18–48 h)	–	–	[100]
AlCrMoTiW	Arc melting	–	1000°C	Up to 48 h	~ 8.4 mg/cm^2	–	–	~ 180 μm (porous), ~ 20 μm (dense)	[101]
AlCrMoTiW	Arc melting	–	1100°C	Up to 48 h		–	–	–	[101]
AlCrMoTiNb	Arc melting	–	1000°C	Up to 48 h	~ 12	–	–	–	[101]
AlCrMoTiNb	Arc melting	–	1100°C	Up to 48 h	~ 8.2	–	–	–	[101]
AlCrMoTaTi	Arc melting	–	1000°C	Up to 48 h	~ 0.5	–	–	–	[101]
AlCrMoTaTi	Arc melting	–	1100°C	Up to 48 h	~ 3	–	–	–	[101]
CrMoTaTi	Arc melting, annealing	BCC, Cr_2Ta	1000°C	Up to 100 h	~ 0.25 mg/cm^2 (after 50 h) ~ 1.1 mg/cm^2 (after 100 h)	–	$CrTaO_4$, TiO_2, unknown phase	287 ± 5 μm (after 50 h)	[103]
CrMoTaTi0.5Y	Arc melting, annealing	BCC, Cr_2Ta	1000°C	Up to 100 h	~ 1.3 mg/cm^2 (after 50 h) ~ 2.2 mg/cm^2 (after 100 h)	–	$CrTaO_4$, TiO_2, unknown phase	148 ± 3 μm (after 50 h)	[103]

(Continued)

TABLE 10.3 *(Continued)*

The Oxidation Temperature, Holding Time, Mass Gain, Oxidation Rate Constant, Phase(s) Formed After Oxidation, and Oxide Layer Thickness of the Alloys Prepared by Different Methods

Alloy System	Preparation Method	Phases/Structure	Oxidation Temperature	Oxidation Time	Mass Gain	Oxidation Rate Constant (Based on Mass Gain)	Phase(s) Formed After Oxidation	Oxide Layer Thickness	Ref.
CrMoTaTi1.0Y	Arc melting, annealing	BCC, Cr_2Ta	1000°C	Up to 100 h	~ 1.1 mg/cm² (after 50 h) ~ 1.2 mg/cm² (after 100 h)	–	$CrTaO_4$, TiO_2, unknown phase	93 ± 3 µm (after 50 h)	[103]
MoNbTaW	Arc melting		800°C	10 h	~ 6 mg/g	–	$Nb_2W_3O_{14}$, $Ta_{16}W_{18}O_{94}$	–	[110]
$(MoNbTaW)_{99.5}B_{0.5}$	Arc melting		800°C	10 h	~ 7 mg/g	–		–	[110]
$(MoNbTaW)_{98}B_2$	Arc melting		800°C	10 h	~ 3 mg/g	–	$Nb_2W_3O_{14}$, $Ta_{16}W_{18}O_{94}$	–	[110]
$(MoNbTaW)_{99.5}C_{0.5}$	Arc melting		800°C	10 h	~ 9 mg/g	–		–	[110]
$(MoNbTaW)_{98}C_2$	Arc melting		800°C	10 h	~ 2.8 mg/g	–	$Nb_2W_3O_{14}$, $Ta_{16}W_{18}O_{94}$	–	[110]
CoCrFeNi	Vacuum arc melting	FCC	1000°C	60 h	2.13 mg/cm²	0.083 mg²/(cm⁴h)	FCC, Laves, Cr_2O_3, Fe_3O_4, CoFe	–	[102]
$CoCrFeNiMo_{0.2}$	Vacuum arc melting	FCC	1000°C	60 h	1.71 mg/cm²	0.058 mg²/(cm⁴h)	FCC, Laves, Cr_2O_3, Fe_3O_4, CoFe	–	[102]
$CoCrFeNiMo_{0.2}Nb_{0.1}$	Vacuum arc melting	FCC + Laves	1000°C	60 h	1.55 mg/cm²	0.045 mg²/(cm⁴h)	FCC, Laves, Cr_2O_3, Fe_3O_4, CoFe, Nb_2O_5	–	[102]
$CoCrFeNiMo_{0.2}Nb_{0.2}$	Vacuum arc melting	FCC + Laves	1000°C	60 h	1.11 mg/cm²	0.023 mg²/(cm⁴h)	FCC, Laves, Cr_2O_3, Fe_3O_4, CoFe, Nb_2O_5	–	[102]

$Hf_{24}Nb_{23}Ti_{38}V_{15}$	Magnetic levitation melting	BCC	700 – 1300°C	1 h	~ 23 mg/cm² (1200°C) ~ 35 mg/cm² (1300°C)	–	$TiNb_2O_7$, HfO_2, $V_3Ti_6O_{17}$, and Nb_2O_5 (1000 °C/60 min)	–	[91]
HfNbTaTiZr	Vacuum arc melting	–	700, 900, 1100, 1300°C	100 h	~ 75 mg/cm² (1100°C/100 h) ~ 250 mg/cm² (1300°C/100 h)	19.97 mg $cm^{-2}s^{-1}$ (1100°C/1–100 h) 26.35 mg $cm^{-2}s^{-1}$ (1300°C/1–10 h), 183.34 mg $cm^{-2}s^{-1}$ (1300°C/10–100 h)	–	–	[93]
$Al_{0.3}$HfNbTaTiZr	Vacuum arc melting	–	700, 900, 1100, 1300°C	100 h	~ 68 mg/cm² (1100°C/100 h) ~ 250 mg/cm² (1300°C/100 h)	12.63 mg $cm^{-2}s^{-1}$ (1100°C/1–100 h) 23.97 mg $cm^{-2}s^{-1}$ (1300°C/1–10 h), 236.02 mg $cm^{-2}s^{-1}$ (1300°C/10–100 h)	–	–	[93]
$Al_{0.5}$HfNbTaTiZr	Vacuum arc melting	–	700, 900, 1100, 1300°C	100 h	~ 60 mg/cm² (1100°C/100 h) ~ 250 mg/cm² (1300°C/100 h)	10.34 mg $cm^{-2}s^{-1}$ (1100°C/1–100 h) 24.09 mg $cm^{-2}s^{-1}$ (1300°C/1–10 h), 240.04 mg $cm^{-2}s^{-1}$ (1300°C/10–100 h)	–	–	[93]
$Al_{0.75}$HfNbTaTiZr	Vacuum arc melting	–	700, 900, 1100, 1300°C	100 h	~ 57 mg/cm² (1100°C/ 100 h) ~ 250 mg/cm² (1300°C/100 h)	8.07 mg $cm^{-2}s^{-1}$ (1100°C/1–100 h) 22.53 mg $cm^{-2}s^{-1}$ (1300°C/1–10 h), 69.70 mg $cm^{-2}s^{-1}$ (1300°C/10–100 h)	–	–	[93]
$Al_{1.0}$HfNbTaTiZr	Vacuum arc melting	–	700, 900, 1100, 1300°C	100 h	~ 50 mg/cm² (1100°C/100 h) ~ 250 mg/cm² (1300°C/100 h)	6.83 mg $cm^{-2}s^{-1}$ (1100°C/1–100 h) 23.37 mg $cm^{-2}s^{-1}$ (1300°C/1–10 h), 47.93 mg $cm^{-2}s^{-1}$ (1300°C/10–100 h)	–	–	[93]

(Continued)

TABLE 10.3 *(Continued)*

The Oxidation Temperature, Holding Time, Mass Gain, Oxidation Rate Constant, Phase(s) Formed After Oxidation, and Oxide Layer Thickness of the Alloys Prepared by Different Methods

Alloy System	Preparation Method	Phases/Structure	Oxidation Temperature	Oxidation Time	Mass Gain	Oxidation Rate Constant (Based on Mass Gain)	Phase(s) Formed After Oxidation	Oxide Layer Thickness	Ref.
17.6Al-Cr-20.3 Mo-15.2Nb-2.9Si-13.4Ta-5.4Ti	Cold crucible levitation melting	BCC, Laves, tetragonal	1200, 1300, 1400°C	100 h	6.00 mg/cm^2 (1200°C/100 h), 9.91 mg/cm^2 (1300°C/100 h), and 18.43 mg/cm^2 (1400°C/100 h)	–	Al_2O_3, Cr_2O_3, Rutile-type oxide (1200°C/ 100 h), Al_2O_3, Rutile-type oxide, mullite (1400°C/ 100 h)		[94]
$Al_{23}Cr_{20}Nb_{15}Ti_{32}Zr_{10}$	Vacuum arc melting	B2, Laves	800, 900, 1000°C	100 h	~ 10 mg/cm^2 (800°C/ 100 h) ~ 20 mg/cm^2 (1000°C/ 10 h)	0.58 mg^2/cm^{4}h (800°C/100 h), 5.03 mg^2/cm^{4}h (900°C/100 h), 39.24 mg^2/cm^{4}h (1000°C/100 h)	TiO_2, Al_2O_3, $Ti_{0.4}Al_{0.3}Nb_{0.3}O_2$, $Zr_{0.5}Al_{0.5}O_2$, CrO_2	–	[95]
$Al_{28}Cr_{20}Nb_{15}Ti_{27}Zr_{10}$	Vacuum arc melting	B2, Laves	800, 900, 1000°C	100 h	–58 mg/cm^2 (800°C/100 h), –14 mg/cm^2 (1000°C/10 h)	–	–	–	[95]
$Al_{33}Cr_{20}Nb_{15}Ti_{22}Zr_{10}$	Vacuum arc melting	B2, Laves	800, 900, 1000°C	100 h	–18 mg/cm^2 (800°C/100 h), –30 mg/cm^2 (1000°C/10 h)	–	–	–	[95]

In order to assess the performance of HEAs as an alternative to Ni-based superalloys in medium to high-temperature applications, it is essential to understand the oxidation behavior of both TMHEA and RHEA. However, the oxidation performance of HEA is reported at a wide range of temperature [88–90]. For example, Zhu et al. [88] evaluated the oxidation resistance of AlCoCrFeNi HEA at 800–1000°C and compared with commercially available 825 Ni-based alloy, and 2205 duplex stainless steel. The mass gain was relatively low (~ 2.5 mg/cm^2) even after high-temperature exposure (1000°C) for an extended duration (100 hours). Despite similar Cr content, AlCoCrFeNi HEA offered better oxidation resistance compared to its counterparts, owing to the enrichment of Cr-oxide in the outer oxide scale. The oxidation resistance of CoCrFeMnNi was studied in detail by Kim et al. [90]. The mass gain after 24 hours of oxidation at 900, 1000, and 1100°C was 1.76, 4.45, and 9.08 mg/cm^2 and the outer oxide layer was composed of the protective oxides of Cr and Mn. Ouyang et al. [91] evaluated the oxidation performance of $Ti_{38}V_{15}Nb_{23}Hf_{24}$ RHEA, prepared by magnetic levitation melting process, at 700–1300°C (with an interval of 100°C) for 1 hour in air. The formation of a dense, composite oxide layer was observed up to 1000°C, which was decomposed beyond 1000°C, owing to volatilization of V_2O_5. As a consequence, the diffusivity of oxygen into the bulk alloy increased above 1000°C, leading to internal oxidation and subsequent formation of needle-like HfO_2.

Many a time, the addition of strong oxide-forming elements is an attractive avenue to improve the oxidation behavior of HEA. For instance, Wang et al. [92] evaluated the oxidation resistance of $Al_xCoCrFeNiTi_{0.5}$ HEA (x = 0.5–1.5) with different Al content. In another work, Chang et al. [93] assessed the oxidation performance of Al-incorporated HfNbTaTiZr RHEA in the temperature range of 700–1300°C with isothermal holding up to 100 hours. In the base alloy (HfNbTaTiZr), considerable mass gain was noticed with increasing temperature. Furthermore, pesting was observed in HfNbTaTiZr RHEA during oxidation at 700–900°C. In contrast, Al addition was beneficial to arrest the pesting of the base alloy. The alloys offered sub-parabolic/near-cubic oxidation behavior at 1100°C, which changed to almost linear behavior when oxidized at 1300°C. In spite of Al addition, the oxidation protectiveness was less at 1100°C and the oxidation resistance was poor at 1300°C, owing to the formation of porous oxide layer, which was incapable of arresting the inward diffusion of oxygen to the bulk, thereby leading to inferior oxidation resistance. Lo et al. [94] analyzed the oxidation resistance of 17.6Al-Cr-20.3Mo-15.2Nb-2.9Si-13.4Ta-5.4Ti RHEAs prepared by cold crucible levitation melting at 1200, 1300, and 1400°C for 100 hours in air. The mass gain per unit area was found to increase with oxidation temperature and the maximum mass gain of 18.43 mg/cm^2 was recorded in the sample oxidized at 1400°C for 100 hours. With increasing oxidation temperature, there was substantial change in oxidation behaviour of RHEA. For instance, parabolic oxidation behaviour was found upon oxidation at 1200°C; which changed to single power-law behavior at 1400°C, owing to the volatilization of Cr_2O_3 and formation of mullite. The phase analysis of the sample oxidized at 1400°C revealed the formation of Al_2O_3, rutile-type oxide, and mullite. Lately, Yurchenko et al. [95] evaluated the oxidation performance of hypoeutectic, eutectic and hypereutectic $Al_xCr_{20}Nb_{15}Ti_{(55-x)}Zr_{10}$ (at. %) RHEA, where x = 23, 28, and 33 at. % for hypoeutectic, eutectic, and hypereutectic system, respectively. The oxidation resistance of the alloys was evaluated at 800–1000°C. The authors found that mass gain is reduced at 800–900°C with enhancing the Al/Ti ratio. However, upon oxidation at 1000°C for 1 hour, spallation of the oxide layer was noticed in alloys with higher Al content (x = 28 and 33 at. %). None of the alloys could survive oxidation at 1000°C for 25 hours. Müller et al. [96] evaluated the oxidation mechanism of different RHEAs (AlCrMoTaTi, AlCrMoNbTi, AlCrMoNb, and AlCrMoTa) at 1000°C in air. Owing to the formation of stable oxide layer composed of Al_2O_3, Cr_2O_3, and $CrTaO_4$, TaMoCrTiAl RHEAs offered better oxidation resistance compared to Nb-containing systems. During oxidation, the formation of rutile also enhances the oxidation resistance of the Ti-containing RHEA system (TaMoCrTiAl). Figure 10.6(a) shows the mass gain of different alloy systems with respect to isothermal holding time at 1000°C, revealing that the RHEA undergoes rapid oxidation even after relatively short duration of exposure. Figure 10.6(b) and (c) respectively show the variation of oxidation rate with specific stiffness and specific strength of different material systems viz., RHEA, TMHEA, Ni-superalloys, Fe-alloys, and Ti-alloys. Many a time, the RHEA also experiences negative mass gain, which is ascribed to multiple factors, including formation of volatile oxide, peeling off the oxide scale due to thermal expansion mismatch, or pesting of oxide layer. As a consequence, the widespread applicability of RHEA in the oxidizing environment at medium to high temperature is a matter of concern [79].

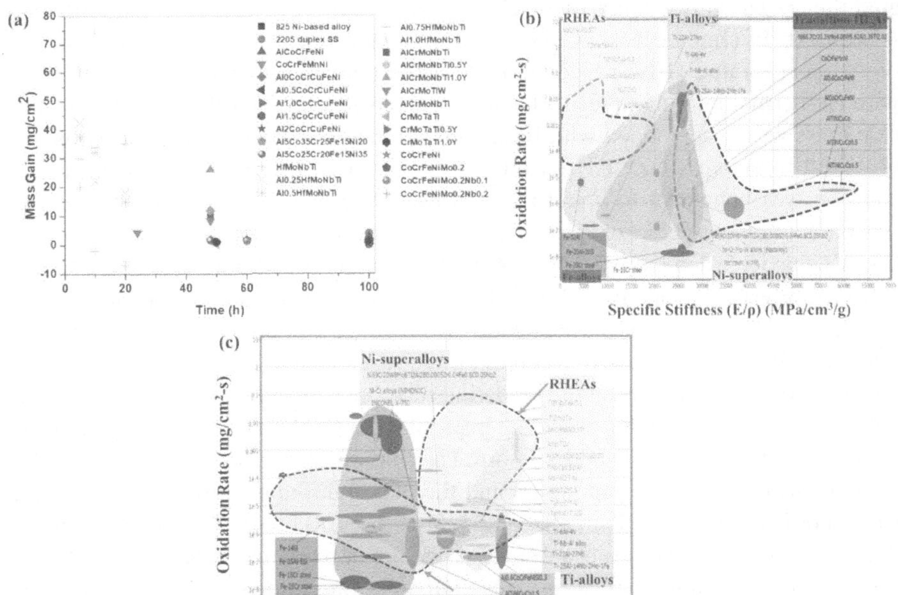

FIGURE 10.6 (a) The variation in mass gain with respect to isothermal holding time during oxidation at 1000°C (Data obtained from [88, 93, 97–103]). The comparative oxidation rate RHEA, transition metal HEA, Ni-superalloys, Fe alloys, Ti alloys with (b) specific stiffness and (c) specific strength. ((b) and (c) are reproduced from [104]).

10.8 WEAR RESISTANCE

For high-temperature structural applications, it is imperative to assess the tribological performance of HEA at different temperatures under various loading conditions and compare their performance with existing structural materials. In the last few years, the wear behavior of transition metal and refractory HEA has been illustrated in detail by many researchers [79, 111, 112]. The wear rate depends on several factors, including the composition of the alloy, phase, hardness, normal load, sliding frequency, type of counter body, temperature of wear, etc. Figure 10.7 shows the variation of wear rate of various superalloys and HEA with test temperatures. For $TiZrV_{0.5}Nb_{0.5}$, $TiZrV_{0.5}Nb_{0.5}Al_{0.25}$, $TiZrV_{0.5}Nb_{0.5}Al_{0.5}$, and $TiZrV_{0.5}Nb_{0.5}Al_{0.75}$ RHEAs, the normal load was 10 N. In the case of Inconel 718 and GH4065, the normal load was 20 N. For the rest of the alloys, the normal load was 5 N.

Bhardwaj et al. [113] studied the effect of Al addition on microstructural evolution, mechanical properties, and wear resistance of HfNbTiZr RHEA, prepared by the arc melting route. The authors evaluated the sliding wear behavior of the alloys using GCr15 steel ball as counter media and found that the wear resistance of the alloy improved with the Al addition, owing to the compaction of oxide-rich debris on the wear track. Wei et al. [114] developed the $Hf_{20}Mo_7Nb_{20}Ti_{30}Ta_{10}V_{10}W_3$ RHEA by vacuum arc melting. The alloy exhibited an excellent combination of yield strength (1225 MPa) and ductility (> 45%) at ambient temperature. The authors examined the room temperature and high temperature (400 and 800°C) wear behavior of the RHEA, applying normal load of 5 N using a ball-on-disk test setup. Silicon nitride balls were used as counter media. The wear rate of RHEA after the test at room temperature (400 and 800°C) was $(1.72 \pm 0.04) \times 10^{-4}$, (3.75 ± 0.08)

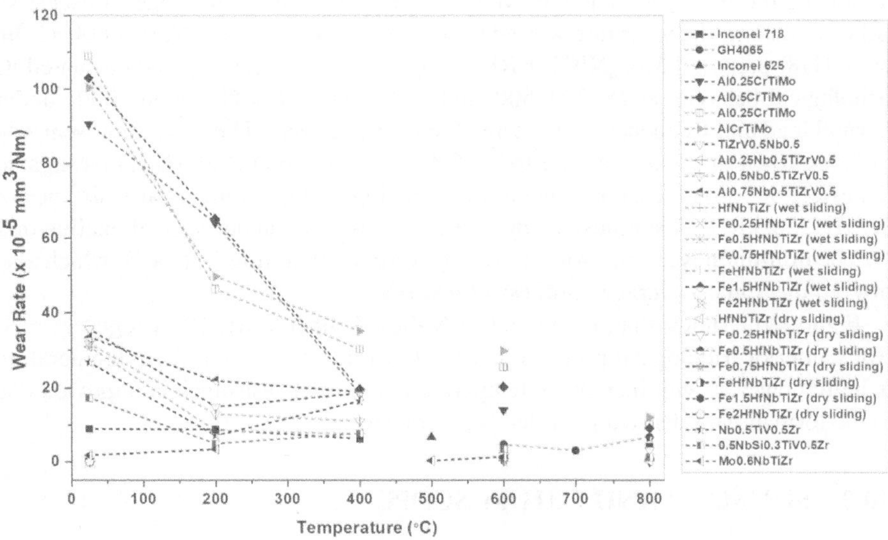

FIGURE 10.7 Wear rate of superalloy and HEA with respect to the test temperature. In all cases, the duration of wear test was 30 min. (Data obtained from [79, 111–118, 120–125].)

$\times 10^{-5}$, and (5.50 ± 0.11) $\times 10^{-6}$ mm^3/Nm, indicating that wear rate decreased with increasing temperature owing to the evolution of dense oxide layer. Furthermore, the width and the depth of wear track also followed a decreasing trend with temperature and the lowest track width of 547.27 μm and depth of 0.93 μm was reported in RHEAs after the wear test at 800°C. In another work, Sadeghilaridjani et al. [115] developed BCC-structured HfNbTaTiZr and HfNbTiZr RHEA by vacuum arc melting with an average grain size of ~ 400 μm and ~ 50 μm, respectively. The reciprocating wear test was performed at 25–300°C applying a normal load of 5 N at two different frequencies (5 and 50 Hz). The ambient temperature wear rate of HfNbTiZr and HfNbTaTiZr RHEA was 1.7 $\times 10^{-4}$ mm^3/Nm and 0.5 $\times 10^{-4}$ mm^3/Nm. Subsequent to a wear test at 300°C, the respective wear rate of HfNbTiZr and HfNbTaTiZr RHEA was 0.7 $\times 10^{-4}$ mm^3/Nm and 1.7 $\times 10^{-4}$ mm^3/Nm and the wear rate was found to increase with reciprocating frequency. The decrease in wear rate in HfNbTiZr at 300°C was due to the formation of dense and protective oxide layer. Pole et al. [116] assessed the reciprocating wear behavior of TaTiVWZr and HfTaTiVZr RHEA at 25, 150, 300, and 450°C using Si_3N_4 counter body and compared with SS 304. In the temperature range of 25–500°C, the hardness of TaTiVWZr was the maximum, which was followed by HfTaTiVZr and SS304. The wear rate of HfTaTiVZr, TaTiVWZr, and SS304 at 150°C was ~ 7 $\times 10^{-4}$, ~ 8 $\times 10^{-4}$ and ~ 3.5 $\times 10^{-4}$ mm^3/Nm, which respectively decreased to ~ 2.5 $\times 10^{-4}$, ~ 1 $\times 10^{-4}$, and ~ 0.8 $\times 10^{-4}$ mm^3/Nm during a wear test at 450°C due to a protective oxide layer formation. The oxidative wear was found to be the dominant wear mechanism in these systems at 450°C. Alvi et al. [117] developed equiatomic CuMoTaVW using spark plasma sintering and assessed its wear behavior against alloy steel and Si_3N_4 counter body under an applied load of 5 N. A relatively high wear rate of 2.3 $\times 10^{-2}$ mm^3/Nm was found at 200°C due to galling wear. Owing to oxidation of Cu to CuO at 400°C, the wear rate was relatively low (5 $\times 10^{-3}$ mm^3/Nm). Due to excessive tribo-oxidation, the wear rate was slightly high (4.5 $\times 10^{-2}$ mm^3/Nm) at 600°C. Jin et al. [118] prepared $Mo_{0.6}NbTiZr$ RHEA by vacuum arc melting and evaluated its tribological behavior at 25, 200, 500, and 800°C against Si_3N_4 counter body under normal load and frequency of 10 N and 5 Hz, respectively. The minimum wear rate of 2.4 $\times 10^{-6}$ mm^3/Nm was reported at 500°C. Lately, Wan et al. [119] investigated the effect of boron incorporation on room and high-temperature wear resistance of TiZrHfTa0.5 RHEA synthesized by vacuum arc melting under normal loading of 5 N. The minimum wear rate was offered by RHEA containing 1 at. % B, which also offered the highest average hardness of 438 HV.

Figure 10.7 shows that, compared to Ni-based superalloys, HEA registers relatively inferior tribological properties. The wear rate of RHEA at room temperature is high; however, with increasing temperature, the wear resistance is improved due to the formation of dense and protective oxide layers.

10.9 SUMMARY AND FUTURE SCOPE

Since the first work of RHEAs in 2010, researchers have focused on the development of new RHEAs by varying compositions. This development leads to the fabrication of RHEAs consisting of single BCC phase, dual (B_2+BCC phases or

BCC1+BCC2), and multiphase systems (BCC+B2+Laves phases +precipitates). Altogether, RHEAs have shown remarkable room temperature and high-temperature mechanical properties compared to Ni-based superalloys. In addition, researchers have developed wear and corrosion-resistant RHEAs. However, for further development of RHEAs towards their applicability at high temperatures, a few areas need constant attention.

One of the major limitations of the use of RHEAs and the characterization of RHEAs is the fabrication of large-sized specimens. Due to the relatively high melting point of each element in RHEAs, they are majorly prepared by the arc melting route. As the elements of RHEAs are prone to oxidation, fabrication involves constraints of using vacuum or inert atmosphere. Therefore, the preparation of large-sized specimens becomes a challenge. As a result, hardly any literature on the tensile and creep properties of RHEAs was reported. Also, based on limited literature, the creep properties of reported HEAs and RHEAs are found to be poor compared to superalloys. On the other hand, the technology for the preparation of large-sized superalloys is in a much-advanced stage owing to the lower melting point of superalloys (< 1400°C). Processing of superalloys using vacuum induction melting, electro-slag remelting, directional solidification followed by aging as well as deformation through hot-forging, hot-rolling, and extrusion to get wrought form is well established and vastly understood. However, it is too early to draw the same conclusion for RHEAs. The RHEAs are largely cast followed by homogenization to break the cast structures. Many groups have also employed hot isostatic pressing and cold isostatic pressing before heat treatment. The main aim of these processes is to produce polycrystalline RHEAs from dendritic grains and to yield precipitate in the microstructure, which would make the phase stable at high temperatures. Another advantage of having precipitate is that it acts as a barrier to dislocation motion and, therefore, increases the high-temperature strength. At this point, it is important to note that RHEAs are primarily exhibiting the BCC solid solution phase as opposed to the FCC matrix of superalloys. As the FCC crystal structure is a close-packed system and slip systems are along close-packed planes and directions, these systems more often provide better ductility and toughness compared to BCC systems. So far, Hf containing RHEAs such as HfNbTaTiZr, HfNbTiVZr, and their derived alloys show an excellent combination of strength and ductility matching to that of superalloys.

Synthesis of RHEAs by the powder metallurgy route also poses a challenge. Mechanical alloying by ball milling to produce RHEAs is energy-intensive and preparation is not as straightforward as other counterpart TMHEAs. A minor content of impurity can affect the formation of a solid solution phase in RHEAs. Also, consolidation of RHEA powders requires advanced sintering techniques, such as spark plasma sintering, hot pressing, etc., and more often, the solid-solution phase decomposes to two or more enriched regions having precipitates. This decomposition may become beneficial or detrimental to the evolved mechanical properties. Here, we would like to point out that a small variation of composition due to induced impurity in the production of RHEAs by mechanical alloying might result in different microstructure upon sintering. There are a few studies on RHEAs that have been reported and have a lot to explore owing to the vast design possibilities of

RHEAs. Recently, additive manufacturing (AM) techniques have been employed largely to fabricate complex, innovative components. These techniques are widely used and well established for Inconel systems than RHEAs. Attributing to superior properties at elevated temperatures and the brittle nature of RHEAs, these systems can be used as thin films, deposited by sputtering techniques on various substrates, which would help protect the components facing a high temperature. Recently, thin films of MoNbTaVW and AlMoNbTaVW RHEA coatings were grown on SS-304 substrate to improve the corrosion resistance of stainless steel. Also, the CuMoTaW RHEA coating on stainless steel has shown excellent wear resistance. Therefore, in the future, researchers must focus attention on developing thin coatings of RHEAs on different substrates to understand the behavior of RHEA coatings and to explore the potential utility of the RHEAs.

The lower oxidation resistance of RHEAs is a pressing issue. Many researchers have incorporated oxide former elements, such as Al, Ti, Cr, etc., and studied the oxidation performances. Nevertheless, it is found that the oxidation properties are not as good as those of superalloys. The wear characteristics of RHEAs is no different. Another concern in using RHEAs is that they inherently have a higher density than superalloys. Therefore, researchers are incorporating relatively lighter elements, such as Al, Ti, Cr, V, Zr, Mo, etc., in order to reduce the density. With this strategy, the density of RHEAs can become lower than superalloys, and as low as 5.6 gcm^{-3} is possible (refer Table 10.2). However, these alloying additions often cause a reduction in ductility. Therefore, in the near future, strategies must be made to get a good combination of strength and ductility in lightweight RHEAs.

In the end, it is imperative to state that the development of RHEAs is relatively immature compared to that of superalloys. Among larger possibilities of RHEAs, only a few have been explored. One of the striking differences between superalloys and RHEAs is the phase system of these alloys. Superalloys have $L1_2$ precipitates or carbides in the FCC matrix, whereas RHEAs exhibit BCC and B2 as solid solution phases. Some RHEAs also exhibit the Laves phase and other intermetallics in addition to the BCC matrix phase. The strengthening of RHEAs is mainly contributed by solid solution strengthening, while the contribution of precipitation strengthening is limited. In the case of superalloys, however, precipitation strengthening is the main strengthening mechanism through the coherency of γ / γ' phases. Moreover, this coherency between BCC/B2 or Laves phase is not well established in RHEAs. Therefore, in the future, researchers must focus on developing HESAs, whose microstructure is quite similar to superalloys in addition to novel RHEAs.

REFERENCES

1. G. Gudivada, A.K. Pandey, Recent developments in nickel-based superalloys for gas turbine applications: Review, J. Alloys Compd. 963 (2023) 171128. https://doi.org/10.1016/j.jallcom.2023.171128
2. P. Sengupta, I. Manna, Advanced High-Temperature Structural Materials in Petrochemical, Metallurgical, Power, and Aerospace Sectors—An Overview, in: Future Landscape of Structural Materials in India, Springer Nature Singapore, Singapore, 2022: pp. 79–131. https://doi.org/10.1007/978-981-16-8523-1_5

3. D.V.V. Satyanarayana, N.E. Prasad, Nickel Based Superalloys, in: R. Rasad, N., Wanhill (Ed.), Aerospace Materials and Material Technology, Indian Institute of Metals Series, Springer, Singapore, 2017: p. 586. https://doi.org/10.1007/978-981-10-2134-3
4. T.M. Pollock, Alloy design for aircraft engines, Nat. Mater. 15 (2016) 809–815. https://doi.org/10.1038/nmat4709
5. M.H. Tsai, J.W. Yeh, High-entropy alloys: A critical review, Mater. Res. Lett. 2 (2014) 107–123. https://doi.org/10.1080/21663831.2014.912690
6. D.B. Miracle, J.D. Miller, O.N. Senkov, C. Woodward, M.D. Uchic, J. Tiley, Exploration and development of high entropy alloys for structural applications, Entropy 16 (2014) 494–525. https://doi.org/10.3390/e16010494
7. J.W. Yeh, S.K. Chen, S.J. Lin, J.Y. Gan, T.S. Chin, T.T. Shun, C.H. Tsau, S.Y. Chang, Nanostructured high-entropy alloys with multiple principal elements: Novel alloy design concepts and outcomes, Adv. Eng. Mater. 6 (2004) 299–303. https://doi.org/10.1002/adem.200300567
8. B. Cantor, I.T.H. Chang, P. Knight, A.J.B. Vincent, Microstructural development in equiatomic multicomponent alloys, Mater. Sci. Eng. A 375–377 (2004) 213–218.
9. P. Sreeramagiri, G. Balasubramanian, Directed energy deposition of multi-principal element alloys, Front. Mater. 9 (2022) 1–15. https://doi.org/10.3389/fmats.2022.825276
10. D.B. Miracle, O.N. Senkov, A critical review of high entropy alloys and related concepts, Acta Mater. 122 (2017) 448–511. https://doi.org/10.1016/j.actamat.2016.08.081
11. S. Praveen, H.S. Kim, high-entropy alloys: Potential candidates for high-temperature applications – An overview, Adv. Eng. Mater. 20 (2018) 1–22. https://doi.org/10.1002/adem.201700645
12. N. Kumar, P. Sengupta, M. Kumar, M. Debata, S. Bajpai, D. Debasish, A. Panigrahi, Microstructure, thermal expansion, and high-temperature oxidation behavior of spark plasma sintered AlCoCrSiNi high entropy alloy, Mater. Today Commun. 40 (2024) 110063. https://doi.org/10.1016/j.mtcomm.2024.110063
13. J.-W. Yeh, Alloy design strategies and future trends in high-entropy alloys, JOM 65 (2013) 1759–1771. https://doi.org/10.1007/s11837-013-0761-6
14. O.N. Senkov, G.B. Wilks, D.B. Miracle, C.P. Chuang, P.K. Liaw, Refractory high-entropy alloys, Intermetallics 18 (2010) 1758–1765. https://doi.org/10.1016/j.intermet.2010.05.014
15. Y.J. Zhou, Y. Zhang, Y.L. Wang, G.L. Chen, Solid solution alloys of AlCoCrFeNi Tix with excellent room-temperature mechanical properties, Appl. Phys. Lett. 90 (2007). https://doi.org/10.1063/1.2734517
16. C.J. Tong, M.R. Chen, S.K. Chen, J.W. Yeh, T.T. Shun, S.J. Lin, S.Y. Chang, Mechanical performance of the AlxCoCrCuFeNi high-entropy alloy system with multiprincipal elements, Metall. Mater. Trans. A Phys. Metall. Mater. Sci. 36 (2005) 1263–1271. https://doi.org/10.1007/s11661-005-0218-9
17. O.N. Senkov, G.B. Wilks, J.M. Scott, D.B. Miracle, Mechanical properties of Nb25Mo25Ta25W25 and V20Nb20Mo20Ta20W20 refractory high entropy alloys, Intermetallics 19 (2011) 698–706. https://doi.org/10.1016/j.intermet.2011.01.004
18. M. Shahwaz, P. Nath, I. Sen, A critical review on the microstructure and mechanical properties correlation of additively manufactured nickel-based superalloys, J. Alloys Compd. 907 (2022) 164530. https://doi.org/10.1016/j.jallcom.2022.164530
19. D. Liu, Q. Ding, Q. Zhou, D. Zhou, X. Wei, X. Zhao, Z. Zhang, H. Bei, Microstructure, mechanical properties and thermal stability of Ni-based single crystal superalloys with low specific weight, Crystals 13 (2023) 610. https://doi.org/10.3390/cryst13040610
20. Y. Hu, X. Lin, Y. Li, S. Zhang, Q. Zhang, W. Chen, W. Li, W. Huang, Influence of heat treatments on the microstructure and mechanical properties of Inconel 625 fabricated by directed energy deposition, Mater. Sci. Eng. A. 817 (2021) 141309. https://doi.org/10.1016/j.msea.2021.141309

21. E. Hosseini, V.A. Popovich, A review of mechanical properties of additively manufactured Inconel 718, Addit. Manuf. 30 (2019) 100877. https://doi.org/10.1016/j.addma.2019.100877
22. C.-C. Juan, M.-H. Tsai, C.-W. Tsai, C.-M. Lin, W.-R. Wang, C.-C. Yang, S.-K. Chen, S.-J. Lin, J.-W. Yeh, Enhanced mechanical properties of HfMoTaTiZr and HfMoNbTaTiZr refractory high-entropy alloys, Intermetallics 62 (2015) 76–83. https://doi.org/10.1016/j.intermet.2015.03.013
23. O.N. Senkov, C.F. Woodward, Microstructure and properties of a refractory NbCrMo0.5Ta0.5TiZr alloy, Mater. Sci. Eng. A 529 (2011) 311–320. https://doi.org/10.1016/j.msea.2011.09.033
24. O.N. Senkov, J.M. Scott, S.V. Senkova, F. Meisenkothen, D.B. Miracle, C.F. Woodward, Microstructure and elevated temperature properties of a refractory TaNbHfZrTi alloy, J. Mater. Sci. 47 (2012) 4062–4074. https://doi.org/10.1007/s10853-012-6260-2
25. O.N. Senkov, J.M. Scott, S.V. Senkova, D.B. Miracle, C.F. Woodward, Microstructure and room temperature properties of a high-entropy TaNbHfZrTi alloy, J. Alloys Compd. 509 (2011) 6043–6048. https://doi.org/10.1016/j.jallcom.2011.02.171
26. Y. Long, X. Liang, K. Su, H. Peng, X. Li, A fine-grained NbMoTaWVCr refractory high-entropy alloy with ultra-high strength: Microstructural evolution and mechanical properties, J. Alloys Compd. 780 (2019) 607–617. https://doi.org/10.1016/j.jallcom.2018.11.318
27. J. Málek, J. Zýka, F. Lukáč, M. Vilémová, T. Vlasák, J. Čížek, O. Melikhova, A. Macháčková, H.-S. Kim, The effect of processing route on properties of HfNbTaTiZr high entropy alloy, Materials (Basel). 12 (2019) 4022. https://doi.org/10.3390/ma12234022
28. Y. Cao, Y. Liu, Y. Li, B. Liu, A. Fu, Y. Nie, Precipitation behavior and mechanical properties of a hot-worked TiNbTa0.5ZrAl0.5 refractory high entropy alloy, Int. J. Refract. Met. Hard Mater. 86 (2020) 105132. https://doi.org/10.1016/j.ijrmhm.2019.105132
29. L. Raman, K. Guruvidyathri, G. Kumari, S.V.S. Narayana Murty, R.S. Kottada, B.S. Murty, Phase evolution of refractory high-entropy alloy CrMoNbTiW during mechanical alloying and spark plasma sintering, J. Mater. Res 34 (2019) 756–766. https://doi.org/10.1557/jmr.2018.483
30. H. Zhang, W. Xu, Y. Xu, Z. Lu, D. Li, The thermal-mechanical behavior of WTaMoNb high-entropy alloy via selective laser melting (SLM): Experiment and simulation, Int. J. Adv. Manuf. Technol. 96 (2018) 461–474. https://doi.org/10.1007/s00170-017-1331-9
31. H. Dobbelstein, E.P. George, E.L. Gurevich, A. Kostka, A. Ostendorf, G. Laplanche, Laser metal deposition of refractory high-entropy alloys for high-throughput synthesis and structure-property characterization, Int. J. Extrem. Manuf 3 (2020). https://doi.org/10.1088/2631-7990/abcca8
32. S. Alvi, D.M. Jarzabek, M.G. Kohan, D. Hedman, P. Jenczyk, M.M. Natile, A. Vomiero, F. Akhtar, Synthesis and mechanical characterization of a CuMoTaWV high-entropy film by magnetron sputtering, ACS Appl. Mater. Interfaces. 12 (2020) 21070–21079. https://doi.org/10.1021/acsami.0c02156
33. S.K. Bachani, C.J. Wang, B.S. Lou, L.C. Chang, J.W. Lee, Microstructural characterization, mechanical property and corrosion behavior of VNbMoTaWAl refractory high entropy alloy coatings: Effect of Al content, Surf. Coatings Technol. 403 (2020) 126351. https://doi.org/10.1016/j.surfcoat.2020.126351
34. V. Shankar, K. Bhanu Sankara Rao, S. Mannan, Microstructure and mechanical properties of Inconel 625 superalloy, J. Nucl. Mater. 288 (2001) 222–232. https://doi.org/10.1016/S0022-3115(00)00723-6
35. T. Bhujangrao, F. Veiga, A. Suárez, E. Iriondo, F.G. Mata, High-temperature mechanical properties of IN718 alloy: Comparison of additive manufactured and wrought samples, Crystals 10 (2020) 689. https://doi.org/10.3390/cryst10080689

36. W. Xia, X. Zhao, L. Yue, Z. Zhang, Microstructural evolution and creep mechanisms in Ni-based single crystal superalloys: A review, J. Alloys Compd. 819 (2020) 152954. https://doi.org/10.1016/j.jallcom.2019.152954
37. N. Yurchenko, E. Panina, M. Tikhonovsky, G. Salishchev, S. Zherebtsov, N. Stepanov, Structure and mechanical properties of an in situ refractory Al20Cr10Nb15Ti20V25Zr10 high entropy alloy composite, Mater. Lett. 264 (2020) 127372. https://doi.org/10.1016/j.matlet.2020.127372
38. A. Katz-Demyanetz, I.I. Gorbachev, E. Eshed, V.V. Popov, V.V. Popov, M. Bamberger, High entropy Al0.5CrMoNbTa0.5 alloy: Additive manufacturing vs. casting vs. CALPHAD approval calculations, Mater. Charact. 167 (2020) 110505. https://doi.org/10.1016/j.matchar.2020.110505.
39. L.H. Rettberg, P.G. Callahan, B.R. Goodlet, T.M. Pollock, Rejuvenation of directionally solidified and single-crystal nickel-base superalloys, Metall. Mater. Trans. A. 52 (2021) 1609–1631. https://doi.org/10.1007/s11661-021-06150-7
40. A. Sengupta, S.K. Putatunda, L. Bartosiewicz, J. Hangas, P.J. Nailos, M. Peputapeck, F.E. Alberts, Tensile behavior of a new single crystal nickel-based superalloy (CMSX-4) at room and elevated temperatures, J. Mater. Eng. Perform. 3 (1994) 664–672. https://doi.org/10.1007/BF02645265.
41. C. Körner, M. Ramsperger, C. Meid, D. Bürger, P. Wollgramm, M. Bartsch, G. Eggeler, Microstructure and mechanical properties of CMSX-4 single crystals prepared by additive manufacturing, Metall. Mater. Trans. A. 49 (2018) 3781–3792. https://doi.org/10.1007/s11661-018-4762-5
42. O.N. Senkov, S.V. Senkova, C. Woodward, Effect of aluminum on the microstructure and properties of two refractory high-entropy alloys, Acta Mater. 68 (2014) 214–228. https://doi.org/10.1016/j.actamat.2014.01.029
43. H. Chen, A. Kauffmann, S. Laube, I.-C. Choi, R. Schwaiger, Y. Huang, K. Lichtenberg, F. Müller, B. Gorr, H.-J. Christ, M. Heilmaier, Contribution of lattice distortion to solid solution strengthening in a series of refractory high entropy alloys, Metall. Mater. Trans. A 49 (2018) 772–781. https://doi.org/10.1007/s11661-017-4386-1
44. N.D. Stepanov, D.G. Shaysultanov, G.A. Salishchev, M.A. Tikhonovsky, Structure and mechanical properties of a light-weight AlNbTiV high entropy alloy, Mater. Lett. 142 (2015) 153–155. https://doi.org/10.1016/j.matlet.2014.11.162
45. É. Fazakas, V. Zadorozhnyy, L.K. Varga, A. Inoue, D.V. Louzguine-Luzgin, F. Tian, L. Vitos, Experimental and theoretical study of Ti20Zr20Hf20Nb20X20 (X = V or Cr) refractory high-entropy alloys, Int. J. Refract. Met. Hard Mater. 47 (2014) 131–138. https://doi.org/10.1016/j.ijrmhm.2014.07.009
46. Y. Mu, H. Liu, Y. Liu, X. Zhang, Y. Jiang, T. Dong, An ab initio and experimental studies of the structure, mechanical parameters and state density on the refractory high-entropy alloy systems, J. Alloys Compd. 714 (2017) 668–680. https://doi.org/10.1016/j.jallcom.2017.04.237
47. Y. Liu, Y. Zhang, H. Zhang, N. Wang, X. Chen, H. Zhang, Y. Li, Microstructure and mechanical properties of refractory HfMo0.5NbTiV0.5Six high-entropy composites, J. Alloys Compd. 694 (2017) 869–876. https://doi.org/10.1016/j.jallcom.2016.10.014
48. N.N. Guo, L. Wang, L.S. Luo, X.Z. Li, Y.Q. Su, J.J. Guo, H.Z. Fu, Microstructure and mechanical properties of refractory MoNbHfZrTi high-entropy alloy, Mater. Des. 81 (2015) 87–94. https://doi.org/10.1016/j.matdes.2015.05.019
49. Y. Zhang, Y. Liu, Y. Li, X. Chen, H. Zhang, Microstructure and mechanical properties of a refractory HfNbTiVSi0.5 high-entropy alloy composite, Mater. Lett. 174 (2016) 82–85. https://doi.org/10.1016/j.matlet.2016.03.092
50. M.C. Gao, B. Zhang, S. Yang, S.M. Guo, Senary refractory high-entropy alloy HfNbTaTiVZr, Metall. Mater. Trans. A 47 (2016) 3333–3345. https://doi.org/10.1007/s11661-015-3105-z

51. S. Maiti, W. Steurer, Structural-disorder and its effect on mechanical properties in single-phase TaNbHfZr high-entropy alloy, Acta Mater. 106 (2016) 87–97. https://doi.org/10.1016/j.actamat.2016.01.018
52. H.W. Yao, J.W. Qiao, J.A. Hawk, H.F. Zhou, M.W. Chen, M.C. Gao, Mechanical properties of refractory high-entropy alloys: Experiments and modeling, J. Alloys Compd. 696 (2017) 1139–1150. https://doi.org/10.1016/j.jallcom.2016.11.188
53. M. Todai, T. Nagase, T. Hori, A. Matsugaki, A. Sekita, T. Nakano, Novel TiNbTaZrMo high-entropy alloys for metallic biomaterials, Scr. Mater. 129 (2017) 65–68. https://doi.org/10.1016/j.scriptamat.2016.10.028
54. C. Chattopadhyay, A. Prasad, B.S. Murty, Phase prediction in high entropy alloys – A kinetic approach, Acta Mater. 153 (2018) 214–225. https://doi.org/10.1016/j.actamat.2018.05.002
55. Y. Zhang, Y.J. Zhou, J.P. Lin, G.L. Chen, P.K. Liaw, Solid-Solution phase formation rules for multi-component alloys, Adv. Eng. Mater. 10 (2008) 534–538. https://doi.org/10.1002/adem.200700240
56. X. Yang, Y. Zhang, Prediction of high-entropy stabilized solid-solution in multi-component alloys, Mater. Chem. Phys 132 (2012) 233–238.
57. S. Guo, Q. Hu, C. Ng, C.T. Liu, More than entropy in high-entropy alloys: Forming solid solutions or amorphous phase, Intermetallics 41 (2013) 96–103. https://doi.org/10.1016/j.intermet.2013.05.002
58. A.B. Melnick, V.K. Soolshenko, Thermodynamic design of high-entropy refractory alloys, J. Alloys Compd. 694 (2017) 223–227. https://doi.org/10.1016/j.jallcom.2016.09.189
59. S. Guo, C. Ng, J. Lu, C.T. Liu, Effect of valence electron concentration on stability of fcc or bcc phase in high entropy alloys, J. Appl. Phys. 109 (2011) 103505. https://doi.org/10.1063/1.3587228
60. M. Behera, A. Panigrahi, M. Bönisch, G. Shankar, P.K. Mishra, Structural stability and thermal expansion of TiTaNbMoZr refractory high entropy alloy, J. Alloys Compd. 892 (2022) 162154. https://doi.org/10.1016/j.jallcom.2021.162154
61. T. Vlasák, J. Čížek, O. Melikhova, F. Lukáč, D. Preisler, M. Janeček, P. Harcuba, M. Zimina, O. Srba, Thermal stability of microstructure of high-entropy alloys based on refractory metals Hf, Nb, Ta, Ti, V, and Zr, Metals (Basel). 12 (2022) 394. https://doi.org/10.3390/met12030394
62. F. Cverna, ASM Ready Reference: Thermal Properties of Metals. ASM International, Ohio (USA) (2002).
63. A.S. Hakeem, F. Patel, N. Minhas, A. Malkawi, Z. Aleid, M.A. Ehsan, H. Sharrofna, A. Al Ghanim, Comparative evaluation of thermal and mechanical properties of nickel alloy 718 prepared using selective laser melting, spark plasma sintering, and casting methods, J. Mater. Res. Technol. 12 (2021) 870–881. https://doi.org/10.1016/j.jmrt.2021.03.043
64. A.K. Rai, H. Trpathy, R.N. Hajra, S. Raju, S. Saroja, Thermophysical properties of Ni based super alloy 617, J. Alloys Compd. 698 (2017) 442–450. https://doi.org/10.1016/j.jallcom.2016.12.183
65. H. Song, F. Tian, D. Wang, Thermodynamic properties of refractory high entropy alloys, J. Alloys Compd. 682 (2016) 773–777. https://doi.org/10.1016/j.jallcom.2016.04.320
66. O.U. Tukac, A. Ozalp, E. Aydogan, Development and thermal stability of Cr10Mo25Ta25Ti15V25 refractory high entropy alloys, J. Alloys Compd. 930 (2023) 167386. https://doi.org/10.1016/j.jallcom.2022.167386
67. H. Huang, Y. Wu, J. He, H. Wang, X. Liu, K. An, W. Wu, Z. Lu, Phase-transformation ductilization of brittle high-entropy alloys via metastability engineering, Adv. Mater. 29 (2017) 1–7. https://doi.org/10.1002/adma.201701678

68. S. Sheikh, S. Shafeie, Q. Hu, J. Ahlström, C. Persson, J. Veselý, J. Zýka, U. Klement, S. Guo, Alloy design for intrinsically ductile refractory high-entropy alloys, J. Appl. Phys. 120 (2016). https://doi.org/10.1063/1.4966659
69. X.W. Nie, M.D. Cai, S. Cai, Microstructure and mechanical properties of a novel refractory high entropy alloy HfMoScTaZr, Int. J. Refract. Met. Hard Mater. 98 (2021) 105568. https://doi.org/10.1016/j.ijrmhm.2021.105568
70. M.M. de Oliveira, A.A. Couto, G.F.C. Almeida, D.A.P. Reis, N.B. de Lima, R. Baldan, Mechanical behavior of Inconel 625 at elevated temperatures, Metals (Basel). 9 (2019) 301. https://doi.org/10.3390/met9030301.
71. T. Sugahara, A.A. Couto, M.J.R. Barboza, F.P. Neto, R.J. Takahashi, D.A.P. Reis, Creep and mechanical behavior study of Inconel 718 Superalloy, Mater. Res. 25 (2022) 1–9. https://doi.org/10.1590/1980-5373-MR-2021-0280
72. W. Tucho, A. Sletsjøe, N. Sayyar, V. Hansen, Optimizing tensile properties and hardness of inconel 718 by cold rolling, Metals (Basel). 14 (2024) 455. https://doi.org/10.3390/met14040455
73. R. Wang, J. Wang, T. Cao, R. Zhao, X. Lu, W. Guan, H. Tao, S. Shuai, X. Songzhe, W. Xuan, C. Panwisawas, C. Chen, Z. Ren, Microstructure characteristics of a René N5 Ni-based single-crystal superalloy prepared by laser-directed energy deposition, Addit. Manuf. 61 (2023) 1–14. https://doi.org/10.1016/j.addma.2022.103363
74. A. Basak, S. Holenarasipura Raghu, S. Das, Microstructures and microhardness properties of CMSX-4® additively fabricated through scanning laser epitaxy (SLE), J. Mater. Eng. Perform. 26 (2017) 5877–5884. https://doi.org/10.1007/s11665-017-3008-9
75. H.W. Yao, J.W. Qiao, M.C. Gao, J.A. Hawk, S.G. Ma, H.F. Zhou, Y. Zhang, NbTaV-(Ti,W) refractory high-entropy alloys: Experiments and modeling, Mater. Sci. Eng. A 674 (2016) 203–211. https://doi.org/10.1016/j.msea.2016.07.102
76. G.E. Dieter, Mechanical Metallurgy, 3rd ed., McGraw Hill Education (India) Pvt. Ltd., 2013.
77. C. Gadelmeier, Y. Yang, U. Glatzel, E.P. George, Creep strength of refractory high-entropy alloy TiZrHfNbTa and comparison with Ni-base superalloy CMSX-4, Cell Reports Phys. Sci. 3 (2022) 100991. https://doi.org/10.1016/j.xcrp.2022.100991
78. U. Glatzel, F. Schleifer, C. Gadelmeier, F. Krieg, M. Müller, M. Mosbacher, R. Völkl, Quantification of solid solution strengthening and internal stresses through creep testing of ni-containing single crystals at 980°C, Metals (Basel). 11 (2021). https://doi.org/10.3390/met11071130
79. O.N. Senkov, D.B. Miracle, K.J. Chaput, J.-P. Couzinie, Development and exploration of refractory high entropy alloys—A review, J. Mater. Res 33 (2018) 3092–3128. https://doi.org/10.1557/jmr.2018.153
80. T. Cao, J. Shang, J. Zhao, C. Cheng, R. Wang, H. Wang, The influence of Al elements on the structure and the creep behavior of AlxCoCrFeNi high entropy alloys, Mater. Lett. 164 (2016) 344–347. https://doi.org/10.1016/j.matlet.2015.11.016
81. C.M. Cao, J. Xu, W. Tong, Y.X. Hao, P. Gu, L.M. Peng, Creep behaviour and microstructural evolution of AlxCrMnFeCoNi high-entropy alloys, Mater. Sci. Technol. (United Kingdom) 35 (2019) 1283–1290. https://doi.org/10.1080/02670836.2019.1619306
82. Y. Bin Kang, S.H. Shim, K.H. Lee, S.I. Hong, Dislocation creep behavior of CoCrFeMnNi high entropy alloy at intermediate temperatures, Mater. Res. Lett. 6 (2018) 689–695. https://doi.org/10.1080/21663831.2018.1543731
83. F. Dobeš, H. Hadraba, Z. Chlup, A. Dlouhý, M. Vilémová, J. Matějíček, Compressive creep behavior of an oxide-dispersion-strengthened CoCrFeMnNi high-entropy alloy, Mater. Sci. Eng. A 732 (2018) 99–104. https://doi.org/10.1016/j.msea.2018.06.108

84. T.K. Tsao, A.C. Yeh, C.M. Kuo, K. Kakehi, H. Murakami, J.W. Yeh, S.R. Jian, The high temperature tensile and creep behaviors of high entropy superalloy, Sci. Rep. 7 (2017) 1–9. https://doi.org/10.1038/s41598-017-13026-7
85. G.A. Greene, C.C. Finfrock, Oxidation of Inconel 718 in air at high temperatures, Oxid. Met 55 (2001) 505–521. https://doi.org/10.1023/a:1010359815550
86. A.M. de Sousa Malafaia, R.B. de Oliveira, L. Latu-Romain, Y. Wouters, R. Baldan, Isothermal oxidation of Inconel 625 superalloy at 800 and 1000°C: Microstructure and oxide layer characterization, Mater. Charact. 161 (2020) 110160. https://doi.org/10.1016/j.matchar.2020.110160.
87. H.T. Mallikarjuna, N.L. Richards, W.F. Caley, Effect of alloying elements and microstructure on the cyclic oxidation performance of three nickel-based superalloys, Materialia 4 (2018) 487–499. https://doi.org/10.1016/j.mtla.2018.11.004.
88. J. Zhu, S. Lu, Y. Jin, L. Xu, X. Xu, C. Yin, Y. Jia, High-temperature oxidation behaviours of AlCoCrFeNi high-entropy alloy at 1073–1273 K, Oxid. Met. 94 (2020) 265–281. https://doi.org/10.1007/s11085-020-09991-6
89. T.M. Butler, M.L. Weaver, Oxidation behavior of arc melted AlCoCrFeNi multicomponent high-entropy alloys, J. Alloys Compd. 674 (2016) 229–244. https://doi.org/10.1016/j.jallcom.2016.02.257.
90. Y.K. Kim, Y.A. Joo, H.S. Kim, K.A. Lee, High temperature oxidation behavior of Cr-Mn-Fe-Co-Ni high entropy alloy, Intermetallics 98 (2018) 45–53. https://doi.org/10.1016/j.intermet.2018.04.006
91. D. Ouyang, Z. Chen, H. Yu, K.C. Chan, L. Liu, Oxidation behavior of the Ti38V15Nb23Hf24 refractory high-entropy alloy at elevated temperatures, Corros. Sci. 198 (2022) 110153. https://doi.org/10.1016/j.corsci.2022.110153
92. S. Wang, Z. Chen, P. Zhang, K. Zhang, C.L. Chen, B.L. Shen, Influence of Al content on high temperature oxidation behavior of AlxCoCrFeNiTi0.5 high entropy alloys, Vacuum 163 (2019) 263–268. https://doi.org/10.1016/j.vacuum.2019.01.053
93. C. Chang, M.S. Titus, J. Yeh, Oxidation behavior between 700 and 1300°C of refractory TiZrNbHfTa high-entropy alloys containing aluminum, Adv. Eng. Mater. 20 (2018) 1–8. https://doi.org/10.1002/adem.201700948
94. K.-C. Lo, H. Murakami, J.-W. Yeh, A.-C. Yeh, Oxidation behaviour of a novel refractory high entropy alloy at elevated temperatures, Intermetallics 119 (2020) 106711. https://doi.org/10.1016/j.intermet.2020.106711
95. N. Yurchenko, E. Panina, S. Zherebtsov, N. Stepanov, Oxidation behaviour of eutectic refractory high-entropy alloys at 800–1000°C, Corros. Sci. 205 (2022) 110464. https://doi.org/10.1016/j.corsci.2022.110464.
96. F. Müller, B. Gorr, H.-J. Christ, J. Müller, B. Butz, H. Chen, A. Kauffmann, M. Heilmaier, On the oxidation mechanism of refractory high entropy alloys, Corros. Sci. 159 (2019) 108161. https://doi.org/10.1016/j.corsci.2019.108161
97. Y.Y. Liu, Z. Chen, Y.Z. Chen, J.C. Shi, Z.Y. Wang, S. Wang, F. Liu, Effect of Al content on high temperature oxidation resistance of AlxCoCrCuFeNi high entropy alloys (x=0, 0.5, 1, 1.5, 2), Vacuum 169 (2019) 108837. https://doi.org/10.1016/j.vacuum.2019.108837
98. R. Gawel, Ł. Rogal, J. Dąbek, M. Wójcik-Bania, K. Przybylski, High temperature oxidation behaviour of non-equimolar AlCoCrFeNi high entropy alloys, Vacuum 184 (2021) 109969. https://doi.org/10.1016/j.vacuum.2020.109969
99. X. Yang, Z. An, Y. Zhai, X. Wang, Y. Chen, S. Mao, X. Han, Effect of Al content on the thermal oxidation behaviour of AlHfMoNbTi high-entropy alloys analysed by in situ environmental TEM, Corros. Sci. 191 (2021) 109711. https://doi.org/10.1016/j.corsci.2021.109711
100. F. Müller, B. Gorr, H.-J. Christ, H. Chen, A. Kauffmann, M. Heilmaier, Effect of Y additions on the oxidation behaviour of novel refractory high-entropy alloy NbMoCrTiAl at 1000°C in air, Oxid. Met. 94 (2020) 147–163. https://doi.org/10.1007/s11085-020-09983-6

101. B. Gorr, F. Müller, M. Azim, H.J. Christ, T. Müller, H. Chen, A. Kauffmann, M. Heilmaier, High-temperature oxidation behavior of refractory high-entropy alloys: Effect of alloy composition, Oxid. Met. 88 (2017) 339–349. https://doi.org/10.1007/s11085-016-9696-y
102. T. Wu, L. Yu, G. Chen, R. Wang, Y. Xue, Y. Lu, B. Luan, Effects of Mo and Nb on the microstructure and high temperature oxidation behaviors of CoCrFeNi-based high entropy alloys, J. Mater. Res. Technol. 27 (2023) 1537–1549. https://doi.org/10.1016/j.jmrt.2023.10.058
103. Y. Du, D. Ding, L. Lai, S. Xiao, N. Guo, B. Song, S. Guo, Effect of Y on the high-temperature oxidation behavior of CrMoTaTi refractory high entropy alloy, Int. J. Refract. Met. Hard Mater. 103 (2022) 105755. https://doi.org/10.1016/j.ijrmhm.2021.105755
104. P. Kumar, T.-N. Lam, P.K. Tripathi, S.S. Singh, P.K. Liaw, E.-W. Huang, Recent progress in oxidation behavior of high-entropy alloys: A review, APL Mater. 10 (2022). https://doi.org/10.1063/5.0116605
105. C.M. Younes, G.C. Allen, J.A. Nicholson, High temperature oxidation behaviour of single crystal superalloys RR3000 and CMSX-4, Corros. Eng. Sci. Technol. 42 (2007) 80–88. https://doi.org/10.1179/174327807X159925
106. N. Mu, J. Liu, J.W. Byeon, Y.H. Sohn, Y.L. Nava, Long-term oxidation and phase transformations in aluminized CMSX-4 superalloys, Surf. Coatings Technol. 188–189 (2004) 27–34. https://doi.org/10.1016/j.surfcoat.2004.07.091
107. S. Sheikh, M.K. Bijaksana, A. Motallebzadeh, S. Shafeie, A. Lozinko, L. Gan, T.K. Tsao, U. Klement, D. Canadinc, H. Murakami, S. Guo, Accelerated oxidation in ductile refractory high-entropy alloys, Intermetallics 97 (2018) 58–66. https://doi.org/10.1016/j.intermet.2018.04.001
108. S. Sheikh, L. Gan, A. Ikeda, H. Murakami, S. Guo, Alloying effect on the oxidation behavior of a ductile Al0.5Cr0.25Nb0.5Ta0.5Ti1.5 refractory high-entropy alloy, Mater. Today Adv. 7 (2020) 100104. https://doi.org/10.1016/j.mtadv.2020.100104
109. A. Anupam, A.S.M. Ang, K. Guruvidyathri, M. Abbas, D. Sivaprahasam, P. Munroe, C.C. Berndt, B.S. Murty, R.S. Kottada, Evaluating the influence of microstructural attributes: Fraction, composition, size and spatial distribution of phases on the oxidation behaviour of high-entropy alloys, Corros. Sci. 184 (2021) 109381. https://doi.org/10.1016/j.corsci.2021.109381
110. L. Shao, Z. Liu, Y. Zou, Effects of boron and carbon on the oxidation behavior of a NbMoTaW refractory high entropy alloy, J. Alloys Compd. 927 (2022) 166946. https://doi.org/10.1016/j.jallcom.2022.166946
111. X. Pei, Y. Du, X. Hao, H. Wang, Q. Zhou, H. Wu, H. Wang, Microstructure and tribological properties of TiZrV0.5Nb0.5Al refractory high entropy alloys at elevated temperature, Wear 488–489 (2022) 204166. https://doi.org/10.1016/j.wear.2021.204166
112. W. Xiong, A.X.Y. Guo, S. Zhan, C.T. Liu, S.C. Cao, Refractory high-entropy alloys: A focused review of preparation methods and properties, J. Mater. Sci. Technol. 142 (2023) 196–215. https://doi.org/10.1016/j.jmst.2022.08.046
113. V. Bhardwaj, Q. Zhou, F. Zhang, W. Han, Y. Du, K. Hua, H. Wang, Effect of Al addition on the microstructure, mechanical and wear properties of TiZrNbHf refractory high entropy alloys, Tribol. Int. 160 (2021) 107031. https://doi.org/10.1016/j.triboint.2021.107031
114. Q. Wei, A. Zhang, J. Han, B. Xin, B. Su, X. Wang, Q. Ma, J. Meng, Development of a Ti30Hf20Nb20Ta10V10Mo7W3 refractory high entropy alloy with excellent mechanical properties and wear resistance, J. Alloys Compd. 966 (2023) 171571. https://doi.org/10.1016/j.jallcom.2023.171571
115. M. Sadeghilaridjani, M. Pole, S. Jha, S. Muskeri, N. Ghodki, S. Mukherjee, Deformation and tribological behavior of ductile refractory high-entropy alloys, Wear 478–479 (2021) 203916. https://doi.org/10.1016/j.wear.2021.203916

116. M. Pole, M. Sadeghilaridjani, J. Shittu, A. Ayyagari, S. Mukherjee, High temperature wear behavior of refractory high entropy alloys based on 4-5-6 elemental palette, J. Alloys Compd. 843 (2020) 156004. https://doi.org/10.1016/j.jallcom.2020.156004
117. S. Alvi, F. Akhtar, High temperature tribology of CuMoTaWV high entropy alloy, Wear 426–427 (2019) 412–419. https://doi.org/10.1016/j.wear.2018.12.085
118. C. Jin, X. Li, H. Li, Q. Li, H. Wang, Tribological performance of a TiZrNbMo0.6 refractory high entropy alloy at elevated temperatures, J. Alloys Compd. 920 (2022) 165915. https://doi.org/10.1016/j.jallcom.2022.165915
119. Q. Wan, K. Hua, Z. Zhou, F. Zhang, H. Wu, Q. Zhou, H. Wang, Revealing the B addition on tribology performance in TiZrHfTa0.5 refractory high-entropy alloy at ambient and elevated temperature, J. Alloys Compd. 931 (2023) 167521. https://doi.org/10.1016/j.jallcom.2022.167521
120. R. Zhang, K. Tulugan, A. Zhang, J. Meng, J. Han, Effect of aluminum content on the tribological properties of AlxCrTiMo refractory high-entropy alloys, J. Mater. Eng. Perform. 31 (2022) 984–993. https://doi.org/10.1007/s11665-021-06243-9
121. W. Wang, K. Yang, Q. Wang, P. Dai, H. Fang, F. Wu, Q. Guo, P.K. Liaw, N. Hua, Novel Ti-Zr-Hf-Nb-Fe refractory high-entropy alloys for potential biomedical applications, J. Alloys Compd. 906 (2022) 164383. https://doi.org/10.1016/j.jallcom.2022.164383
122. X. Pei, Y. Du, H. Wang, M. Hu, H. Wang, W. Liu, Investigation of high temperature tribological performance of TiZrV0.5Nb0.5 refractory high-entropy alloy optimized by Si microalloying, Tribol. Int. 176 (2022) 107885. https://doi.org/10.1016/j.triboint.2022.107885
123. L. Bai, S. Wan, G. Yi, Y. Shan, S.T. Pham, A.K. Tieu, Y. Li, R. Wang, Temperature-mediated tribological characteristics of 40CrNiMoA steel and Inconel 718 alloy during sliding against Si3N4 counterparts, Friction 9 (2021) 1175–1197. https://doi.org/10.1007/s40544-020-0455-3
124. Y. Wang, S. Zhao, Z. Jia, J. Ji, D. Liu, T. Guo, Y. Ding, Study on friction and wear behavior of Inconel 625 superalloy during hot extrusion, Adv. Mater. Sci. Eng. 2020 (2020). https://doi.org/10.1155/2020/5453703
125. K. Feng, T. Shao, Tribo-induced surface structure of nickel-based superalloy at elevated temperature, Appl. Surf. Sci. 571 (2022) 151290. https://doi.org/10.1016/j.apsusc.2021.151290

11 Aerospace Applications of High Entropy Alloys

Sandeep Kumar Yadav, Ayush Sourav, Sudeep Kumar T and Shanmugasundaram T.

High entropy alloys (HEAs) have emerged as a revolutionary class of materials for aerospace applications, offering unprecedented mechanical properties and thermal stability under extreme conditions. Compared to conventional alloys and Ni-based superalloys, HEAs exhibit superior strength, oxidation resistance, and exceptional wear performance at elevated temperatures, making them ideal candidates for critical aerospace components subjected to harsh environments. The unique multi-component design of HEAs enables the tailoring of specific alloy compositions, providing enhanced microstructural stability and resistance to thermal degradation. Moreover, HEA coatings have shown excellent protection against high-temperature wear and oxidation, outperforming traditional coating systems. HEAs hold immense potential to revolutionize aerospace materials with their capability to reduce component weight, improve fuel efficiency, and enhance sustainability.

11.1 INTRODUCTION

As the aerospace industry strives for increased capacity, higher performance, fuel efficiency, and cost reduction, there is a drive towards improving the life of existing materials or developing energy-efficient next-generation materials. Developing materials that can operate at higher temperatures and sustain for longer periods of time would solve some of the problems mentioned above. Owing to their ability to operate at elevated temperature conditions, nickel, cobalt, and titanium alloys are used extensively in the aerospace industry, nuclear reactors, chemical industries, and automotive high-temperature (HT) applications [1]. Nickel-based superalloys are widely used in gas turbine applications owing to their high-temperature strength, ductility, fatigue, and corrosion resistance properties [2]. Gas turbine components such as combustion liners, nozzles, and blades are mostly made of nickel-based superalloys to perform efficiently over a longer duration of time at elevated temperatures [3]. Turbine blades experience very high temperatures as the hot gases pass over the surfaces, causing erosion, corrosion, oxidation, and other possible hazards [4].

The outer body of an aircraft is mostly made of aluminum alloys, whereas most of the engine components are made of superalloys. Owing to their ability to operate at elevated temperature conditions, nickel, cobalt, and titanium alloys are used extensively in the aerospace industry. There is a continued effort towards improving

DOI: 10.1201/9781003664093-15

the life of existing materials or developing energy-efficient next-generation materials. Developing materials that can operate at higher temperatures and sustain for longer periods of time would solve some of the problems associated with a high-temperature environment since the various components of the engine are subjected to different temperatures. Recently, high entropy alloys (HEAs), made of five or more elements in equi-atomic/near equi-atomic compositions, have shown excellent high-temperature strength, oxidation, and wear resistance due to their four core effects. This chapter outlines the recent developments and various properties such as high temperature strength, creep and wear resistance, and coating of various HEAs associated with aerospace applications.

11.2 HIGH ENTROPY ALLOYS FOR AEROSPACE APPLICATIONS

Thrust-to-weight ratio (TWR) is a critical parameter in aerospace design, influencing both performance and efficiency. Achieving a higher TWR can be accomplished either by increasing engine thrust or by reducing aircraft weight. While increasing thrust typically necessitates larger engines and more fuel, which in turn adds weight, a more sustainable approach involves reducing the density of structural components. Traditionally, aluminum (Al) alloys are favored for fabricating the airframe of an airplane due to their low density and excellent specific strength. However, Al alloys are not suitable for components within the engine that must endure extremely high temperatures, where material properties such as high-temperature strength and creep resistance are crucial.

Titanium (Ti) alloys have emerged as ideal candidates for intermediate-temperature components, such as compressors, due to their balance of strength, weight, and thermal stability. However, in the higher temperature sections of aeroengines, such as turbine blades, materials with even greater structural stability are required. Here, nickel-based (Ni-based) and cobalt-based (Co-based) superalloys are typically used, with Ni-based superalloys being more prevalent due to the relatively higher cost of Co. These superalloys are specifically designed to withstand the severe thermal and mechanical stresses encountered in these environments.

Ni-based superalloys derive their exceptional high-temperature strength from γ/γ' precipitation strengthening. In these alloys, the γ phase represents the face-centered cubic (FCC) solid-solution matrix, while the γ' phase consists of ordered, coherent precipitates (such as Ni_3Al, which has an L12 structure). The interplay between these phases is critical to the alloy's performance, providing strength through precipitation hardening. One notable feature of many Ni-based superalloys is that their strength increases with temperature, peaking around 800°C. However, beyond this temperature, the tensile strength typically decreases [5].

Despite this, these alloys maintain excellent high-temperature stability, albeit at the cost of increased density. Most Ni-based superalloys have a density ranging between 8.5 to 9.2 g/cc [6], which is more than twice that of typical Ti and Al alloys. This high density negatively impacts the TWR and overall fuel efficiency, making the search for lower-density alternatives crucial. Furthermore, Ni-based superalloys are expensive due to their complex compositions, which often include rare and refractory metals. This high cost, combined with their density, drives the ongoing

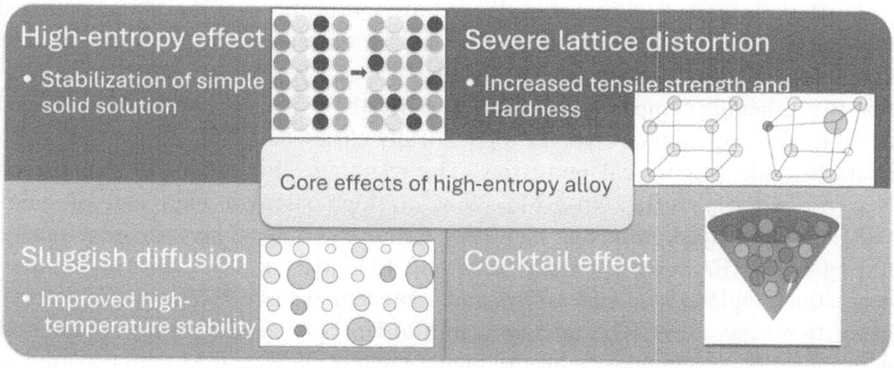

FIGURE 11.1 The core effect of HEAs

search for materials that can meet the demanding high-temperature property portfolio with lower weight and cost. HEAs have emerged as a promising solution, offering a potential alternative for high-temperature aerospace applications.

It has been hypothesized that HEAs show such superior properties through 4 'core effects' present in these alloys: the high entropy effect, the lattice distortion effect, sluggish diffusion, and the cocktail effect [7] (Figure 11.1). The high entropy effect is the signature concept of these alloys. High entropy enables these alloys to have a single-phase structure and thus retain the ductility at increased strength. The lattice distortion effect means that the atoms in the material would be arranged so that the building structure of bonds would be seen as distorted. It also results in an increase of strength without losing ductility. The sluggish diffusion means that the movement of atoms in these alloys would be very slow, resulting in superior properties at high temperatures. The final cocktail effect simply means that the result is unpredictable and more significant than the sum of the parts.

11.2.1 High-Temperature Strength and Strengthening Mechanisms in HEAs

One of the earliest efforts to develop low-density (less than 8 g/cc) HEAs for high-temperature applications was conducted by Yeh et al. [8]. Utilizing thermodynamic design principles, particularly the CALculation of PHAse Diagrams (CALPHAD) method, they designed the $Ni_{48.6}Co_{17}Cr_{7.5}Al_{10.3}Ti_{5.8}Ta_{0.6}Fe_{9}Mo_{0.8}W_{0.4}$ high entropy superalloy (HESA) alloy. This HEA, with a density lower than 8 g/cc, exhibited higher γ' solvus temperatures compared to traditional superalloys. The γ and γ' phases in HESA possess a higher degree of randomness in their ordered structures than those found in CM247LC and RR2101, a defining characteristic of HEAs. Moreover, the overall cost of HESA is reduced due to its minimal content of costly refractory or precious metals. The alloy maintains a stable γ/γ' microstructure from intermediate to high temperatures and demonstrates superior hardness compared to the conventional CM247LC alloy. HESA offers better mechanical properties than CM247LC at only 80% of the cost, with the added benefit of lower density.

Further research by Chen et al. [9] led to modifications of the aforementioned alloy, specifically by increasing the Mo and W content while removing Fe. The hypothesis behind this modification was that Mo and W provide greater solid solution strengthening compared to Fe. Additionally, the Ti and Nb content was reduced to avoid the formation of complex topologically close-packed (TCP) phases, which could degrade mechanical properties. The resulting alloy, with the composition $Ni_{56.11}Co_{16}Cr_8Al_{10}Ti_{3.2}Ta_{1.7}Nb_{0.65}Mo_{1.3}W_{2.7}C_{0.34}$ high strength high entropy alloy (HSHEA), displayed promising high-temperature mechanical performance. Tensile testing of HSHEA reveals a yield strength of 1025 MPa at 850°C, surpassing first-generation single-crystal superalloys and approaching the performance of second-generation superalloys. The microstructure of HSHEA consists of an FCC matrix with L12-type precipitates, with the higher content of W and Mo contributing to enhanced solid-solution strengthening, as these elements typically have higher strengthening coefficients than Co, Cr, and Fe [10].

Figure 11.2 shows the comparative graph of variation of tensile yield strength with respect to temperature for various HEAs and conventional superalloys. It can be seen that the HEA (Refractory HEA-RHEA) outperforms the conventional superalloy at almost all temperatures. While single-crystal HEAs (HESA-2) have shown promising high-temperature properties, their high manufacturing costs are a significant limitation. Consequently, achieving high-temperature softening resistance in polycrystalline structures is highly desirable. This goal requires the arrest of grain growth at elevated temperatures.

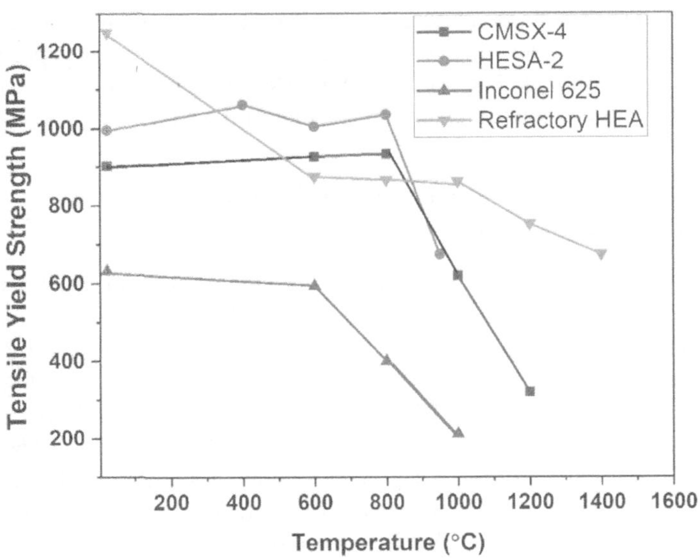

FIGURE 11.2 The comparison of tensile yield strength of CMSX-4 single-crystal superalloy [11], high entropy superalloy (HESA-2) [11], refractory high entropy alloy [12], and Inconel 625 [11].

A novel directionally solidified HEA with the nominal composition $Ni_{47.9}Al_{10.2}Co_{16.9}Cr_{7.4}Fe_{8.9}Ti_{5.8}Mo_{0.9}Nb_{1.2}W_{0.4}C_{0.4}$ (at.%), and a configurational entropy (ΔSmix) of 1.60 R, was designed by Tsao et al. The tensile yield stress of this alloy at high temperatures outperformed many HEAs and was comparable to the single-crystal CMSX-2, though it was lower than the more advanced CMSX-4 and CMSX-10 superalloys. However, the tensile yield strength of the alloy remains remarkably stable from room temperature to 800°C, a highly desirable feature for superalloys intended for gas turbine engine applications. The alloy's creep rate was also comparable to that of the CMSX-2 alloy. The alloy can provide mechanical properties similar to Ni-based superalloys while offering better cost performance due to the relatively lower content of refractory elements [13]. Table 11.1 depicts the High-Temperature yield strength of various reported high entropy alloys.

A similar HEA with the composition $Ni_{48.3}Co_{16.9}Al_{10.2}Fe_{8.9}Cr_{7.4}Ti_{5.8}Nb_{1.2}Mo_{0.9}W_{0.4}$ (at.%) and a configurational entropy of 1.5R was fabricated as a single crystal (SC) [11]. The as-cast single crystal underwent heat treatment at 1500 K for 20 hours, followed by air cooling (HT-1), and subsequent aging at 1023 K for 20 hours, followed by water quenching (HT-2). The alloy exhibits a unique hierarchical microstructure with FCC and L12 phases, where L12 phases are present in the FCC matrix, and disordered FCC particles are observed within the ordered L12 phase. The presence of FCC particles within the L12 phase attracts dislocations, resulting in a wavy slip character [11]. The mechanical properties of the HEA in the HT-2 heat-treated condition surpass those of many advanced superalloys. For instance, the tensile yield strength of the HEA [HT-2] is 110 MPa greater than that of CMSX-4, with three times greater tensile elongation at 1023 K. The cost-specific yield strength of HEA [HT-2] is 1.6 times that of CMSX-4, demonstrating the alloy's superior mechanical performance at a lower cost.

Traditionally, superalloys feature a disordered matrix with ordered precipitates. However, Yang et al. explored a different approach by designing a $Ni_{43.9}Co_{22.4}Fe_{8.8}A$

TABLE 11.1
Comparison of High-Temperature Yield Strength of Various Reported High Entropy Alloys

	Temperature (°C)	Yield Stress (MPa)	Phases	Ref
NiCoCrAlTi	600	1060	L12	[14]
NiCoCrAlTiTaNbMoW C	850	1025	FCC+L12	[9]
$AlCoCrFeTi_{0.5}Ni_{2.5}$	600	1000 (C)	FCC+BCC	[15]
$Al_{20}Co_{20}Cr_{20}Fe_{20}Ni_{20}$	600	1075 (C)	A2+B2	[16]
NbMoTaW(HfN)x	1000	1192 (C)	BCC+HfN	[17]
$W_{30}Ta_5(FeNi)_{65}$	800	1000 (C)	BCC+FCC	[18]
$Ni_{41.98}Co_{23.7}Fe_8Cr_{10}Ti_5Al_{8.6}Ta_{1.0}Nb_{0.5}Mo_{0.5}W_{0.5}Hf_{0.1}Zr_{0.02}B_{0.1}$	700	1045	FCC+l12	[19]
$Co_{43.1}Ni_{28.8}Cr_{10.2}Al_{10.0}Ti_{2.1}Mo_{2.2}$Ta2.1$Nb_{1.5}$	800	755	L12	[20]
$Ni_{41.5}Co_{31}Cr_8Ti_4Al_{10}W_{0.5}Mo_2Ta_3$	800	873	L12	[21]

$l_{10.7}Ti_{11.7}B_{2.5}$ alloy with nanoscale interfacial disordering (NDI-SMs) [22]. This alloy features a microstructure with micrometer-scale ordered superlattice grains and nanometer-scale disordered interfaces, providing excellent strength-ductility synergy. The alloy exhibited an ultrahigh yield strength (σy) of ~1040 ± 61 MPa with an appreciable 25% tensile elongation. It is nearly 2.6 times stronger than the conventional binary Ni_3Al alloy. Additionally, the alloy has shown negligible reduction in hardness up to 800°C and exceptional resistance to grain coarsening at high temperatures, likely due to the configurational entropy of the disordered nanolayer interface lowering the Gibbs energy of grain boundaries, thereby reducing the driving force for grain growth.

Another low-density (7.71 g/cm^3) HEA superalloy with the composition Fe-33.2Ni-13.4Co-3.1Al-2.2Ti-1.1Nb-13.8Cr-1.7Mo-0.03C-0.015B-0.03Zr (wt. %) displayed exceptional high-temperature strength, low coarsening rate of precipitates, and creep resistance [23]. The addition of trace amounts of C, B, and Zr helped refine the grain size, and the alloy displayed a sluggish coarsening rate of K = 1.7 × 10–29 m^3/s at 1023 K, significantly lower than that of many conventional superalloys. This slow coarsening rate of γ' precipitates was attributed to a lower lattice misfit (~ –0.25%) between the γ and γ' phases. Additionally, another L12-strengthened HEA with the composition $Ni_{33.3}Co_{33.3}Cr_{23.4}Al_5Ti_5$ has demonstrated impressive mechanical properties at both room and elevated temperatures [14]. At room temperature, the alloy exhibited a yield stress of approximately 1300 MPa, an ultimate tensile stress of 1610 MPa, and a ductility of 14%. Remarkably, these mechanical properties were largely retained at 600°C, where the alloy displayed a yield stress of around 1060 MPa, an ultimate tensile stress of 1271 MPa, and a ductility of 25% [14]. Microstructural analysis revealed that multiple strengthening mechanisms contributed to these outstanding properties. The enhanced strength and ductility were primarily attributed to the presence of dislocations and stacking fault-based deformation substructures. The complex interaction between the L12 precipitates and hierarchical stacking faults was particularly significant, providing the alloy with robust high-temperature mechanical performance.

One of the defining features of a precipitation-hardened HEA is the presence of strong, yet ductile multicomponent intermetallic nanoparticles (MCNIPs) instead of brittle conventional precipitates that are present in conventional superalloys [24]. The carefully designed experiments have shown that the MCNIPs present in $(FeCoNi)_{86}$-Al_7Ti_7 (Al7Ti7) have higher strength and ductility than the simple Ni_3Al phase, exhibiting a ductility of ~40%, a yield strength of ~600 MPa, and also an excellent stain hardening ability of about ~2.1 GPa. The excellent strength-ductility combination of the Al7Ti7 alloy results from a higher Ti/Al ratio which increases antiphase boundary energy and a higher work hardening. Also, small valence electron concentration values (~8.0) of the MCINP prevent the transformation from the ductile ordered cubic phase to the brittle hexagonal phase (such as η-Ni3Ti).

Figure 11.3 shows the schematic diagram of various mechanisms reported in HEAs which makes them distinct from conventional alloys and offers better high-temperature properties. Figure 11.3 (a) shows the evolution hierarchical microstructure [11] where there is precipitation of disordered FCC particles inside the ordered $L1_2$ phase. Figure 11.3 (b) represents the innovative approach to design a

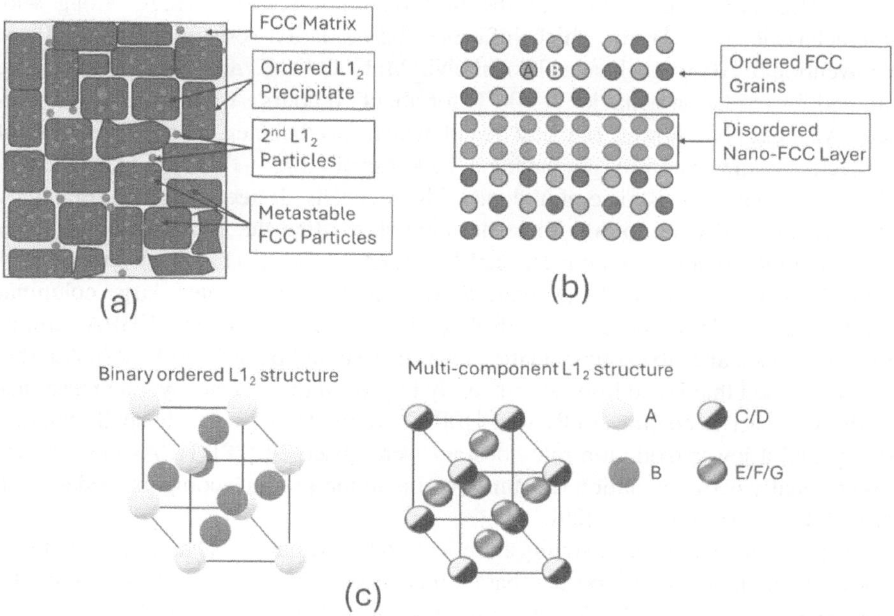

FIGURE 11.3 Schematic diagram showing the phenomenon of (a) hierarchical precipitation in a high entropy superalloy [11], (b) nano-disordered layer between ordered FCC grains [22], and (c) the multicomponent intermetallic precipitate in a HEA [24].

high-temperature alloy where a nano-disordered layer is introduced in between the ordered grains [22]. Figure 11.3 (c) shows the schematic of the presence of multi-component intermetallic precipitates which are not only strong, but also more ductile than the conventional intermetallic particles [24].

11.3 OXIDATION RESISTANCE

Prior to the development of HEAs, there were many conventional alloys studied for oxidation behavior. The oxidation behavior of various classes of conventional bond coat materials, such as (Ni, Pt) aluminides, NiCrAlY, CoCrAlY, and NiCoCrAlY, has been investigated by various researchers over the decades. MCrAlY is one of the commonly used oxidation-resistant alloys. However, it has relatively low strength at high temperatures, which restricts its application as a structural component. In search of newer materials for better efficiency, research on the oxidation behavior of HEAs has been carried out by many researchers, and the results have proven to be potential alternatives for existing bond coat materials as they exhibit promising characteristics for high-temperature applications, such as the slow growth rate of the thermally grown oxides (TGO) layer due to sluggish diffusion behavior and good mechanical properties even at high temperatures.

Several systems of HEAs and RHEAs have been studied for oxidation performance at different temperatures and time duration and reported by various researchers.

Table 11.2 summarizes some of the best oxidation-resistant HEAs along with characteristic oxide layers which influence the oxidation resistance of the alloys. Conventional HEAs such as AlCoCrFeNi, MnCoCrFeNi, and SiCoCrFeNi were studied for oxidation behavior at 900°C for about 50 hours. Among the alloys studied, AlCoCrFeNi showed the best result which is attributed to the formation of protective oxide layer of Al_2O_3 having a thickness of f 2.15 ± 0.15 mm [25]. The isothermal oxidation resistance of Y-doped $AlCoCrFeNi_{2.1}$ Eutectic high entropy alloy (EHEA) at 1100°C up to 500 hours duration shows that the alloy exhibits a lower oxidation rate than that of conventional NiCoCrAlY alloys. This is attributed to the rapid formation of the α-Al_2O_3 scale as well as the oxide scales' large columnar grain sizes [26]. When co-doped with Y-Hf elements, $AlCoCrFeNi_{2.1}$ EHEA exhibits improved resistance to cyclic oxidation as well as spallation at 1200°C [27]. Another study revealed that Hf addition significantly improved the thermal cycling resistance of thermal barrier coatings (TBCs) at 1100°C, where decreased scale spallation tendency and a lower oxidation rate constant were observed [28]. HEAs have shown exceptionally better oxidation resistance owing to their unique complex oxides, such as $CrTaO_4$ oxide in RHEAs [29].

Aluminum content has a significant effect on the oxidation behavior of the alloys. For example, it has been reported that an increase in Al content increases the oxidation resistance of AlxCoCrCuFeNi alloy when tested at 1000°C for 100 hours [30]. Various researchers studied the Al effect on the oxidation resistance in $Al_xCoCrFeNi_2$ HEA and found that a lower Al content is detrimental to the oxidation behavior of the alloy [31, 32]. Some HEA systems, like $FeCoCrNi_2Al$, $FeCoCrNiAl_{0.3}$, $FeCoNiAlTi_{0.4}$, and NiCoCrAlSi, were prepared by the mechanical alloying method and studied for the possible bond coat materials for the TBC system. The sintered samples were subjected to isothermal oxidation at 1050°C for 300 hours in a static air atmosphere. Because of the presence of the α-alumina TGO layer, the $FeCoCrNi_2Al$ HEA alloy was found to have better oxidation resistance than the other alloys. The alloy also exhibits a reasonable coefficient of thermal expansion and good mechanical properties [33]. AlCoCrFeNi family of HEAs are widely studied for oxidation behavior at various temperatures [32, 34, 35]. The presence of Al and Cr elements have a positive effect on the oxidation behavior of the $Al_xCoCr_yFeNi_2$ ($x = 1, 1.5; y = 1, 1.5$) HEA [31, 36].

While the isothermal oxidation studies provide valuable insights into material behavior, cyclic oxidation testing is essential for accurately assessing a material's suitability for real-world applications. The more aggressive nature of cyclic oxidation, characterized by repeated heating and cooling cycles, often results in accelerated material degradation compared to isothermal conditions. Cyclic oxidation yields a significantly higher mass gain than isothermal oxidation [46]. This finding underscores the importance of conducting laboratory-scale cyclic oxidation experiments to accurately simulate the complex operational conditions encountered in real-world applications [47].

The comparison of oxidation resistance of AlCoCrFeNi family of HEAs [31, 48] with that of Ni superalloy [47] and commercially available bond coat materials [49, 50] is shown in Figure 11.4. The HEAs have shown better performance during the cyclic oxidation test which makes them potential materials for high-temperature applications [31]. A recent study revealed that the oxidation resistance can be enhanced by the heat treatment of the HEAs by altering the microstructure [51].

TABLE 11.2
Mass Gain Values of Various HEAs After Oxidation at Different Temperatures and Time

Composition	Temp (^{0}C)	Specific Weight Gain (mg/cm^2) ~ 50 h	100 h	Remarks	Ref
AlCoCrFeNi	900	0.02		Better oxidation resistance, predominant Al_2O_3 oxide formation	[25]
$Ni_2FeCoCr$	900	0.0464		Better oxidation resistance, Cr_2O_3 oxide formation	[32]
$Ni_2FeCoCrAl_{0.5}$	900	0.0330		Better oxidation resistance, Al_2O_3, Cr_2O_3 oxides formation	[32]
$Ni_2FeCoCrAl$	900	0.0217		Better oxidation resistance, dominant Al_2O_3 oxide formation	[32]
$FeCoCrNi_2Al$	1050	0.15	0.20	Better oxidation resistance, dominant Al_2O_3 oxide formation	[33]
$FeCoCrNiAl_{0.3}$	1050	–0.772	–0.52	Al_2O_3, Cr_2O_3 oxides formation	[33]
$FeCoNiAlTi_{0.4}$	1050	1.11	1.68	TiO_2 oxide formation	[33]
NiCoCrAlSi	1050	0.17	0.47	Better oxidation resistance, Al_2O_3 oxide formation	[33]
$CrMo_{0.5}NbTa_{0.5}TiZr$	1000	70	120	Poor oxidation resistance	[37]
AlCrMoTaTi	900–1100	0.6		$CrTaO_4$ complex oxide formation	[38]
AlCrMoNbTi	1000	9		Better oxidation resistance	[39]
AlCrMoNbTi-1Si	1000	6.5		Better oxidation resistance	[39]
AlCrMoTiW	1000	8		Better oxidation resistance	[40]
NbTiZrCr	1000	80	100	Poor oxidation resistance $Nb_2Zr_6O_{17}$ complex oxide formation	[41]
TiZrNbHfTa	1100		77	Poor oxidation resistance	[42]
$(Al_{10}CrNbTiZr)$-Mo-0	1000	24		Poor oxidation resistance	[43]
Al20Nb30Ta10Ti30Zr10	1000	37.5	50	Poor oxidation resistance	[44]
Cr17.6Al20.3Mo15.2Nb2.9Si13.4Ta5.4Ti NV1 alloy	1200		6	Protective $CrTaO_4$ oxide formation	[45]

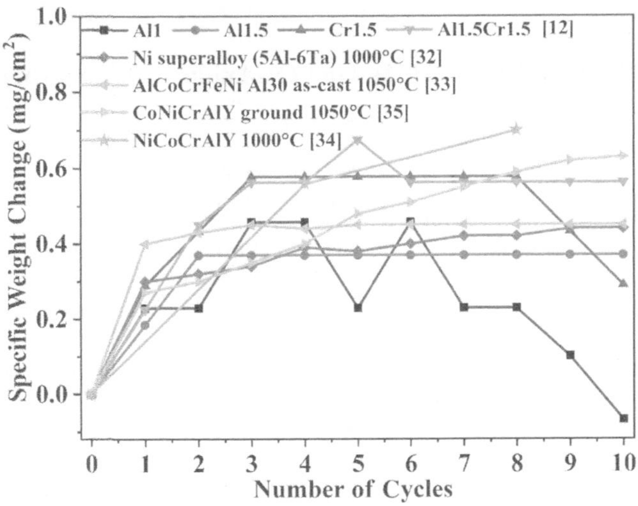

FIGURE 11.4 Specific weight change of various HEAs and conventional alloys after 10 cycles of cyclic oxidation at 1050°C up to 55 hours [31].

Oxidation kinetics is an important parameter in understanding the oxidation phenomenon because it tells us if it follows linear, parabolic, or logarithmic law. Generally, all the metals follow the parabolic law of kinetics, which indicates that initially, the rapid mass gain occurs due to the formation of oxides followed by steady-state oxidation. The equation to determine the oxidation rate constant following the parabolic law is expressed as follows:

$$\left(\frac{\Delta \mathrm{m}}{A}\right)^2 = K_p.\mathrm{t} \tag{11.1}$$

where Δm is mass change ($\mathrm{mg.cm^{-2}}$), A is the total area of the sample subjected to oxidation ($\mathrm{cm^2}$), Kp is the oxidation rate constant ($\mathrm{g^2.cm^{-4}.s^{-1}}$), and t is the oxidation time (h).

11.4 HIGH-TEMPERATURE WEAR IN AEROSPACE

High-temperature wear is a critical phenomenon in aerospace applications where materials are exposed to extreme conditions that significantly affect their performance and longevity. In aerospace environments, components such as turbine blades, engine nozzles, and heat shields are routinely subjected to temperatures exceeding 1000°C. These conditions can induce severe material degradation through mechanisms such as oxidation, corrosion, and mechanical wear, potentially leading to catastrophic failure if not properly managed.

Understanding and mitigating high-temperature wear is crucial for the safety, efficiency, and cost-effectiveness of aerospace operations. As aircraft engines and space vehicles push the boundaries of performance, the demand for materials that can withstand these harsh environments without significant degradation is more critical than ever. Traditional materials like nickel-based superalloys have been the backbone of high-temperature aerospace applications due to their excellent mechanical properties [52–54].

11.4.1 Mechanisms of High-Temperature Wear

High-temperature wear in aerospace components is governed by several complex mechanisms, each contributing to material degradation in unique ways. The primary mechanisms include oxidative wear, adhesive wear, abrasive wear, and thermal fatigue, all of which are influenced by the high-temperature environment and the materials involved.

1. **Oxidative Wear:** Oxidative wear occurs when high temperatures accelerate the formation of oxide layers on the surface of materials. While thin oxide layers can provide some protective benefits, thicker layers often spall off due to mechanical stresses, leading to material loss and further exposure to wear. This cycle of oxidation and spallation can severely compromise the integrity of aerospace components. Oxidative wear is particularly problematic in environments where components are exposed to fluctuating temperatures, such as during take-off and landing [55].
2. **Adhesive Wear:** Adhesive wear arises from the interaction of contacting surfaces at elevated temperatures, where material transfer occurs between the surfaces due to adhesion. At high temperatures, the material softens, making it easier for asperities to adhere and transfer material between the contacting surfaces. This wear mechanism can result in significant surface damage and is particularly critical in aerospace applications where precise tolerances and surface finishes are required [56].
3. **Abrasive Wear:** Abrasive wear occurs when hard particles or asperities on one surface remove material from another surface during relative motion. At high temperatures, the hardness of the materials can decrease, making them more susceptible to abrasive wear. This mechanism is often exacerbated in aerospace environments where dust, debris, or oxidation products act as abrasive particles, leading to accelerated material loss [57].
4. **Thermal Fatigue:** Thermal fatigue is the result of cyclic thermal stresses that occur due to repeated heating and cooling cycles. These stresses can lead to the initiation and propagation of cracks, which eventually result in material failure. Thermal fatigue is a significant concern in aerospace components like turbine blades, which experience rapid temperature changes during operation. The high stresses combined with elevated temperatures make thermal fatigue a dominant wear mechanism in many aerospace applications [58].
5. **Delamination:** Delamination wear at elevated temperatures occurs due to repeated mechanical stresses, leading to subsurface cracks that propagate parallel to the surface. These cracks result in the peeling or separation of material layers, causing significant wear. At elevated temperatures, thermal softening reduces material strength, accelerating crack formation and

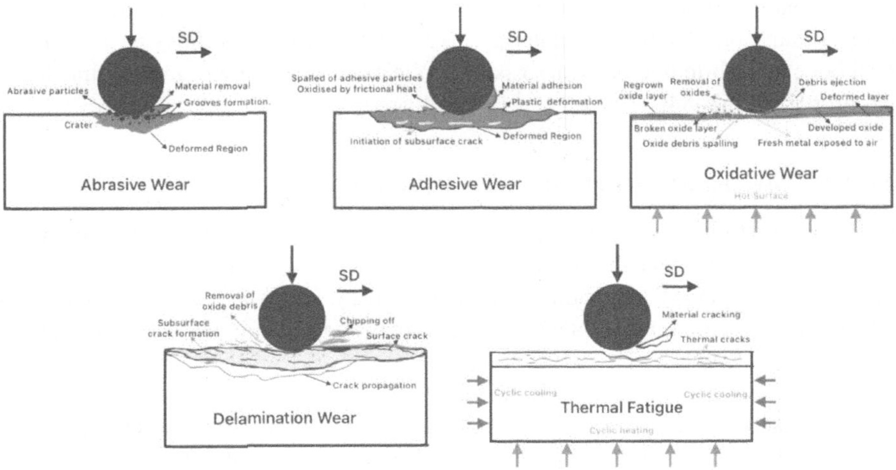

FIGURE 11.5 Schematic of various wear mechanisms.

growth. This process is particularly problematic in aerospace applications, where high operational temperatures exacerbate the wear. The combination of mechanical and thermal stresses weakens the material's surface, leading to premature failure. Mitigating delamination wear requires advanced material designs, such as coatings or high-performance alloys, to maintain surface integrity in extreme environments [59].

Figure 11.5 shows the schematic of transition of wear mechanism from abrasive and adhesive to oxidative, delamination and fatigue under the constant applied load. Understanding these wear mechanisms is essential for the development of materials and coatings that can withstand the harsh conditions encountered in aerospace environments. HEAs with their unique microstructures and superior high-temperature properties, have shown promise in resisting these wear mechanisms, making them strong candidates for aerospace applications.

11.4.2 High-Temperature Wear in HEAs

Conventional materials used in high-temperature aerospace applications, such as nickel-based superalloys and titanium alloys, have been the industry standard due to their excellent mechanical properties and oxidation resistance. However, these alloys have limitations. Nickel-based superalloys, for example, suffer from phase instability and oxidation-induced wear at very high temperatures, often necessitating protective coatings. Similarly, titanium alloys, while lightweight and strong, are prone to oxidation and embrittlement at elevated temperatures, which limits their applicability in extreme environments.

The inherent stability of HEAs at high temperatures makes them particularly suitable for aerospace applications, where materials must withstand extreme environments. The wear resistance of HEAs at elevated temperatures is one of their most promising attributes. For instance, studies have shown that HEAs such as CoCrFeMnNi exhibit exceptional wear resistance due to their unique microstructures, which include a single-phase face-centered cubic (FCC) structure that is both tough and resistant to wear. Furthermore, some HEAs form stable oxide layers during high-temperature exposure, which further enhances their wear resistance by reducing direct metal-to-metal contact.

Oxidation resistance is critical for materials operating at high temperatures, as oxidation can significantly accelerate wear processes. HEAs have demonstrated superior oxidation resistance compared to traditional alloys. For example, the AlCoCrFeNi HEA forms a continuous and adherent oxide layer when exposed to high temperatures, providing effective protection against further oxidation and wear. This property is attributed to the high entropy effect, which stabilizes the oxide scales and prevents their spallation under thermal cycling conditions. Experimental data further supports the superior performance of HEAs in high-temperature wear scenarios. For instance, a study by Joseph et al. (2019) found that CoCrFeMnNi HEA exhibited a lower wear rate at 800°C compared to conventional Ni-based superalloys, attributed to the uniform distribution of elements and the formation of a protective oxide tribolayer [60]. Additionally, Yu et al. studied the tribological properties of $AlCoCrFeNiTi_{0.5}$ HEA under dry sliding and reported that the wear resistance was better than Ti-46Al-2Cr-2Nb, $Fe_{77}Ni_{23}$ alloy, and 3.72 times better when compared to AISI 52100 steel [61]. Moreover, HEAs can also form protective oxide layers on their surface when exposed to high temperatures, acting as a barrier against the wear process. Jithin et al. studied the wear performance of the AlCoCrFeNi system and reported that the wear performance of the alloy surpassed that of Inconel 718 at 900°C because of the formation of a compact oxide layer at the contact region [60]. Verma et al. studied high-temperature wear in CoCrFeNiCux HEA and observed that Cu present in dendritic and interdendritic regions forms a self-lubricating layer at an elevated temperature [62].

The CoCrFeNi system was reported to have better tribological properties over a wide temperature range up to 800°C, attributed to its ability to form a compact oxide layer at elevated temperatures [79]. Furthermore, the incorporation of solid lubricants such as graphite, molybdenum disulfide, sulfur, and copper led to further enhancements in the high-temperature wear resistance of the CoCrFeNi system [80–82]. Yu et al. [83] Reported that niobium addition increased the hot hardness of CoCrFeNi HEA, enhancing high-temperature wear resistance due to dense oxide tribo-layers at 800°C. The laser-cladded FeNiCoAlCu HEA coating demonstrated excellent thermal stability and outstanding wear resistance from room temperature to 800°C, owning to oxide film formation [84]. Figure 11.6 shows an Ashby plot comparing the wear rate of conventional super alloys and various HEAs across various temperatures. It can be clearly observed that the performance of many HEAs is far better than Inconel and Hayes superalloys at elevated temperatures. The formation of a stable, protective, and self-lubricating multi-element tribo-oxide layer is unique in HEAs, which offers outstanding wear resistance at elevated temperatures. These multi-element oxides grow uniformly on the surface of the HEA and stabilize, which protects further diffusion of oxygen into the metal, resisting oxidation. It also acts as protective lubricating film between the sliding surfaces, resisting wear-induced

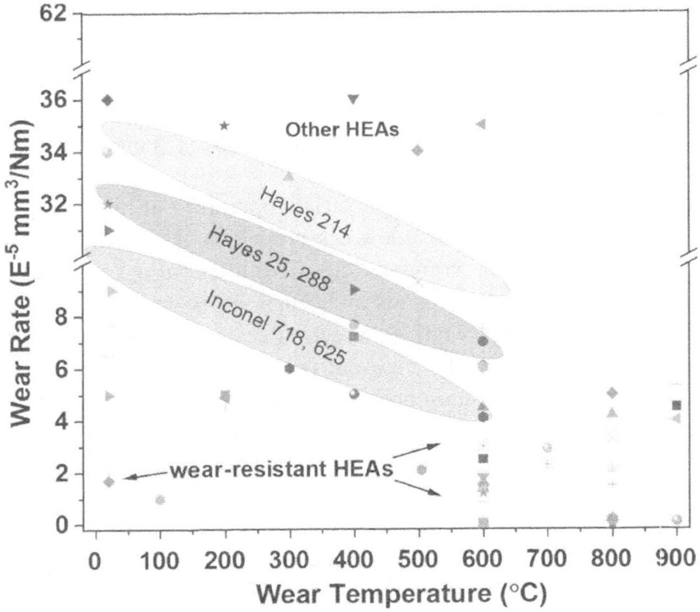

FIGURE 11.6 Wear rate comparison of reported HEAs and conventional superalloys [60, 62–78].

damage by avoiding metal-to-metal contact. Various studies have reported the formation of mixed oxides, such as Fe_2NiO_4, $CoFe_2O_4$, $CoAl_2O_4$, $AlFe_2O_4$, Al_2O_3, CoO, and NiO, known for their wear and oxidation-resistant properties and superior stability at high temperatures [14, 20–25]. Figure 11.7 shows a scanning electron microscopy (SEM) of the surface of a HEA sample that underwent a wear test at 700°C. The figure depicts the formation of a self-protective, self-lubricating, multicomponent tribo-oxide layer on the surface, which is confirmed by corresponding peaks of X-ray photospectroscopy (XPS). This oxide layer acts as self-lubricant and protects the material from wear-induced damage. Table 11.3 shows the summary of

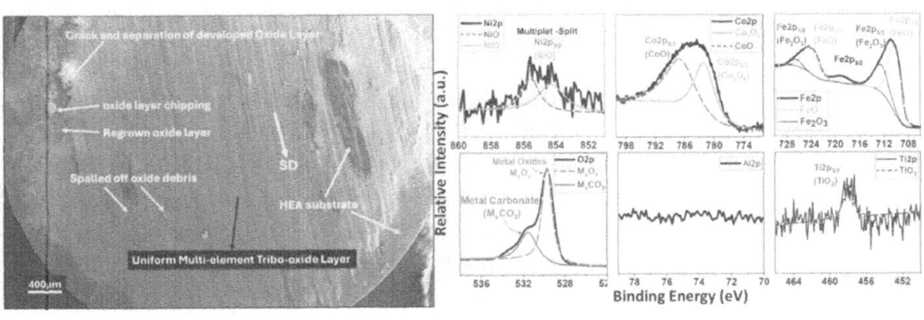

FIGURE 11.7 High temperature post wear SEM depicting formation of multi-element tribo-oxide layer (LHS) on the sample surface confirmed by XPS analysis (RHS).

TABLE 11.3
Wear Rates of Various HEAs at Elevated Temperatures, Their Wear Mechanism, and Processing Technique

Alloy Composition	Temperature (°C)	Wear Rate × E^{-5} mm^3/Nm	Wear Mechanism	Processing Technique	Ref.
CoCrFeNiCu0	600	2.5	Abrasive and Adhesive	Arc Melting	[62]
CoCrFeNiCu0.2	600	1.5			
CoCrFeNiCu0.4	600	1.35			
CoCrFeNiCu0.6	600	1.4			
CoCrFeNiCu0.8	600	1.5			
CoCrFeNiCu1	600	1.3			
CoCrFeNb0.5Ni	20	4.1	Abrasive, Adhesive, and Oxidative	Arc Melting	[85]
	200	6.1			
	400	7.5			
	600	3			
	800	2			
CoCrFeNb0.65Ni	20	1.87			
	200	5			
	400	7.66			
	600	4.5			
	800	0.02			
CoCrFeNb0.8Ni	20	1.7			
	200	4.9			
	400	9			
	600	6			
	800	0.01			
CoCrFeNi+Ag+BaF2/CaF2	20	40	Abrasive, Adhesive, and Oxidative	SPS	[60]
	200	27			
	400	24			
	600	24			
	800	12			
CoCrFeNi	20	6.5	Abrasive, Adhesive, and Oxidative	SPS	[79]
	200	4.5			
	400	7.2			
	600	7			
	800	4.2			
	500	59			
TiZrV0.5Nb0.5Al0	20	36	Abrasive with Plastic Deformation	VAM	[66]
	800	0.1	Adhesive, Delamination, and Oxidative		

(Continued)

TABLE 11.3 ***(Continued)***
Wear Rates of Various HEAs at Elevated Temperatures, Their Wear Mechanism, and Processing Technique

Alloy Composition	Temperature (°C)	Wear Rate × E^{-5} mm^3/Nm	Wear Mechanism	Processing Technique	Ref.
TiZrV0.5Nb0.5Al0.25	20	31	Abrasive with Plastic Deformation		
	800	0.1	Adhesive, Delamination, and Oxidative		
TiZrV0.5Nb0.5Al0.5	20	32	Abrasive with Plastic Deformation		
	800	0.17	Adhesive, Delamination, and Oxidative		
TiZrV0.5Nb0.5Al0.75	20	34	Abrasive with Plastic Deformation		
	800	0.1	Adhesive, Delamination, and Oxidative		
Al0.3CoCrFeNi	25	25	Abrasive, Oxidative, and Delamination	VAM	[67]
	300	12			
	600	2			
	800	1.5			
	900	4.5			
Al0.6CoCrFeNi	25	22.5	Abrasive, Oxidative, and Delamination	VAM	[67]
	300	12			
	600	1.8			
	800	5			
	900	4			
AlCoCrFeNi	25	9	Abrasive, Oxidative, and Delamination	VAM	[68]
	300	6			
	600	1.2			
	800	0.2			
	900	0.1			
AlCoCrFeNiTi	25	7.7	Abrasive, Adhesive, and Oxidative	SPS	[69]
	500	9.3			
	700	2.3			
	900	5.3			

(Continued)

TABLE 11.3 *(Continued)*
Wear Rates of Various HEAs at Elevated Temperatures, Their Wear Mechanism, and Processing Technique

Alloy Composition	Temperature (°C)	Wear Rate × E⁻⁵ mm³/Nm	Wear Mechanism	Processing Technique	Ref.
Al0.25CoCrFeNi	20	14	Delamination, Abrasive, Adhesive, and Oxidative	VAM	[70]
	100	23			
	200	27			
	400	36			
	500	34			
	600	35			
Al0.6CoCrFeNi	20	5	Delamination, Abrasive, Adhesive, and Oxidative	VAM	[84]
	100	1			
	200	35			
	300	45			
	400	5			
	500	52			
	600	52			
AlCoCrFeNi + TiC reinforced (Ti0)	600	0.054	Abrasive, Adhesive, and Oxidative	Laser Cladding	[71]
AlCoCrFeNi + TiC reinforced (Ti0.2)	600	0.043			
AlCoCrFeNi + TiC reinforced (Ti0.4)	600	0.036			
AlCoCrFeNi + TiC reinforced (Ti0.6)	600	0.023			
AlCoCrFeNi + TiC reinforced (Ti0.8)	600	0.01			
AlCoCrFeNi + TiC reinforced (Ti1.0)	600	0.007			
CuMoTaWV	25	400	Adhesive and Oxidative	SPS	[72]
	200	2300			
	400	500			
	600	4500			
Alx(TiZrHfNb)100-x	500	2.09	Abrasive and Oxidative	VAM	[73]
AlCoCrFeNi2.1	900	55.5	Abrasive	Arc Melting and Electrochemical Boronizing	[75]
AlCoCrFeNi2.1	500	10.8	Oxidative	SLM	[74]
	700	2.9			

high-temperature wear resistance of various reported HEAs with their wear mechanisms and processing techniques.

Some recently reported the (FeCoCrNi)Al_xTi_y HEA system has shown excellent trifecta of strength, wear, and oxidation resistance, along with being lightweight, outperforming superalloys in every domain. Unlike Ni-based superalloys, these HEAs can maintain their oxidation and wear resistance without the need for external coatings. This intrinsic property reduces the material's overall weight and simplifies maintenance. These HEAs often maintain their mechanical properties over a broader temperature range, exhibiting high thermal stability and resistance to softening at temperatures where conventional superalloys would fail. These attributes make HEAs a compelling alternative to traditional materials in aerospace applications, especially in components subjected to high wear and oxidation at elevated temperatures.

11.5 COATING OF HIGH ENTROPY ALLOYS

A thermal barrier coating (TBC) system is used in high-temperature applications to prevent the degradation of load-bearing component materials due to thermal stress. TBCs can operate at elevated temperatures due to their high melting point [4]. The TBC system comprises a ceramic topcoat and a metallic bond coat. The most used ceramic topcoat in TBC coating is zirconium oxide (ZrO_2), in which 6–8% yttrium oxide is commonly added to partially stabilize the zirconia tetragonal phase. It is commonly referred to as yttria stabilized zirconia (YSZ) coating. YSZ is generally coated on nickel-based superalloys for thermal insulation [86]. In recent developments, HEAs have been shown to exhibit excellent oxidation resistance and have the potential to be an alternative for bond coat materials. There are various methods by which the HEAs can be coated such as thermal spraying, laser cladding, electrodeposition, magnetron sputtering, and others.

The bond coat is the intermediate layer between the ceramic topcoat and the superalloy substrate. The bond coat is primarily designed to provide bonding between the topcoat and substrate. In addition, the bond coat material should also fulfill other requirements, such as protecting the substrate material from oxidation and bridging the coefficient of thermal expansion (CTE) difference between the topcoat and the substrate [87]. The bond coat provides oxidation resistance to the substrate by forming a protective TGO layer between the topcoat and the bond coat. The TGO layer primarily consists of an Al_2O_3 oxide layer [88]. Commonly used bond coat materials are aluminides (Ni/Ti,Al) or MCrAlY coatings (where M = Ni/Co), which are commercially available. Although NiAl has excellent oxidation resistance, it is not suitable for bond coat application due to its brittleness and poor strength at high temperatures [89]. Huang et al. started the exploration of HEAs for coating application with $AlCoCrFeMo_{0.5}NiSiTi$ and $AlCrFeMo_{0.5}NiSiTi$ alloys on 304 stainless-steel substrates. The coatings resulted in improved oxidation and wear resistance [90]. Since then, various studies have been reported on the coating of HEAs on substrates for high-temperature applications such as Ni/Co/Ti based alloys. Commonly used HEAs consist of Al, Co, Cr, Fe, and Ni elements, as these elements are referred to as 3d transition elements. Owing to their low conductivity properties, HEAs are beneficial as a coating material to protect the superalloys from high temperatures

TABLE 11.4
Various HEAs Used as Coating Materials for Aerospace Applications

High Entropy Coatings (HEC)	Composition	Substrate	Coating Process	Phase/Structure	Temp (°C)	Properties Exhibited	Ref
3d Transition Metal Coatings	Cu-Ni-Fe-Mo-xCr	Medium carbon steel	Laser cladding	Laves phase	500	Good oxidation resistance	[94]
	$Al_{1.4}Co_{2.1}Cr_{0.7}Ni_{2.4}Si_{0.2}Ti_{0.14}$	Ni-super alloy	HVOF	FCC, BCC	1050	Excellent oxidation resistance	[95]
	$CuAl_xNiCrFe$	Inconel 718	Laser Cladding	FCC, BCC	1100	High thermal stability, high oxidation resistance	[96]
	$NiCo_{0.6}Fe_{0.2}Cr_{1.5}SiAlTi_{0.2}$	Incoloy 800H	Plasma Spraying	BCC, Cr_3Si-phase	1100	High oxidation resistance	[97]
	Y-Hf doped AlCoCrFeNi	Ni-super alloy	Sintering process	FCC, BCC	1100	Low oxidation rate, high oxidation resistance	[98]
	NiCrCoTiVAl	Ti-6Al-4V	Laser Alloying	BCC, Intermetallic	1000	High oxidation resistance, high thermal stability	[99]
	NiCoCrAlTi	Haynes 230	HVOF	FCC, BCC	1150	Excellent oxidation resistance	[100]
	AlCoCrFeNi	Ni-super alloy	Cold Spray	FCC, BCC	1100	High thermal stability, high oxidation resistance	[101]
	CuAlNiCrFe	Inconel 718	Laser Cladding	FCC, BCC	1050	High thermal stability, high oxidation resistance	[102]
Refractory High Entropy Metal Coatings	HfNbTaZr	Inconel 718	HVOF	BCC	–	High hardness, high wear resistance	[21]
	MoNbTaVW	Inconel 718	Air Plasma Spray	BCC	–	High hardness, high wear resistance	[21]
	AlTiVMoNb	Ti–6Al–4V	Laser Cladding	BCC	800	High hardness oxidation resistance	[103]
High Entropy Ceramic Coatings	$RE_2(Ce_{0.2}Zr_{0.2}Hf_{0.2}Sn_{0.2}Ti_{0.2})_2O_7$ (RE = Y, Ho, Er, Yb)	Ni-super alloy	APS	Fluorite	1200	High thermal expansion coefficient, low thermal conductivity	[104]
	$(La_{0.2}Nd_{0.2}Sm_{0.2}Eu_{0.2}Gd_{0.2})_2Zr_2O_7$	Ni-super alloy	APS	Fluorite	1100	High thermal stability, high coefficient of thermal expansion	[105]

[91]. RHEAs are also well known for their excellent wear and abrasion resistance in high-temperature conditions. These alloys are made of refractory elements such as Mo, Ta, Ti, W, and Zr [92]. Satish et al. investigated the HfNbTaZr and MoNbTaVW RHEA coatings on SS321 and Inconel 718 substrates through air plasma spray (APS) and high velocity oxygen fuel (HVOF) thermal spray methods. The RHEA coatings have shown good abrasion resistance and high hardness [93]. Table 11.4 shows the summary of various HEAs used as coating material for aerospace applications.

11.6 CONCLUSION

In conclusion, HEAs represent a transformative development in materials science for aerospace applications. Their high strength, superior oxidation resistance, and exceptional wear performance at elevated temperatures surpass those of conventional and Ni-based superalloys. Furthermore, the microstructural stability of HEAs ensures sustained performance even in extreme conditions, making them ideal for components exposed to high thermal and mechanical stresses. HEA coatings further enhance the surface protection of components, offering remarkable resistance to both wear and oxidation. The design flexibility of HEAs allows for the customization of alloy compositions, tailored to meet specific aerospace requirements. The potential of HEAs to reduce component weight, improve fuel efficiency, and support sustainability in aerospace operations makes them a key candidate for future materials innovation in the industry.

REFERENCES

1. E. Akca, A. Gürsel, A review on superalloys and IN718 nickel-based INCONEL superalloy, Periodicals of Engineering and Natural Sciences (PEN) 3 (2015). https://doi.org/10.21533/pen.v3i1.43
2. S.M. Meier, D.K. Gupta, The Evolution of Thermal Barrier Coatings in Gas Turbine Engine Applications. (1994). http://www.asme.org/about-asme/terms-of-use.
3. Z. Mutasim, Gas Turbines Durability in Harsh Environments. Proceedings of the First Middle East Turbomachinery Symposium February 13-16, 2011, Doha, Qatar
4. R.V. Hillery, Coatings for High-Temperature Structural Materials. National Academies Press (1996). https://doi.org/10.17226/5038.
5. R.C. Reed, The Superalloys: Fundamentals and Applications. Cambridge University Press (2006). pp. 1–372. https://doi.org/10.1017/CBO9780511541285.
6. R.A. MacKay, T.P. Gabb, J.L. Smialek, M.V. Nathal, A new approach of designing superalloys for low density, JOM 62 (2010) 48–54. https://doi.org/10.1007/S11837-010-0011-0
7. D.B. Miracle, O.N. Senkov, A critical review of high entropy alloys and related concepts, Acta Mater 122 (2017) 448–511. https://doi.org/10.1016/J.ACTAMAT.2016.08.081.
8. A.C. Yeh, T.K. Tsao, Y.J. Chang, K.C. Chang, J.W. Yeh, M.S. Chiou, S.R. Jian, C.M. Kuo, W.R. Wang, H. Murakami, Developing new type of high temperature Alloys–High entropy superalloys, International Journal of Metallurgical & Materials Engineering (2015). https://doi.org/10.15344/2455-2372/2015/107
9. H. Chen, X. Yuan, W. Ren, J. Peng, B. Ding, T. Zheng, J. Yu, P.K. Liaw, Y. Zhong, A new single crystal high entropy alloy with excellent high-temperature tensile property, Mater Res Express 7 (2020) 046507. https://doi.org/10.1088/2053-1591/AB8491
10. A.J. Goodfellow, E.I. Galindo-Nava, C. Schwalbe, H.J. Stone, The role of composition on the extent of individual strengthening mechanisms in polycrystalline Ni-based superalloys, Mater Des 173 (2019). https://doi.org/10.1016/j.matdes.2019.107760

11. Y.T. Chen, Y.J. Chang, H. Murakami, T. Sasaki, K. Hono, C.W. Li, K. Kakehi, J.W. Yeh, A.C. Yeh, Hierarchical microstructure strengthening in a single crystal high entropy superalloy, Sci Rep 10 (1) (2020) 1–11. https://doi.org/10.1038/s41598-020-69257-8
12. O.N. Senkov, G.B. Wilks, J.M. Scott, D.B. Miracle, Mechanical properties of Nb25Mo25Ta25W25 and V20Nb20Mo20Ta20W20 refractory high entropy alloys, Intermetallics (Barking) 19 (2011) 698–706. https://doi.org/10.1016/J.INTERMET.2011.01.004
13. T.K. Tsao, A.C. Yeh, C.M. Kuo, K. Kakehi, H. Murakami, J.W. Yeh, S.R. Jian, The high temperature tensile and creep behaviors of high entropy superalloy, Sci Rep 7 (1) (2017) 1–9. https://doi.org/10.1038/s41598-017-13026-7
14. H. Zhang, H. Meng, F. Meng, Y. Tong, P.K. Liaw, X. Yang, L. Zhao, H. Wang, Y. Gao, S. Chen, Magnificent tensile strength and ductility synergy in a NiCoCrAlTi high-entropy alloy at elevated temperature, Journal of Materials Research and Technology 28 (2024) 522–532. https://doi.org/10.1016/J.JMRT.2023.12.038
15. R. Guo, P. Zhang, J. Pan, J. Xu, L. Liu, C. Zhang, L. Liu, Achieving prominent high-temperature mechanical properties in a dual-phase high-entropy alloy: A synergy of deformation-induced twinning and martensite transformation, Acta Mater 264 (2024) 119591. https://doi.org/10.1016/J.ACTAMAT.2023.119591
16. K.R. Lim, K.S. Lee, J.S. Lee, J.Y. Kim, H.J. Chang, Y.S. Na, Dual-phase high-entropy alloys for high-temperature structural applications, J Alloys Compd 728 (2017) 1235–1238. https://doi.org/10.1016/J.JALLCOM.2017.09.089
17. Y. Wan, Y. Cheng, Y. Chen, Z. Zhang, Y. Liu, H. Gong, B. Shen, X. Liang, A nitride-reinforced NbMoTaWHfN refractory high-entropy alloy with potential ultra-high-temperature engineering applications, Engineering 30 (2023) 110–120. https://doi.org/10.1016/J.ENG.2023.06.008
18. T. Li, J.X. Chen, T.W. Liu, Y. Chen, J.H. Luan, Z.B. Jiao, C.T. Liu, L.H. Dai, D022 precipitates strengthened W-Ta-Fe-Ni refractory high-entropy alloy, J Mater Sci Technol 177 (2024) 85–95. https://doi.org/10.1016/J.JMST.2023.07.069
19. Y.H. Zhou, J.Y. Zhang, J. Zhang, X.Y. Yao, J.H. Luan, Q. Li, S.F. Liu, B. Xiao, J. Ju, S.J. Zhao, Y.L. Zhao, Z.Y. Sun, H. Nan, M. Yan, T. Yang, A strong-yet-ductile high-entropy alloy in a broad temperature range from cryogenic to elevated temperatures, Acta Mater 268 (2024) 119770. https://doi.org/10.1016/J.ACTAMAT.2024.119770
20. B.X. Cao, H.J. Kong, Z.Y. Ding, S.W. Wu, J.H. Luan, Z.B. Jiao, J. Lu, C.T. Liu, T. Yang, A novel L12-strengthened multicomponent Co-rich high-entropy alloy with both high γ'-solvus temperature and superior high-temperature strength, Scr Mater 199 (2021) 113826. https://doi.org/10.1016/J.SCRIPTAMAT.2021.113826.
21. W. Yang, S. Lin, Q. Wang, C. Liu, J. Qin, J. Zhang, Accelerated design of L12-strengthened single crystal high entropy alloys based on machine learning and multi-objective optimization, Mater Adv 5 (2024) 5772–5780. https://doi.org/10.1039/D4MA00189C.
22. T. Yang, Y.L. Zhao, W.P. Li, C.Y. Yu, J.H. Luan, D.Y. Lin, L. Fan, Z.B. Jiao, W.H. Liu, X.J. Liu, J.J. Kai, J.C. Huang, C.T. Liu, Ultrahigh-strength and ductile superlattice alloys with nanoscale disordered interfaces, Science (1979) 369 (2020) 427–432. https://doi.org/10.1126/SCIENCE.ABB6830
23. C. Liu, Y. Li, J. Li, Z. Wang, Q. Wang, C. Dong, P.K. Liaw, A novel low-density and high-strength Fe-Ni-base high-entropy superalloy with stable γ/γ' coherent microstructure at 1023 K, Scr Mater 252 (2024) 116236. https://doi.org/10.1016/J.SCRIPTAMAT.2024.116236
24. T. Yang, Y.L. Zhao, Y. Tong, Z.B. Jiao, J. Wei, J.X. Cai, X.D. Han, D. Chen, A. Hu, J.J. Kai, K. Lu, Y. Liu, C.T. Liu, Multicomponent intermetallic nanoparticles and superb mechanical behaviors of complex alloys, Science 362 (2018) 933–937. https://doi.org/10.1126/SCIENCE.AAS8815

25. W. Kai, C.C. Li, F.P. Cheng, K.P. Chu, R.T. Huang, L.W. Tsay, J.J. Kai, Air-oxidation of FeCoNiCr-based quinary high-entropy alloys at 700–900°C, Corros Sci 121 (2017) 116–125. https://doi.org/10.1016/j.corsci.2017.02.008
26. J. Lu, H. Zhang, Y. Chen, L. Li, X. Liu, W. Xiao, N. Ni, X. Zhao, F. Guo, P. Xiao, Y-doped AlCoCrFeNi2.1 eutectic high-entropy alloy with excellent oxidation resistance and structure stability at 1000°C and 1100°C, Corros Sci 180 (2021). https://doi.org/10.1016/j.corsci.2020.109191
27. J. Lu, H. Zhang, L. Li, Y. Chen, X. Liu, X. Zhao, F. Guo, Y-Hf co-doped AlCoCrFeNi2.1 eutectic high-entropy alloy with excellent oxidation and spallation resistance under thermal cycling conditions at 1100°C and 1200°C, Corros Sci 187 (2021). https://doi.org/10.1016/j.corsci.2021.109515
28. C. Jiang, S. Li, H. Liu, Z. Bao, J. Zhang, S. Zhu, F. Wang, Effect of Hf addition in (Ni,Pt)Al bond coat on thermal cycling behavior of a thermal barrier coating system at 1100 °C, Corros Sci 166 (2020). https://doi.org/10.1016/j.corsci.2019.108424
29. K.C. Lo, Y.J. Chang, H. Murakami, J.W. Yeh, A.C. Yeh, An oxidation resistant refractory high entropy alloy protected by CrTaO 4 -based oxide, Sci Rep 9 (2019). https://doi.org/10.1038/s41598-019-43819-x
30. Y.Y. Liu, Z. Chen, Y.Z. Chen, J.C. Shi, Z.Y. Wang, S. Wang, F. Liu, Effect of Al content on high temperature oxidation resistance of AlxCoCrCuFeNi high entropy alloys (x=0, 0.5, 1, 1.5, 2), Vacuum 169 (2019). https://doi.org/10.1016/j.vacuum.2019.108837
31. S. Kumar T., A. Sourav, B.S. Murty, A. Chelvane, S. Thangaraju, Role of Al and Cr on cyclic oxidation behavior of AlCoCrFeNi2 high entropy alloy, J Alloys Compd 919 (2022). https://doi.org/10.1016/j.jallcom.2022.165820
32. W. Kai, F.P. Cheng, Y.R. Lin, C.W. Chuang, R.T. Huang, D. Chen, J.J. Kai, C.T. Liu, C.J. Wang, The oxidation behavior of Ni2FeCoCrAlx high-entropy alloys in dry air, J Alloys Compd 836 (2020). https://doi.org/10.1016/j.jallcom.2020.155518
33. M. Jadhav, S. Singh, M. Srivastava, G.S. Vinod Kumar, An investigation on high entropy alloy for bond coat application in thermal barrier coating system, J Alloys Compd 783 (2019) 662–673. https://doi.org/10.1016/j.jallcom.2018.12.361
34. A. Anupam, A.S.M. Ang, K. Guruvidyathri, M. Abbas, D. Sivaprahasam, P. Munroe, C.C. Berndt, B.S. Murty, R.S. Kottada, Evaluating the influence of microstructural attributes: Fraction, composition, size and spatial distribution of phases on the oxidation behaviour of high-entropy alloys, Corros Sci 184 (2021). https://doi.org/10.1016/j.corsci.2021.109381
35. W. Jiao, T. Li, X. Chang, Y. Lu, G. Yin, Z. Cao, T. Li, A novel Co-free Al0.75CrFeNi eutectic high entropy alloy with superior mechanical properties, J Alloys Compd 902 (2022). https://doi.org/10.1016/j.jallcom.2022.163814
36. T. Liu, C. Wang, H. Shen, W. Chou, N.Y. Iwata, A. Kimura, The effects of Cr and Al concentrations on the oxidation behavior of oxide dispersion strengthened ferritic alloys, Corros Sci 76 (2013). https://doi.org/10.1016/j.corsci.2013.07.004
37. O.N. Senkov, S.V. Senkova, D.M. Dimiduk, C. Woodward, D.B. Miracle, Oxidation behavior of a refractory NbCrMo 0.5Ta 0.5TiZr alloy, J Mater Sci 47 (2012). https://doi.org/10.1007/s10853-012-6582-0
38. B. Gorr, F. Müller, M. Azim, H.J. Christ, T. Müller, H. Chen, A. Kauffmann, M. Heilmaier, High-temperature oxidation behavior of refractory high-entropy alloys: Effect of alloy composition, Oxid Met 88 (2017). https://doi.org/10.1007/s11085-016-9696-y
39. B. Gorr, F. Mueller, H.J. Christ, T. Mueller, H. Chen, A. Kauffmann, M. Heilmaier, High temperature oxidation behavior of an equimolar refractory metal-based alloy [Formula presented] with and without Si addition, J Alloys Compd 688 (2016). https://doi.org/10.1016/j.jallcom.2016.07.219
40. B. Gorr, M. Azim, H.J. Christ, T. Mueller, D. Schliephake, M. Heilmaier, Phase equilibria, microstructure, and high temperature oxidation resistance of novel

refractory high-entropy alloys, J Alloys Compd 624 (2015). https://doi.org/10.1016/j.jallcom.2014.11.012

41. T.M. Butler, K.J. Chaput, J.R. Dietrich, O.N. Senkov, High temperature oxidation behaviors of equimolar NbTiZrV and NbTiZrCr refractory complex concentrated alloys (RCCAs), J Alloys Compd 729 (2017). https://doi.org/10.1016/j.jallcom.2017.09.164
42. C.H. Chang, M.S. Titus, J.W. Yeh, Oxidation behavior between 700 and 1300°C of refractory TiZrNbHfTa high-entropy alloys containing aluminum, Adv Eng Mater 20 (2018). https://doi.org/10.1002/adem.201700948
43. O.A. Waseem, H.J. Ryu, Combinatorial synthesis and analysis of AlxTayVz-Cr20Mo20Nb20Ti20Zr10 and Al10CrMoxNbTiZr10 refractory high-entropy alloys: Oxidation behavior, J Alloys Compd 828 (2020). https://doi.org/10.1016/j.jallcom.2020.154427
44. T.M. Butler, K.J. Chaput, Native oxidation resistance of Al20Nb30Ta10Ti30Zr10 refractory complex concentrated alloy (RCCA), J Alloys Compd 787 (2019). https://doi.org/10.1016/j.jallcom.2019.02.128
45. K.C. Lo, H. Murakami, J.W. Yeh, A.C. Yeh, Oxidation behaviour of a novel refractory high entropy alloy at elevated temperatures, Intermetallics (Barking) 119 (2020). https://doi.org/10.1016/j.intermet.2020.106711
46. M. Salehi Doolabi, B. Ghasemi, S.K. Sadrnezhaad, A. Feizabadi, A. HabibollahZadeh, D. Salehi Doolabi, M. AsadiZarch, Comparison of isothermal with cyclic oxidation behavior of “Cr-aluminide” coating on inconel 738LC at 900°C, Oxid Met 87 (2017). https://doi.org/10.1007/s11085-016-9657-5
47. S.J. Park, S.M. Seo, Y.S. Yoo, H.W. Jeong, H.J. Jang, Effects of Al and Ta on the high temperature oxidation of Ni-based superalloys, Corros Sci 90 (2015) 305–312. https://doi.org/10.1016/j.corsci.2014.10.025
48. T.M. Butler, M.L. Weaver, Oxidation behavior of arc melted AlCoCrFeNi multi-component high-entropy alloys, J Alloys Compd 674 (2016) 229–244. https://doi.org/10.1016/j.jallcom.2016.02.257
49. A. Feizabadi, M. Salehi Doolabi, S.K. Sadrnezhaad, M. Rezaei, Cyclic oxidation characteristics of HVOF thermal-sprayed NiCoCrAlY and CoNiCrAlY coatings at 1000 °C, J Alloys Compd 746 (2018) 509–519. https://doi.org/10.1016/j.jallcom.2018.02.282.
50. D. Mercier, C. Kaplin, G. Goodall, G. Kim, M. Brochu, Parameters influencing the oxidation behavior of cryomilled CoNiCrAlY, Surf Coat Technol 205 (2010) 2546–2553. https://doi.org/10.1016/j.surfcoat.2010.10.008
51. S. Kumar T., A. Sourav, S. Yebaji, L. Chauhan, A. Babu, A. Chelvane, T. Shanmuga sundaram, Effect of heat treatment on the oxidation behavior of an AlCoCrFeNi2 near-eutectic high entropy alloy, Corros Sci 221 (2023). https://doi.org/10.1016/j.corsci.2023.111298
52. J.-W. Yeh, S.-K. Chen, S.-J. Lin, J.-Y. Gan, T.-S. Chin, T. Shun, C.-H. Tsau, S.Y. Chang, Nanostructured high-entropy alloys with multiple principal elements: Novel alloy design concepts and outcomes, Adv Eng Mater 6 (2004) 299–303. https://doi.org/10.1002/adem.200300567
53. B. Cantor, I.T.H. Chang, P. Knight, A.J.B. Vincent, Microstructural development in equiatomic multicomponent alloys, Mater Sci Eng A 375–377 (2004) 213–218. https://doi.org/10.1016/J.MSEA.2003.10.257
54. A. Gali, E.P. George, Tensile properties of high- and medium-entropy alloys, Intermetallics (Barking) 39 (2013) 74–78. https://doi.org/10.1016/J.INTERMET.2013.03.018
55. M.H. Tsai, J.W. Yeh, High-entropy alloys: A critical review, Mater Res Lett 2 (2014) 107–123. https://doi.org/10.1080/21663831.2014.912690
56. T. Pollock, S. Tin, Nickel-based superalloys for advanced turbine engines: Chemistry, microstructure and properties, J Propuls Power 22 (2006) 361–374. https://doi.org/10.2514/1.18239.
57. C. Leyens, M. Peters, Titanium and Titanium Alloys: Fundamentals and Applications. Wiley (2003). https://doi.org/10.1002/3527602119

58. S. Guo, C.T. Liu, Phase stability in high entropy alloys: Formation of solid-solution phase or amorphous phase, Prog Nat Sci Mater Int 21 (2011) 433–446. https://doi.org/10.1016/S1002-0071(12)60080-X
59. N.P. Suh, The delamination theory of wear, Wear 25 (1973) 111–124. https://doi.org/10.1016/0043-1648(73)90125-7
60. J. Joseph, N. Haghdadi, K. Shamlaye, P. Hodgson, M. Barnett, D. Fabijanic, The sliding wear behaviour of CoCrFeMnNi and AlxCoCrFeNi high entropy alloys at elevated temperatures, Wear 428–429 (2019) 32–44. https://doi.org/10.1016/j.wear.2019.03.002
61. Y. Yu, J. Wang, J.S. Li, H.C. Kou, S.Z. Niu, S.Y. Zhu, J. Yang, W.M. Liu, Dry-sliding tribological properties of AlCoCrFeNiTi0.5 high-entropy alloy, Rare Metal 41 (2022) 4266–4272. https://doi.org/10.1007/s12598-016-0815-3
62. A. Verma, P. Tarate, A.C. Abhyankar, M.R. Mohape, D.S. Gowtam, V.P. Deshmukh, T. Shanmugasundaram, High temperature wear in CoCrFeNiCux high entropy alloys: The role of Cu, Scr Mater 161 (2019) 28–31. https://doi.org/10.1016/j.scriptamat.2018.10.007
63. L.H. Tian, W. Xiong, C. Liu, S. Lu, M. Fu, Microstructure and wear behavior of atmospheric plasma-sprayed AlCoCrFeNiTi high-entropy alloy coating, J Mater Eng Perform 25 (2016) 5513–5521. https://doi.org/10.1007/s11665-016-2396-6
64. G. Jin, Z. Cai, Y. Guan, X. Cui, Z. Liu, Y. Li, M. Dong, D. Zhang, High temperature wear performance of laser-cladded FeNiCoAlCu high-entropy alloy coating, Appl Surf Sci 445 (2018) 113–122. https://doi.org/10.1016/j.apsusc.2018.03.135
65. A. Zhang, J. Han, B. Su, J. Meng, A promising new high temperature self-lubricating material: CoCrFeNiS0.5 high entropy alloy, Mater Sci Eng A 731 (2018) 36–43. https://doi.org/10.1016/j.msea.2018.06.030
66. A. Zhang, J. Han, B. Su, J. Meng, A novel CoCrFeNi high entropy alloy matrix self-lubricating composite, J Alloys Compd 725 (2017) 700–710. https://doi.org/10.1016/j.jallcom.2017.07.197
67. L. Tian, Z. Feng, W. Xiong, Microstructure, microhardness, and wear resistance of AlCoCrFeNiTi/Ni60 coating by plasma spraying, Coatings 8 (2018). https://doi.org/10.3390/coatings8030112
68. L.M. Du, L.W. Lan, S. Zhu, H.J. Yang, X.H. Shi, P.K. Liaw, J.W. Qiao, Effects of temperature on the tribological behavior of Al0.25CoCrFeNi high-entropy alloy, J Mater Sci Technol 35 (2019) 917–925. https://doi.org/10.1016/j.jmst.2018.11.023
69. H. Liu, J. Liu, P. Chen, H. Yang, Microstructure and high temperature wear behaviour of in-situ TiC reinforced AlCoCrFeNi-based high-entropy alloy composite coatings fabricated by laser cladding, Opt Laser Technol 118 (2019) 140–150. https://doi.org/10.1016/j.optlastec.2019.05.006
70. S. Alvi, F. Akhtar, High temperature tribology of CuMoTaWV high entropy alloy, Wear 426–427 (2019) 412–419. https://doi.org/10.1016/j.wear.2018.12.085
71. X. Pei, Y. Du, X. Hao, H. Wang, Q. Zhou, H. Wu, H. Wang, Microstructure and tribological properties of TiZrV0.5Nb0.5Alx refractory high entropy alloys at elevated temperature, Wear 488–489 (2022) 204166. https://doi.org/10.1016/J.WEAR.2021.204166
72. H. Jiang, L. Jiang, D. Qiao, Y. Lu, T. Wang, Z. Cao, T. Li, Effect of niobium on microstructure and properties of the CoCrFeNbxNi high entropy alloys, J Mater Sci Technol 33 (2017) 712–717. https://doi.org/10.1016/j.jmst.2016.09.016
73. X. Li, H. Li, Q. Li, C. Jin, K. Hua, H. Wang, The determining role of Al addition on tribology properties and oxidation behavior at elevated temperatures of TiZrHfNb refractory high-entropy alloy, Mater Charact 189 (2022) 111921. https://doi.org/10.1016/J.MATCHAR.2022.111921
74. X. Chen, J. Kong, J. Li, S. Feng, H. Li, Q. Wang, Y. Liang, K. Dong, Y. Yang, High-strength AlCoCrFeNi2.1 eutectic high entropy alloy with ultrafine lamella structure via additive manufacturing, Mater Sci Eng A 854 (2022) 143816. https://doi.org/10.1016/J.MSEA.2022.143816

75. J. Dong, H. Wu, Y. Chen, Y. Zhang, Y. Wu, S. Yin, Y. Du, K. Hua, H. Wang, Study on self-lubricating properties of AlCoCrFeNi2.1 eutectic high entropy alloy with electrochemical boronizing, Surf Coat Technol 433 (2022). https://doi.org/10.1016/j.surfcoat.2022.128082
76. Q. Xing, A.C. Feltrin, F. Akhtar, Processing, microstructure and high temperature dry sliding wear of a Cr-Fe-Hf-Mn-Ti-Ta-V high-entropy alloy based composite, Mater Today Commun 28 (2021). https://doi.org/10.1016/j.mtcomm.2021.102657
77. Y. Geng, J. Chen, H. Tan, J. Cheng, S. Zhu, J. Yang, Tribological performances of CoCrFeNiAl high entropy alloy matrix solid-lubricating composites over a wide temperature range, Tribol Int 157 (2021) 106912. https://doi.org/10.1016/J.TRIBOINT.2021.106912
78. G.R. Fox, H. Liang, Wear mode comparison of high-performance inconel alloys, J Tribol 132 (2010) 1–6. https://doi.org/10.1115/1.4001170
79. A. Zhang, J. Han, B. Su, P. Li, J. Meng, Microstructure, mechanical properties and tribological performance of CoCrFeNi high entropy alloy matrix self-lubricating composite, Mater Des 114 (2017) 253–263. https://doi.org/10.1016/j.matdes.2016.11.072
80. J. Lu, B. Wang, X. Qiu, Z. Peng, M. Ma, Microstructure evolution and properties of CrCuFexNiTi high-entropy alloy coating by plasma cladding on Q235, Surf Coat Technol 328 (2017) 313–318. https://doi.org/10.1016/j.surfcoat.2017.08.019
81. J.B. Cheng, X.B. Liang, B.S. Xu, Effect of Nb addition on the structure and mechanical behaviors of CoCrCuFeNi high-entropy alloy coatings, Surf Coat Technol 240 (2014) 184–190. https://doi.org/10.1016/j.surfcoat.2013.12.053
82. X. Xian, L. Lin, Z. Zhong, C. Zhang, C. Chen, K. Song, J. Cheng, Y. Wu, Precipitation and its strengthening of Cu-rich phase in CrMnFeCoNiCux high-entropy alloys, Mater Sci Eng A 713 (2018) 134–140. https://doi.org/10.1016/j.msea.2017.12.060
83. C.C. Juan, C.Y. Hsu, C.W. Tsai, W.R. Wang, T.S. Sheu, J.W. Yeh, S.K. Chen, On microstructure and mechanical performance of AlCoCrFeMo 0.5Ni x high-entropy alloys, Intermetallics (Barking) 32 (2013) 401–407. https://doi.org/10.1016/j.intermet.2012.09.008
84. M.H. Chuang, M.H. Tsai, W.R. Wang, S.J. Lin, J.W. Yeh, Microstructure and wear behavior of AlxCo 1.5CrFeNi1.5Tiy high-entropy alloys, Acta Mater 59 (2011) 6308–6317. https://doi.org/10.1016/j.actamat.2011.06.041
85. Y. Yu, F. He, Z. Qiao, Z. Wang, W. Liu, J. Yang, Effects of temperature and microstructure on the triblogical properties of CoCrFeNiNbx eutectic high entropy alloys, J Alloys Compd 775 (2019) 1376–1385. https://doi.org/10.1016/j.jallcom.2018.10.138
86. R. Darolia, Thermal barrier coatings technology: Critical review, progress update, remaining challenges and prospects, Int Mater Rev 58 (2013) 315–348. https://doi.org/10.1179/1743280413Y.0000000019
87. D.R.G. Achar, R. Munoz-Arroyo, L. Singheiser, W.J. Quadakkers, Modelling of phase equilibria in MCrAlY coating systems, Surf Coat Technol 187 (2004) 272–283. https://doi.org/10.1016/j.surfcoat.2004.02.018
88. C. Vorkötter, D.E. Mack, O. Guillon, R. Vaßen, Superior cyclic life of thermal barrier coatings with advanced bond coats on single-crystal superalloys, Surf Coat Technol 361 (2019) 150–158. https://doi.org/10.1016/j.surfcoat.2019.01.001
89. B. Gleeson, Thermal barrier coatings for aeroengine applications, J Propuls Power 22 (2006) 375–383. https://doi.org/10.2514/1.20734
90. P.K. Huang, J.W. Yeh, T.T. Shun, S.K. Chen, Multi-principal-element alloys with improved oxidation and wear resistance for thermal spray coating, Adv Eng Mater 6 (2004) 74–78. https://doi.org/10.1002/adem.200300507
91. M. Arshad, M. Amer, Q. Hayat, V. Janik, X. Zhang, M. Moradi, M. Bai, High-entropy coatings (HEC) for high-temperature applications: Materials, processing, and properties, Coatings 12 (2022). https://doi.org/10.3390/coatings12050691

92. O.N. Senkov, G.B. Wilks, D.B. Miracle, C.P. Chuang, P.K. Liaw, Refractory high-entropy alloys, Intermetallics (Barking) 18 (2010) 1758–1765. https://doi.org/10.1016/j.intermet.2010.05.014
93. S. Dixit, S. Rodriguez, M.R. Jones, P. Buzby, R. Dixit, N. Argibay, F.W. DelRio, H.H. Lim, D. Fleming, Refractory high-entropy alloy coatings for high-temperature aerospace and energy applications, J Therm Spray Tech 31 (2022) 1021–1031. https://doi.org/10.1007/s11666-022-01324-0
94. Y. Yang, A. Wang, D. Xiong, Z. Wang, D. Zhou, S. Li, H. Zhang, Effect of Cr content on microstructure and oxidation resistance of laser-clad Cu-Ni-Fe-Mo-xCr alloy coating, Surf Coat Technol 384 (2020). https://doi.org/10.1016/j.surfcoat.2019.125316
95. M. Srivastava, M.S. Jadhav, R.P.S. Chethan, S. Chakradhar Singh, Investigation of HVOF sprayed novel Al1.4Co2.1Cr0.7Ni2.45Si0.2Ti0.14 HEA coating as bond coat material in TBC system, J Alloys Compd 924 (2022). https://doi.org/10.1016/j.jallcom.2022.166388
96. Q.L. Xu, K.C. Liu, K.Y. Wang, L.Y. Lou, Y. Zhang, C.J. Li, C.X. Li, TGO and Al diffusion behavior of CuAlxNiCrFe high-entropy alloys fabricated by high-speed laser cladding for TBC bond coats, Corros Sci 192 (2021). https://doi.org/10.1016/j.corsci.2021.109781
97. W.L. Hsu, Y.C. Yang, C.Y. Chen, J.W. Yeh, Thermal sprayed high-entropy NiCo0.6Fe0.2Cr1.5SiAlTi0.2 coating with improved mechanical properties and oxidation resistance, Intermetallics (Barking) 89 (2017) 105–110. https://doi.org/10.1016/j.intermet.2017.05.015
98. J. Lu, L. Li, Y. Chen, X. Liu, X. Zhao, F. Guo, P. Xiao, Y-Hf co-doped AlCoCrFeNi high-entropy alloy coating with superior oxidation and spallation resistance at 1100°C, Corros Sci 182 (2021). https://doi.org/10.1016/j.corsci.2021.109267
99. Z. Cai, X. Cui, G. Jin, Z. Liu, W. Zheng, Y. Li, L. Wang, Microstructure and thermal stability of a Ni-Cr-Co-Ti-V-Al high-entropy alloy coating by laser surface alloying, Met Mater Int 23 (2017) 1012–1018. https://doi.org/10.1007/s12540-017-6583-2
100. X. Ma, P. Ruggiero, R. Bhattacharya, O.N. Senkov, A.K. Rai, Evaluation of new high entropy alloy as thermal sprayed bondcoat in thermal barrier coatings, J Therm Spray Tec 31 (2022) 1011–1020. https://doi.org/10.1007/s11666-021-01279-8
101. A. Anupam, S. Kumar, N.M. Chavan, B.S. Murty, R.S. Kottada, First report on cold-sprayed AlCoCrFeNi high-entropy alloy and its isothermal oxidation, J Mater Res 34 (2019) 796–806. https://doi.org/10.1557/jmr.2019.38
102. Q.L. Xu, Y. Zhang, S.H. Liu, C.J. Li, C.X. Li, High-temperature oxidation behavior of CuAlNiCrFe high-entropy alloy bond coats deposited using high-speed laser cladding process, Surf Coat Technol 398 (2020). https://doi.org/10.1016/j.surfcoat.2020.126093
103. L. Chen, Y. Wang, X. Hao, X. Zhang, H. Liu, Lightweight refractory high entropy alloy coating by laser cladding on Ti–6Al–4V surface, Vacuum 183 (2021). https://doi.org/10.1016/j.vacuum.2020.109823
104. H. Junjie, H. Guo, L. Jing, T. Jingchao, New class of high-entropy defect fluorite oxides RE2(Ce0.2Zr0.2Hf0.2Sn0.2Ti0.2)2O7 (RE = Y, Ho, Er, or Yb) as promising thermal barrier coatings, J Eur Ceram Soc 41 (2021) 6080–6086. https://doi.org/10.1016/j.jeurceramsoc.2021.05.044
105. L. Zhou, F. Li, J.X. Liu, Q. Hu, W. Bao, Y. Wu, X. Cao, F. Xu, G.J. Zhang, High-entropy thermal barrier coating of rare-earth zirconate: A case study on (La0.2Nd0.2Sm0.2Eu0.2Gd0.2)2Zr2O7 prepared by atmospheric plasma spraying, J Eur Ceram Soc 40 (2020) 5731–5739. https://doi.org/10.1016/j.jeurceramsoc.2020.07.061

12 Novel Electrochemical Sensing Applications of High Entropy Alloys

Lakshmi Krishnaa Suresh, Simi M, Navya Roby, Parvathi J, Santhy A and R. Vinaykumar

This chapter explores the emerging field of electrochemical sensing applications utilizing high entropy alloys (HEAs), which are characterized by their unique composition and structural properties. HEAs, composed of multiple principal elements in near-equal proportions, exhibit exceptional mechanical strength, corrosion resistance, and thermal stability, making them promising candidates for advanced sensing technologies. The chapter reviews various applications of HEAs in detecting environmental pollutants, biomolecules, and industrial chemicals, highlighting their advantages over traditional materials. Furthermore, we delve into the mechanisms behind the electrochemical processes involved, including the role of alloy composition and microstructure in sensor performance. By presenting case studies and recent advancements in the field, this chapter aims to provide insights into the potential of HEAs as next-generation materials for electrochemical sensors, paving the way for innovative solutions in environmental monitoring, healthcare, and industrial applications.

12.1 INTRODUCTION

High entropy alloys (HEAs) have emerged as an important class of materials in electrochemical sensing, particularly over the past decade. These alloys consist of five or more principal elements mixed in near-equiatomic proportions, resulting in unique structural and functional properties [1]. Unlike traditional alloys that are based on a single primary element, HEAs benefit from high configurational entropy. This leads to advantages such as enhanced corrosion resistance, stability, catalytic activity, high hardness, and excellent ductility, making them particularly suitable for electrochemical sensing applications [2].

Firstly, HEAs have chemical environments that are very complicated since they are composed of five or more main elements with almost similar atomic proportions. Its unique composition allows tailoring of its surface properties, enabling the designing of surfaces that are highly selective for specific analytes, which is very important for electrochemical detection [3]. Second, since they are made up of more than one element, HEAs often show synergistic effects that enhance the alloy's catalytic performance better than the sum of its components [4]. Because of this, they work

DOI: 10.1201/9781003664093-16

well with electroanalytical techniques. Electroanalytical techniques hold significant promise when applied to HEAs, materials known for their unique properties due to the random distribution of multiple elements [5, 6]. These techniques are particularly useful in characterizing HEAs because they allow for precise measurements of their electrochemical behavior, including conductivity, corrosion resistance, and catalytic activity [7]. Techniques like cyclic voltammetry, electrochemical impedance spectroscopy, and differential pulse voltammetry can reveal the role of each element in enhancing the alloy's performance for specific applications, such as energy storage, catalysis, and sensor technology [8]. The ability of electroanalytical techniques to monitor changes in electrochemical behavior also facilitates the optimization of HEAs for specific applications, making them invaluable tools for advancing our understanding and use of these versatile materials.

The configurational entropy in HEAs develops highly stable structures that will not rust or corrode, which makes them ideal for the harsh conditions that are common in electrochemical sensing applications [9]. A research article by Yunchao Xie et al. explains how certain HEA compositions, like FeNiCoCrRu, achieve stability and corrosion resistance through high entropy, making them suitable for prolonged use in demanding conditions like seawater splitting and other electrochemical sensing techniques. Several other articles provide comprehensive insights into the mechanisms by which high configurational entropy in HEAs contributes to their stability, preventing corrosion, and enhancing performance in electrochemical environments [10–12]. The stability of HEAs guarantees long-term use, even in acidic or basic environments. Also, nanoporous or nanostructured surfaces can be added to HEAs, which maximizes the active surface area. This improves the electrochemical performance by speeding up the reaction rates and charge transfer, which are important for making sensors very sensitive [13]. Furthermore, HEAs are mechanically robust, which means they maintain their original form even when under a lot of stress, going through temperature changes, and being used for a long time [14]. These characteristics, along with the ability to formulate unique compositions, make HEAs useful and promising types of material for the next wave of electrochemical sensors [15].

In recent years, researchers have been focusing on the design and synthesis of electrocatalysts with high catalytic activity, low cost, and good stability. Since the alloy catalyst gained more attention, the intermetallic alloys were intensively investigated by varying the composition of the alloys in order to optimize the surface electronic structure of the alloy catalysts, which in turn improves the catalytic performance [16]. Jein Wei Yeh and his co-workers were the first to propose the concept of HEAs [17]. They suggested that the presence of five or more elements in nearly equal atomic proportions raised the configurational entropy of the mixing by a sufficient amount to overcome the enthalpy of compound formation, lattice distortion, and slow diffusion caused by the mixing of multiple components, which are all known to be responsible for the high mechanical properties and high thermal stability of HEAs [10].

HEAs were initially defined in the early 2000s, when the research was primarily focused on their mechanical properties and thermal stability [18]. Researchers observed that HEAs exhibited remarkable corrosion resistance and stability under harsh conditions, indicating their potential for electrochemical applications [19]. By the early 2010s, research in the field broadened to explore the electrocatalytic properties of HEAs, particularly for applications such as fuel cells and water electrolysis [20]. Researchers started modifying elemental compositions to enhance catalytic performance, which established a foundation for applying HEAs in electrochemical sensing. Also in 2019, Pengfei Xie et al. reported highly efficient HEA catalysts made of earth-abundant elements which showed great potential for catalyzing chemical reactions [21]. Gradually, HEAs started gaining interest in electrochemical sensing, owing to their elevated surface area, tunable composition, and superior conductivity [22]. Researchers focused on HEAs for applications in gas sensors and biosensors, utilizing their distinctive surface chemistries and catalytic sites [23]. This period marked the initial applications of HEAs in biomolecule detection, including glucose and dopamine, highlighting their versatility [24]. Further, researchers began employing HEAs to detect more complex analytes, including heavy metals, pollutants, and biological molecules. The flexibility in HEA composition allowed fine-tuning for specific target analytes, which improved selectivity and sensitivity. High entropy oxides, derived from HEAs, also gained attention, particularly for heavy metal ion detection due to their improved ionic conductivity [25].

Currently, HEAs are more customizable than ever, as a result of the advancement in nanostructuring and synthesis techniques, which enables the precise control of elemental distribution and morphology. These developments are promoting the use of HEAs in ultra-sensitive and miniaturized sensors for complex matrices, including environmental and clinical samples. At present, researchers are also investigating the potential of HEAs in the context of real-time monitoring systems and wearable sensors.

12.2 VARIOUS TECHNIQUES EMPLOYED FOR THE SYNTHESIS OF HEAs

Several manufacturing processes have been used for the production of HEAs, according to the literature reports [26–28]. The techniques employed in the manufacturing processes play a very important role in tuning the properties of HEAs. The following sections of the chapter will give a brief description about some of the most common methods used for the fabrication of HEAs for different applications.

12.2.1 Mechanical Alloying (MA)

MA serves as a prevalent technique for the generation of pre-alloyed HEA powders. MA, combined with spark plasma sintering, seems to be a promising approach for the production of bulk HEAs. MA is commonly employed to develop HEAs due to its ability to modify any composition. The initial report on MA of HEAs dates back to 2008, and its popularity has increased significantly since that time [29]. It is a solid-state fabrication process that operates under non-equilibrium conditions, utilized for the creation of advanced materials characterized by uniform grain size and microstructure [30]. It also offers enhanced compositional precision and yields uniform nanostructured alloys characterized by a high diffusion rate [31]. Recently, a novel AlCoMoFeNi HEA was synthesized through the MA method [32]. The produced sample exhibits a reduction in size of crystallite and a rise in lattice strain as milling time increases. In another work, ultrafine grains of $Al_{10}Cr_{25}Co_{20}Ni_{25}Fe_{20}$ HEA were synthesized using the MA method [33].

12.2.2 Electrochemical Synthesis

Electrodeposition is a highly efficient technique for the synthesis of nanoparticles (NPs), allowing for precise control for its formation. Furthermore, the electrodeposition production method is straightforward, and the composition may be modified by altering the parameters [34]. Moreover, electrooxidation is an innovative metallurgical technique for producing alloys from solid oxide. In an electrolytic cell, a metal oxide is employed as the cathode, graphite as the anode, and molten salt as the electrolyte. Shape, density, and porosity of the oxide precursor, as well as the working temperature, applied potential, and duration of electrolysis, are some of the most important factors for electrodeposition.

Yao et al. pioneered the preparation of BiFeCoNiMn HEA films through an electrochemical method utilizing DMF-CH_3CN mixed solvents [35]. Amorphous thin films of Al–Cr–Fe–Mn–Ni and Al–Cr–Cu–Fe–Mn–Ni were synthesized by electrochemical deposition at various temperatures in organic solvents and the deposited alloy was crystallized by annealing at 738 K [20]. Recently, a CoCrFeMnNi HEA thin film was synthesized using the electrodeposition method on a copper substrate [36].

12.2.3 Carbothermal Shock (CTS) Synthesis

The task of systematically integrating various immiscible components into a singular nanoparticle through conventional synthetic methods presents a significant challenge. The creation of a synthetic approach that can accurately regulate element composition, particle size, and phase may result in a novel range of alloys and nanostructures with exceptional properties. Yao et al. introduced a CTS method and implemented a two-step approach to rapidly heat and cool the metal precursor on the oxygenated carbon support [37]. HEAs NPs comprising as many as seven metal elements were synthesized using this method. The HEAs NPs were synthesized through CTS, resulting in a narrow and uniform profile that is well dispersed on the carbon support. The synthesis of HEA NPs of PtFeCoNiCu, on CO_2-activated carbon nanofibers further validated this work [38]. A FeCoNiPtRu HEA was recently synthesized on carbon support using the CTS process [39]. The interaction between metal elements significantly improved the catalytic activity. The catalyst's stability and oxidation resistance were enhanced by targeting the HEA on the carbon support. It demonstrated outstanding catalytic performance in the water splitting process. Nonetheless, the CTS method is constrained to the synthesis of HEAs NPs, utilizing metal precursors that are compatible with loading onto the carbon support.

12.2.4 Fast-Moving Bed Pyrolysis (FMBP)

The elevated specific surface area and interfacial interactions contribute to the notable catalytic performance observed in supported nanoparticle catalysts. HEA NPs can be uniformly immobilized on conductive activated carbon fibers using the CTS method, as discussed in the previous section. Nonetheless, the CTS method presents challenges in the immobilization of HEA NPs on granular substrates like Al_2O_3 and zeolite. Alternatively, the FMBP method has demonstrated the ability to produce supported HEA NPs on various substrates. The FMBP method enabled the concurrent pyrolysis of metal precursors while facilitating the formation of small nucleations, due to its rapid heating rate. The FMBP method was used for the synthesis of CuPdSnPtAu HEA NPs' various supports, including γ-A12O3, zeolite, carbon black, and graphene oxide (GO) [40]. Also, this method was used for the synthesis of denary MnCoNiCuRhPdSnIrPtAu HEA NPs on GO support.

12.2.5 Wet Chemical Method

The current leading approach for synthesizing high entropy alloys through wet chemistry is the solvothermal technique. The solvothermal process takes place in a non-aqueous medium at temperatures that exceed the boiling point of the solvent used. The elevated temperature and pressure facilitate the enhancement of multiple metal miscibility. For example, Bondesgaard et al. synthesized quaternary and quinary HEA NPs ($Pt_{0.25}Pd_{0.25}Rh_{0.25}Ru_{0.25}$ and $Pt_{0.20}Pd_{0.20}Ir_{0.20}Rh_{0.20}Ru_{0.20}$), exhibiting a face-centered cubic crystal structure. This was achieved by utilizing the noble metal acetylacetone as a precursor and a solvent composed of an acetone-ethanol mixture in a solvothermal autoclave at 200°C for 24 hours. The utilization of acetylacetone salt is crucial in the synthesis of HEA NPs, as the strong interaction between the metal and acetylacetone effectively decreases the coprecipitation rate. Using the solvothermal method, Wu et al. synthesized HEA NPs comprising six platinum group metals (Ru, Rh, Pd, Os, Ir, Pt) [41]. The as-synthesized Pt group HEA NPs, exhibit a narrow size distribution of 3.1 ± 0.6 nm.

12.3 ELECTROCHEMICAL SENSING APPLICATIONS OF HIGH ENTROPY ALLOYS

Recently, there has been a growing interest in HEAs for their applications in electrochemical sensing of various materials such as biomolecules, gas, heavy metals, and environmental pollutants. Their unique multi-component structure, adjustable surface characteristics, and promising catalytic performance make them an appealing field of study.

A groundbreaking advancement in biosensor technology was emerged with the rapid fabrication of a low-cost, high entropy porous CrO/CrN/C biosensor. This innovative approach, utilizing mechanical milling, enabled the simultaneous detection of dopamine (DA) and uric acid (UA), marking a significant step forward in the field of analytical science by making use of biosensor carriers sourced from the natural waste of jasmine flowers (JFs), which is graphically represented in Figure 12.1 [42].

A biosensing material incorporating dispersed chromium oxide/chromium nitride sites was synthesized via high-energy ball milling of JFs modified with dopamine and chromium nitrate. This process generated numerous active sites, significantly enhancing the biosensor's performance in simultaneously detecting dopamine and uric acid. The resulting HEA-modified electrode effectively measured both DA and

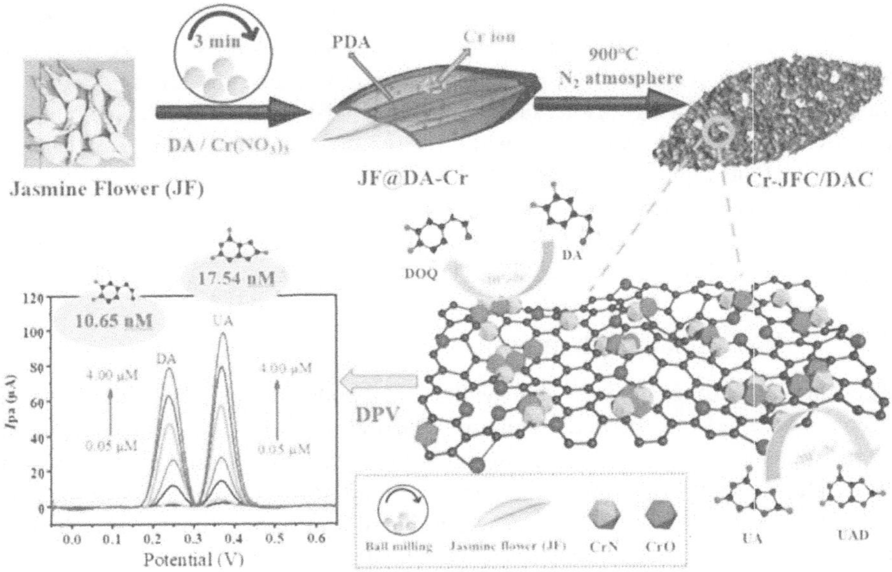

FIGURE 12.1 Graphical representation of high entropy porous CrO/CrN/C biosensor for highly sensitive simultaneous detection of dopamine and uric acid [42].

UA, operating within a range of 0.05 to 4 μM for both analytes, achieving detection limits of 10.65 nM for DA and 17.54 nM for UA. Tests using human serum samples demonstrated excellent recoveries (97.00–104.00%), highlighting the sensor's robust anti-interference properties and stability. The CrO/CrN/C-based HEA electrode shows considerable promise for practical applications in simultaneous DA and UA detection.

Dandan Song and colleagues recently developed a precisely controlled, template-based method for synthesizing ultrathin nanotubes of quinary PtPdRhCuM high entropy alloys (where M represents manganese, cobalt, or nickel). These nanotubes exhibit significant lattice distortion and a high density of defect sites [43]. Combining galvanic exchange, co-reduction, and dealloying, a novel approach was used to create a PtPdRhCuM HEA-based biosensor. The HEA's unique properties, including diverse composition, high mixing entropy, inherent conductivity, and complex surface structure, enable ultrasensitive detection of the carbamate pesticide aldicarb sulfone. The sensor achieved a detection range of 0.45 pM to 45 nM, with a remarkable LOD of 86.5 fM. Furthermore, it successfully detected pesticides in real food samples.

Shashanka Rajendrachari and other researchers have accomplished the successful synthesis of HEAs with the composition 20-Fe-20Cr-20Ni-20Mn-20Ti using a ball-milling technique, conducted for 20 hours after optimizing the milling parameters [44]. They utilized planetary ball-milled HEA powders for the electrochemical investigation of methyl orange (MO), which causes threats to human health and the environment. In 2022, Rajendrachari et al. successfully fabricated HEA (23Fe-21Cr-18Ni-20Ti-18Mn) powders by the ball-milling process [45]. They also investigated

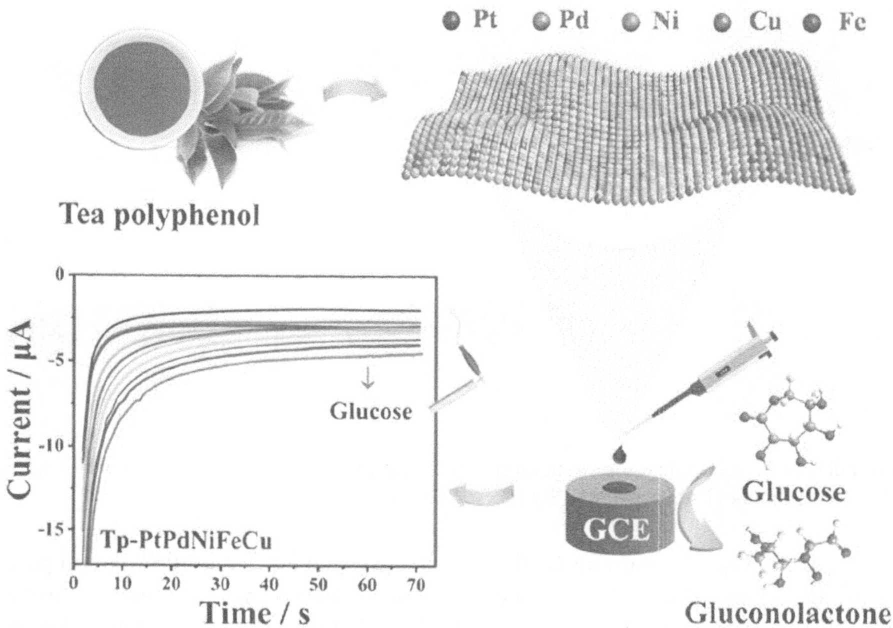

FIGURE 12.2 Green synthesis of PtPdNiFeCu HEA NPs for glucose detection [46].

the electrochemical response of the HEA powders to determine ascorbic acid using cyclic voltammetry utilizing varying concentrations of ball-milled HEA-modified carbon paste electrodes. The developed modified electrodes showed outstanding current sensitivity and selectivity.

The emergence of HEAs marks a significant advancement in material science, captivating the attention of researchers due to their remarkable characteristics, such as a stable microstructure and unique catalytic properties. In pursuit of a safer preparation method, Fengxia Wang et al. unveiled an innovative strategy focused on green synthesis, employing tea polyphenols to create PtPdNiFeCu HEA NPs for the detection of glucose, which is depicted in Figure 12.2 [46].

The HEA-modified electrochemical sensors underwent meticulous examination through transmission electron microscopy and different electrochemical experiments. The examination of physical properties revealed that the NP exhibits enhanced dispersibility, with an average particle size measuring 7.5 nm. The electrochemical findings revealed that PtPdNiFeCu HEA-NPs exhibited remarkable sensitivity at 1.264 $\mu A\ mM^{-1}\ cm^{-2}$, alongside a minimal LOD of 4.503 μM and an extensive detection range for glucose spanning from 0 to 10 mM.

In 2021, Ming Li et al. used high entropy nanomaterial, PtPdRhRuIrNi, as an electrochemical-sensing catalyst to detect serotonin molecules with printed sensor chips, for the first time. The nanomaterial was synthesized as a highly active catalyst for the electro-oxidation of serotonin. Using conductive graphene nanofiber as support, the sensing material PtPdRhRuIrNi exhibited great sensitivity (LOD of 10 nM) and selectivity, effectively detecting serotonin in serum samples [47].

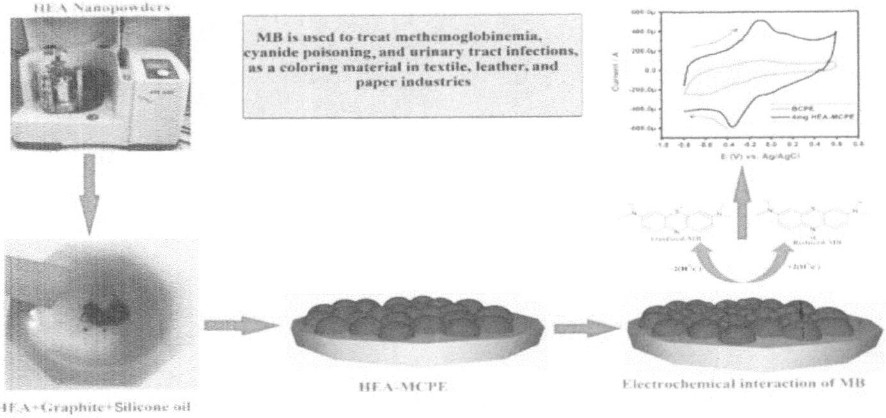

FIGURE 12.3 Graphical representation of the experimental studies carried out followed by the electrochemical studies of MB by HEA-MCPE [48].

Recently, Rajendrachari and colleagues developed a HEA-modified carbon paste electrode for the electrochemical detection of methylene blue dye in wastewater. The HEA powder, composed of 25Fe-19Cr-19Ni-18Ti-19Mn, was synthesized using a planetary ball mill. The HEA-MCPE demonstrated effective MB oxidation within a pH range of 6 to 7.6, exhibiting a substantial current response. The graphical representation of experimental and electrochemical studies of methylene blue dye has been depicted in Figure 12.3 [48].

The newly developed HEA-modified carbon paste electrode demonstrated high sensitivity in electrochemically oxidizing MB. This promising electrode material holds potential for future applications in detecting a variety of substances, including dyes, pharmaceuticals, heavy metals, and pesticides.

12.4 CHALLENGES AND FUTURE PERSPECTIVES IN HEA-BASED SENSING

HEAs, consisting of several principal elements, exhibit significant properties such as enhanced stability, corrosion resistance, and superior mechanical strength, rendering them suitable for applications in electrochemical sensing. Despite this, challenges persist in altering these materials for sensor applications [49]. A significant challenge lies in the optimization of composition; the wide range of potential elements and combinations further complicates the identification of the right combination for the desired electrochemical properties, making the process complex and time-consuming. HEAs necessitate specialized synthesis methods, including arc melting or mechanical alloying, to attain a homogeneous distribution of elements, which can be difficult to control on a large scale. These technical challenges affect the viability of manufacturing HEA-based electrochemical sensors [50].

Another notable challenge pertains to the surface properties of HEAs, which directly affect their sensing capabilities. In electrochemical sensing, the active surface area of HEAs, stability within the electrochemical environment, and the electron transfer kinetics are essential factors [51]. High compositional entropy in HEAs can lead to undesirable surface segregation or phase separation during operational conditions, potentially minimizing sensor sensitivity and longevity. Furthermore, HEAs are under investigation regarding their electrochemical response mechanisms, which differ according to the distinct properties of each alloy. The lack of mechanistic understanding restricts the ability to develop HEAs for targeted sensing applications, especially in the detection of trace analytes where sensitivity and selectivity are critical [52].

The future of HEAs in electrochemical sensing looks promising, due to advancements in materials science and fabrication techniques. Future efforts should prioritize creating computational models to predict the relationships between alloy composition and performance, which will accelerate the identification of HEAs with enhanced properties for sensing applications [11, 53]. Integrating HEAs with nanostructured or composite materials can enhance electrochemical performance through improved surface area and electron transfer efficiency. Scalable synthesis techniques like AM may enhance the practicality of HEA-based sensors for commercial use. HEAs may emerge as versatile materials for next-generation electrochemical sensors as research advances [54, 55].

12.5 CONCLUSION

HEAs exhibit significant potential for enhancing electrochemical sensing owing to their distinctive characteristics, such as improved catalytic activity, corrosion resistance, and structural stability. Nonetheless, other hurdles persist, encompassing concerns regarding synthesis, surface activity, structural integrity, and biocompatibility. By tackling these issues with sophisticated synthesis processes, computational design, and surface engineering, the prospects for HEA-based electrochemical sensors appear promising. Ongoing research indicates that HEAs may significantly contribute to the advancement of next-generation sensing technologies across several applications.

REFERENCES

1. P. Kumari, A. K. Gupta, R. K. Mishra, M. S. Ahmad and R. R. Shahi, "A Comprehensive Review: Recent Progress on Magnetic High Entropy Alloys and Oxides," *J. Magn. Magn. Mater*, vol. 554, p. 169142, 2022. doi: 10.1016/j.jmmm.2022.169142.
2. J.-T. Ren, L. Chen, H.-Y. Wang and Z.-Y. Yuan, "High-Entropy Alloys in Electrocatalysis: from Fundamentals to Applications," *Chem. Soc. Rev*, vol. 52, no. 23, pp. 8319–8373, 2023. doi: 10.1039/D3CS00557G.
3. X. Han *et al.*, "Nanoscale High-Entropy Alloy for Electrocatalysis," *Matter*, vol. 6, no. 6, pp. 1717–1751, 2023. doi: 10.1016/j.matt.2023.03.034.
4. Z. Zhang *et al.*, "Recent Research Progress on High-Entropy Alloys as Electrocatalytic Materials," *J. Alloys Compd*, vol. 918, p. 165585, 2022. doi: 10.1016/j.jallcom.2022.165585.
5. A. Santhy, R. Rejithamol, R. Chandana, P. Sivanand and S. Beena, "Electrochemical Sensor Applications of Mechanically Alloyed Materials," In S. Rajendrachari & R. Mahale (Eds.), *Advancements in Powder Metallurgy*. IGI Global Scientific Publishing. 2024, pp. 33–53. doi: 10.4018/978-1-6684-9385-4.ch002.

6. S. Antherjanam and B. Saraswathyamma, "Simultaneous Electroanalysis of Adenine and Adenosine on a Poly-Nicotinic Acid Modified Pencil Graphite Electrode," *Monatshefte für Chemie - Chem. Mon*, vol. 153, no. 12, pp. 1143–1153, 2022. doi: 10.1007/s00706-022-02985-x.
7. S. Antherjanam, B. Saraswathyamma and S. Murugesan Senthil Kumar, "Simultaneous Electrochemical Determination of the Tumour Biomarkers Homovanillic Acid and Vanillylmandelic Acid Using a Modified Pencil Graphite Electrode," *Microchem. J*, vol. 190, p. 108659, 2023. doi: 10.1016/j.microc.2023.108659.
8. A. S. Murali *et al.*, "Facile Formation of Surface Functionalised α-Fe2O3 Nanoparticles and Their Role in Developing Dual Sensors Towards Cysteamine Quantification," *Mater. Res. Bull*, vol. 174, p. 112734, 2024. doi: 10.1016/j.materresbull.2024.112734.
9. Y. Shi, B. Yang and P. Liaw, "Corrosion-Resistant High-Entropy Alloys: A Review," *Metals (Basel)*, vol. 7, no. 2, p. 43, 2017. doi: 10.3390/met7020043.
10. Z. Ye, C. Li and J. Gu, "Phase Stability in High-Entropy Alloys: The Role of Configurational Entropy," *JOM*, vol. 74, no. 11, pp. 4154–4161, 2022. doi: 10.1007/s11837-022-05307-y.
11. R. Nartita, D. Ionita and I. Demetrescu, "A Modern Approach to HEAs: From Structure to Properties and Potential Applications," *Crystals*, vol. 14, no. 5, p. 451, 2024. doi: 10.3390/cryst14050451.
12. J. Y. Zhang *et al.*, "High-Entropy Alloys: A Critical Review of Aqueous Corrosion Behavior and Mechanisms," *High Entropy Alloy. Mater*, vol. 1, no. 2, pp. 195–259, 2023. doi: 10.1007/s44210-023-00021-z.
13. M. Kogler *et al.*, "Bulk-Independent Surface Oxide Composition Controls the Electrochemical Performance of High-Entropy Alloys," *J. Mater. Chem. A*, vol. 12, no. 34, pp. 22565–22575, 2024. doi: 10.1039/D4TA03619K.
14. W. Li, D. Xie, D. Li, Y. Zhang, Y. Gao and P. K. Liaw, "Mechanical Behavior of High-Entropy Alloys," *Prog. Mater. Sci*, vol. 118, p. 100777, 2021. doi: 10.1016/j.pmatsci.2021.100777.
15. X. Wang, W. Guo and Y. Fu, "High-Entropy Alloys: Emerging Materials for Advanced Functional Applications," *J. Mater. Chem. A*, vol. 9, no. 2, pp. 663–701, 2021. doi: 10.1039/D0TA09601F.
16. J. Zhang, L. Zhang and Z. Cui, "Strategies to Enhance the Electrochemical Performances of Pt-Based Intermetallic Catalysts," *Chem. Commun*, vol. 57, no. 1, pp. 11–26, 2021. doi: 10.1039/D0CC05170E.
17. J. W. Yeh *et al.*, "Nanostructured High-Entropy Alloys with Multiple Principal Elements: Novel Alloy Design Concepts and Outcomes," *Adv. Eng. Mater*, vol. 6, no. 5, pp. 299–303, 2004. doi: 10.1002/adem.200300567.
18. S. Rajendrachari, "An Overview of High-Entropy Alloys Prepared by Mechanical Alloying Followed by the Characterization of Their Microstructure and Various Properties," *Alloys*, vol. 1, no. 2, pp. 116–132, 2022. doi: 10.3390/alloys1020008.
19. B. J. Yeh *et al.*, "Nanostructured High-Entropy Alloys with Multiple Principal Elements : Novel Alloy Design Concepts and Outcomes," *Adv. Eng. Mater.*, vol. 6, no. 5, pp. 299–303, 2004. doi: 10.1002/adem.200300567.
20. V. Soare *et al.*, "Electrochemical Deposition and Microstructural Characterization of AlCrFeMnNi and AlCrCuFeMnNi High Entropy Alloy Thin Films," *Appl. Surf. Sci*, vol. 358, pp. 533–539, 2015. doi: 10.1016/j.apsusc.2015.07.142.
21. P. Xie *et al.*, "Highly Efficient Decomposition of Ammonia Using High-Entropy Alloy Catalysts," *Nat. Commun*, vol. 10, no. 1, p. 4011, 2019. doi: 10.1038/s41467-019-11848-9.
22. W. Zhang *et al.*, "Functional High-Entropy Alloys: Promising Catalysts for High-Performance Water Splitting," *J. Mater. Chem. A*, vol. 12, no. 30, pp. 18705–18732, 2024. doi: 10.1039/D4TA02271H.

23. V. R. Naganaboina, M. Anandkumar, A. S. Deshpande and S. G. Singh, "Single-Phase High-Entropy Oxide Nanoparticles for Wide Dynamic Range Detection of CO 2," *ACS Appl. Nano Mater*, vol. 5, no. 3, pp. 4524–4536, 2022. doi: 10.1021/acsanm.2c00855.
24. R. Sheng *et al.*, "Ultrafine FeCuAgCeGd–based High–Entropy Nanozyme: Preparation, Catalytic Mechanism, and Point–of–Care Detection of Dopamine in Human serum," *Chem. Eng. J*, vol. 485, p. 149913, 2024. doi: 10.1016/j.cej.2024.149913.
25. A. Sarkar, H. Hahn and R. Kruk, "High Entropy Oxides," *Encycl. Mater. Electron*, vol. 1–3, pp. V1-536–V1-545, 2023. doi: 10.1016/B978-0-12-819728-8.00096-6.
26. Y. A. Alshataif, S. Sivasankaran, F. A. Al-Mufadi, A. S. Alaboodi and H. R. Ammar, "Manufacturing Methods, Microstructural and Mechanical Properties Evolutions of High-Entropy Alloys: A Review," *Met. Mater. Int*, vol. 26, no. 8, pp. 1099–1133, 2020. doi: 10.1007/s12540-019-00565-z.
27. P. P. Bhattacharjee *et al.*, "Microstructure and Texture Evolution During Annealing of Equiatomic CoCrFeMnNi High-Entropy Alloy," *J. Alloys Compd*, vol. 587, pp. 544–552, 2014. doi: 10.1016/j.jallcom.2013.10.237.
28. H. Luo, Z. Li, A. M. Mingers and D. Raabe, "Corrosion Behavior of an Equiatomic CoCrFeMnNi High-Entropy Alloy Compared with 304 Stainless Steel in Sulfuric Acid Solution," *Corros. Sci*, vol. 134, pp. 131–139, 2018. doi: 10.1016/j.corsci.2018.02.031.
29. M. Tokarewicz and M. Grądzka-Dahlke, "Review of Recent Research on AlCoCrFeNi High-Entropy Alloy," *Metals*, vol. 11, no. 8, p. 1302, 2021. doi: 10.3390/met11081302.
30. J.-J. Suñol, "Mechanical Alloying: Processing and Materials," *Metals*, vol. 11, no. 5, p. 798, 2021. doi: 10.3390/met11050798.
31. K. R. Ramkumar, N. Radhika, S. Sivasankaran and H. S. Kim, "Evolution of Microstructure and Mechanical Properties of [Cu–10Ni]–Si3N4 Nanocomposites Developed Using Mechanical Alloying and Spark Plasma Sintering," *J. Alloys Compd*, vol. 899, p. 163319, 2022. doi: 10.1016/j.jallcom.2021.163319.
32. P. K. Verma, A. Singh and A. Kumar, "Microstructure Characterization and Phase Evolution of Equiatomic AlCoMoFeNi High Entropy Alloy Synthesized by Mechanical Alloying," *Mater. Chem. Phys*, vol. 318, p. 129325, 2024. doi: 10.1016/j.matchemphys.2024.129325.
33. D. Jeyasimman, V. Vijayaraghavan and S. Venkateshwara, "Synthesis and Characterization Study of Al10Cr25Co20Ni25Fe20 High-Entropy Alloy Powders through Mechanical Alloying," *J. Mater. Eng. Perform*, 2024. doi: 10.1007/s11665-024-09667-1.
34. B. Sajeevan, M. G. Gopika, A. S. Murali, S. Murugesan and B. Saraswathyamma, "An Electrochemical Sensor Based on a Pencil Graphite Electrode Modified with Poly-Riboflavin for 4-Nitrophenol Quantification," *Mater. Chem. Phys*, vol. 301, p. 127568, 2023. doi: 10.1016/j.matchemphys.2023.127568.
35. C.-Z. Yao *et al.*, "Electrochemical Preparation and Magnetic Study of Bi–Fe–Co–Ni–Mn High Entropy Alloy," *Electrochim. Acta*, vol. 53, no. 28, pp. 8359–8365, 2008. doi: 10.1016/j.electacta.2008.06.036.
36. A.-M. J. Popescu *et al.*, "Electrodeposition, Characterization, and Corrosion Behavior of CoCrFeMnNi High-Entropy Alloy Thin Films," *Coatings*, vol. 11, no. 11, p. 1367, 2021. doi: 10.3390/coatings11111367.
37. Y. Yao *et al.*, "Carbothermal Shock Synthesis of High-Entropy-Alloy Nanoparticles," *Science (80-.)*, vol. 359, no. 6383, pp. 1489–1494, 2018. doi: 10.1126/science.aan5412.
38. J. Chen *et al.*, "Well-defined Nanostructures of High Entropy Alloys for Electrocatalysis," *Exploration*, 2024. doi: 10.1002/EXP.20230036.
39. M. Geng *et al.*, "Carbothermal Shock Synthesis of FeCoNiPtRu High-Entropy Alloy for Dual-Function Water Splitting in Alkaline media," *J. Alloys Compd*, vol. 1005, p. 176180, 2024. doi: 10.1016/j.jallcom.2024.176180.

40. S. Gao *et al.*, "Synthesis of High-Entropy Alloy Nanoparticles on Supports by the Fast Moving Bed Pyrolysis," *Nat. Commun*, vol. 11, no. 1, p. 2016, 2020. doi: 10.1038/s41467-020-15934-1.
41. D. Wu *et al.*, "Platinum-Group-Metal High-Entropy-Alloy Nanoparticles," *J. Am. Chem. Soc*, vol. 142, no. 32, pp. 13833–13838, 2020. doi: 10.1021/jacs.0c04807.
42. W. Gong *et al.*, "A Low-Cost High-Entropy Porous CrO/CrN/C Biosensor for Highly Sensitive Simultaneous Detection of Dopamine and Uric Acid," *Microchem. J*, vol. 175, p. 107188, 2022. doi: 10.1016/j.microc.2022.107188.
43. D. Song *et al.*, "Rational Construction of Ultrathin PtPdRhCuM (M=Mn, Co or Ni) High Entropy Alloy Nanotubes with Rich Defects for Enhanced Electrochemical Activity: Electrochemical Aldicarb Sulfone Sensing," *Sensors Actuators B Chem*, vol. 405, p. 135337, 2024. doi: 10.1016/j.snb.2024.135337.
44. S. Rajendrachari, R. Shrinivas Mahale, V. Adimule, M. Gulen, V. Mahesh and N. Basavegowda, "Fabrication of Mechanically Alloyed High Entropy Alloys (20-Fe-20Cr-20Ni-20Mn-20Ti) Modified Carbon Paste Electrode Sensor for the Cyclic Voltammetric Determination of Methyl Orange," *Inorg. Chem. Commun*, vol. 164, p. 112428, 2024. doi: 10.1016/j.inoche.2024.112428.
45. S. Rajendrachari, V. Adimule, M. Gulen, F. Khosravi and K. K. Somashekharappa, "Synthesis and Characterization of High Entropy Alloy 23Fe-21Cr-18Ni-20Ti-18Mn for Electrochemical Sensor Applications," *Materials*, vol. 15, no. 21, p. 7591, 2022. doi: 10.3390/ma15217591.
46. F. Wang, X. Feng, Y. Gao, X. Ding, W. Wang and J. Zhang, "Green Synthesis of PtPdNiFeCu High-Entropy Alloy Nanoparticles for Glucose Detection," *ACS Omega*, vol. 8, no. 50, pp. 47773–47780, 2023. doi: 10.1021/acsomega.3c06122.
47. M. Li, X. Wang, Y. Cheng, L. Su, P. Xu and X. Li, "High-Entropy Alloy Nanoparticles As Catalyst For Nanomolar-Level Detection Of Neurotransmitter Serotonin In Serum," in *2021 IEEE 34th International Conference on Micro Electro Mechanical Systems (MEMS)*, Jan. 2021, pp. 314–317. doi: 10.1109/MEMS51782.2021.9375269.
48. S. Rajendrachari, V. M. Adimule, G. K. Jayaprakash and A. Pandith, "Electrochemical Oxidation of Methylene Blue Dye in Wastewater Using Mechanically Alloyed High Entropy Alloy Modified Carbon Paste Electrode Using Cyclic Voltammetry," *Mater. Res. Express*, vol. 10, no. 5, p. 054003, 2023. doi: 10.1088/2053-1591/acd3d7.
49. D. B. Miracle, "High-Entropy Alloys: A Current Evaluation of Founding Ideas and Core Effects and Exploring 'Nonlinear Alloys," *JOM*, vol. 69, no. 11, pp. 2130–2136, 2017. doi: 10.1007/s11837-017-2527-z.
50. Y. Zhang and R. Li, "New Advances in High-Entropy Alloys," *Entropy*, vol. 22, no. 10, p. 1158, 2020. doi: 10.3390/e22101158.
51. G. Gao *et al.*, "High Entropy Alloy Electrocatalysts," *J. Energy Chem*, vol. 99, pp. 335–364, 2024. doi: 10.1016/j.jechem.2024.07.049.
52. R. Ashwini, M. K. P. Kumar, M. Y. Rekha, M. S. Santosh and C. Srivastava, "High Entropy Alloy Nanoparticle - Graphene (HEA:G) Composite for non-Enzymatic Glucose Oxidation : Optimization for Enhanced Catalytic Performance," *Carbon Trends*, vol. 9, p. 100216, 2022. doi: 10.1016/j.cartre.2022.100216.
53. B. E. MacDonald *et al.*, "Recent Progress in High Entropy Alloy Research," *JOM*, vol. 69, no. 10, pp. 2024–2031, 2017. doi: 10.1007/s11837-017-2484-6.
54. Y. F. Ye, Q. Wang, J. Lu, C. T. Liu and Y. Yang, "High-Entropy Alloy: Challenges and Prospects," *Mater. Today*, vol. 19, no. 6, pp. 349–362, 2016. doi: 10.1016/j.mattod.2015.11.026.
55. X. Gao *et al.*, "FeMnNiAlCr High-Entropy Alloys with High-Efficiency Surface Oxide Solar Absorbers for Concentrating Solar Power Systems," *High Entropy Alloy. Mater*, vol. 2, no. 1, pp. 97–109, 2024. doi: 10.1007/s44210-024-00033-3.

13 Applications of High Entropy Alloys as Catalysts and as Coating Materials

Raghavendra Rao R, Sharath B N, Shashanka R and Pradeep S

A family of advanced materials known as high entropy alloys (HEAs) is distinguished by its extraordinary qualities and distinct composition. HEAs have five or more principal elements, each of which is present in significant amounts (generally between 5% and 35% atomic fraction), in contrast to standard alloys, which are usually based on one or two principal elements. This unorthodox method of alloy design produces a number of unique characteristics. They have been used extensively in a variety of industries, including catalysis, because of their unique characteristics. Because of their superior mechanical, thermal, and chemical qualities, HEAs are being utilised extensively as coating materials. Applications needing exceptional durability, wear resistance, and protection from severe conditions are best suited for coatings manufactured from HEAs.

13.1 INTRODUCTION

Because of their special composition and characteristics, a family of materials known as high entropy alloys (HEAs) has attracted a lot of interest in materials science. HEAs are composed of five or more principal elements in nearly equal proportions, in contrast to standard alloys, which are typically based on one or two principal elements with trace amounts of other components. Single-phase solid-solution formations are stabilised by this high compositional entropy, which frequently produces advantageous mechanical and functional characteristics. Coatings based on HEAs have a lot of potential in a variety of industries. Greater specific strength, excellent mechanical qualities, superconductivity, super-paramagnetic properties, great fracture toughness at cryogenic temperatures, and more were all provided by these alloys. Multiple primary components are present in these alloys, and compared to traditional alloys, a much greater range of HEA compositions are available. As a result, HEAs are now more frequently used as feedstock materials for coating processes, particularly for protecting the surfaces of different materials in different conditions. Thus, the purpose of this book is to give a general review

DOI: 10.1201/9781003664093-17

of HEA coatings, including their varieties, fabrication processes, post-processing techniques, and various applications. Next, a summary of the HEAs-based coating's mechanical and tribological characteristics created by different researchers is provided. Furthermore, the characteristics of HEAs are covered in this work [1, 2]. Various classes of HEA coatings and methods of fabrication are as shown in Figures 13.1 and 13.2, respectively.

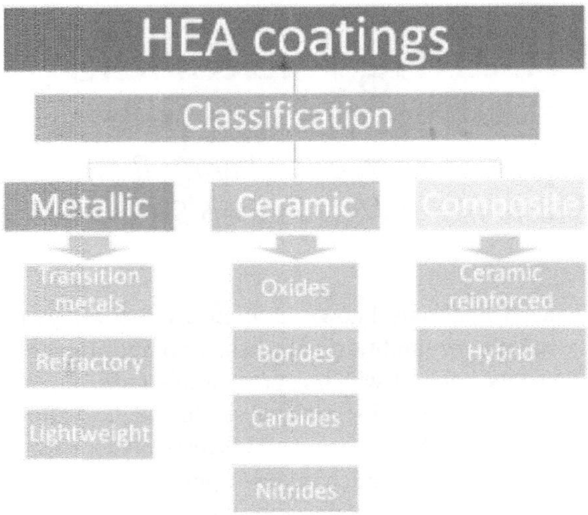

FIGURE 13.1 Various classes of HEA coatings [3].

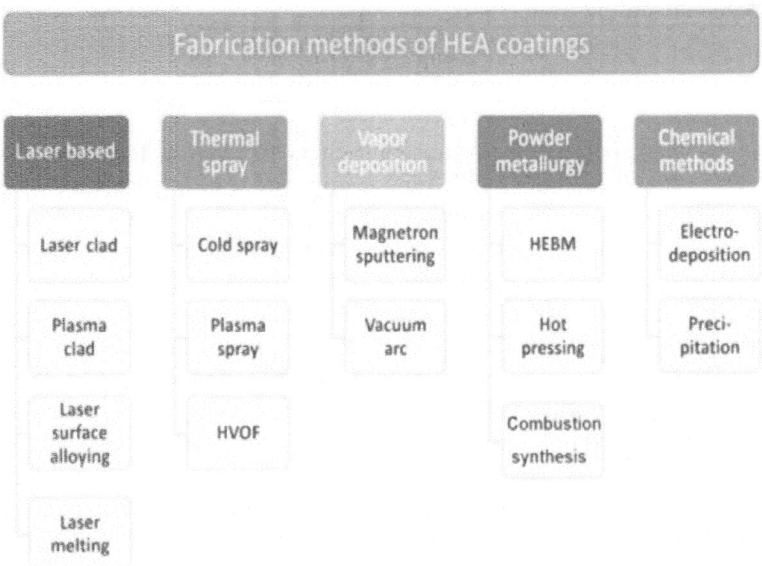

FIGURE 13.2 Various fabrication methods of HEA coatings [3].

13.2 HIGH ENTROPY ALLOYS AS CATALYSTS

HEAs are very promising as catalysts for a variety of chemical reactions, such as hydrogen evolution, oxygen reduction, and CO_2 reduction, because of their distinctive structural and chemical characteristics [4].

The unique characteristics which make them good for use as catalysts are:

i. Structural Complexity: Catalytic activity is increased by the distinct active sites produced by the random atomic distribution.
ii. Synergistic Effects: Reaction kinetics are improved by the many electrical and geometric effects that are contributed by various constituents.
iii. Chemical and Thermal Stability: HEAs are more resilient to harsh environments than traditional catalysts.
iv. Tunable Properties: Catalytic performance for particular reactions can be maximised by varying the constituent composition [5, 6].

13.2.1 Applications of HEAs in Catalysis

13.2.1.1 Hydrogen Evolution Reaction (HER)

Transition metal-based HEAs (such as Co, Fe, Ni, Mo, and W) have good HER activity, which makes them valuable for water electrolysis [7].

For instance, FeCoNiMoW is similar to catalysts based on platinum, and HEA shows excellent stability and activity.

13.2.1.1.1 HER Reaction Mechanism

HER occurs at the cathode and follows two primary reaction pathways in acidic and alkaline media.

In Acidic Medium (H^+ Reduction)

$$2H^+ + 2e^- \rightarrow H_2$$

Volmer Step (Electrochemical adsorption)

$$H^+ + e^- \rightarrow H_2$$

Heyrovsky Step (Electrochemical desorption)

$$H_{ads} + H^+ + e^- \rightarrow H_2$$

Tafel Step (Recombination desorption)

$$H_{ads} + H_{ads} \rightarrow H_2$$

In Alkaline Medium (Water Reduction)

$$2H_2O + 2e^- \rightarrow H_2 + 2OH^-$$

Volmer Step:

$$H_2O + e^- \rightarrow H_{ads} + OH^-$$

Heyrovsky Step:

$$H_{ads} + H_2O + e^- \rightarrow H_2 + OH^-$$

Tafel Step:

$$H_{ads} + H_{ads} \rightarrow H_2$$

The rate-determining step varies depending on the catalyst and reaction conditions.

13.2.1.1.2 HER Catalysts

Since HER requires a catalyst to lower the over-potential (η) and improve efficiency, researchers focus on high-performance catalysts.

A. **Noble Metal Catalysts (Benchmark Materials)**
- Platinum (Pt) – Best HER catalyst due to optimal hydrogen-binding energy (HBE) and high exchange current density.
- Palladium (Pd), Rhodium (Rh), Iridium (Ir) – Show excellent catalytic activity but are expensive.

B. **Non-Noble Metal Catalysts (Cost-Effective Alternatives)**
- Transition Metal–Based Catalysts
  Nickel (Ni), cobalt (Co), molybdenum (Mo), iron (Fe)
  Examples: MoS_2, CoP, NiFe alloys.
- Carbides, Nitrides, and Phosphides
  Mo_2C, WC, Fe_4N, CoP, Ni_2P.
- HEAs
  Multi-element alloys (e.g., FeCoNiMoW) offer tunable catalytic properties.

C. **Metal-Free Catalysts**
  Graphene, carbon nanotubes (CNTs), doped carbon materials
  Examples: N-doped graphene, Co-doped carbon.

13.2.1.1.3 Key Performance Metrics for HER Catalysts

Over-potential (η): Lower values indicate better efficiency.
Exchange Current Density (j_0): Higher values reflect better intrinsic activity.
Tafel Slope (mV/dec): Determines the reaction mechanism (smaller slope = faster kinetics).
Stability: Long-term operation without degradation.
Faradaic Efficiency: Ratio of charge used for H_2 production.

13.2.1.2 Reaction of Oxygen Reduction (ORR)

The hydrogen evolution reaction (HER is an electrochemical process which produces hydrogen gas (H_2) by reducing protons (H^+) in an aqueous solution. In fuel cells, hydrogen synthesis for renewable energy uses and water electrolysis is an essential process. HEAs including Pt, Pd, and Cu improve ORR efficiency and lessen reliance on noble metals when used in fuel cells and metal-air batteries [8, 9].

13.2.1.3 Oxygen Evolution Reaction (OER)

In water-splitting applications, HEAs with oxophilic elements (such as Mn, Fe, Co, and Ni) demonstrate strong catalytic activity in the oxygen evolution reaction (OER).

13.2.1.4 CO_2 Reduction Reaction (CO_2RR)

HEAs provide multifunctional active sites that help convert CO_2 into useful compounds (methanol, hydrocarbons).

13.2.1.5 Ammonia Production and Decomposition

For hydrogen storage applications, HEAs increase the catalytic efficiency of nitrogen fixation and ammonia breakdown.

13.3 HIGH ENTROPY ALLOYS AS COATING MATERIALS

A type of sophisticated material known as an HEA is made up of many primary elements, usually in proportions that are equiatomic or nearly so. These alloys' remarkable qualities, such as their high hardness, resistance to corrosion, resistance to wear, and thermal stability, have drawn a lot of interest in them as coating materials.

Typical High Entropy Coating Alloys:

i. CrMnFeCoNi: One of the most researched HEAs, it is beneficial in wear-resistant applications due to its high ductility and toughness.
ii. AlCoCrFeNi: This alloy is perfect for thermal barrier coatings because of its exceptional hardness and resistance to oxidation.
iii. TiZrNbHfTa: High melting point refractory HEA that is appropriate for harsh settings, such as aerospace engines.
iv. CoCrFeNiMo: It is an excellent choice for maritime and biomedical applications because of its improved corrosion resistance [10].

13.3.1 Benefits of HEAs as Coating Materials

i. Superior mechanical properties: HEAs are great for wear-resistant coatings because of their high hardness, strength, and toughness. They have an outstanding ability to withstand deformation from mechanical stresses.
ii. Oxidation and corrosion resistance: Because of their even distribution of alloying elements, HEAs frequently show exceptional resistance to oxidation and corrosion, even at high temperatures.

iii. Thermal stability: HEAs are appropriate for high-temperature applications and continue to function and retain their structural integrity in harsh environments.
iv. Bonding and adhesion: Strong connections can be formed between HEAs and substrate materials, resulting in dependable and long-lasting performance.
v. Customisable attributes: HEAs can be designed for certain uses, like improved electrical resistivity, thermal conductivity, or hardness, by changing their composition [11].

13.3.2 Applications of HEAs as Coating Materials

i. HEA Coatings Applications in Aerospace: Find applications in coatings that protect engine parts and turbine blades from oxidative conditions and high temperatures.
ii. Power and Energy: Coatings for solar panels, nuclear reactors, and fuel cells because of their superior thermal stability and resistance to corrosion.
iii. Medical Equipment: Coatings that are resistant to wear on biomedical devices, implants, and surgical instruments [12, 13].
iv. Tools for Industry: Coatings to increase durability and decrease wear on drilling, milling, and cutting instruments.
v. Automotive and Defence: Armour and engine parts with HEA coatings to withstand mechanical stress, corrosion, and abrasion.

13.3.3 Deposition Techniques for HEA Coatings

Several innovative processes are employed to deposit HEA coatings.

13.3.3.1 Physical Vapour Deposition (PVD)

One of the best methods for applying HEA coatings is physical vapour deposition, or PVD. It entails turning solid material into vapour, which condenses onto a substrate to create a coating or thin film. Because PVD produces homogeneous, high-purity, and well-adhered coatings, it is widely employed. It is commonly used to produce thin, homogeneous coatings with high adherence [14].

The main advantages of PVD are as follows:

- Precise Composition Control – PVD allows fine-tuning of the HEA coating composition.
- High Purity – The process occurs in a vacuum, preventing contamination.
- Good Adhesion – PVD coatings strongly bond to substrates, reducing peeling or delamination.
- Fine-Grained Microstructure – PVD results in coatings with superior hardness and wear resistance.
- Low Processing Temperature – Ideal for coating temperature-sensitive materials.

13.3.3.2 Magnetron Sputtering

HEA coatings are commonly deposited using the PVD method of magnetron sputtering. It provides fine control over the microstructure, content, and thickness of the film, which makes it a great option for creating high-temperature, corrosion-, and wear-resistant coatings. A vacuum chamber is filled with a target (elemental targets or HEA targets). An intense electric field is applied to create plasma, which is an ionised gas, often argon. Atoms are ejected from the target by the bombardment of plasma ions (sputtering process). A thin layer of HEA is created when these atoms condense onto the substrate. Enhancement of the Magnetic Field: A magnetron, such as an electromagnet or permanent magnet, is positioned behind the target. This improves sputtering speeds and ionisation efficiency by trapping electrons close to the target. It is appropriate for materials that are sensitive to temperature since it lowers substrate heating.

HEA Coatings Magnetron Sputtering Types:

i. Magnetron Sputtering Using Direct Current (DC)
ii. For conductive HEA coatings (such CrMnFeCoNi), this is utilised. High rates of deposition. 2. Magnetron Sputtering at Radio Frequencies (RF)
iii. It works with conductive and non-conductive HEAs and creates coatings that are more dense and more consistent. Improved adhesion as a result of the high energy plasma.
iv. High-Power Impulse Magnetron Sputtering (HiPIMS): Creates denser, flawless HEA coatings; produces pulsed high-energy plasma for higher film quality; excellent for high-temperature and wear-resistant applications.

Magnetron Sputtering Benefits for HEAs:

- Grain boundaries and porosity are lessened by uniform and dense coatings.
- Low Substrate Temperature: Perfect for materials that are sensitive to temperature changes.
- Good Adhesion: Bonding to the substrate is improved by plasma bombardment.
- Controlled Composition: To attain the required composition, one can utilise either individual element targets or pre-alloyed HEA targets.

Applications of Magnetron Sputtering for HEA Coatings:

- Wear-resistant coatings for machine parts and cutting tools.
- Marine and biomedical implants with corrosion-resistant coatings.
- Aerospace and turbine blades with high-temperature coatings.
- Functional coatings for microelectronics and sensors.

13.3.3.3 Thermal Spray Coatings

For high-temperature stability, corrosion prevention, and wear resistance, HEA coatings are applied using thermal spray spraying, a flexible surface engineering method.

Thermal spray is perfect for industrial applications since it can produce coatings that are thicker (tens to hundreds of microns) than PVD processes. It is suitable for applying thick HEA coatings to large surfaces.

HEA Thermal Spray Coating Applications:

i. Aerospace: Thermal barriers, heat shields, and jet engines.
ii. Marine: Corrosion-resistant coatings for ships and offshore constructions.
iii. Industrial Tools: Cutting tools, machine components, and wear-resistant coatings.
iv. Biomedical: HEA coatings for prostheses and implants.
v. Energy Sector: High-temperature parts, fuel cells, and nuclear reactors.

Importance of HEA Coatings Using Thermal Spray:

- Thick, Dense Coatings: Perfect for situations involving high temperatures and wear.
- High Deposition Rate: This method covers vast surfaces more quickly than PVD techniques.
- Broad Material Compatibility: Able to cover composites, metals, and ceramics.
- Cost-effective: Less expensive per unit area than vacuum-based deposition techniques.

Typical Methods for Thermal Spraying HEA Coatings:

i. **High-velocity oxygen fuel spraying (HVOF):**
Oxygen is combined with a fuel (such as kerosene or propane) and ignited. After being injected into the hot gas stream, the HEA powder accelerates at supersonic speeds in the direction of the substrate. High-density, low-porosity coatings are produced by the impact of the molten or semi-molten particles, which also provide excellent adhesion and wear resistance. It is ideal for industrial components such as turbine blades and aircraft.

ii. **Plasma spraying method (APS):**
HEA powder is melted in a high-temperature plasma (10,000–15,000 K). Droplets of molten material accelerate and land on the surface. It is effective with high-melting-point HEAs, such as refractory HEAs. Better purity and oxidation resistance are provided by vacuum plasma spraying (VPS), which is also utilised in high-temperature and jet engine thermal barrier coatings.

iii. **Cold spray coating (CS) method:**
Compressed gas is used to accelerate HEA powder at a high speed. Although it connects through extreme plastic deformation, the powder stays solid and maintains its original material qualities (no melting or oxidation). The method is excellent for biomedical and corrosion-resistant coatings. It is used in additive manufacturing and part repair.

13.3.3.4 Electrochemical Deposition

Electrochemical deposition (electrodeposition) is a promising and cost-effective method for depositing HEA coatings onto various substrates. This technique is especially useful for fabricating corrosion-resistant, wear-resistant, and functional coatings at room temperature and ambient pressure.

Importance of Electrochemical Deposition in HEA Coatings:

- Low-Cost and Scalable: This method works well for large-scale production and only requires basic equipment.
- Room-Temperature Process: No need for vacuum or high-temperature melting.
- Controlled Composition and Thickness: By modifying the bath composition and deposition settings, alloy ratios can be precisely tuned.
- Good Adhesion: Strong coating-substrate interactions are guaranteed via electrochemical bonding.
- Adaptable Substrate Compatibility: Able to cover ceramics, metals, and polymers.

Steps Involved in Electrochemical Deposition for HEA Coatings:

i. Preparation of the Electrolyte: A solution comprising metal salts, such as Ni^{2+}, Fe^{2+}, Cr^{3+}, Co^{2+}, Cu^{2+}, etc., is made.
ii. Placement of the Substrate: In an electrochemical cell, the substance to be coated serves as the cathode.
iii. Current Application: Metal ions are reduced onto the substrate by applying a direct or pulsed current.
iv. HEA Film Formation: A uniform HEA coating is created when several elements are simultaneously deposited.
v. Post-Treatment: To enhance characteristics, annealing, heat treatment, or surface changes may be used [13].

Methods of Electrodeposition for HEA Coatings:

i. **Direct Current (DC) Electrodeposition:** Easy to use and popular for consistent coatings. It works well with conductive substrates. It also may result in composition segregation because of varying rates of metal deposition.
ii. **Electrodeposition via Pulse:** Grain refining and coating homogeneity are enhanced by the use of pulsed current. It aids in regulating growth and nucleation, producing finer microstructures. It lessens coating stress and fractures.
iii. **Alloy Co-Deposition (Simultaneous Plating of Multiple Elements):** Non-noble HEAs can be deposited from aqueous or ionic solutions thanks to alloy co-deposition, which is the simultaneous plating of multiple elements. To prevent phase segregation, careful control of pH, temperature, and voltage is necessary [14].

Applications of Electrochemically Deposited HEA Coatings:

- Marine, chemical industry, and biomedical implants all benefit from corrosion protection.
- Tools, bearings, and industrial machinery are examples of wear-resistant coatings. Heat exchangers, power plants, and aircraft are examples of high-temperature applications.
- Fuel cells, super capacitors, and batteries are examples of electrocatalysts and energy storage.

13.3.3.5 Laser Cladding

Laser cladding is an innovative surface modification method used to provide HEA coatings on a variety of substrates. A layer of material, usually metal, is applied to a substrate via laser cladding, an advanced material processing technique that enhances the substrate's surface qualities like thermal conductivity, corrosion resistance, and wear resistance. A powerful laser is used in the process to melt the substrate's surface and concurrently transfer a feed material (wire or powder) to the molten pool. A metallurgically bonded layer is produced when the laser merges the feed material and substrate as it passes across the surface. This procedure improves corrosion resistance, wear resistance, and stability at high temperatures, which makes it perfect for industrial, automotive, marine, and aerospace applications.

Some of the salient characteristics of laser cladding are:

- Control and Precision: By precisely regulating the heat input, the laser beam reduces thermal distortion and guarantees a consistent coating.
- Minimal Heat-Affected Zone (HAZ): Warping and unintended material deterioration are less likely due to the localised heat input creating a restricted heat-affected zone.
- Broad Variety of Materials: Depending on the use, cladding can be made of a variety of materials, such as metals, ceramics, and composites.
- Decreased Material Waste: Compared to alternative coating techniques, less material is wasted because of the process's high level of precision.

In sectors including manufacturing, energy, automotive, and aerospace, where high-performance components need to belong-lasting, with specially made coatings, laser cladding is frequently utilised.

13.4 CONCLUSION

An innovative method for improving the longevity and performance of aerospace components is the use of HEA coatings. HEAs are positioned to be extremely important in the future of aerospace engineering, especially in jet engines, hypersonic vehicles, and spacecraft thermal protection systems, due to their exceptional high-temperature strength, resilience to corrosion, and resistance to wear. The aerospace sector will employ deposition technology more widely as a result of ongoing research and development. Also, HEAs have shown great promise as catalysts because of their exceptional stability, adjustable electronic structures, and distinctive multi-element composition. They are appropriate for a variety of catalytic applications, such as hydrogen evolution, CO_2 reduction, and biomass conversion, due to their capacity to produce a synergistic catalytic effect, improved adsorption qualities, and resistance to deactivation. The catalytic performance of HEAs can be further improved by their logical design and optimisation through structural and compositional adjustment. To expedite their use in sustainable catalysis, future study should concentrate on comprehending their basic reaction processes, enhancing scalability, and creating affordable synthesis techniques.

REFERENCES

1. Luo, Long, Huimin Han, Liangpan Chen, Dianchen Feng, Lirong Li, Tingting Zhai, Zhiwei Chen, Rihao Gao, Haibin Wu, Suxia Liu, Wenli Pei and Yongzhi Li. "Critical review of high-entropy alloys for catalysts: Design, synthesis, and applications." *International Journal of HydrogenEnergy*, 90, (2024): 885–917. https://doi.org/10.1016/j.ijhydene.2024.09.379
2. Kumar, Santosh. "Comprehensive review on high entropy alloy-based coating." *Surface and Coatings Technology*, 477, (2024): 130327. https://doi.org/10.1016/j.surfcoat.2023.130327
3. Sharma, A. "High entropy alloy coatings and technology." *Coatings* 11 (2021): 372. https://doi.org/10.3390/coatings11040372
4. Shashanka, R. and Debasis Chaira. "Phase transformation and microstructure study of nano-structured austenitic and ferritic stainless steel powders prepared by planetary milling." *Powder Technology*, 259 (2014): 125–136.
5. Shashanka, R. and Debasis Chaira. "Development of nano-structured duplex and ferritic stainless steels by pulverisette planetary milling followed by pressureless sintering." *Materials Characterization*, 99 (2015): 220–229.
6. Shashanka, R. and D. Chaira. "Optimization of milling parameters for the synthesis of nano-structured duplex and ferritic stainless steel powders by high energy planetary milling." *Powder Technology*, 278, (2015): 35–45. https://doi.org/10.1016/j.powtec.2015.03.007
7. Rajendrachari, Shashanka, Debasis Chaira and Orhan Uzun. "Synthesis of nano-structured duplex and ferritic stainless steel powders by dry milling and its comparison with wet milling." *Archives of Metallurgy and Materials*, 65, (2020): 5–14. https://doi.org/10.24425/amm.2019.131091
8. Rajendrachari, Shashanka, Vinayak Adimule, Gururaj Jayaprakash and Anup Pandith. "Electrochemical oxidation of methylene blue dye in wastewater using mechanically alloyed high entropy alloy modified carbon paste electrode using cyclic voltammetry." *Materials Research Express*, 10, (2023): 1–13. https://doi.org/10.1088/2053-1591/acd3d7
9. Rajendrachari, Shashanka, Nagaraj Basavegowda, R. Vinaykumar, Damanapeta Narsimhachary, Prathap Somu and Meng-Jen Lee. "Electrocatalytic determination of methyl orange dye using mechanically alloyed novel metallic glass modified carbon paste electrode by cyclic voltammetry." *Inorganic Chemistry Communications*, 155, (2023): 111010.
10. Rajendrachari, Shashanka, Rayappa Shrinivas Mahale, Vinayak Adimule, Mahir Gulen, V. Mahesh and Nagaraj Basavegowda. "Fabrication of mechanically alloyed high entropy alloys (20-Fe-20Cr-20Ni-20Mn-20Ti) modified carbon paste electrode sensor

for the cyclic voltammetric determination of methyl orange." *Inorganic Chemistry Communications*, 164, (2024): 112428. https://doi.org/10.1016/j.inoche.2024.112428

11. Rajendrachari, Shashanka, R. Vinaykumar, Ramesh Katti, B. G. Koujalagi, Gururaj Kudur Jayaprakash and Ramya Ramkumar. "Cyclic voltammetric determination of rhodamine b dye using a gas-atomized duplex stainless steel powder-modified carbon paste electrode." *Topics in Catalysis*, (2024). https://doi.org/10.1007/s11244-024-02020-9
12. Rajendrachari, Shashanka, Emre Altaş, Sachin Salunkhe, Ilker Dağ, Baris Avar and Debasis Chaira. "Hot isostatic pressing of ball milled novel grade high entropy alloys at different sintering time and the investigation of their mechanical and corrosion resistance properties." *Surface Topography: Metrology and Properties*, 12, (2024): 045006. https://doi.org/10.1088/2051-672X/ad880b
13. Dağ, İ.E., B. Avar and M. Panigrahi et al. "Evaluation of the microstructural, mechanical, and corrosion behavior of BN-, B4C-, and FeB-added distaloy AE composites produced by high-energy ball milling and hot-pressing." *Metallography, Microstructure, and Analysis,* 14, (2025): 210–228. https://doi.org/10.1007/s13632-025-01169-3
14. Gülen, Mahir, Recep Taş, Hamza Dünya and Shashanka Rajendrachari. "Enlargement of the surface area of high-entropy alloys with acid treatment as positive electrode for high specific capacitance supercapacitors." *ChemistryOpen,* 14, (2025): e202400312.

14 Other Advanced Applications and the Future of High Entropy Alloys

Sharath B N, Raghavendra Rao, Hareesha M, Rajesh M and Shashanka R

High entropy materials (HEMs), which include high entropy alloys HEAs, high entropy polymers (HEPs), and high entropy ceramics (HECs), have recently been an attraction for innovative studies because of their remarkable thermal stability, mechanical strength, and resistance to oxidation, among other remarkable qualities. Aerospace, military, medicinal, and chemical engineering are just a few of the sophisticated fields that can benefit greatly from these revolutionary materials. This chapter delves into HEAs, exploring their effects, classifications, and different manufacturing procedures to find optimal alloys for important advanced uses. By outperforming more conventional materials like metals, alloys, and composites, HEAs have established themselves as a viable substitute in industries with strong demand. Additionally, it provides a comprehensive analysis of potential future developments in HEA production and addresses significant issues in this field. Its objective is to enhance the usefulness of HEMs and other advanced applications in various contexts and to unleash their full potential in other innovative sectors.

14.1 INTRODUCTION

The tremendous advances forward in scientific knowledge and technological capability have recently propelled space exploration to new levels. Reliable spacecraft, satellites, and probes have made it possible to conduct revolutionary experiments, explore faraway planets, and gather vital data about our solar system. Advancements in robotics, materials science, and propulsion systems have enhanced the efficiency, safety, and productivity of space missions, paving the way for potential future explorations and human settlements beyond Earth. Engineers meticulously design spacecraft for the challenging space environment to ensure safe and reliable operations, able to withstand harsh conditions during launch, space travel, and reentry into Earth's atmosphere [1]. Rigorous studies and analyses precede any design changes to ensure maximum efficiency, safety, and mission success. From an aeronautical perspective, materials used

DOI: 10.1201/9781003664093-18

in spacecraft must exhibit high strength-to-weight ratios, thermal and corrosion resistance, wear resistance, and the ability to endure extreme environmental conditions. When it comes to engine components, the aerospace industry is always on the lookout for new, innovative materials that may enhance fuel economy, decrease emissions, endure higher temperatures, and protect against space debris' abrasive impacts. While existing materials meet some of these criteria, further research and development are necessary to enhance their performance [2, 3]. One revolutionary approach to these problems is the use of high entropy alloys, or HEAs. In contrast to standard alloys, which usually consist of two or three main elements, HEAs contain five or more main elements in nearly equal proportions. The combination of elements creates a crystal structure that gives this unusual composition remarkable qualities, especially when subjected to cryogenic temperatures. Expanding the notion of high entropy to non-metallic materials beyond alloys led to the development of high entropy ceramics (HECs) and high entropy polymers (HEPs). Various applications in major sectors can now benefit from these materials' expanded choice of high-performance possibilities. However, it still requires overcoming obstacles to fully optimize these materials for use in aerospace and other cutting-edge industries. Tackling these issues opens avenues for additional exploration and development, expanding the frontiers of material science and its practical uses [4, 5].

14.1.1 Overview of High-Entropy Alloys

Historically, experts typically described HEAs as alloys that contain five or more elements in an equimolar or nearly equimolar ratio, with each main component accounting for 5 to 35 atomic percent (at. %) of the total. The primary focus of this first definition was on composition. But a new way of thinking about HEAs, defined by their mixing configurational entropy, came up as the area developed. Entropy, the intrinsic degree of disorder within a closed system, indicates greater unpredictability in this context. The result, according to materials science, is a more disorganized distribution of components inside the alloy. The transition to defining HEAs through mixing entropy considerably expanded the scope of these alloys, enabling greater compositional flexibility and the capacity to tune their properties for specific purposes. The entropy-based definition has the benefit of focusing on the mixing configurational entropy rather than the enthalpy of compound formation, which is usually more important. This process lessens the possibility of intermetallic compound formation and stabilizes solid solutions as they solidify. This implies that HEAs lack a distinct phase structure, as their constituent atoms disperse across crystallographic lattice locations, rendering them indistinguishable from each other. Because of new processes that give HEAs their unique mechanical and physical properties, the high mixing entropy of these alloys makes it easier for them to dissolve in a number of different substances. Because of this quality, HEAs are able to be very flexible and adaptable, which further establishes them as a groundbreaking material class with enormous promise for cutting-edge uses in many fields [6, 7].

14.1.2 Importance and Potential Uses Among Alloys with High Entropy

The outstanding room-temperature mechanical capabilities of HEAs are the result of their amazing ductility and strength-to-weight ratio. Because they have more mixing configurational entropy, they have strange properties like slow diffusion, high entropy effects, cocktail effects, and severe lattice distortion [8–11]. Differences in atomic size, electronic structure, and mixing enthalpy primarily cause the great chemical complexity of HEAs, which sets them apart from more traditional alloys. HEAs are naturally very strong. This is because their complexity creates local structural inhomogeneities on an atomic level. These inhomogeneities raise lattice friction and have a big impact on how they deform [12–15]. HEAs achieve improved mechanical performance through careful heat treatment process control and strategic alloy design. Some important methods for improving strength and flexibility are metastable engineering to create transformation-induced plasticity (TRIP), creating multi-stage coherent precipitates, and adding ordered oxygen complexes (OOCs) by adding atoms between the matrix [16–19]. Because of how these aspects work together, HEAs have the ability to change material design in many cutting-edge industries, and they are also very suitable as high-strength structural materials for many demanding applications. The remarkable performance of HEAs throughout a broad temperature range makes them extremely adaptable for cutting-edge uses. As an example of their exceptional low-temperature capabilities, HEAs are able to overcome low-temperature embrittlement. Cantor alloy stands out among these because of its superior fracture toughness compared to most metallic materials; this is due to its single-phase face-centered cubic (FCC) crystal structure [20]. The Cantor alloy's remarkable tolerance to low-temperature conditions is one of its most remarkable features. As the temperature drops from room temperature to as low as –200°C, the Cantor alloy shows improved strength and ductility, in contrast to typical alloys that frequently experience embrittlement [21]. Because of its high-performance properties at low temperatures, such as its ability to withstand fracture stresses up to 200 MPa·m$^{1/2}$, FCC HEAs are ideal for use in aircraft, space exploration, and polar environments.

HEAs excel in high-temperature performance, with RHEAs being particularly noteworthy. These alloys, predominantly composed of refractory elements such as Nb, Ta, Mo, and W, benefit structurally from the high entropy effect and their high melting temperatures. Resisting high-temperature softening, refractory HEAs keep their single-phase body-centered cubic (BCC) structure stable up to 1400°C and continue to show remarkable strength and stability even at 1600°C, an ultra-high temperature [22–26].

Refractory HEAs are essential for high-temperature molds, ultra-high-temperature coatings, aerospace components, and other demanding applications because of their characteristics. Their dependability, even in the harshest environments, is a testament to their ability to change the way many industries view material performance [27].

HEAs are incredibly desirable for cutting-edge uses in harsh settings due to their exceptional radiation resistance. Numerous primary elements with intrinsic chemical instability interact to modify energy propagation mechanisms and decrease the average mean free path. This significantly lowers the potential for radiation damage to accumulate [28]. When it comes to the structural management of radiation-induced defects like vacancies and interstitial atoms, HEAs' intrinsic severe lattice distortion is crucial. This distortion further enhances the radiation resistance of HEAs by reducing dislocation mobility and enhancing defect recombination. Recent research has shown that HEAs show tremendous promise as neutron irradiation shields, which could make them useful in improved nuclear reactors and other nuclear energy machinery [29, 30]. Not only are HEAs resistant to radiation, but they also have remarkable physical features like superconductivity, intriguing magnetic behaviors, and high corrosion resistance. The aerospace, energy, electronics, and automotive industries are just a few of the many that could benefit from HEAs due to their adaptability and versatility. They have the ability to change the face of material science and meet the rigorous demands of future technological applications thanks to their unique blend of radiation tolerance and varied functional characteristics [31].

14.2 PROPERTIES OF HIGH ENTROPY ALLOYS

The alloy's noticeable lattice distortion is a result of the elements' varied atomic radii. This distortion significantly complicates dislocation motion, thereby enhancing the material's strength and hardness. Furthermore, the intrinsic "cocktail effect" [32] of HEAs amplifies the interactions among the different alloying elements. So, these alloys show how the qualities of each element work together in a harmonious whole. This composite effect enhances the exceptional hardness and strength of the HEAs.

Numerous studies have proven the superior hardness of HEAs. The four types of HEAs that Guo Jianbin [33] created are ZrTiHfVNb, ZrTiHfNbMo, ZrTiHfVTa, and TaMoWNbZr. Results from compression testing showed that the yield strengths of the three materials were 1152 MPa, 1134 MPa, and 1089 MPa, respectively. However, upon raising the testing load, these three alloys exhibited minimal plastic strain and experienced early cracking. Alternatively, ZrTiHfVNb achieved an impressive plastic strain of 55% while having a lower yield strength of 836 MPa. Compared to many commonly used stainless steels, HEAs exhibit better corrosion resistance due to their simpler crystal structure. Several HEAs also contain dendritic segregation structures, which are a blend of basic surface and body structures [34]. These dendritic formations are highly resistant to corrosion because of their nanoscale particles and chaotic configurations. The HEA $Cu_{0.5}NiAlCoCrFeSi$ is one example. It was better at resisting corrosion than 304 stainless steels at room temperature in solutions of 1 mol/L NaCl and 0.5 mol/L H_kSO_4. The field of thermodynamics uses the entropy to describe the degree of disorder present in a system. A higher entropy number more accurately represents chaos. Notably, a more stable system is one that has a higher entropy value. Moreover, HEAs are thermally more stable since rising temperatures lower the system's free energy. As a result, these alloys are highly protected against heat deterioration [34].

Microstructure plays a significant role in determining the mechanical characteristics of HEAs. HEAs with meticulously engineered microstructures and compositions exhibit outstanding mechanical properties such as low-temperature toughness, ductility, and high strength. Aerogenerators and other high-temperature applications could benefit greatly from refractory HEAs due to their exceptional thermal stability [35, 36]. Zr-Hf-Nb HEAs also show enormous promise as advanced nuclear structural materials due to their high radiation resistance. The transportation industry is one that places a premium on minimizing weight, making lightweight HEAs an ideal choice because of their low density. The high-configurational-entropy effect in HEAs changes the compositions of alloys to make it easier to make single solid solution (SS) phases. This is done to get rid of intermetallic (IM) structures that don't work well together. Balance between the SS matrix phase and the strengthening phase enhances their qualities [37]. HEAs with a FCC structure offer advantages such as ease of processing and excellent plasticity, though their strength is relatively low. Researchers have explored various strengthening mechanisms, such as second-phase dispersion strengthening, interstitial atom strengthening, deformation strengthening, and phase transformation, to address this issue. Researchers have demonstrated that cryogenic temperature rolling refines microstructures and significantly enhances mechanical strength, thereby expanding the potential applications of HEAs [38]. In order to improve the mechanical strength and microstructure of HEAs, cryogenic temperature rolling is an essential process. Cryogenic deformation of HEAs reveals an unprecedented combination of strength and ductility, outperforming the ductile-brittle transition seen in conventional alloys. This is a crucial factor to consider when working with aircraft materials, as they are especially well-suited for cryogenic settings [15]. Cryogenic machinability, low thermal conductivity, great strength, and outstanding toughness are essential properties for aeronautical component materials. Because of their exceptional strength-toughness synergy in these environments, HEAs show great promise as a material for many cryogenic uses in the aerospace sector. Their exceptional mechanical characteristics make them promising structural materials for use in cryogenic settings, which bodes well for their future in cutting-edge aerospace applications [39].

14.3 PREPARATION METHOD OF HIGH ENTROPY ALLOYS

The preparation, manufacturing, and selection of process parameters for HEAs significantly influence the final product's characteristics and functionality. Furthermore, the choice of preparation method significantly influences the processing costs, underscoring the importance of assessing and optimizing these factors. To achieve the ideal equilibrium between material qualities and economic efficiency, it is crucial to carefully choose preparation processes and process parameters.

A lot has changed in the manufacturing of HEAs throughout the years. Initially, the vacuum arc melting and casting technology dominated the field, offering a reliable approach to alloy synthesis. Technological developments, such as surface coating techniques, mechanical alloying and others, have presented numerous novel ways. These new techniques allow for more precise control over alloy microstructures and compositions, which in turn allows for the creation of HEAs for niche uses while taking scalability and affordability into account.

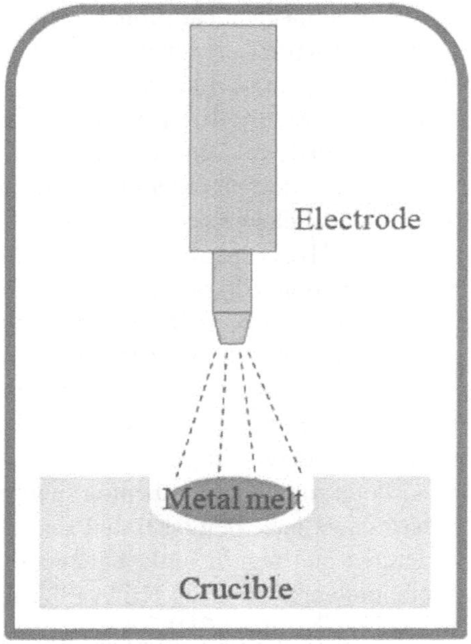

FIGURE 14.1 Schematic diagram of vacuum arc melting method [41].

14.3.1 Vacuum Arc Melting

Figure 14.1 shows the vacuum arc melting procedure, the first preparation technique for HEAs. This process uses an arc discharge to quickly melt the metal components, then swiftly cools and solidifies the liquid metal. Vacuum arc melting has been the method of choice for preparing alloy ingots in most recent studies due to its capacity to melt metal components with high melting points. However, this approach comes with a few limitations. Rapid cooling during solidification causes metal surfaces to shrink significantly, which in turn causes flaws that compromise the material's strength. Therefore, the best use of vacuum arc melting is for preliminary HEA characteristics investigation, not for mass manufacturing or applications that require defect-free materials [40].

14.3.2 Resistance Furnace Melting

When dealing with HEAs that have high melting temperatures or are prone to oxidation, this method performs less effectively than it does with alloys that have a more uniform microstructure and minimal segregation. The open nature of the operation and insufficient control over the melting environment make it unsuitable for preventing oxidation during the melting phase. Moreover, resistance furnaces cannot melt the refractory components of some HEAs at the high temperatures they reach. Therefore, the resistance furnace process does not prepare certain alloys, especially those with demanding structural and purity specifications, to their best potential [7].

14.3.3 Vacuum Induction Melting

One method for melting metals in a vacuum is vacuum induction melting, or VIM for short [42]. Preparing large-size ingots of HEAs is where this method really shines. To keep the material pure and reduce oxidation when melting, VIM provides a controlled, low-contamination environment.

For the advancement of HEA research and applications, the capacity to manufacture large-scale ingots using this method is essential. For in-depth investigations of alloy deformation, microstructure evolution, mechanical properties, and raw material availability, these bigger ingots are more than enough. Thus, VIM facilitates the transition of novel HEAs from the experimental stage to the industrial stage, while also simplifying their development. Research and development in the field of HEAs has favored this technology due to its capacity to create ingots of excellent quality [43].

14.3.4 Vacuum Electron Beam Melting

A focused, high-energy electron beam is used on metal substrates in a very vacuumed chamber to make HEAs. This process is called vacuum electron beam melting. The powerful electron beam heats the metal to a melting point, after which it cools and solidifies in a controlled environment. Among its many other benefits, this approach can improve the mechanical properties of HEAs. Vacuum electron beam melted alloy samples showed a significant improvement in plasticity (almost 20%) and ultimate strength (1400 MPa) over their electric arc furnace-melted counterparts, with only a little drop in total strength.

However, vacuum electron beam melting has its own set of restrictions. The solidification process can lead to uneven material characteristics, often causing severe anisotropy within the microstructure. Furthermore, the high-vacuum environment prohibits the processing of elements with high volatility using this approach. Because these elements can evaporate during melting, the alloy's composition may change. Notwithstanding these obstacles, this technique is nonetheless useful for some applications that need customized mechanical characteristics [44].

14.4 APPLICATIONS OF HIGH ENTROPY ALLOYS

The remarkable mechanical, thermal, and chemical properties of HEAs are leading to their widespread use in various industries. Extreme settings, such as those found in nuclear reactors, power plants, and aircraft, are perfect for their high strength, wear resistance, corrosion resistance, and oxidation resistance. The biomedical industry is investigating HEAs with biocompatible compositions for use in implants and prosthetics due to their exceptional corrosion behavior and wear resistance. Energy applications (Figure 14.2) such as fuel cells, batteries, and hydrogen generation are showing interest in HEAs due to their exceptional thermal stability and distinctive catalytic characteristics. Additionally, researchers are developing lightweight HEAs for airplanes and automobiles to enhance fuel efficiency without compromising structural integrity. HEAs provide even more

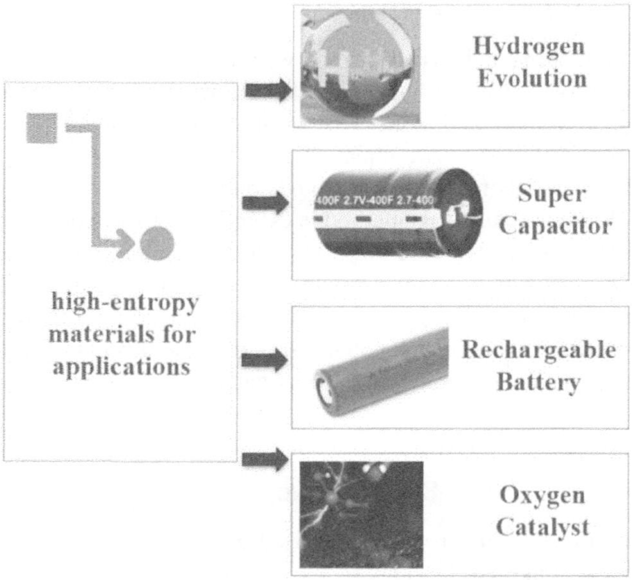

FIGURE 14.2 High entropy materials for applications [41].

adaptability in coatings by increasing surface durability and resilience to severe environmental factors. Researchers anticipate that the range of HEA applications will grow in the future to meet the demands of cutting-edge technology and environmental objectives.

14.5 IDENTIFICATION OF POTENTIAL RESEARCH DIRECTIONS AND AREAS FOR FURTHER DEVELOPMENT

A wide range of scientific and engineering fields have taken an interest in HEAs due to their remarkable mechanical and functional qualities and unusual chemical compositions. Nevertheless, there is a tremendous deal of room for growth in both study and technical implementation in this area, since it is still in the beginning stages. A potential approach is to deepen the theoretical knowledge of HEAs, focusing on the processes that govern their phase stability and deformation. To better understand the kinetic and thermodynamic characteristics of HEAs, state-of-the-art computational methods, including first-principles calculations and predictive modeling based on machine learning, can be employed [42, 45–48]. These endeavors have the potential to improve phase prediction models and uncover new alloys designed for uses. Furthermore, understanding the atomic-scale interactions of HEAs may reveal novel mechanisms behind their exceptional strength, ductility, and corrosion resistance. Aerospace, power-generating, and nuclear applications are good fits for HEAs because of their strength retention and resistance to oxidation at high temperatures. Investigations can center on making these materials more resistant to radiation and high temperatures. The advancement of technologies for space exploration, turbines, and jet engines depends on the creation of refractory HEAs and coatings that can endure high pressures and temperature cycling. HEAs with non-toxic molecules and outstanding corrosion resistance could be beneficial for biomedical applications such as implants and prostheses.

Researchers need to investigate how resistant materials are to wear in order to make more advanced medical devices. Bioinspired HEAs with non-toxic molecules and great corrosion resistance could be useful for biomedical uses like implants and prostheses. Developing sophisticated medical devices requires investigations into their wear resistance, bioinertness, and compatibility with human tissues. Additional investigation into HEAs for complex-geometry, custom medical implants is possible with the use of the additive manufacturing process and compatibility with human tissues. Additional investigation into HEAs for complex-geometry, custom medical implants is possible with the use of the additive manufacturing process. There are promising new avenues for energy-related applications made possible by the exceptional electrical and catalytic characteristics of HEAs. Electrodes and catalysts for fuel cells, batteries, and water-splitting devices are only a few potential applications for HEAs. Their customizable electronic structures and surface chemistries optimize electrochemical systems for greater energy efficiency and durability [49–52].

Scientists are looking into possibilities for developing aluminum, magnesium, and titanium-based lightweight HEAs to meet the increasing demand for such materials in the aerospace and automotive industries. Using these materials to make cars and buildings much lighter could lead to significant improvements in fuel economy and pollution. When combined with HEAs, advanced manufacturing technologies like powder metallurgy and additive manufacturing (AM) can produce materials with highly customizable microstructures and characteristics. Complex and elaborate patterns are now within reach, thanks to AM techniques that make them possible. Better material performance and wider industrial usage could result from additional research into the interaction between HEA powders and AM techniques.

The development of green technologies can benefit from HEAs, which are durable and recyclable, and there is a growing focus on sustainability. Aligning HEA components with environmental aims can be achieved by research that focuses on lowering resource usage during production and promoting recyclability. Furthering worldwide sustainability initiatives, HEAs may also be involved in carbon capture technologies and renewable energy systems. Additionally, AM has the potential to increase material performance and find wider use in industry.

A revolution in the field of materials research, HEAs might have far-reaching effects in fields as diverse as aerospace, healthcare, energy, and ecology. Scientists may bring forth HEAs' full potential by filling in existing information gaps and making use of new processing technologies, breakthroughs in computer modeling, and interdisciplinary collaborations. By collaborating, we can expand the potential applications of HEAs and explore innovative solutions for some of the world's most pressing issues [53].

14.6 CONCLUSION

In materials research and engineering, a major step forward has been the appearance of high entropy materials, specifically HEAs, HEPs, and HECs. The aerospace, military, medical, and chemical engineering industries can benefit greatly from their oxidation resistance, mechanical strength, and thermal stability, making them ideal for demanding applications. Despite these materials' obvious advantages over more traditional options, they have yet to reach their full potential. There are prospects for HEMs because production methods are improving and more is known about their properties. Therefore, as production procedures improve and research continues, these materials pave the way for new technological capabilities and advances in various industries. The best way to ensure they have the most impact is to keep researching ways to improve their properties, find new ways to make them cheap, and find new uses. There will be a long-term, major impact on material choice and design in high-tech engineering applications from this new area of HEMs.

REFERENCES

1. Miracle DB, Senkov ON. A critical review of high entropy alloys and related concepts. Acta Materialia. 2017;122:448–511.
2. Sanchez JM, Vicario I, Albizuri J, Guraya T, Garcia JC. Phase prediction, microstructure and high hardness of novel light-weight high entropy alloys. Journal of Materials Research and Technology. 2019;8(1):795–803.
3. Mohanty S, Maity TN, Mukhopadhyay S, Sarkar S, Gurao NP, Bhowmick S, Biswas K. Powder metallurgical processing of equiatomic AlCoCrFeNi high entropy alloy: Microstructure and mechanical properties. Materials Science and Engineering: A. 2017;679:299–313.
4. Mohanty S, Gurao NP, Biswas K. Sinter ageing of equiatomic Al20Co20Cu20Zn20Ni20 high entropy alloy via mechanical alloying. Materials Science and Engineering: A. 2014;617:211–8.
5. Guruvidyathri K, Murty BS, Yeh JW, Kumar KH. Gibbs energy-composition plots as a tool for high-entropy alloy design. Journal of Alloys and Compounds. 2018;768:358–67.
6. Jien-Wei YE. Recent progress in high entropy alloys. Ann. Chim. Sci. Mat. 2006;31(6):633–48.
7. Yeh JW, Chen SK, Lin SJ, Gan JY, Chin TS, Shun TT, Tsau CH, Chang SY. Nanostructured high-entropy alloys with multiple principal elements: Novel alloy design concepts and outcomes. Advanced Engineering Materials. 2004;6(5):299–303.
8. Zhou YJ, Zhang Y, Wang FJ, Chen GL. Phase transformation induced by lattice distortion in multiprincipal component CoCrFeNiCuxAl1− x solid-solution alloys. Applied Physics Letters. 2008;92(24).
9. Tsai KY, Tsai MH, Yeh JW. Sluggish diffusion in Co–Cr–Fe–Mn–Ni high-entropy alloys. Acta Materialia. 2013;61(13):4887–97.
10. Senkov ON, Senkova SV, Dimiduk DM, Woodward C, Miracle DB. Oxidation behavior of a refractory NbCrMo 0.5 Ta 0.5 TiZr alloy. Journal of Materials Science. 2012;47:6522–34.
11. Yeh JW. Alloy design strategies and future trends in high-entropy alloys. Jom. 2013;65:1759–71.
12. Zhang F, Tong Y, Jin K, Bei H, Weber WJ, Huq A, Lanzirotti A, Newville M, Pagan DC, Ko JY, Zhang Y. Chemical complexity induced local structural distortion in NiCoFeMnCr high-entropy alloy. Materials Research Letters. 2018;6(8):450–5.

13. Sohn SS, Kwiatkowski da Silva A, Ikeda Y, Körmann F, Lu W, Choi WS, Gault B, Ponge D, Neugebauer J, Raabe D. Ultrastrong medium-entropy single-phase alloys designed via severe lattice distortion. Advanced Materials. 2019;31(8):1807142.
14. Ma E. Unusual dislocation behavior in high-entropy alloys. Scripta Materialia. 2020;181:127–33.
15. Li Z, Pradeep KG, Deng Y, Raabe D, Tasan CC. Metastable high-entropy dual-phase alloys overcome the strength–ductility trade-off. Nature. 2016;534(7606):227–30.
16. Huang H, Wu Y, He J, Wang H, Liu X, An K, Wu W, Lu Z. Phase-transformation ductilization of brittle high-entropy alloys via metastability engineering. Advanced Materials. 2017;29(30):1701678.
17. Lei Z, Liu X, Wu Y, Wang H, Jiang S, Wang S, Hui X, Wu Y, Gault B, Kontis P, Raabe D. Enhanced strength and ductility in a high-entropy alloy via ordered oxygen complexes. Nature. 2018;563(7732):546–50.
18. Jiao M, Lei Z, Wu Y, Du J, Zhou XY, Li W, Yuan X, Liu X, Zhu X, Wang S, Zhu H. Manipulating the ordered oxygen complexes to achieve high strength and ductility in medium-entropy alloys. Nature Communications. 2023;14(1):806.
19. Fan L, Yang T, Zhao Y, Luan J, Zhou G, Wang H, Jiao Z, Liu CT. Ultrahigh strength and ductility in newly developed materials with coherent nanolamellar architectures. Nature Communications. 2020;11(1):6240.
20. Chung H, Choi WS, Jun H, Do HS, Lee BJ, Choi PP, Han HN, Ko WS, Sohn SS. Doubled strength and ductility via maraging effect and dynamic precipitate transformation in ultrastrong medium-entropy alloy. Nature Communications. 2023;14(1):145.
21. Cantor B, Chang IT, Knight P, Vincent AJ. Microstructural development in equiatomic multicomponent alloys. Materials Science and Engineering: A. 2004;375:213–8.
22. Li P, Wan X, Su J, Liu W, Guo Y, Yin H, Wang D. A single-phase FeCoNiMnMo high-entropy alloy oxygen evolution anode working in alkaline solution for over 1000 h. ACS Catalysis. 2022;12(19):11667–74.
23. Rost CM, Sachet E, Borman T, Moballegh A, Dickey EC, Hou D, Jones JL, Curtarolo S, Maria JP. Entropy-stabilized oxides. Nature Communications. 2015;6(1):8485.
24. Senkov ON, Wilks GB, Miracle DB, Chuang CP, Liaw PK. Refractory high-entropy alloys. Intermetallics. 2010;18(9):1758–65.
25. Gludovatz B, Hohenwarter A, Catoor D, Chang EH, George EP, Ritchie RO. A fracture-resistant high-entropy alloy for cryogenic applications. Science. 2014;345(6201):1153–8.
26. Senkov ON, Miracle DB, Chaput KJ, Couzinie JP. Development and exploration of refractory high entropy alloys—A review. Journal of Materials Research. 2018;33(19):3092–128.
27. Zhang Y, Stocks GM, Jin K, Lu C, Bei H, Sales BC, Wang L, Béland LK, Stoller RE, Samolyuk GD, Caro M. Influence of chemical disorder on energy dissipation and defect evolution in concentrated solid solution alloys. Nature Communications. 2015;6(1):8736.
28. El-Atwani O, Li N, Li M, Devaraj A, Baldwin JK, Schneider MM, Sobieraj D, Wróbel JS, Nguyen-Manh D, Maloy SA, Martinez E. Outstanding radiation resistance of tungsten-based high-entropy alloys. Science Advances. 2019;5(3):eaav2002.
29. Koželj P, Vrtnik S, Jelen A, Jazbec S, Jagličić Z, Maiti S, Feuerbacher M, Steurer W, Dolinšek J. Discovery of a superconducting high-entropy alloy. Physical Review Letters. 2014;113(10):107001.
30. Han L, Maccari F, Souza Filho IR, Peter NJ, Wei Y, Gault B, Gutfleisch O, Li Z, Raabe D. A mechanically strong and ductile soft magnet with extremely low coercivity. Nature. 2022;608(7922):310–6.
31. Wu P, Gan K, Yan D, Fu Z, Li Z. A non-equiatomic FeNiCoCr high-entropy alloy with excellent anti-corrosion performance and strength-ductility synergy. Corrosion Science. 2021;183:109341.

32. Guo J, Li Y. Study on microstructure and mechanical properties of five-element refractory high entropy alloys: Beijing Institute of Technology; 2015.
33. Yang X. High-entropy alloy with broad application prospects. Advanced Materials Industry. 2009;9:71–4.
34. Yin X, Xu S. Properties and preparation of high entropy alloys. In MATEC Web of Conferences 2018 (Vol. 142, p. 03003). EDP Sciences.
35. Zhang Y, Ting T, Tang Z, Gao MC, Dahmen KA, Liaw PK, Ping Z. Progress in materials science microstructures and properties of high-entropy alloys. Progress in Materials Science. 2014;61:1–93.
36. Shi P, Ren W, Zheng T, Ren Z, Hou X, Peng J, Hu P, Gao Y, Zhong Y, Liaw PK. Enhanced strength–ductility synergy in ultrafine-grained eutectic high-entropy alloys by inheriting microstructural lamellae. Nature Communications. 2019;10(1):489.
37. Fu Z, Jiang L, Wardini JL, MacDonald BE, Wen H, Xiong W, Zhang D, Zhou Y, Rupert TJ, Chen W, Lavernia EJ. A high-entropy alloy with hierarchical nanoprecipitates and ultrahigh strength. Science Advances. 2018;4(10):eaat8712.
38. Zhang Z, Mao MM, Wang J, Gludovatz B, Zhang Z, Mao SX, George EP, Yu Q, Ritchie RO. Nanoscale origins of the damage tolerance of the high-entropy alloy CrMnFeCoNi. Nature Communications. 2015;6(1):10143.
39. Gao X, Chen R, Liu T, Fang H, Qin G, Su Y, Guo J. High-entropy alloys: A review of mechanical properties and deformation mechanisms at cryogenic temperatures. Journal of Materials Science. 2022;57(12):6573–606.
40. Yeh JW, Lin SJ, Chin TS, Gan JY, Chen SK, Shun TT, Tsau CH, Chou SY. Formation of simple crystal structures in Cu-Co-Ni-Cr-Al-Fe-Ti-V alloys with multiprincipal metallic elements. Metallurgical and Materials Transactions A. 2004;35:2533–6.
41. Shashanka R, Chaira D. Phase transformation and microstructure study of nano-structured austenitic and ferritic stainless steel powders prepared by planetary milling. Powder Technol. 2014;259:125–136.
42. Chaira D. Development of nano-structured duplex and ferritic stainless steels by pulverisette planetary milling followed by pressureless sintering. Materials Characterization. 2015;99:220–9.
43. Zhou YJ, Zhang Y, Wang YL, Chen GL. Microstructure and compressive properties of multicomponent Alx (TiVCrMnFeCoNiCu) 100− x high-entropy alloys. Materials Science and Engineering: A. 2007;454:260–5.
44. Fujieda T, Shiratori H, Kuwabara K, Kato T, Yamanaka K, Koizumi Y, Chiba A. First demonstration of promising selective electron beam melting method for utilizing high-entropy alloys as engineering materials. Materials Letters. 2015;159:12–5.
45. Shashanka R, Chaira D. Optimization of milling parameters for the synthesis of nano-structured duplex and ferritic stainless steel powders by high energy planetary milling. Powder Technology. 2015;278:35–45.
46. Shashanka R, Uzun O, Chaira D. Synthesis of nano-structured duplex and ferritic stainless steel powders by dry milling and its comparison with wet milling. Archives of Metallurgy and Materials. 2020;65(1).
47. Rajendrachari S, Adimule VM, Jayaprakash GK, Pandith A. Electrochemical oxidation of methylene blue dye in wastewater using mechanically alloyed high entropy alloy modified carbon paste electrode using cyclic voltammetry. Materials Research Express. 2023;10(5):054003.
48. Rajendrachari S, Basavegowda N, Vinaykumar R, Narsimhachary D, Somu P, Lee M-J. Electrocatalytic determination of methyl orange dye using mechanically alloyed novel metallic glass modified carbon paste electrode by cyclic voltammetry. Inorganic Chemistry Communications. 2023;155:111010.

49. Rajendrachari S, Mahale RS, Adimule V, Gulen M, Mahesh V, Basavegowda N. Fabrication of mechanically alloyed high entropy alloys (20- Fe-20Cr-20Ni-20Mn-20Ti) modified carbon paste electrode sensor for the cyclic voltammetric determination of methyl orange. Inorganic Chemistry Communications. 2024;164:112428.
50. Rajendrachari S, Vinaykumar R, Katti R, Koujalagi BG, Jayaprakash GK, Ramkumar R. Cyclic voltammetric determination of rhodamine B dye using a gas-atomized duplex stainless steel powder-modified carbon paste electrode. Topics in Catalysis. 2024;1–2.
51. Rajendrachari S, Altas E, Salunkhe S, Dag IE, Avar B, Chaira D. Hot isostatic pressing of ball milled novel grade high entropy alloys at different sintering time and the investigation of their mechanical and corrosion resistance properties. Surface Topography: Metrology and Properties. 2024;12(4):045006.
52. Gülen M, Taş R, Dünya H, Rajendrachari S. Enlargement of the surface area of high-entropy alloys with acid treatment as positive electrode for high specific capacitance supercapacitors. ChemistryOpen. 2025;14(5):e202400312.
53. Wang Z, Zhang S. Research and application progress of high-entropy alloys. Coatings. 2023;13(11):1916.

Index

Note: page numbers in **bold** and *italics* refer to tables and figures, respectively.

I

K

L

M

N

R

S

T

U

V

W

X

Y

Z